THE
SOLAR SYSTEM

THE SOLAR SYSTEM

Volume 1

Archaeoastronomy—Jupiter's Ring System

Editors

David G. Fisher and **Richard R. Erickson**
Lycoming College

SALEM PRESS
Pasadena, California   Hackensack, New Jersey

Editor in Chief: Dawn P. Dawson

Editorial Director: Christina J. Moose   *Production Editor:* Joyce I. Buchea
Acquisitions Editor: Mark Rehn   *Page Design:* James Hutson
Manuscript Editor: Jennifer L. Campbell   *Layout:* William Zimmerman
Photo Editor: Cynthia Breslin Beres   *Editorial Assistant:* Dana Garey

Cover photo: (©Andrea Danti/Dreamstime.com)

Library of Congress Cataloging-in-Publication Data

The solar system / editors, David G. Fisher, Richard R. Erickson.
    p. cm.
    Includes bibliographical references and index.
    ISBN 978-1-58765-530-2 (set : alk. paper) — ISBN 978-1-58765-531-9 (v. 1 : alk. paper) — ISBN 978-1-58765-532-6 (v. 2 : alk. paper) — ISBN 978-1-58765-533-3 (v. 3 : alk. paper) — 1. Solar system. 2. Astronomy. I. Fisher, David G. II. Erickson, Richard R., 1945-
    QB501.S627 2010
    523.2—dc22

2009013008

First Printing

Publisher's Note

The Solar System covers 180 features of Earth's solar system, including every major body and phenomenon: from the astrophysics of the Sun and the major features of every planet, their satellites, and small bodies such as comets and asteroids, to scientific methodologies, interplanetary phenomena, and related topics in stellar astronomy and cosmology. Authoritative and essential to the core curriculum in astronomy and Earth sciences, these three volumes offer both students and general library patrons detailed basic information on all major aspects of Earth's solar system. No other general reference dedicated to the solar system is this complete and up to date, incorporating the latest perspectives offered by space telescopes, interplanetary probes, and planetary missions.

Scope and Coverage

The essays are designed to meet the needs of both general readers and students enrolled in courses in the Earth sciences (with 25 topics on Earth's geology, geophysics, and astrophysics alone), as well as astronomy, planetology, and cosmology. An older edition, under Salem's Magill's Choice imprint (1998), now more than a decade old, here receives a complete overhaul. For this new edition, we have added 58 topics and have thoroughly expanded every essay, from the text through the bibliographies, to bring the coverage up to date.

The result is the most thorough reference available on our expanding understanding of Earth's universal neighborhood. All of the original 122 essays, as well as the 58 new topics, were reviewed and updated by Professors David G. Fisher and Richard R. Erickson of Lycoming College's Department of Astronomy. They not only have added key developments of the past decade but also have scrutinized and heavily edited the existing text to ensure current accuracy—often increasing the lengths of old essays by more than a third. Such revision and expansion were essential in order to take into account the many planetary and interplanetary missions that have exponentially increased our

knowledge of the solar system—from Pioneer and Voyager to the data from Galileo, the Hubble Space Telescope, Cassini-Huygens, NASA's New Horizons program, and the latest Mars rovers.

Organization and Format

Coverage is presented in an A-Z format—from "Archaeoastronomy" and "Asteroids" through "Venus's Volcanoes" to "X-Ray and Gamma-Ray Astronomy"—and is supplemented by hundreds of diagrams and photos. For those wishing instant access to essays grouped by planetary system, a Category List appears in the front of every volume.

The set's essays fall into one or more of the following categories: the Cosmological Context (13), Earth (25), the Jovian System (11), Life in the Solar System (5), Mars (11), Mercury (1), Natural Planetary Satellites (23), the Neptunian System (7), Planets and Planetology (62), the Saturnian System (8), Scientific Methods (13), Small Bodies (17), the Solar System as a Whole (7), the Stellar Context (13), the Sun (20), the Uranian System (7), and Venus (5).

Every essay ranges in length from 2,000 to 5,000 words (3 to 7 pages) and offers not only a complete overview of the topic but also an assessment of knowledge gained, methods of study, or applications. Each essay displays standard ready-reference top matter and subsections:

- **Essay Title:** Topic name, from "Archaeoastronomy" and "Asteroids" through "Mars: Possible Life" to "X-Ray and Gamma-Ray Astronomy."
- **Categories:** Lists scientific subdisciplines relevant to standard undergraduate curricula.
- **Significance:** Summarizes the importance of the topic and current state of our knowledge.
- **Overview:** Rehearses the main facts about the topic.
- **Knowledge Gained *or* Methods of Study *or* Applications:** Headed as appropriate, this section details how the topic is investi-

gated, what scientific knowledge we have accumulated, or the uses of the knowledge we have gained.

- **Context:** Addresses the topic from the larger perspective of the history of solar system science and its relevance for humankind.
- **Byline:** All essays are signed by the scholars who wrote them; these experts are also listed, with their academic affiliations, in the front matter to Volume 1.
- **Further Reading:** An annotated selection of the most important print resources for further study.
- **See also:** Lists cross-references to other essays in *The Solar System* on related topics.

Special Features

Several special resources and finding aids enhance coverage and access to the set's contents. Each volume's front matter includes both an Alphabetical List of Contents (A-Z) and a Category List of Contents (with essays arranged by area of the solar system studied). The set is lavishly illustrated with photos, sidebars, diagrams, and tables. A Glossary, a General Bibliography, a list of Web Sites, and full Subject Index round out the set.

Miscellaneous Notes

The editors have endeavored to follow standard stylistics and usage throughout, using NASA sources in most cases. The words "natural satellite," "satellite" (where unambiguous with technology), and "moon" are used interchangeably. It is important to note that several essays refer to observances of the Sun in optical (visible) light, and we remind readers that one should *never* look directly at the Sun with the naked eye, through a camera, or through sunglasses or other such devices, which offer no protection and can cause significant damage to the eyes.

The Editors and Contributors

Salem Press extends its appreciation to all involved in the development of this work. Special thanks go to the set's editors, David G. Fisher and Richard R. Erickson of Lycoming College's Department of Astronomy, along with their assistant, graduate student Jennifer L. Campbell, who performed much of the manuscript editing. Professor Fisher is Professor of Physics and Astronomy as well as an accomplished spaceflight historian. In the latter capacity, Dr. Fisher served as co-editor of Salem Press's three-volume *USA in Space, Third Edition* (2006). He has published extensively on spaceflight history (concerning not only American robotic and crewed space programs but also Soviet/Russian and other international space projects), physics and physics education, and popular science topics. Professor Erickson is Associate Professor of Astronomy and Physics and Director of the Detwiler Planetarium at Lycoming College. Dr. Erickson has professional interests in and has published articles and papers concerning stellar dynamics, nearby stars, variable stars, galactic structure, relativity and cosmology, impacts and extinctions, and the modeling of the physics of impact events.

The essays were written and are signed by more than eighty scientists, scholars, and other experts, a list of whom will be found in the following pages, accompanied by their academic affiliations. Without their contributions, a project of this magnitude would not be possible.

Contributors

Stephen R. Addison
University of Central Arkansas

Arthur L. Alt
College of Great Falls

Michael S. Ameigh
St. Bonaventure University

Victor R. Baker
University of Arizona

Iona C. Baldridge
Lubbock Christian University

Thomas W. Becker
Webster University

Reta Beebe
New Mexico State University

Timothy C. Beers
Michigan State University

Raymond D. Benge, Jr.
Tarrant County College

Alvin K. Benson
Utah Valley University

John L. Berkley
*State University of New York
College at Fredonia*

Larry M. Browning
South Dakota State University

Michael L. Broyles
*Collin County Community
College*

David S. Brumbaugh
Northern Arizona University

Jessica Lynn Bugno
Lycoming College

Jennifer Campbell
Lycoming College

Dennis Chamberland
Science Writer

D. K. Chowdhury
*Indiana University and Purdue
University at Fort Wayne*

John H. Corbet
Memphis State University

John A. Cramer
Oglethorpe University

Robert L. Cullers
Kansas State University

Joseph Di Rienzi
*College of Notre Dame of
Maryland*

Bruce D. Dod
Mercer University

Dave Dooling
D2 Associates

Steven I. Dutch
*University of Wisconsin—
Green Bay*

John J. Dykla
Loyola University of Chicago

Richard R. Erickson
Lycoming College

Dale C. Ferguson
Baldwin-Wallace College

David G. Fisher
Lycoming College

Richard R. Fisher
*National Center for Atmospheric
Research*

Gerald J. Fishman
*National Aeronautics and Space
Administration*

Michael P. Fitzgerald
Vestal, New York

Dennis R. Flentge
Cedarville College

George J. Flynn
*State University of New York
College at Plattsburgh*

John W. Foster
Illinois State University

Donald R. Franceschetti
Memphis State University

Roberto Garza
San Antonio College

Karl Giberson
Eastern Nazarene College

David Godfrey
*National Optical Astronomy
Observatory*

Gregory A. Good
West Virginia University

C. Alton Hassell
Baylor University

Robert M. Hawthorne, Jr.
Marlboro, Vermont

Paul A. Heckert
Western Carolina University

David Wason Hollar, Jr.
Rockingham Community College

Earl G. Hoover
Science Writer

Hugh S. Hudson
*University of California,
San Diego*

Brian Jones
Science Writer

Richard C. Jones
Texas A&M University

Pamela R. Justice
*Collin County Community
College*

Karen N. Kähler
Pasadena, California

Christopher Keating
University of South Dakota

John P. Kenny
Bradley University

Firman D. King
University of South Florida

Richard S. Knapp
Belhaven College

Narayanan M. Komerath
Georgia Institute of Technology

Kristine Larsen
Central Connecticut State University

Joel S. Levine
National Aeronautics and Space Administration

James C. LoPresto
Edinboro University of Pennsylvania

George E. McCluskey, Jr.
Lehigh University

Michael L. McKinney
University of Tennessee, Knoxville

V. L. Madhyastha
Fairleigh Dickinson University

David W. Maguire
C. S. Mott Community College

Randall L. Milstein
Oregon State University

Brendan Mullan
Pennsylvania State University

Theresa A. Nagy
Pennsylvania State University

Anthony J. Nicastro
West Chester University of Pennsylvania

Divonna Ogier
Oregon Museum of Science and Industry

Steven C. Okulewicz
City University of New York, Hunter College

Satya Pal
New York Institute of Technology

Robert J. Paradowski
Rochester Institute of Technology

Jennifer L. Piatek
Central Connecticut State University

George R. Plitnik
Frostburg State University

Howard L. Poss
Temple University

Gregory J. Retallack
University of Oregon

Clark G. Reynolds
College of Charleston

Mike D. Reynolds
University of North Florida

J. A. Rial
University of North Carolina at Chapel Hill

Charles W. Rogers
Southwestern Oklahoma State University

David M. Schlom
California State University, Chico

Stephen J. Shulik
Clarion University of Pennsylvania

R. Baird Shuman
University of Illinois at Urbana-Champaign

Paul P. Sipiera
William Rainey Harper College

Billy R. Smith, Jr.
Anne Arundel Community College

Roger Smith
Willamette University

Joseph L. Spradley
Wheaton College, Illinois

James L. Whitford-Stark
Sul Ross State University

J. Wayne Wooten
Pensacola Junior College

Clifton K. Yearley
State University of New York at Buffalo

Ivan L. Zabilka
Lexington Kentucky Public Schools

Editors' Introduction

Since ancient times, humans have looked up at the sky and felt both fearsome awe and wondrous amazement. The Sun, Moon, and stars have been worshiped as gods. They have inspired poets, artists, and musicians. Scholars and "ordinary folk" have pondered the nature of what they saw in the sky and our relationship to it.

Probably the first thing to be noticed was that the celestial bodies moved across the sky from east to west, suggesting the Earth was at the center of the universe and everything revolved around us. Most cultures recognized that the stars maintained the same patterns, moving across the sky as groups. Many developed grand stories about these patterns or constellations which became part of their creation mythologies.

In time, it was noted that some of the points of light in the night sky did not move quite the same as most stars did; they "wandered" against the background patterns of the so-called "fixed stars" that remained together in the same constellation. These "wandering stars" or planets (from the Greek word for "wanderer") acquired special significance in many cultures. For example, the Mayans were particularly interested in Venus and kept meticulous records of where it was seen in the sky—sometimes in the east before sunrise and other times in the west after sunset. Some cultures developed rites of worship and built temples devoted to the planets. Some regarded the planets and their movements as portents or divine signs of what would happen here on Earth.

Eclipses and comets also evoked feelings of fear and wonder. Because these visually spectacular events occur only occasionally, those who successfully predicted them could acquire respect and power. There is evidence that one of Stonehenge's functions might have been to predict eclipses; if so, its builders had determined the pattern of eclipses several thousand years ago.

Many cultures tried to explain the nature of the universe (at least what they could see of it) and their place in it. How were the heavens constructed? What kept the clockwork timing of most of the heavens so perfect? What made the planets move so differently? How and why did eclipses occur? Attempts to answer questions like these led to the first models of the universe and theories of cosmology.

A few hundred years B.C.E., Greek philosopher-scientists determined that Earth must be spherical, and they calculated a remarkably accurate value for its circumference, using geometry and a few simple observations. They correctly explained why the Moon went through its phases and how eclipses of the Sun and Moon occurred. They even deduced that the Moon is smaller and the Sun is larger than Earth.

Most Greek models, like those of most other cultures, placed the Earth at the center of all things. Not that some Greeks did not consider the possibility that Earth moved, but such an idea seemed to contradict everything they could perceive. Many centuries would pass and much intense scientific debate would occur before it was accepted that while our existence in the universe may be special, our position is not.

The transformation from an Earth-centered universe to our present picture involved many steps. The first step, led by the likes of Copernicus, Galileo, and Kepler during the 1500's and 1600's, was recognizing that Earth is one of several planets orbiting the Sun. It was not until the 1830's that F. G. W. Struve, F. W. Bessel, and Thomas Henderson independently made the first measurements of stellar parallaxes (which is the only direct way to determine the distances of stars) and thereby confirmed what many had suspected, that stars really are things akin to the Sun but much farther away. Ultimately the Sun was found to be a rather average star. The location of the Sun and solar system in the outer part of the Milky Way galaxy's disk was determined by Harlow Shapley in 1917. A few years later, Edwin Hubble demonstrated that the so-called "spiral nebulae" really are galaxies outside the Milky Way. He then discovered that most galaxy spectra are redshifted, which led to the idea of an expanding universe. At the end of the twentieth century, the expansion of the universe was found to be accelerating, instead of slowing down as was expected. Today, there is serious speculation by some scientists that our universe might be just one of a multitude in a multidimensional "multi-verse."

Observations of planetary motions led to the development of heliocentric models and Sir Isaac Newton's theory of gravity. Early advances in telescope design and construction allowed intriguing but unclear details to be seen on the planets. However, by the middle of the twentieth century, solar system studies had become a backwater of astronomy. Few astronomers then were actively engaged in any sort of planetary or minor body research. (A notable exception was the Dutch-American astronomer Gerard Kuiper.) When one of the editors (Erickson) was an undergraduate at the University of Minnesota in the early 1960's, the textbook used in his first astronomy course had only two chapters consisting of 85 pages (out of a total of 536) that dealt with the planets, their satellites, the asteroids, comets, and meteoroids. (In contrast, current introductory astronomy textbooks typically devote about one-third of their pages to these topics.)

When the other editor (Fisher) was just five years old, he received two cherished books as Christmas presents. Both were from Random House, two selections from the "All About" series. *All About Rockets* and *All About the Universe* were read over and over again. These books describe the status of solar system understanding at the start of the 1960's. The books told about a coming age when humans would send spacecraft out into the solar system to take photographs and make direct measurements of the known planets from Mercury to Pluto, and also examine their moons. The books foretold the adventure of trying to unlock the secrets of the planets and of the great missions to come when humans would extend their reach from the Earth to the planets, maybe even sending people to make discoveries out in space themselves.

The book *All About the Universe* painted a picture of beauty and serenity in both the solar system and the universe around it. Pretty pictures showed the rings of Saturn and the craters of the Moon. Venus was perhaps a planet warmer and wetter than Earth, enjoying a lush, swampy ecosystem filled with plant life. Mars was probably cold and dry, but some form of primitive microbes or plant life might have adapted to live in the harsh environment there. Earth's moon was covered with craters probably created by volcanic action. The moons of the other planets were probably uninteresting balls of ice (with the anticipated exception of Titan about Saturn). Pluto was a small planet, perhaps with some terrestrial characteristics, although the planet would be extremely cold. A very peaceful and safe solar system was described, in which nothing very dramatic happened. All of this was about to change, and a more dynamic solar system soon would be revealed.

The dawn of the space age sparked renewed interest in the solar system. Early Luna (Soviet) and Pioneer (American) flyby missions attempted to reach the Moon in 1959. Only two years after Fisher received *All About the Universe* and *All About Rockets*, a spacecraft named Mariner 2 flew relatively close by Venus and determined that Venus was extraordinarily hot. No lush swamps existed, the surface temperature was hot enough to melt lead, and the atmosphere was thick with carbon dioxide. Then, in 1965, our ideas about Mars changed drastically when Mariner 4 flew by the Red Planet and produced a small set of grainy pictures revealing a dry, cratered surface much more like the Moon than a planet where life might exist.

Rapidly the space program grew in capability and complexity. Now (as of 2009), spacecraft have visited and studied at close range all of the planets in our solar system from Mercury to Neptune, many of their satellites, and several asteroids and comets. More information has been gained about the solar system in a few decades by robotic explorers and space-based observatories than had been learned from millennia of observations from Earth's surface, first with the unaided eye and later with telescopes. Much of what had been believed about solar system bodies and the processes of physical evolution throughout the solar system turned out to be wrong. Craters on the Moon were created largely by impact processes. Earth was (and still is) subject to collisions with comets and asteroids. These large impacts have played a major role in the evolution of the Earth, the development of its atmosphere and oceans, and even the course of biological evolution.

Telescopes were responsible for the first major increase in the storehouse of knowledge and level of understanding about the solar system. In the early 1600's Galileo, using his simple telescopes, found that small bodies orbit around Jupiter (the four Galilean satellites or moons), the Moon has mountains and craters, and the Sun has black spots (sunspots) on it. In 1781, Sir William Herschel discovered Uranus, the first planet not known to the ancients. These early telescopic observations provided the impetus for further investigations of the solar system.

Spacecraft provided the next great leap forward, vastly extending the reach of humanity in investigating our nearest neighborhood, the solar system. However, this is not to say that ground-based telescopes have become obsolete; they continue to provide important clues and new discoveries as well. Solar system studies presently require complementary use of ground-based telescopes, observatories in low-Earth orbit, and spacecraft dedicated to fly out to individual bodies even to the farthest reaches of the solar system.

When the articles for the first edition of *The Solar System* were written, several decades of space exploration had revealed much about our solar system. However, since then there has been such an explosion of information and change in our understanding of the solar system that 58 new articles were written for this second edition. The previous articles have been extensively updated and revised. It has been the editors' job to correct statements in articles that are no longer thought to be true and to update what has been learned since the first edition. For example, when the first edition was published, most scientists believed in the possibility of planets in solar systems other than our own, but none had been definitively detected. At the time of the second edition, approximately three hundred exoplanets have been discovered, and the list continues to grow with each passing year. Furthermore, most scientists have thought that

it would be just a matter of technology and time before clear images of planets orbiting other stars could be obtained; now a Jupiter-sized planet orbiting the star Formalhout has been photographed by the Hubble Space Telescope, although it appears only as a tiny dot.

The influx of new information about our solar system and others continues at a furious pace. Only two weeks before this introduction was composed, the French spacecraft COROT detected the closest thing to an Earth-sized planet yet. Named COROT-Exo-7b, this planet's existence was revealed by observing a dip in the planet's star's brightness as the planet transited in front of the star. The data suggest that this exoplanet is between two and four times the size of our Earth. This planet's "year" is less than one Earth day (approximately 20 hours), which means it orbits extraordinarily close to its star; consequently, this planet must be extremely hot (possibly 1,800 kelvins) and would share very little in common with Earth. A spacecraft named Kepler was launched on March 6, 2009. This Earth-orbiting observatory will examine a far greater number of distant stars by the transit methodology to search for planets— both smaller and larger than Earth—orbiting them. The time may well be near when an Earth-like planet will be imaged and its atmosphere spectroscopically investigated. Famed astrophysicist Michio Kaku commented in regard to the COROT-Exo-7b discovery that humanity could experience an existential shock when we first find a planet with conditions similar to those on Earth orbiting a star within its habitable zone, thus demonstrating that there is a place besides Earth where life might have developed.

Closer to home, as a result of discoveries of bodies similar to and farther out in the solar system than Pluto, the definition of "planet" itself was altered at a meeting of the International Astronomical Union in 2006; Pluto was "demoted" to the status of dwarf planet, becoming instead the first member in a new category of objects called plutoids, leaving our solar system (by convention) now with eight recognized planets. The worldview of solar system objects continues to undergo major changes with increasingly ambitious space missions. The Mars

Phoenix Lander provided the first direct evidence of subsurface water in the Red Planet's northern polar region. For several decades the search for possible primitive life on Mars has been based on the paradigm of "follow the water." Shortly after the Mars Phoenix confirmation that subsurface permafrost exists on Mars, a new discovery of methane in the Martian atmosphere held the promise of a different method for looking for life: Methane replenishment at Mars might indicate a biological origin and may shed light on the methane abundance in the atmosphere on Saturn's moon Titan.

At this writing, the issue of global climate change on Earth remains controversial. To truly understand what might be happening on Earth and the implications for life on Earth, changes in our atmosphere and climate need to be addressed in the broader context of comparative planetology. Venus, Earth, and Mars appear to have had very similar conditions in the past. Why did Venus lose most of its water, develop a thick carbon dioxide atmosphere, and heat up to the point where the surface temperature is enough to melt lead? Why did Earth retain its water, and how did the water lead to conditions where life as we know it could develop? How did Mars lose the vast majority of its surface water and develop a dry, arid surface with only a very thin atmosphere of carbon dioxide? Answers to these questions may lead to a better determination of the direction in which climate change on Earth might be headed.

Global climate change is but one possible threat to Earth. Whereas in the past the solar system had been considered to be a serene, quiet place, it is now understood that violent impacts have played a major role in the evolution of planetary surfaces and even on the evolution of life here on Earth. Some large impacts may have helped spread and promote the development of life, while others caused mass extinctions. An asteroid or comet impact remains a real threat today. Insufficient research is being done to detect those objects which might be heading "our" way. Fifty thousand years ago a small body about half as large as a football field created the impressive Meteor Crater near Flagstaff, Arizona. Compared to the asteroid that probably killed off the dinosaurs 65 million years ago, the one that produced Meteor Crater was a baby. While investigation of potential global climate change is important, a greater threat to life on Earth almost certainly is impact of a large body from space. With the promise of technology, such a natural disaster could be avoided as a result of direct intervention, but only if the threat from a specific body is identified well in advance of its impact.

The field of planetary research will surely remain a rich one, and questions raised by today's investigations may be answered in the relatively near future. Also in the years to come, new questions will be asked in response to information being obtained currently. A spacecraft, New Horizons, is on its way toward Pluto, and it will return images of that body and other Kuiper Belt objects. Efforts are now under way to expand the human presence in space beyond low-Earth orbit, to which it has been restricted since the conclusion of the Apollo program in 1972. The technology for a return to the Moon to stay and to eventually send human beings to Mars is being developed. If these plans materialize, humans may well establish bases upon the lunar surface from which extensive investigations of lunar geology or astronomical observations can be performed on a regular basis. If humans venture out to Mars, questions that numerous robotic probes have sought to address might finally be answered. Mars Exploration Rover principal investigator Dr. Steve Squyres, in his book about the plucky Spirit and Opportunity rovers, remarked that he, as a planetary geologist with trained eyes and hands, could have done all the work of the rovers in just a week's worth of his own surface investigations on Mars. Perhaps soon, scientists will make a better determination of the orbit of 99942 Apophis, an asteroid that will pass very close to (but not hit) the Earth in 2029. It is feasible that, by then, technology will be capable of using gravitational interactions to "tractor" that asteroid away, eliminating its threats to Earth and leading to a planetary defense system against future impacts.

Only the future will tell.

David G. Fisher, Lycoming College
Richard R. Erickson, Lycoming College
February 15, 2009

Contents

Units of Measure

Common prefixes for metric units—which may apply in more cases than shown above—include *giga-* (1 billion times the unit), *mega-* (one million times), *kilo-* (1,000 times), *hecto-* (100 times), *deka-* (10 times), *deci-* (0.1 times, or one tenth the unit), *centi-* (0.01, or one hundredth), *milli-* (0.001, or one thousandth), and *micro-* (0.0001, or one millionth).

Unit	Quantity	Symbol	Equivalents
Acre	Area	ac	43,560 square feet 4,840 square yards 0.405 hectare
Ampere	Electric current	A *or* amp	1.00016502722949 international ampere 0.1 biot *or* abampere
Angstrom	Length	Å	0.1 nanometer 0.0000001 millimeter 0.000000004 inch
Astronomical unit	Length	AU	92,955,807 miles 149,597,871 kilometers (mean Earth-Sun distance)
Barn	Area	b	10^{-28} meters squared (approx. cross-sectional area of 1 uranium nucleus)
Barrel (dry, for most produce)	Volume/capacity	bbl	7,056 cubic inches; 105 dry quarts; 3.281 bushels, struck measure
Barrel (liquid)	Volume/capacity	bbl	31 to 42 gallons
British thermal unit	Energy	Btu	1055.05585262 joule
Bushel (U.S., heaped)	Volume/capacity	bbl	2,747.715 cubic inches 1.278 bushels, struck measure
Bushel (U.S., struck measure)	Volume/capacity	bsh *or* bu	2,150.42 cubic inches 35.238 liters
Candela	Luminous intensity	cd	1.09 hefner candle
Celsius	Temperature	C	1° centigrade
Centigram	Mass/weight	cg	0.15 grain
Centimeter	Length	cm	0.3937 inch
Centimeter, cubic	Volume/capacity	cm^3	0.061 cubic inch
Centimeter, square	Area	cm^2	0.155 square inch
Coulomb	Electric charge	C	1 ampere second

Unit	Quantity	Symbol	Equivalents
Cup	Volume/capacity	C	250 milliliters 8 fluid ounces 0.5 liquid pint
Deciliter	Volume/capacity	dl	0.21 pint
Decimeter	Length	dm	3.937 inches
Decimeter, cubic	Volume/capacity	dm^3	61.024 cubic inches
Decimeter, square	Area	dm^2	15.5 square inches
Dekaliter	Volume/capacity	dal	2.642 gallons 1.135 pecks
Dekameter	Length	dam	32.808 feet
Dram	Mass/weight	dr *or* dr avdp	0.0625 ounce 27.344 grains 1.772 grams
Electron volt	Energy	eV	$1.5185847232839 \times 10^{-22}$ Btus $1.6021917 \times 10^{-19}$ joules
Fermi	Length	fm	1 femtometer 1.0×10^{-15} meters
Foot	Length	ft *or* '	12 inches 0.3048 meter 30.48 centimeters
Foot, cubic	Volume/capacity	ft^3	0.028 cubic meter 0.0370 cubic yard 1,728 cubic inches
Foot, square	Area	ft^2	929.030 square centimeters
Gallon (British Imperial)	Volume/capacity	gal	277.42 cubic inches 1.201 U.S. gallons 4.546 liters 160 British fluid ounces
Gallon (U.S.)	Volume/capacity	gal	231 cubic inches 3.785 liters 0.833 British gallon 128 U.S. fluid ounces
Giga-electron volt	Energy	GeV	$1.6021917 \times 10^{-10}$ joule
Gigahertz	Frequency	GHz	—
Gill	Volume/capacity	gi	7.219 cubic inches 4 fluid ounces 0.118 liter
Grain	Mass/weight	gr	0.037 dram 0.002083 ounce 0.0648 gram

Unit	Quantity	Symbol	Equivalents
Gram	Mass/weight	g	15.432 grains 0.035 avoirdupois ounce
Hectare	Area	ha	2.471 acres
Hectoliter	Volume/capacity	hl	26.418 gallons 2.838 bushels
Hertz	Frequency	Hz	$1.08782775707767 \times 10^{-10}$ cesium atom frequency
Hour	Time	h	60 minutes 3,600 seconds
Inch	Length	in *or* ″	2.54 centimeters
Inch, cubic	Volume/capacity	in^3	0.554 fluid ounce 4.433 fluid drams 16.387 cubic centimeters
Inch, square	Area	in^2	6.4516 square centimeters
Joule	Energy	J	$6.2414503832469 \times 10^{18}$ electron volt
Joule per kelvin	Heat capacity	J/K	$7.24311216248908 \times 10^{22}$ Boltzmann constant
Joule per second	Power	J/s	1 watt
Kelvin	Temperature	K	–272.15 Celsius
Kilo-electron volt	Energy	keV	$1.5185847232839 \times 10^{-19}$ joule
Kilogram	Mass/weight	kg	2.205 pounds
Kilogram per cubic meter	Mass/weight density	kg/m^3	$5.78036672001339 \times 10^{-4}$ ounces per cubic inch
Kilohertz	Frequency	kHz	—
Kiloliter	Volume/capacity	kl	—
Kilometer	Length	km	0.621 mile
Kilometer, square	Area	km^2	0.386 square mile 247.105 acres
Light-year (distance traveled by light in one Earth year)	Length/distance	lt-yr	5,878,499,814,275.88 miles 9.46×10^{12} kilometers
Liter	Volume/capacity	L	1.057 liquid quarts 0.908 dry quart 61.024 cubic inches
Mega-electron volt	Energy	MeV	—
Megahertz	Frequency	MHz	—

Unit	Quantity	Symbol	Equivalents
Meter	Length	m	39.37 inches
Meter, cubic	Volume/capacity	m^3	1.308 cubic yards
Meter per second	Velocity	m/s	2.24 miles per hour 3.60 kilometers per hour
Meter per second per second	Acceleration	m/s^2	12,960.00 kilometers per hour per hour 8,052.97 miles per hour per hour
Meter, square	Area	m^2	1.196 square yards 10.764 square feet
Metric. *See* unit name			
Microgram	Mass/weight	mcg *or* µg	0.000001 gram
Microliter	Volume/capacity	µl	0.00027 fluid ounce
Micrometer	Length	µm	0.001 millimeter 0.00003937 inch
Mile (nautical international)	Length	mi	1.852 kilometers 1.151 statute miles 0.999 U.S. nautical miles
Mile (statute or land)	Length	mi	5,280 feet 1.609 kilometers
Mile, square	Area	mi^2	258.999 hectares
Milligram	Mass/weight	mg	0.015 grain
Milliliter	Volume/capacity	ml	0.271 fluid dram 16.231 minims 0.061 cubic inch
Millimeter	Length	mm	0.03937 inch
Millimeter, square	Area	mm^2	0.002 square inch
Minute	Time	m	60 seconds
Mole	Amount of substance	mol	6.02×10^{23} atoms or molecules of a given substance
Nanometer	Length	nm	1,000,000 fermis 10 angstroms 0.001 micrometer 0.00000003937 inch
Newton	Force	N	$6.14124095407198 \times 10^{25}$ atomic weight 0.224808943099711 pound force 0.101971621297793 kilogram force 100,000 dynes

Units of Measure

Unit	Quantity	Symbol	Equivalents
Newton meter	Torque	N·m	0.7375621 foot-pound
Ounce (avoirdupois)	Mass/weight	oz	28.350 grams 437.5 grains 0.911 troy or apothecaries' ounce
Ounce (troy)	Mass/weight	oz	31.103 grams 480 grains 1.097 avoirdupois ounces
Ounce (U.S., fluid or liquid)	Mass/weight	oz	1.805 cubic inch 29.574 milliliters 1.041 British fluid ounces
Parsec	Length	pc	30,856,775,876,793 kilometers 19,173,511,615,163 miles
Peck	Volume/capacity	pk	8.810 liters
Pint (dry)	Volume/capacity	pt	33.600 cubic inches 0.551 liter
Pint (liquid)	Volume/capacity	pt	28.875 cubic inches 0.473 liter
Pound (avoirdupois)	Mass/weight	lb	7,000 grains 1.215 troy or apothecaries' pounds 453.59237 grams
Pound (troy)	Mass/weight	lb	5,760 grains 0.823 avoirdupois pound 373.242 grams
Quart (British)	Volume/capacity	qt	69.354 cubic inches 1.032 U.S. dry quarts 1.201 U.S. liquid quarts
Quart (U.S., dry)	Volume/capacity	qt	67.201 cubic inches 1.101 liters 0.969 British quart
Quart (U.S., liquid)	Volume/capacity	qt	57.75 cubic inches 0.946 liter 0.833 British quart
Rod	Length	rd	5.029 meters 5.50 yards
Rod, square	Area	rd^2	25.293 square meters 30.25 square yards 0.00625 acre
Second	Time	s or sec	$\frac{1}{60}$ minute $\frac{1}{3600}$ hour

Unit	Quantity	Symbol	Equivalents
Tablespoon	Volume/capacity	T *or* tb	3 teaspoons 4 fluid drams
Teaspoon	Volume/capacity	t *or* tsp	0.33 tablespoon 1.33 fluid drams
Ton (gross or long)	Mass/weight	t	2,240 pounds 1.12 net tons 1.016 metric tons
Ton (metric)	Mass/weight	t	1,000 kilograms 2,204.62 pounds 0.984 gross ton 1.102 net tons
Ton (net or short)	Mass/weight	t	2,000 pounds 0.893 gross ton 0.907 metric ton
Volt	Electric potential	V	1 joule per coulomb
Watt	Power	W	1 joule per second 0.001 kilowatt $2.84345136093995 \times 10^{-4}$ ton of refrigeration
Yard	Length	yd	0.9144 meter
Yard, cubic	Volume/capacity	yd^3	0.765 cubic meter
Yard, square	Area	yd^2	0.836 square meter

Alphabetical List of Contents

Volume 1

Volume 2

Volume 3

Category List of Contents

The Saturnian System

Scientific Methods

Small Bodies

The Solar System as a Whole

The Stellar Context

THE
SOLAR SYSTEM

A

Archaeoastronomy

Category: Scientific Methods

Archaeoastronomy is an interdisciplinary field studying how ancient civilizations sought answers to cosmological questions. Observations of sky phenomena influenced many aspects of ancient cultures, including but not limited to their science, religion, architecture, and lifestyles. Archaeoastronomy helps demonstrate how early technologies developed to better observe and understand the heavens. More often than not, the phenomena of greatest interest involved various solar system objects, especially the rising, setting, and apparent motions of the Sun, Moon, and bright planets.

OVERVIEW

Archaeoastronomy combines astronomy and archaeology into a study of ancient civilizations that focuses on the relationship between their observations of sky phenomena and their science, religion, architecture, and cultural practices. In advancing the discipline, astronomers and archaeologists work alongside ethnographers, geographers, anthropologists, mathematicians, historians, and others. However, many of the interpretations of archaeoastronomy are controversial and open to alternate explanations.

Archaeoastronomy dates back to the late 1600's and early 1700's, when the connection between ancient human-constructed structures and astronomical events was first made. Antiquarians like John Aubrey in 1678 and Henry Chauncy in 1700, studied the astronomical alignments of churches. Initial studies dealt with Middle Eastern and European cultures. Suggested by Euan MacKie, the term "archaeoastronomy" was first used in 1973 by Elizabeth Chesley Baity. Today, locations throughout the world are investigated.

The best-known archaeoastronomy site is Stonehenge in Wiltshire, England. The current structure at the site has been dated to 2500 B.C.E. There is evidence of an older stone structure on the same site. An earthen bank and ditch that encompass the stones has been dated to 3100 B.C.E. The circular bank-and-ditch system measures 110 meters in diameter and has a large opening in the northeast and a smaller one to the south. Flint tools and bones of deer and oxen that were found in the ditch date the ring's age.

Large stones of the structure called sarsens were named for the quarry where they were mined, about 40 kilometers away from Stonehenge. Stones were carefully shaped before being positioned. Thirty were erected vertically in a 33-meter circle within the earthworks structure. Thirty lintel stones were placed on top of the sarsens. Stones are held together with tongue-and-groove joints. Each sarsen is about 4.1 meters tall and 2.1 meters wide, and each weighs about 25 tons. Lintel stones are slightly smaller. Inside this circle, five triliths were arranged in a horseshoe pattern with the open end facing northeast; each trilith is a structure composed of a lintel stone and two supporting vertical stones—that is, a structure consisting of three stones (hence the name, from "tri" for "three" and "lith" for "stone"). Triliths increased in size from 6 to 7.3 meters tall in the southwest. A single sarsen of the Great Trilith remains standing. The stone is 6.7 meters tall above ground level, with another 2.4 meters buried below ground.

In 1666, Aubrey was one of the first to study Stonehenge. His work was continued by William Stukeley, who incorrectly associated the site with the Druids. In 1740, John Wood created the most accurate map of Stonehenge. Some distance outside the sarsen circle is a pointed stone called the heel stone. When viewed from the center of the sarsen circle, the Sun rises directly over the heel stone during the summer solstice. The structure also is aligned with the winter solstice sunset and the southernmost moonrise. From outside Stonehenge, the winter sunset is framed by one particular trilith.

A second, wooden structure was built two miles away from Stonehenge around the same time. The Durrington Walls circle overlooks the Avon River and is remarkably similar to the stone structure. The timber circle (or timber henge), however, aligns with the winter sunrise. The avenue leading to the river aligns with the summer solstice sunset. Archaeoastronomer Mike Parker Pearson believes that the two structures were linked and that each played a key role in the culture and lives of their builders. The purpose of these two henges, however, is still unclear. The Stonehenge Riverside Project, led by Pearson, found evidence that Stonehenge was a burial site. Pearson believes that the site was used to bury thirty to forty genera-

tions of royal family members. The project also conducted excavations near the Durrington Walls site, finding at least three hundred homes believed to have been used as seasonal lodgings during celebrations held there.

The most famous Irish archaeoastronomy site is Newgrange. Built between 3300 and 2900 B.C.E., Newgrange is older than both the Giza pyramids (by five hundred years) and the Stonehenge triliths (by one thousand years). The dominant feature of this site is a 12-meter-high, 76-meter-wide mound that covers an acre of ground. The mound contains an 18-meter-long passage ending in a cross-shaped room. The site is believed to be a tomb. Cremated remains have been found inside the burial chamber. A wooden circle to the southeast and a smaller one to the south of the mound were added during the Neolithic period. The larger

Mercury, Venus, Mars, Jupiter, and Saturn can be seen in the night sky over Stonehenge. (Philip Perkins)

circle has five rings of pits, with the outer one composed of wooden posts. The next inner ring was lined with clay and was used to cremate animals buried in the inner three circles of pits. The mound was surrounded by a circle of large, freestanding stones added during the Bronze Age. Newgrange was not discovered until the late seventeenth century because of slippage covering the mound's entrance. Michael O'Kelley and his team excavated and restored the site between 1962 and 1975.

The complexity of the mound's construction becomes obvious on the winter solstice, when sunlight enters the tomb through a roof box above the entrance and shines directly down the passageway onto the chamber floor. Solar alignments have been found in other tombs, but none is as precise as Newgrange, and none includes a roof box.

The only remaining Great Wonder of the ancient world is the Great Pyramid of Giza, located near Cairo, Egypt. That pyramid is also known as the Pyramid of Khufu or the Cheops Pyramid. The base of the structure is nearly a perfect square, each side measuring 230 meters. The Great Pyramid is almost fifty stories high, measuring 146.7 meters. Two slightly smaller pyramids (the Pyramid of Khafre and the Pyramid of Menkaure) also occupy the Giza complex. The three pyramids all follow north-south and east-west lines within a fraction of a degree. Archaeoastronomers propose various explanations for that orientation. In 2000, Egyptologist Kate Spence published an article with a new theory for the age and alignment of the pyramids. Spence believes that pyramid construction began between 2485 and 2475 B.C.E. Her theory also explains why all the pyramids are slightly offline with true north. Spence argues that the builders used two faint stars, Kochab (in the Little Dipper) and Mizar (in the Big Dipper), as guide stars. In 2467 B.C.E. these two stars were aligned directly with north and the celestial pole star. Egyptian astronomers could locate the two stars and use a plumb line to find the northern horizon. Pyramid construction took so long that Earth's precession eventually caused Kochan and Mizar to lose their exact alignment with north. This theory explains the orientation of the pyramids and gives a more ex-

act estimate of their age. Other scientists argue that the three pyramids simply represent the three stars in Orion's belt.

The most famous Mayan city is Chichén Itzá in Mexico. At the center of the city lies El Castillo, a temple to the god Kukulcan (Quetzalcoatl). Built between 1000 and 1200 C.E., El Castillo is a step pyramid with staircases on each side leading up to a central platform at the top. Each side has 91 steps. Counting the top platform, in total there are 365 steps, one for each day in the Haab, a part of the Mayan calendar. Sculptures of serpents adorn the side of the north-facing staircase. At the vernal and autumnal equinoxes, they cast a shadow, making it appear as though a serpent is climbing down the stairs. The western side is also aligned with the setting Sun around May 25, the day that usually marks the transition between dry and rainy seasons.

Another important Mayan site is Uxmal, located on the Yucatan peninsula. The Governor's Palace is a long, low structure built atop a large platform. The palace is aligned with the southernmost rising of Venus, which occurs once every eight years. The temple is also aligned with the pyramid of Cehtzuc on an azimuth of 118°. Standing in the entrance of the palace, Venus's southernmost rising would be seen directly over the Cehtzuc pyramid 4.6 kilometers away. Venus was very important to the Maya. The Governor's Palace is covered with hundreds of glyphs representing Venus, one of the most important Mayan deities. The Maya built observatories and kept very accurate records of Venus. They recognized that Venus alternated between appearing for several months in the eastern sky before sunrise and then for several months in the western sky after sunset. The Maya used these observations and those of other celestial objects to create one of the most accurate ancient calendars.

Medicine wheels are circular stone structures, with "spokes" radiating outward from a central cairn (pile of stones). The wheels were given the name "medicine" from the speculation that they were used by Native Americans in healing rituals. In 1885, when George Dawson published the first paper about medicine wheels, more than 20,000 of these artifacts had

Built between 1000 and 1200 C.E., El Castillo is a step pyramid with staircases on each side leading up to a central platform at the top. There are 365 steps, one for each day in the Haab, a part of the Mayan calendar. Sculptures of serpents adorn the side of the north-facing staircase. At the vernal and autumnal equinoxes, the shadows they cast make it appear as though a serpent is climbing down the stairs. The western side is aligned with the setting Sun around May 25, marking the transition between dry and rainy seasons. (©Pierdelune/Dreamstime.com)

been identified throughout North America. Today, only 135 survive in the United States and Canada. Some wheels are more than forty-five hundred years old, while others were built in later centuries. Medicine wheels vary in type. Some have spokes that extend out of the circle up to 120 meters. Others contain more rings of stones. The origin and true purpose of medicine wheels will never be known conclusively, but they are bear an unmistakable connection to the night and daytime skies.

The southernmost surviving ring is Bighorn Medicine Wheel, high in the Bighorn Mountains of north-central Wyoming, a site tentatively dated between 1200 and 1700 C.E. In the 1970's, John A. Eddy studied medicine wheels and believes that they were astronomically aligned. The Bighorn wheel has twenty-eight spokes radiating from a central cairn, each spoke ending in its own cairn. The number of spokes is the same as the number of days in a lunar cycle. One of the spokes extends out of the ring by 3.96 meters. Eddy discovered that the longer spoke aligned with sunrise on the summer solstice. Further support for his theory is the fact that Bighorn Medicine Wheel, at an elevation of 9,956 feet, is almost constantly under snow, with the exception of two months in the summer around the solstice. Another spoke aligns with the solstice sunset. Eddy linked other cairns with alignments to the rising of the stars Sirius (in Canis Major), Aldebaran (in Taurus), and Rigel (in Orion). Since these stars hold significance for several Native American tribes, Eddy believes that the wheel was used as a calendar or timepiece.

Eddy and other scientists have found similar alignments with other medicine wheels. Opponents to Eddy's theory argue that Native Americans used medicine wheels for religious or spiritual ceremonies. Having no evidence other than the wheels themselves, archaeoastronomers may never solve all of their mysteries.

METHODS OF STUDY

There are two main methodologies used by archaeoastronomers. Green archaeoastronomy relies only on alignments and is used when little or nothing is known about the builders of a site. Some scientists criticize this approach because it postulates that the ancients were interested in astronomy but offers no insights into why or what role it played in their cultures. Brown archaeoastronomy, by contrast, is often compared with cultural history or the history of astronomy. Brown archaeoastronomy studies historical and ethnographic records to learn about the roles astronomy played in ancient cultures. (The names of these two methods have no import other than their reference to the colors of the covers of the first published texts proposing and explaining the nature of each technique; green came first, followed by brown.)

At some sites, both methods are used. Alexander Thom developed many of the methods used in green archaeoastronomy. He examined British sites, looking for evidence of astronomy in the builders' society. Thom looked for alignments with the Sun (on solstices and equinoxes), the Moon, and the stars. One significant alignment, he believed, could be the result of chance, but more than one, he postulated, indicated deliberate attention to celestial objects on the part of the ancient builders. Euan MacKie, one of Thom's strongest supporters, studied British and Mayan sites, looking for proof of a link between the two societies.

Scientists used brown archaeoastronomy to interpret the Mayan city of Chichén Itzá. They studied Mayan records trying to determine what astronomical bodies were important in their culture. Once archaeoastronomers learned the importance of Venus to the Maya, they evaluated the site and found that many buildings were aligned with the rising or setting of the planet.

Anthony Aveni uses the Incan capital Cuzco as an example of the value of brown methods. Records of Incan life and legends were written by the Spanish and provide the majority of information known about the culture. These records explain that Cuzco was planned extensively before it was constructed. Most structures have astronomical alignments—with sunrise or sunset on the solstices, for example.

CONTEXT

Although controversial, archaeoastronomy—the study of the beliefs, religions, science, architecture, and cosmology of ancient cultures—is an increasingly important scientific field. The astronomical alignments of buildings, structures, and cities not only provide a new way of thinking about ancient societies but also elucidate what was known about the solar system before modern astronomy arose in the sixteenth century. The organizational, engineering, scientific, and mathematical skills of these cultures are showcased by their archaeoastronomy sites. Archaeoastronomy also expands the study of the history of science and religion further into the past than has been possible by the study of written records.

However, many questions remain unanswered. What caused societies in Russia, China, Egypt, Spain, and in other parts of the world to build pyramid-shaped structures? What led these cultures to study and worship the heavens? The study of archaeoastronomy can also reveal some things about ourselves. Many people today pay no attention to the night sky. Most cannot explain what an eclipse or solstice is. Interest in the exploration of the solar system has declined since the 1960's even as the technology to discover more about the solar system and the universe beyond has expanded greatly. Less than one-third of all Americans and Europeans can see the Milky Way from their own backyards because of urban light pollution. Today's societies lack a connection with the heavens, something that was a main component of ancient cultures. One can only wonder how the builders of Stonehenge or the Mayan temples would view this modern attitude.

Jennifer L. Campbell

5

FURTHER READING

Aveni, Anthony. *Stairways to the Stars: Skywatching in Three Great Ancient Cultures*. New York: Wiley, 1997. An informative work focusing on the Mayan Great Pyramid, the Incan capital Cuzco, and Stonehenge. Aveni speculates about how these ancient structures were built and discusses the various theories for their usage and origins.

Chamberlain, Von Del, John B. Carlson, and M. Jane Young. *Songs from the Sky: Indigenous Astronomical and Cosmological Traditions of the World*. College Park, Md.: Ocarina Books, 2005. A collection of papers written by scientists from various fields dealing with cultures of the early Americas and the Near East. Astronomy and cosmology in these societies are investigated, including their roles in architecture.

Fountain, John, and Rolf Sinclair. *Current Studies in Archaeoastronomy: Conversations Across Time and Space*. Durham, N.C.: Carolina Academic Press, 2005. A collection of papers from the fifth Oxford Conference on Archaeoastronomy focusing on sociology, astronomy, and culture of various ancient civilizations.

Kelley, David, and Eugene Milone. *Exploring Ancient Skies: An Encyclopedia Survey of Archaeoastronomy*. New York: Springer, 2004. A comprehensive look at archaeoastronomy, focusing on observational astronomy and its history. Examines the ways ancient cultures watched celestial objects and used them to keep time and develop calendars. Discusses origins of astronomy in cultures throughout the world, including Mesoamerica, Egypt, Africa, China, Korea, India, Japan, Greece, and Europe.

Magli, Giulio. *Mysteries and Discoveries of Archaeoastronomy: From Giza to Easter Island*. New York: Springer, 2009. A comprehensive study of all areas of modern archaeoastronomy. Takes the reader on a tour of the world's greatest ancient sites and discusses their astronomical alignments and cultural astronomy. Asserts that these great structures were built as symbols and the foundation of power. Discusses the Giza pyramid complex in great detail and hypothesizes that

the Cheops and Chephren pyramids were built as part of the same structure.

See also: Coordinate Systems; Earth System Science; Optical Astronomy; Planetary Orbits; Telescopes: Ground-Based.

Asteroids

Category: Small Bodies

Asteroids are minor bodies of a wide variety of sizes in orbit around the Sun, primarily but not exclusively located between the orbits of Mars and Jupiter. They provide important clues regarding the early history of the solar system, including the effect of their collisions on the surfaces of planets or their satellites. A class popularly referred to as near-Earth asteroids threaten to impact Earth.

OVERVIEW

Although discovery of the first asteroid was accidental, it came as no surprise to the astronomical community of the day. In 1766, German astronomer Johann Titius (1729-1796) observed that the positions of the planets could be approximated very closely by a simple empirical rule. Adding 4 to each number in the sequence {0, 3, 6, 12, 24, 48 . . .} and dividing the sum by 10 yields the mean planetary distances from the Sun in astronomical units (the distance from the Earth to the Sun is one astronomical unit, or 1 AU). The exception to this rule is the fifth element in that purely mathematical sequence, where an apparent gap occurs at 2.8 AU. It must be noted that this is just an empirical observation with no known physical basis. This rule was publicized by Johann Bode (1747-1826) and led to a search for a missing planet in the gap between Mars at 1.5 AU and Jupiter at 5.2 AU.

On January 1, 1801, the Sicilian astronomer-monk Giuseppe Piazzi (1746-1826) accidentally discovered a moving object during a routine star survey. He named it Ceres, for the patron goddess of Sicily. Soon its orbit was calculated by Carl Friedrich Gauss (1777-1855). At 2.77 AU,

Ceres was found, coincidentally, to conform closely to the Titius-Bode rule. However, since Ceres seemed to be too small to be classified as a planet, the search continued.

In March, 1802, German astronomer Heinrich Olbers (1758-1840) found a second minor body at the same predicted distance. He named it Pallas. In 1803, Olbers proposed that meteorites come from an exploded planet near 2.8 AU. This possibility led to a continued search resulting in the discovery of Juno in 1804 and Vesta in 1807. The latter discovery again was made by Olbers. It took quite some time for a fifth small body to be discovered (in 1845), but by 1890 the total had reached three hundred. These bodies came to be called "asteroids" for their faint, starlike images. In 1891, German astronomer Max Wolf (1863-1932) began using a long-exposure camera to detect asteroids. Since then, thousands of asteroids have been registered in the official catalog of the Institute of Theoretical Astronomy in Leningrad.

Asteroids are usually referred to officially by both a number and a name, such as 3 Juno or 1,000 Piazzi. About one hundred newly numbered asteroids are cataloged each year. Sky surveys indicate as many as 500,000 asteroids that are large enough to appear on telescopic photographs. Most asteroids are found within the main asteroid belt, which extends from 2.1 to 3.4 AU, and about half are between 2.75 and 2.85 AU. Asteroids revolve around the Sun in the same direction as the planets but tend to have more elliptical orbits. Their orbits are inclined up to 30° to the ecliptic plane, but they are far less eccentric than comet orbits. The smallest asteroids are a few kilometers wide; the largest, 1 Ceres (now considered a "dwarf planet"), about 1,000 kilometers wide. In 1867, American astronomer Daniel Kirkwood (1814-1895) discovered gaps in the asteroid belt where relatively few asteroids are found. These so-called Kirkwood gaps occur where asteroids have orbital periods that are simple fractions of the twelve-year revolution period of the giant planet Jupiter about the Sun, resulting in periodic gravitational influences called resonances. Such depletions occur, for example, at about 3.3 AU (where the periods have a six-year, 1:2 resonance with Jupiter) and 2.5 AU (a four-year, 1:3 resonance); other resonances, however, act to stabilize certain asteroids, such as the Hilda

An artist's rendering of one of two asteroid belts of Epsilon Eridani, a solar system located in the constellation Eridanus. (NASA/JPL-Caltech)

Facts About Selected Asteroids

Name	Diameter (km) or Dimensions	Mass (10^{15} kg)	Rotational Period (hrs)	Orbital Period (yrs)	Distance from Sun (AU)
Ceres (dwarf planet)	960 × 932	870,000	9.075	4.60	2.767
Palas	570 × 525 × 482	318,000	7.811	4.61	2.774
Juno	240	20,000	7.210	4.36	2.669
Vesta	530	300,000	5.342	3.63	2.362
Eugenia	214	6,100	5.699	4.49	2.721
Siwa	110	1,500	18.5	4.52	2.734
Ida	58 × 23	100	4.633	4.84	2.861
Mathilde	66 × 48 × 46	103.3	417.7	4.31	2.646
Eros	33 × 13 × 13	7.2	5.270	1.76	1.458
Gaspra	19 × 12 × 11	10	7.042	3.29	2.209
Icarus	1.4	0.001	2.273	1.12	1.078
Geographos	2.0	0.004	5.222	1.39	1.245
Apollo	1.6	0.002	3.063	1.81	1.471
Chiron	180	4000	5.9	50.70	13.633
Shipka	—	—	—	5.25	3.019
Rodari	—	—	—	3.25	2.194
McAuliffe	2-5	—	—	2.57	1.879
Mimistrobell	—	—	—	3.38	2.249
Toutatis	4.6 × 2.4 × 1.9	0.05	130	3.98	2.512
Nereus	2	—	—	1.82	1.490
Castalia	1.8 × 0.8	0.0005	—	1.10	1.063
Otawara	5.5	0.2	—	3.19	2.168
Braille	2.2 × 1.0	—	—	3.58	2.341

Source: Data are from the National Aeronautics and Space Administration/Goddard Space Flight Center, National Space Science Data Center.

group at 4 AU (2:3 resonance), which is named for 153 Hilda.

Some asteroids have orbits departing greatly from the main belt. In 1772, French mathematician Joseph Lagrange (1736-1813) showed that points in Jupiter's orbit 60° ahead of and behind the planet are gravitationally stable (1:1 resonance). In 1906, Max Wolf discovered the first so-called Trojan asteroid, 588 Achilles, at the Lagrangian point 60° ahead of Jupiter. Subsequent discoveries have revealed several hundred Trojan asteroids. Those ahead of Jupiter are named for Greek heroes, and those behind are named for Trojan heroes; there is one Greek spy (617 Patroclus asteroid) in the Trojan group, and one Trojan spy (624 Hektor) in the Greek group. Hektor is the largest known Trojan asteroid, at about 150 by 300 kilometers, and is the most elongated of the more massive

asteroids. At least two objects have orbits that extend beyond Jupiter: 944 Hidalgo, which may be a burned-out cometary nucleus, and 2060 Chiron, whose orbit extends beyond Saturn.

Some asteroids depart from the main belt over only part of their orbit. Mars-crossing Amor group bodies have elongated orbits that carry them inside Mars's orbit but still keep them well outside Earth's orbit. The Martian satellites Phobos and Deimos have long been suspected by many to be captured asteroids, perhaps from this group. Apollo group members come inside Earth's orbit. (The groups were named for their first examples, discovered in 1932.) Estimates indicate about thirteen hundred Apollos ranging in from 0.4 to 10 kilometers across, with an estimated average Earth-collision rate of about one in 250,000 years. The closest known approaches were Hermes, in

1937, at about 780,000 kilometers, and 1566 Icarus, in 1968, at about 6 million kilometers. Smaller Apollos may be an important source of meteorites, and 100-meter objects capable of making a 1-kilometer crater strike Earth about every two thousand years. Aten-type asteroids are Earth-crossers with elliptical orbits smaller than Earth's. Some asteroids appear to be grouped in families that may be the fragments resulting from an earlier collision between asteroids.

Chemical and physical characteristics of asteroids are mostly determined by remote-sensing techniques that study electromagnetic radiation reflected off their surfaces. More than five hundred asteroids have been studied by remote sensing and radar astronomical techniques. These studies have indicated asteroidal compositions similar to those of meteorites. Comparison with reflected light from meteorites suggests several classes. Rare E-type asteroids possess the highest albedo (23 to 45 percent reflection). They appear to be related to enstatite (a magnesium silicate mineral) chondrites, and are concentrated near the inner edge of the main belt. About 10 percent of asteroids are S-type; they have relatively high albedos (7 to 23 percent) and appear reddish in color. They likely are related to stony chondrites, are found in the inner to central regions of the main belt, and generally range in size from 100 to 200 kilometers. The largest S-type is 3 Juno, at about 250 kilometers in diameter. Much smaller Apollo asteroids are also in this category. A few asteroids in the middle belt are classified as M-type, since their reflected light (7 to 20 percent) reveals evidence of large amounts of nickel-iron metals on their surface, similar to iron or stony-iron meteorites.

About three-quarters of all asteroids are C-type, having relatively low albedos (2 to 7 percent) and grayish colors similar to that of the Earth's moon. They are found in the outer belt and among the Trojans. They resemble carbonaceous chondrite meteorites, containing water-bearing silicate-based and carbon-based

minerals along with some organic compounds (about 1 percent). The largest of all the asteroids, 1 Ceres (now considered a dwarf planet), is in this category. Some evidence supports the claim that Ceres has a mixture of ice and carbonaceous minerals on its surface. Dark reddish, D-type asteroids are found in the same regions and have similar albedos.

About 10 percent of asteroids remain unclassified and are designated as U-type. In general, asteroids with low-temperature volatile materials lie farther from the Sun, whereas those in the inner part of the main belt are richer in high-temperature minerals, displaying little evidence of volatile water and carbon compounds.

Many asteroids exhibit periodic variations in brightness that suggest irregular shapes and rotation. Their measured rotational periods range from about three to thirty hours. There is some evidence that S-type asteroids rotate faster than C-type asteroids but more slowly than M-type asteroids. Large asteroids (greater than 120 kilometers) rotate more slowly with

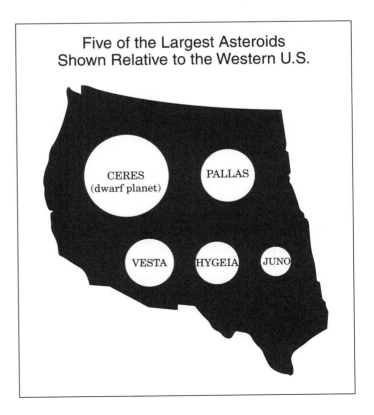

Five of the Largest Asteroids Shown Relative to the Western U.S.

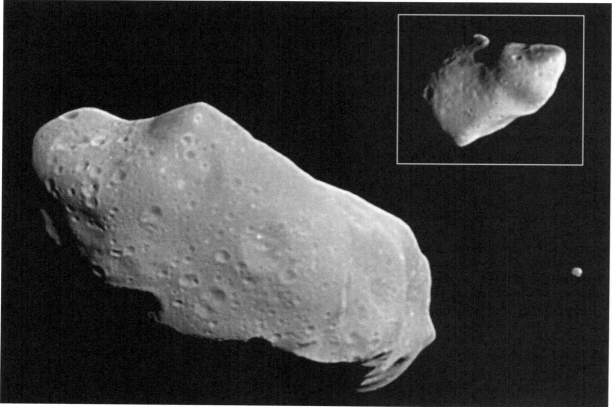

Asteroid Ida, so named for its potato shape, as imaged by the Galileo spacecraft. Ida's satellite Dactyl appears in the inset. (NASA/JPL/USGS)

increasing size, but small asteroids rotate more slowly with decreasing size, suggesting that large asteroids may be primordial bodies, while smaller ones may be fragments produced by collisions. Calculations show that rotation rates longer than two hours produce centripetal forces weaker than gravity, which indicates that loose debris can exist on the surface of even the fastest known rotating asteroid, the Apollo object 1566 Icarus, which has a 2.25-hour rotation rate. Polarization studies of light reflected from asteroids indicate that many do have dusty surfaces.

Named after the Greek god of love, Eros is an S-type asteroid belonging to the Amor group. As the second-largest of the near-Earth asteroids, it is larger than the asteroid generally accepted to have been responsible for the extinction of the dinosaurs that impacted Earth near the Yucatán peninsula 65 million years ago. Eros is

13 by 13 by 33 kilometers in size. It was the first asteroid recognized to approach inside the orbit of Mars, and in 1975 it became the first asteroid to be studied with Earth-based radars.

Computer models suggest the possibility that larger asteroids have a deep layer of dust and rock fragments (or regolith) similar to that on the surface of the Moon. Asteroids with diameters larger than 100 kilometers are believed to have undergone a process of differentiation in which heavier metals sank to the core, leaving a stony surface of lighter materials later pulverized by collisions to form a layer of dust.

Asteroid elongations can be estimated from the change in brightness, which can vary by a factor of three or more. For example, radar evidence indicates that the unusual Trojan asteroid 624 Hektor (150 by 300 kilometers) may be a dumbbell-shaped double asteroid. Kilometer-scale asteroids have been observed with lengths up to six times greater than their width. Main-belt asteroids tend to be less elongated than Mars-crossers of the same size, perhaps because of greater erosion from collisions in the belt. Asteroids larger than about 400 kilometers tend to be more spherical, since their gravitational attractions exceed the strength of their rocky materials, causing deformation and plastic flow into a more symmetric shape.

An asteroid's size occasionally can be determined quite accurately by timing its passage in front of a star, that is, in a stellar occultation. In a few cases, stellar light has been occulted more than once in a single passage, indicating that asteroids may possess satellites. Radar-based studies have confirmed this theory. Also, as the Galileo spacecraft flew through the main belt on its way to enter orbit in the Jupiter system, it imaged a satellite revolving about an asteroid. The irregularly shaped asteroid Ida was discovered to have a small satellite later named Dactyl. Ida is a member of the Koronis family and of S-type, which is 56 by 24 by 21 kilometers in size and rotates once around its own axis every 278 minutes. Dactyl is only 1.2 by 1.4 by 1.6 kilometers in size and is also of S-type. This strongly suggests that it was created when a larger asteroid smashed into Ida. Previously, the Galileo spacecraft had also provided the first close-up images of an asteroid, when it passed within five thousand kilometers of the 19- by 12- by 11-kilometer-sized S-type asteroid Gaspra on October 29, 1991. Gaspra has an irregular shape, one resembling a potato.

The distribution of asteroid sizes and masses supports the idea that many have undergone a process of fragmentation. Typical relative velocities of encounter, about 5 kilometers per second in the main belt, are quite adequate to fragment most asteroids. Ceres contains nearly half the mass of all the asteroids, but it is more than three times smaller than the Moon and about fifty times less massive. About 80 percent of the total mass of all asteroids is contained in the four largest ones, and only about ten are larger than 300 kilometers. Studies suggest that the main belt was several times more massive in the past but that in the process of fragmentation, the smallest dust particles were removed by radiation pressure from the Sun.

Interest in asteroids increased when strong evidence was advanced to solve the mystery of the demise of the dinosaurs 65 million years ago. Physicist Luis Alvarez and his geologist son Walter sampled the worldwide clay layer that marks the end of the Cretaceous period and the start of the Tertiary period (the so-called K-T boundary, which essentially marks the demarcation between the age of dinosaurs and the rise of mammals within the fossil record). This thin layer of clay is enriched in the rare elements of iridium and osmium, having levels more akin to asteroids than Earthly materials. Thus, the impact theory for killing off the dinosaurs was proposed, and largely accepted except by certain portions of the paleontology community. That is, until a crater dated to 65 million years was discovered off the coast of the Yucatán peninsula. Some still insist that more than an asteroid impact was necessary to account for the observed diminishment of dinosaur species leading up to the extinction event 65 million years ago. However, the majority of the scientific community has come to accept the asteroid impact theory, at least as the principal cause of the sudden mass extinction at the end of the Cretaceous period. Since this event marks the boundary between the Cretaceous and Tertiary periods, it is often referred to as the K-T event.

This spurred interest in asteroid and comet impacts causing extreme environmental damage to the Earth at other times in the past, along with a desire to search for near-Earth asteroids that might represent a threat in the future. Twenty-five years after the proposal that an asteroid impact killed the dinosaurs received initial lukewarm acceptance by paleontologists; some researchers proposed that an even bigger asteroid (or comet) impact was responsible for the so-called Great Dying, the mass extinction at the end of the Permian period that closed out the Paleozoic era. At the end of the Permian 248 million years ago, more than 95 percent of all species died off rather suddenly; life nearly did not make it into the Mesozoic era, during which the dinosaurs eventually arose to dominance.

Researchers point to a large crater in the Antarctic (1.5 kilometers under the ice pack that dates to the time of the Permian mass-extinction event) as well as heavily jumbled areas in Siberia (known as the Siberian Traps), that might have received tremendous seismic energy after impact energy would have undergone antipodal focusing off Earth's core. The Siberian Traps also was an area of tremendous volcanic activity at the end of the Permian period. Was this coincidental or the result of an impact with antipodal focusing of seismic energy? In 2008 this theory remained highly speculative, rather than enjoying the widespread acceptance of the K-T event that killed the dinosaurs. However, if the theory is correct, such an event underscores the danger posed by asteroid and comet impacts on Earth.

Impact of even a small asteroid could pose a tremendous threat to human civilization. Throughout the 1990's and the early twenty-first century, a number of newly discovered near-Earth asteroids were thought to have a significant chance of hitting Earth in the quite near future. However, in each of those cases, additional observations refined the asteroid's orbit to the point where it was clear it would not hit Earth after all. There remained one major exception, however. Discovery of the asteroid 99942 Apophis, a member of the Aten group, led to major concern beginning in late 2004 that this 350-meter-across rock had a relatively worrisome potential to impact Earth in 2029. Precise observations of Apophis's orbit, ranging from 0.746 AU to 1.099 AU, dramatically lowered the probability that it would strike the Earth. However, Apophis would indeed pass within the altitude of geosynchronous satellites, less than 36,000 kilometers from Earth's surface. If Apophis passed within a special corridor only 400 meters across, gravitational influences could cause it to return and strike the Earth on Friday, April 13, 2036.

The Torino scale assesses the relative impact hazard of an asteroid impact. For a time after its discovery, Apophis rated a level 4 on the Torino scale; that is the highest level of threat. Further orbit refinements lowered the threat assessment to a level 0 threat, but, after realization of the possible return in 2036, it was raised to Level 1. Although Apophis will come very close to Earth in 2029, refinement of available orbital data has since determined that the chances of Apophis hitting the Earth in 2036 are more comforting: less than 1 in 45,000. The 2036 encounter will set up another close encounter the following year, but the chances that this would result in an Earth impact are calculated to be less than 1 in 12.3 million. Nevertheless, Apophis points out the absolute requirement for close monitoring of asteroids, particularly the near-Earth ones, and the development of means whereby asteroids could be deflected or destroyed in order to preserve Earth's biosphere and save human civilization. This sort of natural megadisaster is one of the few that humans have the potential to mitigate or prevent if action is taken sufficiently early once the threat is identified.

Hollywood has even taken notice of the asteroid or comet impact threat to Earth. Several scientifically incorrect action movies were produced, some of which were popularly received. Many of these movies, such as *Armageddon* (1998), portray the use of some type of nuclear device as the only viable way to avert an asteroid impact. In many real cases, nuclear explosions detonated within, on the surface of, or close to asteroids either would be insufficient or could merely fragment it so badly that an even worse situation, a swarm of impacting bodies, might ensue.

METHODS OF STUDY

Studies of asteroids hold the potential for expanding our understanding of the formation of bodies of sizes between the smallest objects and full planets, and also could lead to development of technology to prevent an impact that might devastate life on Earth and even wipe out civilization.

The Galileo spacecraft passed near enough to two asteroids to photograph them directly. The NEAR spacecraft orbited Eros and later landed on its surface, providing close-up photographs of an asteroidal surface. For the most part, however, indirect methods of remote sensing must be used to determine asteroidal properties by studying the reflected electromagnetic radiation that comes from their surfaces. These methods include photometry, infrared radiometry, colorimetry, spectroscopy, polarimetry, and radar detection. They can be augmented by

The Torino Asteroid Impact Hazard Scale

Scale	Description
0	EVENT HAVING NO LIKELY CONSEQUENCES: The likelihood of a collision is zero, or low enough to be effectively zero. This designation also applies to any small object that, in the event of a collision, is unlikely to reach the Earth's surface intact.
1	EVENT MERITING CAREFUL MONITORING: The chance of collision is extremely unlikely, about the same as a random object of the same size striking the Earth within the next few decades.
2	EVENT MERITING CONCERN: A somewhat close but not unusual encounter. Collision is very unlikely.
3	EVENT MERITING CONCERN: A close encounter, with 1 percent or greater chance of a collision capable of causing localized destruction.
4	EVENT MERITING CONCERN: A close encounter, with 1 percent or greater chance of a collision capable of causing regional devastation.
5	THREATENING EVENT: A close encounter, with a significant threat of a collision capable of causing regional devastation.
6	THREATENING EVENT: A close encounter, with a significant threat of a collision capable of causing a global catastrophe.
7	THREATENING EVENT: A close encounter, with an extremely significant threat of a collision capable of causing a global catastrophe.
8	CERTAIN COLLISION: A collision capable of causing localized destruction. Such an event occurs somewhere on Earth between once per 50 years and once per 1,000 years.
9	CERTAIN COLLISION: A collision capable of causing regional devastation. Such an event occurs between once per 1,000 years and once per 100,000 years.
10	CERTAIN COLLISION: A collision capable of causing a global climatic catastrophe. Such an event occurs once per 100,000 years or less often.

comparative studies with meteorites, whose composition and structure can be analyzed by direct methods in the laboratory. Such methods include chemical, spectroscopic, and microscopic analysis, and processes of fragmentation can be studied by producing high-speed collisions between comparable materials in the laboratory. Such comparative studies must recognize various differences between meteorites and asteroids. The masses of only the three largest asteroids have been determined from their gravitational effects on other bodies; their densities are between 2.3 and 3.3 grams per cubic centimeter.

Photometry is the study of how light is scattered by various surfaces. The varying brightness of reflected sunlight from asteroids can be measured by photoelectric observations to determine their rotation periods and approximate shapes. One test of this method was made in 1931, when the Amor asteroid 433 Eros came close enough (23 million kilometers) for scientists to observe the tumbling motion of this elongated object (7 by 19 by 30 kilometers) and to confirm its 5.3-hour rotation. The size of an asteroid can be estimated from its brightness together with its distance, orbital position, and albedo. The albedo is important, since a bright, small object may reflect as much light as a dark, large object. Since a dark object absorbs more heat than a light object, albedos can be determined by comparing reflected light with thermal radiation measured by infrared radiometry. Photometric measurements also give information on surface textures. Colorimetry involves measuring the range of wavelengths in the reflected light to determine surface colors. Most asteroids are either fairly bright, reddish objects (with albedos of up to 23 percent) composed largely of silicate-type materials or grayish objects, at least as dark as the Moon (11 percent albedo), composed of carbonaceous materials.

Spectroscopy is the spectral analysis of light and can be used to infer the composition of many asteroids. Optical and infrared reflectance spectra exhibit absorption bands at characteristic frequencies for given materials. An asteroid's surface composition is determined by comparing its spectrum with the spectra of light reflected from meteorites of known composition. Examples of this method applied to U-type (unclassified) asteroids include the identification of the silicate mineral pyroxene in the infrared spectrum of Apollo asteroid 1685 Toro, and the matching of the surface of Vesta with a basaltic achondrite that resembles lava. Most asteroids appear to have unmelted surfaces with little or no evidence of lava eruptions. About two-thirds of the Trojans are D-type asteroids with no known meteorite counterparts because of their distance from Earth. Their spectra have been matched with the spectra of coal-tar residues, suggesting possible organic compounds.

Polarimetry uses measurements of the alignment of electric field vibrations of the reflected sunlight and its variation with direction to estimate albedos. Polarization measurements have also been interpreted as evidence for dust-covered surfaces, but they leave uncertainty about the depth of the dust layer. Radar observations of Eros during a close approach to Earth in 1975 were made at a wavelength of 3.8 centimeters and indicated that the surface must be rough on a scale of centimeters. Since optical polarimetry suggests that Eros is dusty, the radar results imply that the dust must be too thin to smooth rock outcrops of more than a few centimeters. Radar measurements also provided independent estimates of the size of Eros, confirming photometric estimates of its dimensions. The NEAR spacecraft confirmed these observations.

As spacecraft results such as this demonstrate, the best method to study asteroids is by means of a space probe. When Pioneers 10 and 11 passed through the asteroid belt, scientists found that it has no more dust than any other part of the solar system. The Galileo probe encountered Gaspra in 1991 and Ida in 1993, both S-type asteroids. The probe determined the masses, sizes, and shapes. The Cassini spacecraft on its way toward orbit about Saturn flew through the asteroid belt and passed asteroid 2685 Masursky at a distance of 1.6 million kilometers. Named after the famed planetary scientist Hal Masursky, this body was a little understood 15- by 20-kilometer asteroid prior to Cassini's encounter.

Before the Pioneer 10 and 11 passages there were serious concerns that spacecraft might not be able to pass safely through the main asteroid belt. Much has been learned about the density of material in the belt since the space age began. Thus far, no spacecraft sent into the belt has experienced serious damage from an impact with asteroidal material or an actual asteroid body. Minor hits on dust detectors have been recorded, however. Robotic spacecraft investigations have provided much information about the nature of the various types of asteroids, as has analysis of meteorites found on Earth that are believed to have come from certain asteroids.

The NEAR spacecraft was launched on February 17, 1996, and was directed toward a rendezvous with the asteroid Eros three years later. Eighteen months out from Earth, NEAR flew by the asteroid Mathilde. It successfully reached Eros, and for well over a year NEAR orbited Eros at varying altitudes providing high-resolution images of the surface of this S-type asteroid. After completing its primary mission, NEAR gently touched down on Eros on February 12, 2001. A total of 69 high-resolution images of the asteroid's surface were taken on the way down during a soft-contact landing. The final picture was taken at an altitude of 130 meters and covered an area of 6 meters by 6 meters. Within that final frame was a portion of a 4-meter-wide boulder, as well as evidence of a dusty surface pocked with small rocks and tiny craters. Much to the surprise of the Johns Hopkins University Applied Physics Laboratory research team controlling the spacecraft, NEAR survived its landing and transmitted data back to Earth for two weeks before falling silent. The team was lucky in that the spacecraft's antenna pointed toward Earth and the solar arrays faced partially toward the Sun after impact.

The next step in spacecraft-based investigations of asteroids is the Dawn mission, a robotic probe designed to orbit two different bodies. Dawn's mission is to visit the two largest asteroids, Ceres and Vesta. By comparison with Ceres, the asteroid upon which NEAR settled was a tiny speck. Ceres is a spherical body with a diameter of 960 kilometers. Indeed under an official review of classification for solar-system objects, Ceres is now officially designated a dwarf planet—a characterization it shares (much to the displeasure of many in the scientific community) with Pluto, which was demoted from full planet status to that of a dwarf. To accomplish its mission on a minimum of propellant, Dawn is outfitted with ion propulsion similar to that demonstrated by the Deep Space 1 spacecraft. To achieve its science goals, Dawn is outfitted with a framing camera, a mapping spectrometer, and a gamma-ray and neutron spectrometer. The goal is to image the surface of these two large asteroids and to determine their composition.

Dawn launched on September 27, 2007, and was set up for a gravity assist from Mars in early 2009. Arrival at Vesta was planned for September, 2011. The ion propulsion system would then break Dawn out of Vesta orbit in April, 2012, and send the spacecraft toward a rendezvous with Ceres in February, 2015. Assuming the spacecraft remains healthy and propellant is available when the primary mission ends in July, 2015, Dawn could be redirected to other asteroids within reach.

Samples of rocks believed to come from various portions of the asteroid belt fall on Earth regularly and have been subjected to intense study. The next step in asteroid investigation would be to return pristine samples of asteroids so the asteroid samples are not altered on their outer layers by passage through Earth's atmosphere. A robotic mission to collect and then return samples from any asteroid is possible with contemporary technology.

Perhaps the greatest potential for insight into the nature of asteroids would be a human expedition to such a body. Shortly after the adoption of the Vision for Space Exploration in 2004, National Aeronautics and Space Administration (NASA) entertained the possibility of sending a crewed Orion Crew Exploration vehicle into deep space for a rendezvous with an asteroid. In terms of propulsion requirements, it is slightly less intensive to send a piloted spacecraft to a near-Earth asteroid than to the Moon. Such a mission could take at least six months to reach a target and up to another year to return to Earth. It could return to Earth large amounts of carefully selected asteroid samples for detailed analysis. The potential for gathering information that might be used someday to divert an asteroid that threatened to impact Earth would be tremendous.

CONTEXT

Asteroids usually cannot be seen with the unaided eye, but they provide important clues for understanding planet formation: They can have major effects on the Earth and, in fact, have had such effects during the planet's history. At one time, it was assumed that the asteroid belt was formed by the breakup of a planet between Mars and Jupiter. However, the combined mass in the belt is much less than that of any planet (only 0.04 percent Earth's mass), and the observed differences in the composition of asteroids at different locations in the belt make it unlikely that they all came from the same planet-sized object. It now appears that asteroids are original debris that was left over after planet formation and that has undergone complex processes such as collisions, fragmentation, and heating. Apparently, strong tidal forces caused by Jupiter's large mass prevented small bodies between it and Mars from combining to form a single planet in their region.

It appears, therefore, that asteroids are among the oldest objects in the solar system, left over from the time immediately before planet formation concluded. Studies of these objects should provide clues to the structure and composition of the primitive solar nebulae. Different types of asteroids found in different regions of the solar system support the theory of planetesimal origin through a sequence of condensation from a nebular disk around the Sun. Asteroids farther from the Sun, beyond the main belt, may have contained more ice; those that formed closer, within the belt, may have been primarily stony or stony-iron materials. Some of these planetesimal precursors of asteroids were probably perturbed during close passes by neighboring planets into elongated Apollo-like orbits that cross Earth's orbit. Other objects on similar orbits may have been comets that remained in the inner solar system long enough to lose their volatile ices by evaporation. Processes of collision and fragmentation among these objects provide direct evidence about the earliest forms of matter.

Special interest in Apollo asteroids arises from their potential for Earth collisions. Objects as small as 100 meters hit Earth about once every two thousand years, and the 30 percent that fall on land can produce craters a kilometer in diameter. Such impacts would devastate much wider areas by their shock waves, and dust thrown into the upper atmosphere could have marked effects on climate. Growing evidence suggests that asteroid collisions in the past might have contributed to major extinctions of species, such as the dinosaurs, and perhaps even caused reversals of Earth's magnetism. Thin layers of iridium, often found in meteorites, have been identified in Earth's crust at layers corresponding to such extinctions. Satellite photography has revealed about one hundred apparent impact craters on Earth with diameters up to 140 kilometers. It is likely that many more succumbed to processes of erosion. Knowledge of Apollo orbits might make it possible to avoid such collisions in the future.

Asteroids also offer the possibility of recovering resources with great economic potential. Some contain great quantities of nickel-iron alloys and other scarce elements; others may yield water, hydrogen, and other materials useful for space-based construction. Estimates of the economic value of a kilometer-sized asteroid reach as high as several trillion dollars. A well-designed approach to space mining might someday help to take pressure off Earth's ecosystem by providing an alternative to dwindling resources, and space-borne manufacturing centers might alleviate pollution on Earth.

Joseph L. Spradley and David G. Fisher

FURTHER READING

Barnes-Svarney, Patricia. *Asteroid: Earth Destroyer or New Frontier?* New York: Basic Books, 2003. In-depth coverage of technical issues about asteroids necessary to understand the danger that an impact on Earth represents. Makes connections to science-fiction stories involving such disasters and allows the reader the ability to determine what is often incorrectly portrayed in doomsday documentaries and fiction with asteroid impact themes.

Beatty, J. Kelly, Carolyn Collins Petersen, and Andrew Chaikin, eds. *The New Solar System.* 4th ed. Cambridge, Mass.: Sky, 1999. A richly illustrated summary of early space-age discoveries that radically revised knowledge of the solar system. Discusses the various types of asteroids.

Bobrowsky, Peter T., and Hans Rickman, eds. *Comet/Asteroid Impacts and Human Society: An Interdisplinary Approach.* New York: Springer, 2007. Suitable for an interdisciplinary college course at the introductory level about the science and societal issues related to an impact on Earth by a near-Earth asteroid or comet.

Gehrels, Tom, ed. *Asteroids.* Tucson: University of Arizona Press, 1979. A classic, authoritative and comprehensive book on asteroids available in English. It contains about fifty articles on every aspect of asteroid research, including extensive references to original research papers. Most articles are technical, but the first seventy-five pages provide a readable introductory survey. Tabulations in the last section provide data of various kinds on all asteroids that have been studied.

Harland, David H. *Jupiter Odyssey: The Story of NASA's Galileo Mission.* New York: Springer Praxis, 2000. Provides virtually all of NASA's press releases and science updates during the first five years of the Galileo mission in a single work, including Galileo's encounters with asteroids. Includes an enormous number of diagrams, tables, lists, and photographs.

Hartmann, William K. *Moons and Planets.* 5th ed. Belmont, Calif.: Thomson Brooks/Cole, 2005. An updated version of a classic text that covers all aspects of planetary science. Results for the entire NEAR mission are presented. Additional material relating to asteroids is included in chapters on comets, meteorites, planetary evolution, and cratering. An appendix on planetary data includes some asteroid data for comparison, and an extensive bibliography includes about seventy entries on asteroids.

Lewis, John S. *Rain of Iron and Ice: The Very Real Threat of Comet and Asteroid Bombardment.* New York: Basic Books, 1997. A comprehensive survey of meteorites, impact-cratering processes, and the concept of cataclysm. About the latter, the book provides a historical look at the change in scientists' belief in uniformitarianism to their recognition of catastrophism. Plans for preventing a major impact are discussed.

Time-Life Books. *Comets, Asteroids, and Meteorites.* Alexandria, Va.: Author, 1990. Heavily illustrated in color, offers an excellent collection of photographs of comets, pictures of meteorites, and descriptions of asteroids.

See also: Ceres; Comet Halley; Comet Shoemaker-Levy 9; Comets; Dwarf Planets; Eris and Dysnomia; Impact Cratering; Kuiper Belt; Meteorites: Achondrites; Meteorites: Carbonaceous Chondrites; Meteorites: Chondrites; Meteorites: Nickel-Irons; Meteorites: Stony Irons; Meteoroids from the Moon and Mars; Meteors and Meteor Showers; Oort Cloud; Planetary Interiors; Planetary Orbits; Planetary Orbits: Couplings and Resonances; Pluto and Charon; Solar System: Origins.

Auroras

Category: Earth

Auroras, commonly called the northern and southern lights, are caused by geomagnetic activity taking place in a planet's atmosphere. By understanding auroras, scientists can gauge the effects of solar activities on planetary environments.

OVERVIEW

Auroral phenomena were first observed on Earth. Only later were such phenomena detected on other planets in the solar system. "Aurora" is a general term for the light produced by charged particles interacting with the upper reaches of the Earth's atmosphere. The term "aurora borealis" specifically refers to the northern dawn, or northern lights; "aurora australis" refers to the southern lights. Auroras appear in an oval girdling the Earth's geomagnetic poles, where magnetic field lines become nearly perpendicular to the surface. In this region, the Earth is not shielded from the space environment as it is at lower latitudes, where magnetic field lines can be almost parallel to the surface. Thus, electrons and ions moving along magnetic field lines can strike the atmosphere directly. Normally, the auroral oval is located about 23° from the north magnetic pole and 18° from the south magnetic pole. Because the north magnetic pole is located in Greenland, the oval is offset toward Canada and away from Europe. Generally, auroras appear at altitudes between 100 and 120 kilometers high, in sheets 1 to 10 kilometers thick and several thousand kilometers long.

The auroral oval is a product of the Earth's magnetic field and is driven by the Sun's output of charged particles. The oval can be enlarged as far north or south as 20° latitude; its normal range is around 55-60°. These variations in range and intensity have been correlated with sunspots, showing that solar activity is the engine that drives auroras and other geomagnetic disturbances. Additionally, scientists usually describe auroral activity in terms of local time relative to the Sun rather than the geographic point over which it occurs. Thus, the Earth can be considered to be rotating beneath auroral events (even though the shape of the oval remains skewed). The first indication that auroral displays might be linked to solar activity came in 1859, when Richard Carrington observed an especially powerful solar flare in white light. A few hours later, he observed a strong auroral display and suspected that the two might be linked.

Electrons impinge upon the upper atmosphere from this field-aligned current, moving

17

in a helical path about Earth's magnetic field lines. The helix of electrons trapped in the Earth's magnetic field will become more pronounced as they approach the poles, until finally their direction is reversed at the "mirror point" and they are reflected back to the opposite pole. Motion back and forth is quite normal. If the electrons are accelerated down into the ionosphere, they encounter oxygen atoms and nitrogen molecules. These collisions will release Bremsstrahlung (braking) radiation. These X rays are absorbed by the atmosphere or radiated into space. The oxygen is dissociated from molecular oxygen and then ionized and electrons freed. Those oxygen ions radiate light when neutralized by free electrons. Nitrogen either is excited and radiates when it returns to the "ground" state or is dissociated and excited. As the atmosphere dwindles gradually into the vacuum of space, starting at about 60 kilometers above the surface of the Earth, the atmosphere forms an electrified layer called the ionosphere, where oxygen and nitrogen molecules are dissociated by sunlight. Because many of these free atoms and molecules are also ionized by sunlight, electric fields and currents move freely, although the net electrical charge is zero.

Auroral structure varies widely. Three major forms have been discerned: quiet or homogeneous arcs, rayed arcs, and diffuse patches. Homogeneous arcs appear as "curtains" or bands across the sky. They sometimes will occur as pairs, and rarely as sets of parallel arcs. They have also been described as resembling ribbons of light. The lower edge of the arc will be sharply defined as it reaches a certain density level in the atmosphere, but the upper edge usually simply fades into space. Pulsating arcs vary in brightness, as energy is pumped in at different rates. Also in the category of quiet arcs are diffuse luminous surfaces, which appear like clouds and have no defined structure; they may also appear as a pulsating surface. Finally, the weakest homogeneous display is a feeble glow, which actually is the upper level of an auroral display just beyond the horizon.

Auroras with rays appear as shafts of light, usually in bundles. A rayed arc is similar to a homogeneous arc but comprises rays rather than evenly distributed light. The formation and dissipation of individual rays may produce the illusion that rays are moving along the length of a curtain. Among rayed arcs, the "drapery" most resembles a curtain and is most active in shape and color changes. If the viewer is directly below the zenith of an auroral event, then it appears as a corona, with parallel rays appearing to radiate from a central point. Drapery displays are often followed by flaming auroras that move toward the zenith.

A controversial aspect of auroras is whether they produce any sound. Many observers from antiquity to the present have reported hearing auroras; however, sensitive sound-recording equipment has yet to capture such sounds. This leaves open the question of whether or not the sound is a psychological perception, an electrostatic discharge, or some other phenomenon.

A view of an aurora, caused by a coronal mass ejection, from the International Space Station. (NASA)

Auroral colors—pink, red, green, and blue-green—are distinct and correspond with specific chemistry rather than being a continuous spectrum typical of thermal radiation. Major emissions come from atomic oxygen (at 557.7- and 630-nanometer wavelengths) and molecular nitrogen (at 391.4, 470, 650, and 680 nanometers). These emissions come from distinct altitudes. The green oxygen line (557.7 nanometers), which peaks at 100 kilometers in altitude, is caused by an excited energy state that relaxes in 0.7 second. The red oxygen line (630 nanometers), which peaks at about 300 kilometers, comes from an excited energy state that relaxes in 200 seconds. While oxygen is energized to this level at lower altitudes, its excited energy will be lost to collisions with other gases long before it can relax naturally. From such comparisons, geophysicists have deduced some of the vertical structure of the atmosphere. X rays and ultraviolet light are also emitted but cannot be detected from the ground. The Dynamics Explorer 1 satellite recorded auroras at 130 nanometers in hundreds of images taken several Earth radii above the North Pole. Numerous Earth-sensing satellites continue to observe and study ionospheric physics and auroral phenomena.

Auroral brightness can vary widely. Four levels of international brightness coefficients (IBCs) are assigned, ranging from IBC I, which is comparable to the brightness of the Milky Way, to IBC IV, which equals the illumination received from a full moon. Auroras usually are eighty times brighter in atomic oxygen than in ionized nitrogen molecules, indicating their origins higher in the atmosphere. Doppler shifting is commonly recorded in the spectra around 656.3 nanometers (hydrogen-alpha), indicating the motion of protons that are neutralized and reionized as they accelerate up or down magnetic field lines. It is theorized that only a small fraction (about 0.5 percent) of the energy that goes into auroras actually produces visible light. The remainder goes into radio waves, ultraviolet rays, and X rays, and into heating the upper atmosphere.

Images from the Dynamics Explorer satellite confirmed the indication by ground-based camera chains that auroras are uneven in density and brightness. One image, for example, shows that auroras thin almost to extinction on the dayside but expand to several hundred kilometers in thickness between about 10:00 P.M. and 2:00 A.M. local time. "Theta" auroras have been recorded in which a straight auroral line crosses the oval in the center, giving the appearance of the Greek letter Θ. This phenomenon may be caused by the splitting of the tail of plasma sheets which extends well into the tail of the magnetosphere, or by the solar wind's magnetic field when it has a direction opposite to that of the Earth.

Photographs taken by spin-scan auroral imagers aboard the Dynamics Explorer 1 satellite show that auroral substorms start at midnight, local time, and expand around the oval. Observations of hundreds of substorms show that they have the same generalized structure but that no two are alike. The satellite imager also showed expansions and contractions in the aurora in response to changes in the interplanetary magnetic field and solar wind. As solar wind plasma meets the Earth's magnetosphere, a shock wave is formed, and the wind is diverted around the Earth. This diversion compresses Earth's magnetic field on the sunward side, while it extends like a comet's tail on the nightside. When the field of the solar wind is oriented toward the south, its field lines reconnect with the field lines of the Earth and allow protons and electrons to enter the magnetosphere; they are normally blocked when the field is oriented to the north.

Auroral activity is strongly driven by the solar wind. If the magnetic field of the wind points north—aligned with the Earth's magnetic field—then the auroral oval is relatively small, and its glow is hard to see. When the solar wind's magnetic field reverses direction, a substorm occurs. The oval starts to brighten within an hour, and bright curtains form within it. At its peak, the oval will be thinned toward the noon side and will be quite thick and active on the midnight side. As the storm subsides and starts to revert to normal about four hours after the field is reversed, the aurora dims and curtains form. Finally, a large, diffuse glow covering the pole may be left as the field becomes stronger in the northward direction.

Two curtain-patterned "dueling auroras," as seen over the Yukon in October, 2001. (©Phil Hoffman/Courtesy, NASA)

Electrojets also form in auroras at low altitudes from an effect known as "E-cross-B drift" (written E × B). At high altitudes, electrons and protons flow freely because there is low gas density and no net current change. At lower altitudes, around 100 kilometers, protons are slowed by collisions with atoms and molecules, but electrons continue to move unopposed. The result is a pair of electrojets, eastward (evening) and westward (morning), which flow toward midnight, then cross the polar cap toward noon. These electrojets heat the ionosphere, especially during active solar periods, when auroras are more intense. This E × B drift in auroral ovals appears to be a major source of plasma for the magnetosphere. It appears that positively charged ions are accelerated upward along the same magnetic field lines, whereas negatively charged electrons precipitate downward. Hydrogen, helium, oxygen, and nitrogen ions compose this flow. Each ion has the same total energy, so their paths vary according to mass. The net effect is that of an ion fountain blowing upward from the auroras which spreads by a wind across the poles.

A little-known subset of the aurora is the subauroral red (SAR) arcs, which appear at the midlatitudes; the magnetic field lines on which they occur are different from those on which auroras appear. SAR arcs, which always emit at 660 nanometers (from oxygen atoms), are dim and uncommon. Modern instrumentation has shown that the SAR arcs are a phenomenon separate from the polar auroras. These arcs may be caused by cold electrons in the plasmasphere interacting with plasma waves or with energetic ions. SAR arcs are believed to originate at an altitude of approximately 19,000-26,000 kilometers during especially strong geomagnetic storms, although the arcs themselves appear at altitudes of around 400 kilometers as the energy from the storm leaks or is forced downward.

Auroras also "appear" in the radio spectrum. Studies in the twentieth century showed that auroras could be sounded by radar at certain

The flow of the solar wind past the magnetosphere generates massive electrical currents, which flow mostly from one side of the magnetosphere to the other. Some of the currents, however, connect down the Earth's magnetic field, into and through the auroral oval. Because an electric current is caused by the flow of charged particles, in the process electrons are brought directly into the ionosphere around the poles. Primary currents enter around the morning side and exit around the evening side. Secondary currents flow in the opposite direction. Changes in electrical potential of the magnetosphere, as when it is pumped up by particles arriving in the solar wind, will force the electrons through the mirror point. They are then accelerated deeper into the ionosphere. This auroral potential structure is thin but extends around the auroral oval for thousands of kilometers even to the point of closing in on itself.

frequencies. Satellites in the 1970's started recording bursts of energy in the low end of the AM radio spectrum. This radiation is called auroral kilometric radiation (AKR), because its wavelength is up to 3 kilometers, reflected outward by the ionosphere. Bursts can release 100 million to 1,000 million watts at a time, making them far more powerful than conventional broadcasts by humans. Bursts originate in a region of the sky about 6,400-18,000 kilometers high in the evening sector of the auroral oval. Because the radiation is polarized, it is likely that AKR is caused directly by electrons spiraling along magnetic field lines.

Earth is not the only planet to display auroras. Earthly auroral activity has a power rating of approximately 100 billion watts. Auroral displays on Jupiter were detected by the Voyager 1 spacecraft in 1979. During the 1990's the Hubble Space Telescope performed several investigations of this phenomenon. Auroral activity on Jupiter is hundreds of times stronger than on Earth and also appears always to be energized rather than intermittent, as on Earth. In 2007 the Chandra X-Ray Observatory and Hubble Space Telescope conducted a coordinated study of Jovian auroral activity, seeing the phenomenon in both visible and ultraviolet (Hubble) and X-radiation (Chandra) simultaneously. Observing in multiple wavelengths provides clues to the basic physical process involved that seeing an aurora only in visible light cannot.

Saturnian auroras presented a problem with the current understanding of how auroral displays are produced. In 2005 the Hubble Space Telescope and Cassini spacecraft performed coordinated observations of Saturn's auroral activity. Hubble observed in ultraviolet and visible wavelengths, whereas the Cassini probe in orbit about Saturn recorded radio emissions tied to the auroras. An oddity of Saturn's auroras is that, whereas Jupiter's are not affected very much by the solar wind, Saturn's appears to be. Another is that Saturn's auroras brighten on the portion of the planet where darkness leads to sunlight as the storm increases in intensity, which is not the case for Earthly or Jovian auroras.

Uranus's auroras were detected by the Voyager 2 spacecraft's ultraviolet spectrometer.

Contemporary studies of Uranian auroras have been performed by the Hubble Space Telescope. The auroral displays of Uranus mimic those on Earth. Uranus's rings have swept clean much of the region that otherwise would have been a rich collection of trapped charged particles that could be taken down along magnetic field lines into Uranus's magnetosphere to generate auroras. Uranus displays both auroras that are centered about its magnetic poles and the subauroral red arcs seen on Earth. Indeed, on Uranus the SAR arcs are more prevalent than auroras that are centered about the magnetic poles. The latter variety are believed to result from currents that connect Uranus's rather unusual satellite Miranda to the gas giant's magnetic pole. Uranian auroras generate only weak radio signals.

Voyager 2 detected auroral activity on Neptune. Studies of these displays reveal that Neptune's auroral activity is only half as energetic as that on Earth, despite the disparity in size of the two planets. Also, due to the complexity of Neptune's magnetic field, auroral activity is found on Neptune over areas of the planet far from the magnetic poles.

Ironically, even though the planet has no significant magnetic field, Mars also appears to have auroral displays. Data from Mars Global Surveyor and the European spacecraft Mars Express indicated hundreds of aurora-like displays with less dramatic color variations than those observed on other planets in the solar system. Since a planetary magnetic field is not responsible, some researchers suggest that primordial magnetism associated with patches of the planet's crust, particularly in the southern hemisphere, may be involved in this auroral phenomenon. Martian auroral displays show up mostly in the ultraviolet range, with little or no visible counterpart.

METHODS OF STUDY

The space age in some small measure owes its birth to a fascination with auroral phenomena. It was the desire to study and understand the Earth-space interface around the globe at high altitude that resulted in launching the first satellites, during the International Geophysical Year in 1957-1958. Until then, ground-based photography and instrumentation were almost the only methods of studying auroras. Aircraft and rockets played lesser roles. Ground-based instrumentation in the 1940's and 1950's confirmed that auroras were linked to the geomagnetic field, for studies showed that auroras occurred in a circle around the north magnetic pole. Photography of auroral displays has always been difficult because the activity is dynamic, sometimes changing from second to second. Not until the 1950's were electronic devices available to analyze the entire auroral spectrum visible from the ground.

Satellites in the 1970's and 1980's expanded the array of instruments available to investigators. While fields and particles instrumentation

An aurora, caused by both solar plasma and debris from Comet Swift-Tuttle, appeared over the observatory at Mount Megantic in Quebec, Canada, in August, 2000. (Sébastien Gauthier/Courtesy, NASA)

has been used to analyze gases and plasmas, imaging instruments have been equally revealing. Notable cameras of various sorts have been carried by Dynamics Explorer 1, the U.S. Air Force HiLat (high latitude) satellite, and the Swedish Viking satellite. In addition, some imaging was performed by polar-orbit weather satellites, but with lower spectral and spatial resolutions. The Skylab crews observed some auroral activity. The space shuttle-based Spacelab 3 crew in 1985 photographed auroras from above the atmosphere. Combining photographs taken a few seconds apart allowed formation of stereo imagery so that the structure could be studied better. In other experiments, small electron guns have been carried into space aboard rockets, on spacecraft, and within the payload bays of space shuttle orbiters. These fire electrons back into the atmosphere in an attempt to generate artificial auroras. An electron gun flew on the first Spacelab (on the space shuttle STS-9 mission) and produced some interesting results. Another such project, on STS-45, was part of the National Aeronautics and Space Administration's (NASA's) INSPIRE program to involve secondary students and undergraduate students in this type of auroral research; unfortunately, although the response of school groups worldwide was strong, the electron gun failed early in the mission.

A key finding from satellite-based research was that auroras are often more active on Earth's dayside, although sunlight and sky completely overwhelm it. Large quantities of radiation are generated in the ultraviolet. This radiation is not seen at the Earth's surface because the atmosphere selectively absorbs such light.

Other planets with magnetospheres also display auroral activity. Such displays are observed almost exclusively by spacecraft either in Earth orbit, like the Hubble Space Telescope, or from spacecraft that either fly by or orbit other planets. The Galileo orbiter routinely picked up auroral activity on Jupiter, and so did the Cassini orbiter on Saturn. Voyager 2 flew past Uranus (1986) and Neptune

(1989) and detected auroral activity on these planets as well.

CONTEXT

Auroras are the most visible manifestation of the interaction between the Earth and space. The study of plasmas has been enhanced by observations made of them. A clear understanding of auroras will provide a means of diagnosing activities in the magnetosphere and the effects of solar activities on the terrestrial environment.

Auroras also serve as a means to study the magnetohydrodynamics of stars and planets. The physics of planetary auroras are essentially the same as that of Earth's auroral activity, although the energies and chemistries involved may be vastly different. Thus, terrestrial auroras can serve as a laboratory for testing basic theories. Planets with magnetic fields also have auroral activity. Much of Jupiter's radio noise is caused by auroral kilometric radiation, and the Einstein Observatory recorded X rays that apparently came from Bremsstrahlung radiation in the Jovian atmosphere. The Voyager 1 spacecraft observed a 29,000-kilometer-long aurora on the nightside of Jupiter, as well as lightning pulses in and above the clouds coincident with the auroral activity. Voyager 2 did too, and it continued on to the rest of the gas giants in the solar system to do the same. Follow-on spacecraft investigations of these planets, and also on Mars, continued to devote considerable effort to characterizing and understanding the nature of auroral activity production. Each planet's auroral activity provides insight into the nature of that planet's magnetic field and also the plasma environment about it.

Dave Dooling

FURTHER READING

Akasofu, Syun-Ichi. "The Dynamic Aurora." *Scientific American* 260 (May, 1989). A detailed, college-level treatment of auroras, written by a physicist who is generally accepted as a world expert.

Bone, Neil. *The Aurora: Sun-Earth Interaction.* New York: John Wiley, 1996. One volume in the Ellis Horwood Library of Space Science and Space Technology. Devoted to describing the electrodynamics of the Sun-Earth environment that produce auroral displays.

Bothmer, Volker, and Ioannis A. Daglis. *Space Weather: Physics and Effects.* New York: Springer Praxis, 2006. A selection from Springer Praxis's excellent Environmental Sciences series, this is an overview of the Sun-Earth relationship and provides a historical and technological survey of the subject. Projects the future of space weather research through 2015 and includes information about contemporary spacecraft.

Delobeau, Francis. *The Environment of the Earth.* New York: Springer, 1971. A technical description of the terrestrial environment, written as a reference for space scientists. Although the work is dated by subsequent discoveries, its description of auroral chemistry is still valid.

Dooling, Dave. "Satellite Data Alters View on Earth-Space Environment." *Spaceflight* 29, suppl. no. 1 (July, 1987): 21-29. An article focusing on the exploration of the magnetosphere by the Dynamics Explorer satellites, with details on auroral imaging and radiation.

Moldwin, Mark. *An Introduction to Space Weather.* Cambridge, England: Cambridge University Press, 2008. This text introduces space weather, the influence the Sun has on Earth's space environment, to the nonscientist. Discusses both the scientific aspects of space weather and issues of technological and societal import.

Savage, Candace. *Aurora: The Mysterious Northern Lights.* New York: Firefly Books, 2001. Provides a history of scientific investigation of auroral phenomena. Heavily illustrated with auroral displays.

See also: Coronal Holes and Coronal Mass Ejections; Earth-Moon Relations; Earth-Sun Relations; Earth's Atmosphere; Earth's Magnetic Field: Origins; Earth's Magnetic Field: Secular Variation; Earth's Magnetic Field at Present; Earth's Magnetosphere; Eclipses; Greenhouse Effect; Interplanetary Environment; Neptune's Magnetic Field; Planetary Magnetospheres; Saturn's Magnetic Field; Solar Flares; Solar Wind; Uranus's Magnetic Field; Van Allen Radiation Belts; Venus's Atmosphere.

Big Bang

Category: The Cosmological Context

The big bang theory was developed to explain the origin of the expanding universe, uniting cosmology with general relativity and elementary particle physics. About 13 to 14 billion years ago, an explosion called the big bang created energy and matter, space and time. Ever since, space has been expanding with time, carrying matter and electromagnetic radiation with it. As space has expanded, its contents have evolved.

OVERVIEW

Sir Isaac Newton's law of universal gravitation led him to suggest that a static universe with a finite distribution of stars would collapse, but that an infinite universe could be stable. The possibility of an expanding universe, however, is contained within Albert Einstein's general theory of relativity, which he published in 1915. In 1917, Einstein himself found that his general theory in its original form would not permit a static universe. Because the scientific consensus then was that the universe on the large scale is static and unchanging, Einstein added an arbitrary constant (later called the cosmological constant) to his field equations to allow static solutions. Physically, the cosmological constant represents a long-distance repulsion that would balance gravitational attraction on a cosmic scale and thus permit a static universe.

Just five years later, in 1922, the Russian mathematical physicist Alexander Alexandrovich Friedmann found two solutions to the original general relativistic field equations (without the cosmological constant) in which the universe initially expands with time. In one (called "open"), the universe continues to expand forever. In the other (called "closed"), the universe expands to some maximum size, after which it contracts.

In 1927, the Belgian priest and cosmologist Abbé Georges Lemaître independently derived the same two solutions to the field equations of general relativity that Friedmann had obtained earlier. However, Lemaître went further, speculating about the origin of the expansion. Extrapolating backward in time, he realized that everything in the universe would come together at the same time in the distant past, thus pointing to a unique beginning of the universe. He envisioned all matter and space compressed into a "primeval atom" that split into all the atoms of all the elements present in the universe. An enormous explosion initiated the expansion of space and its fragmented matter. As he described the aftermath,

> The evolution of the world could be compared to a display of fireworks just ended—some few red wisps, ashes, and smoke. Standing on a well-cooled cinder, . . . we try to recall the vanished brilliance of the origin of the worlds.

Today we know that the chemical elements could not have been created the way Lemaître proposed. However, Lemaître's basic idea was prophetic. Many years later, the explosive origin of the universe was dubbed the "big bang." Just before his death in 1966, Lemaître learned of the discovery of the cosmic microwave background, which is greatly redshifted radiation emitted just a few hundred thousand years after the big bang—the "vanished brilliance of the origin of the worlds" about which he had speculated so many years earlier.

Observational confirmation that the universe actually is expanding came in 1929, when Edwin Hubble, assisted by Milton Humason, showed there is a correlation between galaxy distances and the redshifts of their spectra; the farther away a galaxy is, the more its spectrum is redshifted. The cause of this redshift, termed cosmological, is the expansion of the universe. As the universe expands, wavelengths of electromagnetic radiation are stretched by the expansion, so visible light is shifted toward longer, redder wavelengths. (The term "redshift" has come to be applied to a shift to longer wavelengths of any part of the electromagnetic spectrum.)

Starting in 1935, Friedmann's student George Gamow began work on more rigorously developing Lemaître's hypothesis of an explosive origin. Gamow proposed that the very dense initial state would have been very hot, and the universe cooled as it expanded. In 1946, he suggested that the primordial substance, which he called "ylem," had consisted of neutrons at a temperature of about 10 billion degrees, some of which decayed during the early stages of expansion to form protons and electrons. Successive interactions of the neutrons

and protons then led to the formation of all the chemical elements by nuclear fusion reactions while the early universe still was very hot and dense. Gamow worked out the details of this nucleosynthesis of all the chemical elements with his colleague Ralph A. Alpher at George Washington University. Before they published their results in 1948, Gamow persuaded Hans Albrecht Bethe, the physicist who first described nuclear fusion reactions in stars, to allow them to add his name to their paper to make the list of authors "Alpher, Bethe, Gamow," a wordplay on the first three letters of the Greek alphabet. This came to be referred to as the alpha-beta-gamma theory of the origin of the universe and its chemical elements. (Today we know that the early universe cooled too quickly

George Gamow: Physicist, Cosmologist, Geneticist

Born March 4, 1904, in Odessa, Russia, George Gamow started his scientific career as a boy, when his father gave him a telescope for his thirteenth birthday. Little did his father know that his son would one day become one of the greatest scientists of the twentieth century.

After graduating from the University of Leningrad in 1926, Gamow went to Göttingen, a center for the study of the new quantum mechanics. At this time, natural radioactivity was the focus of research of many of the great physicists of the day, from the Curies to Lord Rutherford, and Gamow was particularly interested in its relationship to the atomic nucleus. In 1928, he made his first great contribution when he described quantum tunneling of alpha particles to explain the radioactive process of alpha decay. His investigation of the atomic nucleus would take him to Copenhagen, where he worked under Niels Bohr laying the theoretical groundwork for nuclear fusion and fission.

During the 1930's, Gamow taught at universities in Copenhagen, Leningrad, Cambridge, Paris, and the United States. In Washington, D.C., he and Edward Teller worked on the theory of beta decay. He also turned his attention to astrophysics and the origin of the elements. This work led to his 1948 proposal of the "big bang" theory of the universe, for which he is best known.

Gamow was more than a theoretical physicist, however: Known for his sense of humor and revered by his students, he was also devoted to education. His "Mr. Tompkins" series used science fiction to explain difficult science in a way that anyone—including Tompkins, whose attention span was notoriously short—could understand. In 1954, inspired by the Watson-Crick DNA model, he theorized that the order of the DNA molecules determined protein structure. The problem, as he saw it, was to determine how the four-letter "alphabet" of nucleic acid bases could be formed into "words." His "diamond code" paved the way for Marshall W. Nirenberg to crack the genetic code in 1961.

In 1956, Gamow settled in Boulder to teach at the University of Colorado. That year, he received UNESCO's Kalinga Prize for his popularization of science, and two years later he was married (a second time) to Barbara "Perky" Perkins, who initiated the George Gamow Lecture Series after his death, in 1968.

(AP/Wide World Photos)

for most of the chemical elements to have formed then; almost all the atoms heavier than helium were formed later by nuclear fusion reactions in stars.)

Gamow and his associates tried to work out other physical processes that would have occurred in the intensely hot, compressed fireball from which the universe expanded. In the same year 1948, Alpher and Robert C. Herman (another of Gamow's colleagues) published a further analysis that predicted a cosmic background radiation left over as a kind of relic from the early hot, dense universe. Because of the expansion of the universe and the corresponding redshift of this radiation, they predicted that it would have cooled from an initial high temperature to only about 5 kelvins at the present time. Since such radiation would be in the microwave part of the spectrum, they had no way of detecting it then, and their prediction was forgotten until the 1960's.

In the 1960's, a team of physicists at Princeton University—Robert H. Dicke, P. J. E. Peebles, P. G. Roll, and David T. Wilkinson— began planning to build an instrument to detect the cosmic background radiation predicted by Alpher and Herman. However, it was accidentally discovered first by Arno A. Penzias and Robert W. Wilson at Bell Telephone Laboratories in 1965. They were using a large radio horn antenna as part of a communication satellite program when they detected microwave radiation coming uniformly from all directions and corresponding to a temperature of about 3 kelvins. Since the signal was so uniform, they thought it might be due to some instrumental noise. Pigeons roosted in the antenna, and Penzias described "a white sticky dielectric substance coating the inside of the antenna." Chasing away the pigeons and cleaning out their droppings did not get rid of the signal. Eventually it was identified by Dicke and his colleagues at Princeton University as the relic radiation from the primeval fireball predicted by Alpher and Herman. In 1978, Penzias and Wilson received the Nobel Prize in Physics for their serendipitous discovery that provided convincing confirmation of a big bang origin for the universe.

Measurements from Earth-orbiting space-craft such as the Cosmic Background Explorer (COBE), launched in 1989, and the Wilkinson Microwave Anisotropy Probe (WMAP), launched in 2003, as well as high-altitude balloons launched from Antarctica, including the Balloon Observations of Millimetric Extragalactic Radiation and Geomagnetics (BOOMERANG) project, have shown that the microwave background spectrum perfectly fits a thermal radiator at a temperature of 2.73 kelvins. The shape of the spectrum is exactly what would be expected for radiation from the early universe, when matter and radiation were in thermal equilibrium; the shape of the original blackbody spectrum has been preserved during the subsequent expansion and cooling. However, the background radiation is not precisely uniform in all directions. There are temperature variations of up to about 0.00001 kelvin over regions with an angular size of about one degree of arc. These variations are thought to represent slight differences in density in the early universe that ultimately produced the "lumpy" universe of clusters of galaxies that we observe today.

The equations that describe the expansion of the universe can be extrapolated back to very early times of incredibly high densities and temperatures, but only to about 10^{-43} second (called the Planck time). Before that time, conditions were so extreme (for example, temperatures in excess of 10^{32} kelvins) that the current understanding of physics breaks down. Scientists believe that, at that time of high temperatures and correspondingly high energies, the four fundamental forces of nature—gravity, strong nuclear, weak nuclear, and electromagnetic— were indistinguishable from one another, or "unified." However, physicists have no workable "theory of everything" (TOE) to describe this unification of forces. At about the Planck time, gravity would have separated (or "frozen out") from the other three forces as temperature and energy decreased. After about 10^{-35} second, when the temperature had decreased below about 10^{28} kelvins, the strong nuclear force separated from the other two. The energy released by this "freeze out" of the strong nuclear force may have initiated a brief period of cosmic inflation, during which the universe increased in size by a factor of 10^{50} in 10^{-32} second.

This artist's rendering shows how the universe might have looked shortly after the big bang, when matter began to form into stars. (NASA/JPL-Caltech/R. Hurt, SSC)

High-energy elementary particle physics has been employed to work out more details of the early development of the universe and its contents. The first particles and antiparticles that would have materialized from energy according to ideas about mass-energy equivalence (as expressed by Einstein's famous equation $E = mc^2$) and particle-antiparticle pair creation are not well understood, although unified field theories are beginning to suggest their possible properties. At the very high temperatures of the early universe, particles moved so fast that they avoided any interaction, but as the universe cooled they could interact to produce new forms of matter, leading to an era dominated by quarks. The known laws of physics can account for the particles that would have existed after about 10^{-12} second at a temperature of about 10^{16} kelvins. At that time, space was filled with photons, quarks, and leptons (electrons, neutrinos, and the like), along with their antiparticles.

After 10^{-6} seconds, the universe had cooled enough so that no more quark-antiquark pairs could be created. From then on, quarks and antiquarks mutually annihilated each other,

producing a brilliant fireball of gamma-ray photons. Equal numbers of quarks and antiquarks had been produced, but some asymmetry resulted in a slight excess of quarks over antiquarks by about one part in a billion. It seems that the asymmetry may occur in the weak nuclear force, which provides a way for antiquarks to decay but no equivalent way for quarks. After the quark-antiquark annihilations were over, all the antiquarks and most of the quarks were gone, but about one quark in a billion had survived; they combined to form protons and neutrons, which went on to become the matter in the universe today.

Neutrons decayed into protons by emitting electrons and antineutrinos, and protons combined with electrons to form neutrons and neutrinos. They were kept nearly equal in number by thermal equilibrium as long as electrons were abundant. When the universe was a few seconds old and the temperature fell to about 6 billion kelvins, photons no longer had enough energy to produce electron-positron pairs. Soon, all positrons and all but one electron out of a billion had mutually annihilated in another burst of gamma-ray photons. With so few electrons remaining, no new neutrons were formed, and the number of neutrons declined as they decayed into protons.

Before all the neutrons decayed, some joined with protons to form nuclei of deuterium (also called heavy hydrogen). However, while the universe was hot enough, gamma-ray photons could break deuterium nuclei apart. After about three to four minutes, when the temperature had dropped below about 1 billion kelvins, photons no longer had enough energy to break up deuterium nuclei, so they could survive. In rapid succession, the deuterium nuclei then collided with protons and neutrons to form helium nuclei, and soon almost all the remaining neutrons combined to form helium. When this nucleosynthesis began, there was about one neutron for every six protons. Using almost all the neutrons to form ordinary helium nuclei resulted in one helium nucleus (two protons and two neutrons) for every ten hydrogen nuclei (each just a single proton), and this very closely matches the cosmic abundance of helium and hydrogen observed today. A few lithium nuclei and even fewer beryllium nuclei also formed, but the temperature dropped too quickly for there to be time to form heavier nuclei. After about fifteen minutes, the temperature had dropped below 400 million kelvins, and nucleosynthesis ended. (The heavier elements eventually formed at much later times through nuclear fusion reactions in stars.) Throughout the early universe, the radiation density exceeded the matter density, but radiation density decreased more rapidly than matter density. After several thousand years, the two densities were equal, and from that time on, matter has been dominant.

When nucleosynthesis ended, one electron remained for each free or bound proton in hydrogen and helium nuclei, but the universe was too hot for electrons to combine with nuclei to form neutral atoms. Free electrons are very effective at scattering photons, so the universe was opaque to electromagnetic radiation. The universe expanded for several hundred thousand years before it was cool enough (about 3,000 kelvins) for electrons to combine with nuclei to form neutral atoms. When this happened, the lack of free electrons made the universe transparent to electromagnetic radiation, and photons were free to travel through space. This decoupling of matter and radiation was the source of the cosmic microwave background radiation. As the universe continued to expand, it stretched the wavelengths and effectively cooled the primeval "relic" radiation until it reached the present temperature of 2.73 kelvins.

APPLICATIONS

The cosmological interpretation of redshifts attributes these spectral shifts to the stretching of wavelengths of electromagnetic radiation as space expands. The greater the redshift, the more space has expanded since the electro-

magnetic radiation was emitted, and the farther back in time one can observe. The Hubble law expresses basically the same idea; the greater the redshift, the farther away the sources and the greater the travel time of electromagnetic radiation to reach Earth. This ability to look back in time aids our understanding of distant objects with large redshifts. Such objects are seen as they were billions of years ago during early stages in the evolution of the universe.

For example, quasars (quasi-stellar radio sources) are mostly unresolved sources with very small angular sizes having very large redshifts and often rapid and erratic changes in brightness. If their redshifts are cosmological (a few astronomers dispute this assertion), then they are very distant. This conclusion, coupled with their apparent brightness, implies that they are incredibly luminous. Their rapid changes in brightness mean they are relatively small in actual size. All of this taken together suggests that they likely are extremely energetic compact phenomena in young galaxies, possibly supermassive black holes forming at the centers of developing galaxies.

The cosmic microwave background radiation is extremely uniform in all directions, but it does have a small asymmetry, being slightly warmer systematically in one half of the sky compared to the other half by a maximum of about 0.007 kelvin. This is interpreted as due to the motion of our solar system through the background radiation field; the temperature difference of 0.007 kelvin in thermal radiation corresponds to a spectral Doppler shift produced by a speed of 380 kilometers per second. Presumably this speed is a combination of the movements of our solar system in the Milky Way galaxy, the Milky Way galaxy in the Local Group of galaxies, the Local Group relative to the Virgo supercluster of galaxies, and maybe other motions as well.

The measured temperature of the cosmic microwave background radiation makes it possible to calculate the expected cosmic abundances of light elements: about 74 percent (by mass) hydrogen, 26 percent helium, a thousandth of a percent deuterium, and a millionth of a percent lithium. All these match the measured abundances within the observational uncertainties. Since no other source for deuterium production is known, these measurements provide additional confirmation for the standard big bang model.

The uniformity of the cosmic background radiation implies thermal equilibrium throughout the universe, even in regions so far apart that electromagnetic radiation has not had time to travel from one to the other. In the early 1980's, Alan Guth proposed "inflation" as a solution to this "horizon problem." He suggested a very early period of rapid expansion at an exponential rate, when the universe increased in size by a factor of 10^{50} in 10^{-32} second. Before inflation, the universe would have been small enough for electromagnetic radiation to travel between all parts of it.

Uniformity also raises a "galaxy problem." How could galaxies and stars form in such a uniform universe? Fine-scale sky maps of the cosmic background radiation made with data from sensitive detectors on the COBE and WMAP spacecraft and BOOMERANG balloons show small temperature (and hence density) variations in the early universe that could have grown by gravity to develop into clusters and superclusters of galaxies. The density variations probably are due to small random quantum fluctuations in the very early universe.

The discovery of the W and Z particles in 1983 by Carlo Rubbia and Simon van der Meer provided support for the electroweak theory, which predicts the electromagnetic force and the weak nuclear force become "unified" or indistinguishable from each other at temperatures above about 10^{15} kelvins. The quark theory predicts a weakening of the strong nuclear force at even higher temperatures; grand unified theories (GUTs) propose that the strong nuclear force is unified with the electromagnetic and weak nuclear forces at temperatures above about 10^{27} kelvins. Theories of everything (TOEs) go still further and suggest the unification of gravity with the other three forces at temperatures above about 10^{32} kelvins.

The high temperatures required to unify the forces occurred shortly after the big bang, making the very early universe a high-temperature laboratory in which it may be possible to test these theories. Gravity would have decoupled at the Planck time, 10^{-43} second. The decoupling of the strong nuclear force at about 10^{27} kelvins would have occurred after about 10^{-35} second and may have released the energy that drove the sudden inflationary expansion of the universe. The last decoupling of the weak nuclear and electromagnetic forces would have occurred at 10^{-11} second, when the temperature had cooled to 10^{15} kelvins.

CONTEXT

Several competing theories have attempted to avoid the creation implications of the big bang theory, but they have not been able to sustain successful alternatives. One of the first was Einstein's early attempt to obtain a solution for a static universe and his introduction of an arbitrary cosmological constant to balance gravitational attraction. When it was later shown that his field equations of general relativity without the cosmological constant were compatible with the observed expansion of the universe, Einstein is reported by Gamow to have remarked that the cosmological constant was the greatest blunder of his life. It is ironic that the early era of cosmic inflation and the recently discovered acceleration of the expansion both involve repulsive forces similar to Einstein's original cosmological constant.

The most serious attempt to defeat the big bang theory was the steady state theory of the universe, introduced in 1948. About the same time that Gamow and his colleagues were working out the details of a Lemaître-type explosive origin, the British cosmologists Hermann Bondi, Thomas Gold, and Fred Hoyle were developing an alternative—the steady state continuous creation model. It was Hoyle who coined the term "big bang" as a derogatory name for the Lemaître-type primordial explosion. However, the name is short and catchy, and it was quickly adopted by most astronomers and physicists no matter which side (if either) they supported.

The steady state theory did not invoke a moment of creation for the entire universe but assumed instead the continuous creation of new matter throughout space at a rate that keeps the mean density of the universe constant for all times as the universe expands. Continuous creation would occur so gradually that it could not be observed until enough matter had been created to form stars and galaxies. Such a steady state universe would be infinite and eternal. Ironically, it required the religious idea of creation *ex nihilo* (from nothing) to avoid another religious idea of a unique creation event in the remote past (the big bang).

Although the steady state theory provided the main competition for the big bang theory during the 1950's and early 1960's, it did not stand the test of time. Since stars and galaxies would form throughout space from the continuous creation of new matter, young and old galaxies should exist side by side. This is contrary to the evidence that galaxies all formed at about the same time, and the galaxies that seem to be much younger (such as quasars) are observed only at great distances and hence at great times in the past. The steady state theory was virtually abandoned after the 1965 discovery of the cosmic background radiation, the relic radiation predicted from the big bang fireball. Even Hoyle, chief spokesman for the steady state theory, helped work out some details of the standard model of the big bang in 1967.

One other attempt to avoid a finite age for the universe was the idea of an oscillating universe. If the density of matter in the universe were large enough eventually to reverse its expansion by gravitational attraction, the universe would collapse toward a "big crunch." The oscillating universe theory proposed that another big bang might follow each big crunch, giving rise to a series of oscillations between successive big bangs, extending indefinitely into the past and future. However, such speculation was laid to rest by the discovery in the 1990's that the expansion of the universe is accelerating, so no contraction seems possible.

Joseph L. Spradley and Richard R. Erickson

FURTHER READING

Barrow, John D., and Joseph Silk. *The Left Hand of Creation*. New York: Basic Books, 1983. A readable account of the origin and evolution of the expanding universe by two astronomers with a good grasp of cosmology. Discusses many theories and problems associated with the big bang model, and includes a good glossary of astrophysical terms and an index.

Chaisson, Eric, and Steve McMillan. *Astronomy Today*. 6th ed. New York: Addison-Wesley, 2008. Very well written college-level textbook for introductory astronomy courses. Two chapters provide a thorough description of the big bang and the evolution of the universe afterward.

Fraknoi, Andrew, David Morrison, and Sidney Wolff. *Voyages to the Stars and Galaxies*. Belmont, Calif.: Brooks/Cole-Thomson Learning, 2006. A well-written, thorough college textbook for introductory astronomy courses. One chapter contains a good description of the big bang and its aftermath.

Freedman, Roger A., and William J. Kaufmann III. *Universe*. 8th ed. New York: W. H. Freeman, 2008. College-level introductory astronomy textbook, thorough and well-written. One chapter contains a good description of the big bang and its aftermath.

Hawking, Stephen W. *A Brief History of Time: From the Big Bang to Black Holes*. New York: Bantam Books, 1988. A very popular and readable account of the development of cosmology and the big bang theory. Includes a helpful glossary and an index.

Jastrow, Robert. *God and the Astronomers*. New York: Warner Books, 1978. A brief but interesting history of the discovery of the expanding universe and development of the big bang theory. Contains many historical photographs of the originators of the theory and supplements on its theological implications.

Lang, Kenneth R., and Owen Gingerich, eds. *A Source Book in Astronomy and Astrophysics, 1900-1975*. Cambridge, Mass.: Harvard University Press, 1979. This volume contains many of the original articles that established the ideas of the expanding universe and the big bang theory, with good introductory sections for each. Contributors include Einstein, Hubble, Friedmann, Lemaître, and Gamow. Some articles are technical, but much can be understood by the general reader.

Schneider, Stephen E., and Thomas T. Arny. *Pathways to Astronomy*. 2d ed. New York: McGraw-Hill, 2008. Very thorough college textbook for introductory astronomy courses.

Divided into short units on specific topics. Several units provide a thorough discussion on the big bang and the evolution of the universe afterward.

Silk, Joseph. *The Big Bang*. Rev. ed. New York: W. H. Freeman, 1989. A good introduction to the standard model of the big bang theory. Includes a good glossary, index, and a thirty-five-page section on mathematical details.

Trefil, James S. *The Moment of Creation: Big Bang Physics from Before the First Millisecond to the Present Universe*. New York: Charles Scribner's Sons, 1983. A good introduction to the big bang theory. Includes a discussion of grand unification and inflationary theories.

Weinberg, Steven. *The First Three Minutes*. New York: Bantam Books, 1977. An excellent introduction to the details of the standard model of the big bang by a leading theoretical physicist and Nobel laureate.

See also: Cosmic Rays; Cosmology; Electromagnetic Radiation: Nonthermal Emissions; Electromagnetic Radiation: Thermal Emissions; General Relativity; Interstellar Clouds and the Interstellar Medium; Milky Way; Novae, Bursters, and X-Ray Sources; Solar System: Element Distribution; Space-Time: Distortion by Gravity; Space-Time: Mathematical Models; Stellar Evolution; Supernovae; Universe: Evolution; Universe: Expansion; Universe: Structure.

Brown Dwarfs

Category: The Stellar Context

Between the giant planets such as Jupiter, in which no nuclear reactions occur, and the small red dwarf stars, in which nuclear reactions produce energy, objects exist whose mass is almost great enough to have initiated a few nuclear reactions but which mostly just radiate the heat that nearly ignited them. Known as brown dwarfs because of the feeble infrared light they emit, the first of these was unequivocally identified only in 1995.

OVERVIEW

Brown dwarfs are defined as objects with masses intermediate between stars and planets—not massive enough to fuse ordinary hydrogen nuclei (consisting of single protons, H^1) into helium in their cores as "real" stars do at some stage, but massive enough to generate energy briefly by nuclear fusion of deuterium (heavy hydrogen consisting of a proton and a neutron, H^2). Theoretical calculations indicate that the upper mass limit is about 7 to 8 percent ($\frac{1}{14}$ to $\frac{1}{14}$) the Sun's mass, or about 70 to 80 times Jupiter's mass; above this, ordinary hydrogen fusion occurs. The lower mass limit is estimated to be about 1.0 to 1.7 percent ($\frac{1}{100}$ to $\frac{1}{60}$) the Sun's mass, or about 10 to 17 times Jupiter's mass; below this, no nuclear reactions of any sort can occur and the object simply is a large planet, a "super-Jupiter." Some astronomers prefer to reserve the term "star" for objects massive enough to initiate ordinary hydrogen fusion, and these astronomers call brown dwarfs "failed stars" or "almost stars" or "substellar objects."

Because of their low mass, brown dwarfs have low temperatures by stellar standards. Their surface temperature is 2,000 kelvins (degrees above absolute zero) or less. In contrast, the Sun—by no means a very hot star—has a surface temperature of about 6,000 kelvins. They are called brown dwarfs because, due to their low surface temperature, most of their electromagnetic radiation is in the infrared part of the spectrum; at visible wavelengths they glow faintly with a dim, dark red color. Their diameter is about $\frac{1}{10}$ the Sun's diameter, which makes them about the same size as Jupiter. Their surface area and its temperature determine their luminosity, which ranges from about $\frac{1}{10,000}$ down to $\frac{1}{1,000,000}$ (10^{-4} to 10^{-6}) of the Sun's luminosity.

METHODS OF STUDY

Astronomer Jill Tarter coined the name "brown dwarf" in 1975 for hypothetical objects in between stars and planets. In subsequent years, other astronomers predicted their appearance and physical properties, postulating that our Milky Way galaxy contained many of them because slightly more massive red dwarf stars are so abundant.

Tarter's speculation touched off a search for brown dwarfs by many of the world's major observatories. The problems in identifying such objects were formidable. Brown dwarfs are very cool and faint, emitting very weak electromagnetic radiation primarily at near-infrared wavelengths of a few microns (10^{-6} meters). Predicted spectral signatures included absorption bands due to water (H_2O) and methane (CH_4), since such stars would be cool enough for these compounds to form; in both compounds the bonds between the hydrogen atoms and the central oxygen or carbon atom absorb energy in a narrow band of wavelengths within the near-infrared part of the spectrum.

Even more conclusive spectral evidence is provided by the element lithium. Small amounts of the isotope lithium 7 (Li^7) were produced in the big bang. Li^7 can undergo nuclear reaction when bombarded with a proton; its nucleus splits, and two atoms of helium 4 (normal helium, He^4) are formed. This happens, however, only at the temperatures found in "real" stars. A brown dwarf is cool enough for lithium to be consumed only very slowly, if at all. A spectrum showing an absorption feature at the wavelength characteristic of lithium, 0.67 micron, is almost certain confirmation that the object in question is a brown dwarf.

The first bodies to be identified as brown dwarfs were Teide 1 in the Pleiades and Gliese 229B, part of a binary star system with the true star Gliese 229, located in the constellation Lepus (the Hare). Since Gliese 229's distance from the Earth was known, the brown dwarf's distance was also known: about 19 light-years. Teide 1 is much farther away (about 400 light-years) and harder to observe. Although Teide 1 was discussed in the research literature before Gliese 229, it had to await final confirmation while the identification of Gliese 229B became fully established. In 1995 Gliese 229B was the subject of two papers: one published in *Science*, which provided methane spectral evidence, and the other in *Nature*, which provided lithium data. Because the methane absorption is so strong, Gliese 229B is considered to be surrounded by a thick methane atmosphere, somewhat like Jupiter. The lithium absorption was the final piece of evidence, confirming that Gliese 229B is a brown dwarf.

Enough brown dwarfs have now been discovered that they have been assigned spectral types L and T as an extension of the existing sequence of stellar spectral types O, B, A, F, G, K, and M. Type M refers to the "real" stars with the coolest surfaces, down to about 2,000 kelvins. Type L is applied to brown dwarfs with surface temperatures between 2,000 and 1,300 kelvins. Their spectra are characterized by absorption bands and lines due to water, carbon monoxide, metal hydrides, sodium, potassium, cesium, and rubidium. Type T refers to brown dwarfs with temperatures from 1,300 kelvins on down (perhaps to about 700 kelvins). Their spec-

An artist's conception of the brown dwarf OTS 44, about fifteen times the size of Jupiter, at the center of a protoplanetary disk. (NASA/JPL-Caltech)

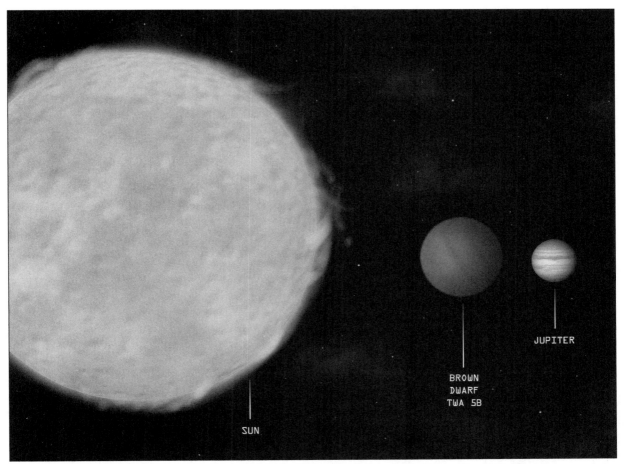

The brown dwarf TWA 5B (center) is shown with the Sun on the left and Jupiter on the right for size comparison. (NASA/CXC/M. Weiss)

tra are characterized by absorption bands of water and methane. Teide 1 is an example of type L, and Gliese 229B is an example of type T. Our understanding of the cooling rate of brown dwarfs is that they start out as type L, and after no more than about 1 billion to 2 billion years they have cooled down to type T.

CONTEXT

The actual detection of brown dwarfs after their existence was predicted helped fill in theories about the formation of stars and planets. Red dwarf stars (also called red main sequence stars, of spectral type M and luminosity class V) with masses down to about 7 or 8 percent of the Sun's mass are the least massive "real" stars, slowly fusing hydrogen into helium in their cores. Many such stars are known; they are the most common type of star found in our solar neighborhood, and presumably throughout the Galaxy. Then there are the giant planets like Jupiter and Saturn, with masses less than 0.1 percent of the Sun's mass. They have nearly the same elemental abundance as young stars (mostly hydrogen, most of the rest helium, and small amounts of other elements), but they lack the mass to have generated high enough temperatures by gravitational contraction to have initiated hydrogen fusion. Theory suggested that between red dwarf stars and giant planets like Jupiter and Saturn there should exist a class of intermediate objects—objects that generated substantial heat as they first contracted, perhaps enough to fuse deuterium (heavy hydrogen), but not enough to fuse protons (ordinary hydrogen nuclei) into helium. The successful identification of brown dwarfs and the continuing discovery of increasing numbers of them suggest a continuity in the mass distribution function (the number of objects as a function of mass) from stars down to planets. In fact, brown dwarfs may outnumber ordinary stars in the Galaxy.

Robert M. Hawthorne, Jr.

FURTHER READING

Chaisson, Eric, and Steve McMillan. *Astronomy Today*. 6th ed. New York: Addison-Wesley, 2008. College-level textbook for introductory astronomy courses. Has more thorough discussion of brown dwarfs than most.

Fraknoi, Andrew, David Morrison, and Sidney Wolff. *Voyages to the Stars and Galaxies*. Belmont, Calif.: Brooks/Cole-Thomson Learning, 2006. A well-written, thorough college textbook for introductory astronomy courses. Several pages refer to brown dwarfs.

Freedman, Roger A., and William J. Kaufmann III. *Universe*. 8th ed. New York: W. H. Freeman, 2008. College-level introductory astronomy textbook. Thorough and well-written with several pages discussing brown dwarfs.

Marley, M. S., et al. "Atmospheric, Evolutionary, and Spectral Models of the Brown Dwarf Gliese 229 B." *Science* 272 (June 28, 1996): 1919-1921. A fairly technical article, but it provides the actual data upon which the identification of Gliese 229B is based.

Mayor, Michel, and Disier Queloz. "Many Brown Dwarfs Being Found by Lithium Signature." *Sky and Telescope* 93 (February, 1997): 17-23. More discoveries, as in the preceding citation.

_____. "Swiss Find Ten New Brown Dwarfs." *Astronomy* 25 (February, 1997): 24-29. Part of the rush of discovery that followed Teide-1 and Gliese 229B.

Nakajima, T., et al. "Discovery of a Cool Brown Dwarf." *Nature* 378 (November 30, 1995): 463-465. Data on such aspects of Gliese 229B as mass, temperature, and luminosity and a demonstration that this brown dwarf is as far from Earth as its parent star. A technical article.

Reid, Neil, and Suzanne Hawley. *New Light on Dark Stars: Red Dwarfs, Low-Mass Stars, Brown Stars*. 2d ed. New York: Springer Praxis, 2005. A technical description of stars and other objects which are not very luminous. Discusses recent discoveries of brown dwarfs and extrasolar planets.

Rosenthal, Edward D., Mark A. Gurwell, and Paul T. P. Ho. "Efficient Detection of Brown Dwarfs Using Methane-Band Imaging." *Nature* 384 (November 21, 1996): 243-244. Methane data following on Nakajima's article.

Schneider, Stephen E., and Thomas T. Arny. *Pathways to Astronomy*. 2d ed. New York: McGraw-Hill, 2008. Very thorough college textbook for introductory astronomy courses. Divided into lots of short sections on specific topics. Has several pages referring to brown dwarfs.

Tyson, Neil de Grasse. "When a Star Is Not Born." *Natural History* 105 (March, 1996): 62-63. A popular discussion of brown dwarfs and their formation, this article touches on the theoretical question of whether brown dwarfs fit into existing star-formation mechanisms.

See also: Comets; Extrasolar Planets; Extrasolar Planets: Detection Methods; Gamma-Ray Bursters; Infrared Astronomy; Main Sequence Stars; Novae, Bursters, and X-Ray Sources; Nuclear Synthesis in Stars; Protostars; Pulsars; Red Dwarf Stars; Red Giant Stars; Stellar Evolution; Supernovae; Thermonuclear Reactions in Stars; White and Black Dwarfs.

C

Callisto

Categories: The Jovian System; Natural Planetary Satellites

Study of Callisto, Jupiter's outermost natural satellite, has led to insights into the formation of the solar system, the possibilities for extraterrestrial life, and the protection from comet impacts that Jupiter gives to the inner planets of the solar system.

OVERVIEW

Callisto is the outermost of the four major satellites of the "gas giant" planet Jupiter. It was discovered with one of the earliest telescopes by Galileo Galilei in 1610. Hence, it is often referred to as one of the Galilean satellites. Callisto is one of the largest satellites in the solar system, ranking third behind Jupiter's Ganymede and Saturn's Titan. With a diameter of 4,800 kilometers (2,985 miles), it is nearly the size of the planet Mercury. Callisto is also tidally locked to Jupiter, meaning that its "day" is the same length as its month, 16.82 Earth days. As a result, the same side of the satellite always faces Jupiter, just as the Moon always presents the same face toward Earth.

If the Galilean satellites had personalities, Callisto would be a frail old man. Unlike the young and vibrant Io, Callisto has neither volcanoes nor large mountains anywhere on its surface. In fact, its total lack of geological activity, both above and below the surface, means that its surface most likely resembles what the satellite looked like during its formation. This is at least partly due to the lack of tidal forces from nearby Jupiter. The lack of squeezing and pulling from Jupiter's gravity reduced the heat and energy within the satellite, leading to a relatively tranquil geology. This unique surface gives astronomers and geologists a glimpse of not only the primordial Jovian system but also the primordial solar system.

Callisto's surface is twice as bright as Earth's moon but still much darker than the surfaces of its Jovian siblings. The first few kilometers of the surface layer is primarily ice, with a darker material having leaked in at some point. Callisto's surface is uniformly covered in craters and is thought to be the most cratered satellite in the solar system. These impacts are the primary force that has shaped the planet, and sometimes great rings appear around the impact craters. The two largest features, Valhalla and Asgard, are respectively 3,000 kilometers (1,865 miles) and almost 1,600 kilometers (1,000 miles) in diameter. While impacts have been the primary force in shaping Callisto's surface, data from the Galileo space probe in the late 1990's showed that some minor erosion has occurred. This erosion is thought to be carbon dioxide sublimating through cracks in the surface ice.

Along with these large impact craters, there are numerous crater chains, or catenae. After the 1979 Voyager flybys, the catenae were thought to be the result of debris from asteroid impacts. This idea was called into question after the spectacular impact of Comet Shoemaker-Levy 9 into Jupiter's atmosphere during late May, 1994. This comet had come within a special distance from Jupiter, known as the Roche limit, and been broken up by the force of gravity. What was once one large comet was now a series of fragments traveling in formation. This event gave credibility to the idea of comets colliding with planets and satellites and has helped to explain Callisto's pockmarked surface.

While the surface has given scientists relatively overt information about the satellite's past, Callisto's interior remains shrouded in mystery and conjecture. With a density of 1.86 grams/centimeter^3, Callisto's density is the smallest of the major Jovian satellites. Scientists at the National Aeronautics and Space Administration (NASA) believe that Callisto is made up of roughly equal parts rock and ice, but the exact internal structure is unclear. Early observations led Galileo scientists to believe that Callisto is undifferentiated, meaning it has the same composition throughout.

Most rocky bodies in the solar system, such as Earth, have multiple layers that form during their creation. Molten materials tend to separate out, or differentiate, due to density. Within Earth, for instance, there is a dense core of iron and some nickel. Moving away from the core are different layers of decreasing density. Initial readings from Galileo showed that this process had not taken place in Callisto. Newer data, from subsequent flybys, do not directly contradict this hypothesis but have made planetary scientists less certain. Further evidence for an undifferentiated interior comes from data showing that Callisto also lacks its own magnetic field, suggesting a lack of a metallic core.

Curiously enough, however, Callisto does alter Jupiter's magnetic field within its vicinity. Because this perturbation in the field arises from increased conductivity within the planet, scientists speculate that a subsurface ocean may exist. Only an ocean with the salinity similar to Earth's oceans could explain the readings.

Callisto also has an extremely thin atmosphere composed primarily of carbon dioxide. With a pressure millions of times lower than Earth's, the atmosphere appeared, based on data from the Galileo flybys of 1998-1999, to have formed relatively recently. These data led scientists to believe that the atmosphere was no more than four years old and due to a combination of processes known as photoionization and magnetospheric sweeping. Photoionization takes place when ultraviolet rays (the same rays that cause sunburns) come in contact with individual carbon dioxide (CO_2) molecules; each CO_2 molecule ejects an electron, similar to the way a solar calculator generates current. Removal of an electron causes the molecule to become charged. Since charges interact with magnetic fields, Jupiter's enormous magnetic field acts like a giant broom and sweeps these ionized particles away from Callisto. Left unchecked, this process would eventually cause Callisto's atmosphere to fade away.

If the atmosphere is not transient, the carbon dioxide gas must be replenished on a continual basis. The obvious source of CO_2 gas is Callisto's icy surface. This ice would have to be located in a region that is permanently shadowed, away from direct light and protected from ionization. It has also been suggested that much of the carbon dioxide that exists on the satellite's surface, as well as this tenuous atmosphere, comes from the comet impacts that Callisto has sustained.

KNOWLEDGE GAINED

The vast majority of Callisto data comes from the Voyager flybys of the late 1970's and the multiple flybys of the Galileo spacecraft during the late 1990's. Before that, the satellite was, at best, a foggy image in

Jupiter's pockmarked moon Callisto, as imaged from the Galileo spacecraft in 2001. (NASA/JPL/DLR)

ground-based professional telescopes and a minuscule, but predictable, pinprick of light in backyard telescopes. Even Hubble Space Telescope images taken in October of 1995 showed a blurry surface. Only uncrewed space probes would produce the information needed to gain further understanding.

Both Voyagers 1 and 2, which took images on their way to the outer solar system, revealed a relatively dead world, battered by impact craters. Two decades later, Galileo returned to focus purely on the Jovian system. Its more sophisticated instruments offered higher-resolution imagery, magnetometric information, and spectroscopic information.

Galileo's most significant discovery about Callisto was the possibility of an underground ocean, similar to Earth's oceans. The discovery of water in the solar system is always a major event because it is thought to be

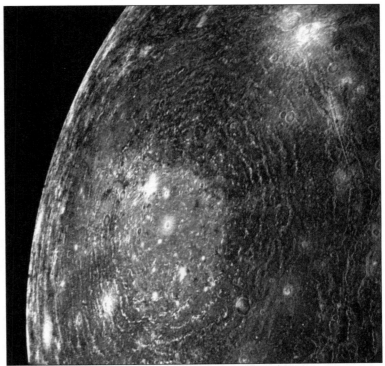

Taken in 1979 by one of the Voyager spacecraft from about 200,000 kilometers, this image of Callisto shows a multiple-ring basin. (NASA/JPL)

an essential ingredient for life. Water was already thought to exist on nearby Europa, and great efforts were made to ensure that Galileo would not contaminate the surface. This included deliberately driving the probe into Jupiter's atmosphere at the conclusion of the mission. Water on Callisto was a much bigger surprise. Could Callisto now be added to the small, but growing, list of potentially fertile worlds within our solar system?

The possibility of a subsurface ocean arises from data on the local magnetic field around Callisto. Callisto does not possess an interior magnetic field but orbits well within the boundary of Jupiter's magnetic field. During multiple flybys, Galileo measured this magnetic field and detected fluxuations in its intensity. The local magnetic environment around Callisto is similar to an electromagnet. Whereas electromagnets have magnetic fields that are induced by the flow of electrons through a looped wire, Jupiter's magnetosphere does the opposite, cap-

turing charged particles from the solar wind and creating electric currents in space. Galileo's instruments showed that this magnetic field was altered by increased conductivity from the satellite itself. While surface ice would not have any effect, the phenomenon could be explained by a subsurface ocean with a salinity level similar to Earth's oceans, conduction of current due to the presence of dissolved salts. This hypothesis is supported by the fact that similar data were taken at Europa, where planetary scientists are more confident that water exists below the surface.

More controversial is the continuing debate over Callisto's differentiation, or lack thereof. This controversy arose from data regarding Callisto's moment of inertia, a measurement of mass that indirectly comes from a body's rotation. This is the phenomenon that controls an ice skater's rotation, increasing it if the arms are brought close to the body and decreasing it when the arms are extended outward. Plane-

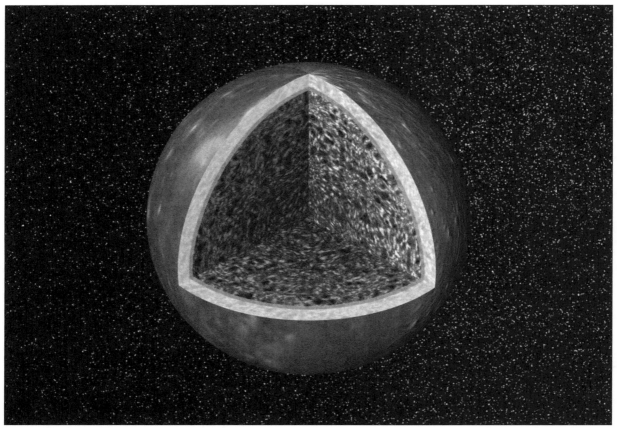

The Galileo spacecraft returned data from Callisto that revealed that the Jovian moon may have a salty ocean underlying its icy crust, as shown in this artist's rendering. (NASA/JPL)

tary scientists take this information one step further to determine the composition of a planet or satellite. A moment of inertia of 0.40 would mean that Callisto is totally undifferentiated. Data from multiple passes by Galileo showed a moment of just 0.38, within one standard deviation of theoretical uniformity. This debate is likely to continue for many years, until another spacecraft is sent. Regardless of the answer, the idea that Callisto is not as differentiated as Ganymede, a satellite similar in size and in distance from Jupiter, hints at an interesting beginning of the Jovian system. Answering the question of Callisto's interior will give scientists insight into planet and satellite formation.

While its innards will remain a mystery, Callisto's surface has helped astromoners understand more about comets, comet impacts, and Jupiter's role as protector of the solar sys-

tem's inner planets (those between it and the Sun). Before the discovery of Comet Shoemaker-Levy 9, the idea of comets impacting planets was not universally accepted. Watching the comet slam into giant Jupiter, and the subsequent "bruises" it temporarily left behind, made the idea of cometary impacts more acceptable. Scientists also learned that it was Jupiter that caused the comet to split into fragments in the first place, leading many to believe that the gas giant has done this in the past. The fact that crater chains exist on the Jupiter-facing hemisphere of Callisto is evidence of past impacts and further evidence that Jupiter is the vacuum cleaner of the solar system, keeping the inner planets safe from dangerous debris.

Finally, studying Callisto may reveal much about the future of humankind, specifically the possibilities of colonizing the solar system. Project HOPE, or Human Outer Planet Exploration, is a futuristic concept mission put forth by NASA. Part of this exploration would include a crewed mission to Jupiter, with a landing on Callisto. Callisto is an optimal choice for a human landing for two reasons. The first is its icy surface, which would provide both a source of water, allowing astronauts to "live off the land," and an opportunity for a first-rate study of impact geology. Second, Callisto's orbit places it in a region of low radiation from Jupiter. This remote, icy outpost would make an excellent location from which to study the Jovian system's past, present, and future.

CONTEXT

Callisto is a wonderful example of how taking a second look leads to a different perception. The Voyager images offered snapshots of Callisto while racing through the solar system's highway. The Galileo probe effectively pulled over and took a look around. Missions like Galileo, which observed the Jovian system from late 1995 to 2003, and Cassini, which began observing Saturn in 2004, offer a chance to understand the distant gas giant planets along with their rocky satellites. Data from Galileo have pointed to the possibility of water on Callisto and have produced debates over its internal structure and its trace of atmosphere—all from a world previously thought dead. Callisto has shown that every object in the solar system has a distinct and complicated personality, arising from a mysterious past, and that we have a long way to go when it comes to understanding our fellow travelers around the Sun.

Michael P. Fitzgerald

FURTHER READING

Bagenal, Fran, Timothy E. Dowling, and William B. McKinnon, eds. *Jupiter: The Planet, Satellites, and Magnetosphere*. New York: Cambridge University Press, 2007. A collection of articles provided by recognized experts in their fields of study, this volume offers a comprehensive look at the biggest planet in the solar system. Excellent repository of photography, diagrams, and figures about the Jovian system and the various spacecraft missions that unveiled its secrets.

Carlson, Robert W. "A Tenuous Carbon Dioxide Atmosphere on Jupiter's Moon Callisto." *Science* (February 5, 1999): 283ff. A discussion of Galileo data regarding CO_2 in Callisto's atmosphere.

Cole, Michael D. *Galileo Spacecraft: Mission to Jupiter*. New York: Enslow, 1999. Provides a full description of the Galileo spacecraft, its mission objectives, and science returns through the primary mission. Particularly good at describing mission objectives and goals. Suitable for a younger audience.

Harland, David H. *Jupiter Odyssey: The Story of NASA's Galileo Mission*. New York: Springer Praxis, 2000. Provides virtually all of NASA's press releases and science updates during the first five years of the Galileo mission in a single volume, along with an enormous number of diagrams, tables, lists, and photographs. Also provides a preview of the Cassini mission. Although the book's coverage ends before completion of the Galileo mission, what is missing can easily be found on numerous NASA Web sites.

Khurana, K. K., et al. "Induced Magnetic Fields as Evidence for Subsurface Oceans in Europe and Callisto." *Nature* 395 (October 22, 1998). This article is the resource for all discussions of the possible subsurface ocean on Callisto.

Leutwyler, Kristin, and John R. Casani. *The Moons of Jupiter*. New York: W. W. Norton, 2003. Casani was the original Galileo program manager, and this book offers a heavily illustrated discussion of the Galilean satellites as well as a number of the lesser known Jovian satellites. The authors attempt to accompany their scientific findings with an artful text, which may please the tastes of some readers more than others.

McKinnon, William B. "Mystery of Callisto: Is It Undifferentiated?" *ICARUS* 130 (1997): 540-543. This article explains why the question of Callisto's differentiation does not have a definitive answer.

Melosh, H. J., and P. Schenk. "Split Comets and the Origin of Crater Chains on Ganymede and Callisto." *Nature* 365 (October 21, 1993). Discusses the hypothesis that crater chains on Callisto come from previous comets similar to Shoemaker-Levy 9.

Showman, Adam P., and Renu Malhotra. "Galilean Satellites." *Science* 286 (October 1, 1999). An excellent overview of Jupiter's four largest satellites.

See also: Enceladus; Eris and Dysnomia; Europa; Ganymede; Iapetus; Io; Jovian Planets; Jupiter's Satellites; Lunar Craters; Lunar History; Lunar Interior; Lunar Maria; Lunar Regolith Samples; Lunar Rocks; Lunar Surface Experiments; Mars's Satellites; Miranda; Neptune's Satellites; Planetary Formation; Planetary Satellites; Pluto and Charon; Saturn's Satellites; Titan; Triton; Uranus's Satellites.

Ceres

Category: Small Bodies

Discovered in 1801, Ceres, named for the Roman goddess of agriculture, is the largest of the main-belt asteroids. For a short time it was believed to be the eighth planet in the solar system, but discovery of additional large main-belt asteroids influenced astronomers to revoke its planetary status. Discoveries of Pluto-sized objects in the Kuiper Belt have once again brought Ceres back into the discussion of what constitutes the definition of a planet.

OVERVIEW

On the night of January 1, 1801, the Italian astronomer Giuseppe Piazzi was observing the heavens when he noted a faint object that did not appear on his star charts. At first he thought it might be a comet, but it did not have the typical "fuzzy" appearance associated with comets. If not a comet, what could it be? By observing its motion over the next several weeks, Piazzi was able to determine that its orbital speed was greater than that of Mars but slower than that of Jupiter. This suggested to him that the object must lie between the orbits of Mars and Jupiter.

Additional help came from the German mathematician Carl Friedrich Gauss, who had perfected a means of calculating orbital motion based on limited observations. When he applied his method to Piazzi's observations, Gauss was able to calculate where and when this mysterious object should next appear, and it did just as he predicted. Within one year of Piazzi's discovery, Heinrich Olbers and Franz von Zach were able to relocate Ceres and refine its 4.6 Earth-year orbit. Later scientists were able to determine that it has a spherical shape with a 930-kilometer diameter. In comparison to the Moon, Ceres is one-third its size, but with only less than 2 percent of its mass, giving it a much lower density of 2.1 grams per cubic centimeter.

Piazzi's observations and the calculations of Gauss led many of the leading scientists of that time to believe that a new planet had been discovered. This conclusion seemed logical, based on an earlier idea first suggested by Johann Daniel Titius of Wittenberg and later championed by Johann Elert Bode. In 1792, Bode pointed out an apparent mathematical relationship between the distances of the various planets to the Sun. He suggested that the planets were positioned at specific distances from each other based on a mathematical ratio that would later be referred to as Bode's law. This worked reasonably well for all the planets from Mercury through Uranus, with the exception of an apparent gap between Mars and Jupiter. When Piazzi found his mystery object positioned in this gap where Bode suggested a planet should be, this seemed to be the observational confirmation of Bode's law. Even though modern science treats Bode's law as more of an interesting coincidence rather than a scientific law, it did serve a purpose at that time and contributed to the eventual discovery of Neptune.

Although initially proclaimed the eighth planet in 1801, Ceres did not long retain its planetary status. The excitement created in the astronomical community by the discovery of Ceres led to a systematic search of the heavens, which centered on the plane of the ecliptic. Scientists believed that many new and interesting objects would soon be found, and they were right. Within the next six years, three new asteroids—Pellas, Juno, and Vesta—were found within the same general vicinity as Ceres. With four minor bodies now occupying the same region of space, scientists concluded that no one planet would fill the gap in Bode's law. A new theory would have to be created to explain the presence of so many small bodies occupying planetary positions.

Since that time, many theories have been created to explain the presence of the main-belt asteroids. One of the more popular, but incorrect, theories described a large planet exploding and creating a huge number of smaller bodies ranging from the largest, Ceres, down to extremely small meteoroid-sized fragments. Perhaps the most widely accepted theory now describes a "planet that never formed." This theory can be supported by the generally accepted nebular hypothesis of planetary formation, which envisions a final stage of accretion during which a huge number of smaller bodies are attracted to each other and form into a much larger object. In the case of Ceres and the other main-belt asteroids, that final stage was interrupted and they never fully accreted into a single large object.

Beginning in the late twentieth century, the study of Ceres and its family of asteroids was no longer regulated by the limitations of Earth-based telescopic observations. The Hubble Space Telescope operating well above the Earth's atmosphere revealed details never before seen by surface-based telescopes. In addition, modern astronomers have the opportunity to send their spacecraft-borne instruments directly to the asteroids to get close-up views of their surfaces and analyze their mineralogical compositions. Several flyby spacecraft missions have investigated a number of smaller asteroids revealing previously unimagined surface conditions. One in particular, the NEAR Shoemaker probe, first orbited and then actually landed on Eros, giving scientists their first detailed images from the surface of an asteroid. The Japanese probe Hayabusa is believed to have landed on the surface of the asteroid Itokawa and collected a sample for return to Earth. Data returned by these missions have rewritten the textbooks on what is known about asteroids. The Dawn spacecraft, en route to both Ceres and Vesta, was designed first to orbit Vesta in 2011 and then leave orbit and go on to rendezvous with and orbit Ceres in 2015. Dawn's mission was to collect sufficient data to help scientists gain a better understanding of the conditions present at the initial accretion stage of planetary formation in the solar system.

Ceres is very different from Vesta. Studies based on a comparison of densities and surface reflectivity have shown that Ceres may have what is considered to be a "wet" surface, composed of water-bearing minerals, as opposed to the "dry" surface minerals of Vesta. Some scientists speculate that Ceres may have a total amount of water locked up in its interior that could rival that of the Earth's surface, while

An artist's conception of the interior of Ceres, formerly classified as an asteroid and now the only dwarf planet in the inner solar system. (NASA/ESA/A. Field, STScI)

Vesta is more comparable to
Earth's moon. Other Dawn experiments should shed light on
the internal structure, shape,
composition, and mass of these
two primordial bodies. Scientists have been able to determine that a particular class of
meteorites, the HED achondrites, is probably derived from
Vesta. It is believed that most
meteorites are fragments of
crustal material that was
blasted off an asteroid's surface
during the accretion process or
from later impacts. This is
based on reflectivity studies of
minerals present on the surface
of Vesta and from radioisotope
chronology studies of the HED
meteorites. All evidence seems
to point to Vesta. No such data
exists for Ceres. This is why a
spacecraft mission to Ceres is so
important.

KNOWLEDGE GAINED

The discovery and later scientific study of the asteroid
Ceres had a profound effect
both on our early view of the so-
lar system and on our subsequent understanding of the origin and nature of planetary bodies.
At the time of its discovery (1801), Ceres represented another "new" object in the heavens that
was unknown to the ancient astronomers. It
had been less than fifty years since the return of
Halley's comet (1758) had literally galvanized
the concept of gravity and its effects on motion,
and only twenty years after the discovery of
Uranus (1791). The discovery of three other asteroids would soon follow, but after 1807, no
other asteroid was detected until 1845.

The following year Neptune was discovered,
and astronomers had definitely moved into the
"modern era" with bigger and better telescopes,
technology that included spectroscopy and photography, and a more scientific perspective of
the universe. With Neptune recognized as the
eighth planet, the asteroids fell into their

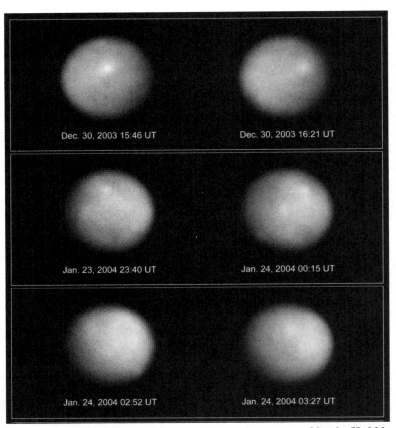

Released in 2005, these six images of Ceres were captured by the Hubble Space Telescope between December, 2003, and January, 2004. (NASA/ J. Parker et al., Southwest Research Institute)

proper place within the structure of the solar
system. The study of asteroids remained within
the domain of observational astronomers until
scientists and engineers could develop the technology to send their scientific instruments to
the planets and minor bodies. Once this happened, the outpouring of data changed our view
of the nature and origin of the planets.

To the early astronomers Ceres was only a
faint speck of light in the night sky. Today modern astronomers see it quite differently, primarily as a result of the observations made by the
Hubble Space Telescope's Advanced Camera for
Surveys. In 2005 astronomers observed Ceres
through a complete nine-hour revolution, taking 267 photographic images. From these observations astronomers were able to determine
that Ceres has a spherical shape with a diameter slightly wider at the equator than at the

poles. This suggests that it has a differentiated internal structure with denser materials forming a core and lighter materials closer to the surface. Scientists suspect that, because Ceres' density is much lower than Earth's, large amounts of water ice may exist either on the surface of Ceres or buried within its crust. This surmise is supported by spectral evidence for water-bearing minerals that may be present on the surface that are not representative of Ceres' crystal rocks. Additional microwave studies suggest that this surface material might be dry clay. All considered, Ceres could turn out to be the most Earth-like body in the solar system; it may even be a haven for primitive forms of life.

CONTEXT

Clearly the asteroids hold many vital clues to unraveling the mysteries surrounding the formation of the planets. On the basis of their respective sizes, densities, and chemical compositions, the planets in our solar system are divided into three major groups: the terrestrial (Earth-like) planets, the Jovian (Jupiter-like) planets, and the dwarf planets. The third group can be further divided into rocky objects like Ceres and into ice bodies like Pluto. These dwarf planets, rocky or ice, most likely represent a fundamental primordial stage in the formation of planets.

By studying these early remnants of planetary formation, scientists can achieve a clearer picture of their formative processes. Each group will have its own distinctive secrets to reveal. The rocky dwarf planets positioned between Mars and Jupiter formed under conditions of higher temperature, higher density, and higher velocity than the icy worlds at the edge of the solar system in the Kuiper Belt. It is believed that at this distance from the Sun, these objects have remained essentially unchanged over the last 4.6 billion years. In 2015, the New Horizons spacecraft will visit Pluto and send back images and data giving science its first close-up look at this unknown world. In that same year the Dawn spacecraft will orbit Ceres. Perhaps then, with both worlds under study, science will be able to fill in many of the gaps in our understanding of planetary formation.

Paul P. Sipiera

FURTHER READING

Bell, Jim, and Jacqueline Mitton, eds. *Asteroid Rendezvous: NEAR Shoemaker's Adventures at Eros*. Cambridge, England: Cambridge University Press, 2002. A collection of nine scientific articles that provide the reader with an overview of an asteroid rendezvous mission and what to expect from the anticipated Dawn mission to Ceres and Vesta. Suitable for a wide range of readers.

Bottke, William F., Jr., Albertoi Cellino, Paolo Paolicchi, and Richard P. Binzel, eds. *Asteroids III*. Tucson: University of Arizona Press, 2002. This comprehensive work is a compilation of scientific papers that cover virtually every aspect of asteroid research. The paper on Piazzi and Ceres is especially relevant. Best suited for the graduate student and professional scientist.

Hartmann, William K. *Moons and Planets*. 5th ed. Belmont, Calif.: Thomson Brooks/Cole, 2005. An excellent reference for a variety of topics in planetary science, including asteroids. Suitable for readers of high school-level and above.

Kowal, Charles T. *Asteroids: Their Nature and Utilization*. Chichester, England: Ellis Harwood, 1988. A good basic reference source for general information about asteroids, including both their potential for commercial use and their potential as a hazard to Earth. Suitable for both astronomy enthusiasts and students at the undergraduate and graduate levels.

Lang, Kenneth R. *The Cambridge Guide to the Solar System*. Cambridge, England: Cambridge University Press, 2003. A concise yet comprehensive book, containing a wealth of information on the members of the solar system. Excellent for a wide range of readers.

Reedy, Francis. "The Tenth Planet." *Astronomy* 33, no. 11 (2005): 68-69. A good basic article for a general readership, describing the scientific controversy over the definition of what constitutes a planet.

Sipiera, P. P. "Dawn Mission." In *USA in Space*. 3d ed. Edited by Russell R. Tobias and David G. Fisher. Pasadena, Calif.: Salem Press, 2006. A concise yet comprehensive article describing the current space mission to the asteroids Ceres and Vesta. Suitable for a wide range of readers.

See also: Comet Halley; Comet Shoemaker-Levy 9; Comets; Dwarf Planets; Eris and Dysnomia; Kuiper Belt; Meteorites: Achondrites; Meteorites: Carbonaceous Chondrites; Meteorites: Chondrites; Meteorites: Nickel-Irons; Meteorites: Stony Irons; Meteoroids from the Moon and Mars; Meteors and Meteor Showers; Oort Cloud; Planetary Classifications; Pluto and Charon.

Comet Halley

Category: Small Bodies

Halley's comet is the brightest, most famous of the known periodic comets. Definitive records of sightings go back more than two thousand years. The comet travels around the Sun roughly once every seventy-six years in a highly eccentric retrograde orbit inclined 20° to the ecliptic plane. Its orbital period has enabled many observers to see Halley's comet twice during their lifetimes.

OVERVIEW

For many years, the idea that comets were "dirty snowballs" has generally been accepted by astronomers. First proposed by Fred L. Whipple in 1950, this was one of a number of different ideas about the makeup of comets. The most popular idea was that they were "flying sandbanks," or collections of interstellar dust and gas accreted as the Sun and planets periodically passed through vast clouds of interstellar matter in their journey through the galaxy. The Sun's gravity then drew in the material that eventually collected to form individual bodies. This idea was popular during the first half of the twentieth century and was championed by British astronomers R. A. Lyttleton and Fred Hoyle. Since the middle of the nineteenth century, meteor streams have been associated with comets, and supporters of the "flying sandbank" model of cometary nuclei suggested that the

particles within meteor streams arose from material escaping from comets as they moved through the solar system.

It is now widely believed that cometary nuclei are composed of material that condensed from the solar nebula at the same time as did the Sun and its planets. The European Space Agency's (ESA's) and other spacecraft that intercepted and studied Halley's comet in March, 1986, detected copious amounts of carbon, nitrogen, and oxygen. The materials given off by the comet signify that these objects were formed in the outer regions of the solar system, where the extremely low temperatures necessary for them to solidify prevailed. Giotto revealed that the nucleus of Halley's comet is a tiny, irregularly shaped chunk of ice coated by a layer of very dark material measuring some 15 kilometers long by 8 kilometers wide. This layer is thought to be composed of carbon-rich compounds and has a very low albedo, reflecting merely 4 percent of incident light. This low reflectivity makes the nucleus of Halley's comet one of the darkest objects known. However, various bright spots were seen on the nucleus. A hill-type feature was found near the terminator, along with one resembling a crater located near a line of vents. The vents seen on the nucleus appear to be fairly long-lived. Dust jets detected by the Russian Vega 1 and Vega 2 probes appear to have emanated from these vents, two of which were also identified by Giotto. Possibly some of the larger vents have survived successive perihelion passages.

Gas and dust that cause all the cometary activity seen (including the coma and tail) emanate from the nucleus via localized vents or fissures in the outer dust layer. These vents cover approximately 10 percent of the total surface area of the nucleus. They become active when exposed to the Sun and cease to expel material when plunged into darkness as the nucleus rotates. The force of these jets of material escaping from the nucleus plays an important role in the comet's motion around the Sun, affecting its orbital speed. Halley's comet was several days late in reaching perihelion during the last apparition in 1986, a result of the jetlike effects of the matter being expelled, as a consequence of Newton's second law of motion. The late arrival of

Halley's comet was one of the factors examined by Swedish astronomer Hans Rickman, who attempted to calculate the mass of the nucleus from the amount of ejected material. Linking the ejection rate to the delay in perihelion, he judged the volume of the nucleus to be between 50 and 130 cubic kilometers. Measurements obtained through spacecraft imagery, however, revealed a volume closer to 500 cubic kilometers. The only conclusion was that the nucleus is markedly porous and far less dense than first anticipated, with an average density of no more than a quarter that of ice. This porosity meshes with the belief that comets formed in the outer regions of the solar nebula, where material coming together would remain loosely bound rather than compacting.

The fact that the nucleus of Halley's comet rotates is not in doubt. What does remain unresolved is the period of rotation. Using photographs of the comet taken during its apparition in 1910, astronomers calculated the rotation period to be 2.2 days around an axis that was fairly well aligned with the poles of the comet's orbit around the Sun. Results obtained by the Giotto, Vega, and Japanese Suisei probes appeared to support this value. Ground-based observations carried out during 1986, however, indicated a rotation period of 7.4 days. This value was supported by other ground-based observations, together with results from the American Pioneer Venus orbiter, which examined Halley's comet when it neared perihelion. Controversy ensued over these differing values, although a possible explanation has been suggested. The nucleus of Halley's comet could actually display both periods of rotation; one being spinning around its axis and another the precession of the axis of rotation. The combination of rotation and precession is still contested by certain astronomers, to some extent because of the porosity of the nucleus. Any precessional properties would

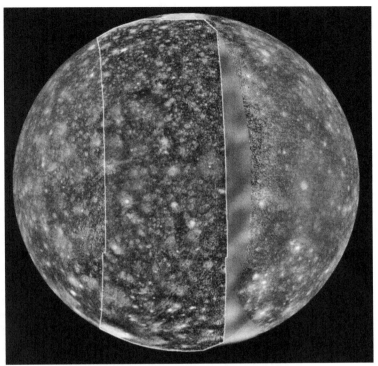

Comet Halley's nucleus. (NASA/GSFC)

quickly disappear unless the nucleus were fairly rigid.

Comets give off copious amounts of gas and dust that spread out in tails across large areas of space. Investigation of this material can reveal much about the composition of cometary interiors. Many of the investigations carried out by the European, Japanese, and Soviet space probes were directed toward a survey of the material ejected by Halley's comet. These investigations were supplemented by observations both from ground-based astronomers and from the American Pioneer Venus, International Cometary Explorer (ICE), and International Ultraviolet Explorer (IUE) spacecraft. As is the case with the surface of Halley's nucleus, the dust thrown off by the comet was found to be very dark and may have emanated from the surface itself rather than the interior. Giotto and Vega carried out analyses of the dust. They found a mixture of different materials, including the lighter elements oxygen, hydrogen, nitrogen, and carbon, and the heavier elements silicon, iron, and magnesium. The amount of carbon found during the investigations coincides quite well with the observed abundance of this material elsewhere in the galaxy, indicating that comets are made of interstellar material.

More than three-quarters of the gas ejected from the nucleus was found to be water vapor, which also appears to constitute more than 80 percent of the nucleus. The rate of production varied during the interval the comet was examined by these space probes. Vega 2 found approximately 16 tons of water coming away from the nucleus during its flyby, while Vega 1 detected double that rate. These large changes are reflected in the fact that Comet Halley's brightness sometimes varied by a factor of two or three from night to night. The velocity of the ejected vapor was found to be between 0.8 and 1.4 kilometers per second. This was the first time that water had been positively identified in a comet, in spite of the fact that cometary nuclei were widely thought to consist of a mixture of dust and water ice. Carbon monoxide and carbon dioxide were also detected, although methane was not found at all. This is strange, in that either any methane which existed in the comet may have been altered chemically during the period since the formation of the comet or methane was lacking in the cloud of material from which the comet formed. If there is methane in Halley's nucleus, it must constitute a very tiny percentage of the total makeup.

Processes involved in the release of gas from the nucleus may have played a prominent role in the evolution of its surface. It has been suggested that, as a comet approaches the Sun after spending its time in temperatures of approximately 40 kelvins in the outer regions of the Sun's influence, warming effects of the Sun can cause the ice within the nucleus to expand. This would result in heat generation and release of trapped gas. Some of this gas may collect in pockets, which eventually explode, producing craterlike features similar to that imaged by Giotto.

METHODS OF STUDY

Halley's comet is unusual (though not unique) in that it is named for the astronomer who first calculated its orbital path rather than the person who discovered it. Edmond Halley observed a bright comet in 1682, the impression of this sighting staying with him and eventually expanding into a deeper interest in comets. In 1705, Halley began a study of a number of bright comets seen between 1337 and 1698. Using methods developed by Sir Isaac Newton, he carried out work on the orbital motions of some twenty-four comets seen during this period. He noticed from his results that there were many similarities between the orbits of the comets observed in 1531 and 1607 and the bright comet he had seen in 1682. The intervals between the sightings were also roughly identical at around seventy-six years. This led Halley to predict that these sightings were of the same comet, and that it would reappear in 1758.

Halley died in 1743, although astronomers began a search for the returning comet as the date forecast by Halley drew near. French astronomer and mathematician Alexis-Claude Clairaut, with the help of Joseph-Jérome de Lalande and Madame Nicole Lepaute, attempted to calculate its orbital path in more detail. Taking into account gravitational effects of Jupiter and Saturn, they calculated that the comet would reach perihelion on April 13, 1759, and published ephemerides (detailed star maps and charts) to help astronomers with their search. Many famous astronomers joined in, although it was the amateur astronomer Johann Georg Palitzsh from Dresden who first spotted the comet on Christmas Day, 1758. The reappearance was quickly confirmed, and the comet was named for Halley in honor of the fact that he had correctly predicted its return. Once a number of observations had been obtained, a revised orbit was calculated. It was found that Clairaut's calculated perihelion date was in error by thirty-two days. Scientists were at a loss to explain this error, although they did not know about the existence of the two giant planets Uranus and Neptune, which were not to be discovered until 1781 and 1846, respectively.

Since the 1758 appearance, Halley's comet has been seen on three occasions: in 1835, in 1910, and in 1985-1986. Times of previous visits of the comet have been calculated by taking into account the gravitational effects of other bodies of the solar system and plotting the comet's

The Adventurous Edmond Halley

In 1684, Edmond Halley was a young scientist who had already made a name for himself as a precocious astronomer: He was the first to observe that the Sun rotated on an axis, during a trip to St. Helena in the South Seas. In 1680, during his Grand Tour of Italy and France, he had observed the comet that would bear his name. He had produced star catalogs and tidal tables, and he was trying to determine why Kepler's laws worked the way they did. Then, in April, his father's disfigured corpse was discovered near a riverbank; he had been missing for more than a month. Edmond's attention was redirected toward a bitter battle with his stepmother over the family estate.

Four months later, Halley was visiting Isaac Newton, who had solved the problems with Kepler's laws but had "misplaced" the solutions, supposedly worked out when Cambridge had been shut down during the plague of 1665. Halley began a campaign of diplomacy to get the eccentric and overly sensitive Newton to publish his results before someone else (Robert Hooke) derived the inverse square law and beat him to it. This was the genesis of Newton's *Principia* of 1687, published at Halley's expense.

In the meantime, Halley was supporting himself as a clerk at the Royal Society and working on a diverse array of projects, from determining the causes of the biblical Flood (which he unorthodoxly and dangerously placed earlier than the accepted date of 4004 B.C.E.) to making the connection between barometric pressure and altitude above sea level. He even calculated the height of the atmosphere, at a remarkably accurate 45 miles. Motivated by his persistent lack of money, Halley also designed various nautical instruments: a prototype diving bell, a device for measuring the path of a ship, and another device for measuring the rate of evaporation of seawater. He even prepared life-expectancy tables that became the basis for modern life insurance. Between 1696 and 1698, he became the deputy comptroller of the Royal Mint at Chester, a post offered him by Newton, who was then the warden of the Mint. Administration did not prove to be one of Halley's many talents, however, and Newton found himself having to defend his friend against the Lord's Commissioners.

(NASA)

In 1698, Halley set out on another expedition to the South Seas to study the magnetic variations of the Earth's compass. The journey was abandoned (with the ship's first lieutenant facing a court-martial on their return), but Halley tried again a year later with more success. He also went on a secret mission in 1701, about which little is known, traveling to France for the Admiralty on the pretext of yet another scientific expedition. In 1703, Halley became a member of the Council of the Royal Society in recognition of his work, and in the same year, he was appointed to the Savilian Chair of Geometry at Oxford, where he conducted his study of comets. It was around this time that he made the observation for which he became famous:

> Many considerations incline me to believe that the comet of 1531 observed by Apianus is the same as that observed by Kepler and Longomontanus in 1607 and which I observed in 1682. . . . I would venture confidently to predict its return, namely in the year 1758. If this occurs there will be no further reason to doubt that other comets ought to return also.

In 1719, on the death of John Flamsteed, Halley succeeded to the post of Astronomer Royal, a position he held until his death in 1742. The practicality and range of his interests made him a celebrity whose achievements far exceeded those for which he is remembered today. He did not live to see his comet, which was sighted on Christmas, 1758.

orbital course backward through time. Dates calculated for previous apparitions have been substantiated by checking against ancient astronomical records, primarily those of Chinese astronomers. The first definite appearance of Halley's comet took place in 240 B.C.E., although the 12 B.C.E. appearance is the first about which detailed information is available. The most famous return was that of 1066, which was interpreted as a bad omen by the Saxons and, in particular, by Harold, the last of the Saxon kings. William of Normandy, who viewed the apparition as a good sign, invaded England, following which Harold died at the Battle of Hastings in October of that year. The Bayeux Tapestry depicts the comet suspended above Harold, who is seen tottering on his throne as his courtiers look on in awe and terror.

The 1531 appearance is important because it was one of two apparitions studied by Halley (the other being that of 1607) prior to his two deductions: that these historical sightings were of the same object, and that the comet is a regular visitor to this region of the solar system. A comprehensive set of observations of the 1531 appearance was made by astronomer Peter Apian, who published his results in 1540. The 1607 appearance was observed and recorded by many astronomers, including Johannes Kepler. This was the last apparition of Halley's comet before the introduction of the telescope.

After the comet's reappearance in 1758 and the discovery of Uranus in 1781, astronomers were able to plot its orbit with even greater accuracy. Long before its scheduled return in 1835, many attempts were made to calculate the expected date of perihelion passage. The consensus was that Halley's comet would pass closest to the Sun in November, 1835. The search for the returning comet started as early as December, 1834, almost a year before it was due to sweep through the inner solar system. The first sighting was not made, however, until August 6, 1835, by Father Dumouchel and Francisco di Vico at the Collegio Romano Observatory. Confirmation came via Friedrich Georg Wilhelm von Struve, who saw the comet on August 21. Perihelion occurred on November 16.

Prominent among the astronomers who studied the comet during the 1835 apparition was Sir John Frederick Herschel, who was then based at a temporary observatory near Cape Town, South Africa. He was in the process of completing the sky survey started by his father, Sir William Herschel, and had moved to South Africa in order to survey the southern stars that were visually inaccessible from England. John Herschel made his first attempt to locate the comet in late January, 1835, although he did not see it until October 28. The 1835 apparition was remarkable in that much activity was seen to occur in the comet. Prior to its temporary disappearance in the Sun's rays as it rounded the Sun, a number of changes were observed in the tail. These disturbances continued after its reappearance. The tail was seen to vary noticeably in length. The head also altered in appearance, at times appearing almost as a point of light, while at others taking on a nebulous form. It was noticed that the coma expanded while undergoing a reduction in brightness, eventually becoming so dim that it merged into the surrounding darkness. Herschel's final observation of Halley's comet in mid-May, 1836, was the last that any astronomer made until the 1910 return. All data scientists have about the 1835 apparition are in the form of sketches and visual descriptions. Photography had not yet made an impact on astronomy, although the appearance in 1910, through the use of the camera, provided the most comprehensive and detailed study of Halley's comet up to that time.

The third predicted return in 1910 was awaited eagerly by astronomers all over the world. The interval between the 1835 and 1910 visits had been littered with numerous bright comets, notable among which were the Great Comet of 1843, Donati's comet of 1858, and the Great September Comet of 1882. The latter is particularly significant in that it was the subject of the first successful attempt to photograph a comet. A good image was obtained by Sir David Gill in South Africa. Observation of Comet Morehouse in 1908 demonstrated that a series of photographs was an ideal means of monitoring cometary structural changes. (Comet Morehouse itself underwent a number of prominent changes that, coupled with the fact that Halley's comet had suffered in a similar fashion three-quarters of a century before, whetted the appetites of astronomers who were gearing up for the forthcoming apparition.) The prolonged period of cometary activity following its last visit had allowed astronomers to perfect their observing techniques and paved the way for observations of the return of Halley's comet.

The comet had passed aphelion in 1872, after which it once more began its long journey toward the inner solar system. The first astronomer to detect the returning visitor was astrophysics professor Max Wolf at Heidelberg, Germany. A photographic plate was exposed on the night of September 11-12 and recorded the comet close to its expected position. It did not become visible to the naked eye until well into 1910. Prior to this, another bright comet made an unexpected appearance. The Great Daylight Comet was first spotted by diamond miners in Transvaal, South Africa, in the early morning sky on January 13, 1910. Confirmation of the discovery was made four days later, and news of this spectacular discovery was distributed to the world's observatories. Unlike Halley's comet, which was to appear later that year, the Great Daylight Comet became a brilliant evening object for observers in the Northern Hemisphere. Its tail attained a maximum length of 30° or more by the end of January. The comet became so bright that it was visible to the naked eye even in broad daylight (hence its name).

The Great Daylight Comet was widely mistaken for Halley's comet by many people who had been expecting its return at about this time, although Halley's comet did not put on as grand a show. Bad weather together with the fact that a full moon occurred at what should have been the best time for observation meant that astronomers north of the equator were disappointed. Yet, even working against these odds, they did obtain many useful photographs and were able to study the comet spectroscopically. The best results, however, were obtained from observatories in the Southern Hemisphere, notably at Santiago in Chile. From mid-April to mid-May, 1910, Halley's comet was in the same area of the morning sky as Venus, the two objects together forming a marvelous visual spectacle in the constellation of Pisces. Much activity was noted in both the nucleus and the tail of the comet. Sequences of photographs showed marked changes in the head, including material being ejected from the nucleus and halos expanding out from the nucleus. The tail also underwent violent changes, with material being seen to condense in various regions. On April 21, the day following perihelion, the previously smooth northern edge of the tail became irregular and distorted. Material seemed to be thrown out in

various directions, and parts of the tail seemed to be ejected into space, an event clearly visible on photographs obtained at the time. For some days following perihelion, a jet of material from the nucleus seemed to be refueling the northern section of the tail. Once this activity ceased, the tail's southern section increased in brightness. A few weeks after perihelion, the two types of cometary tail appeared, a straight and distinct gas tail contrasting with the fainter, more diffuse and curved dust tail. Halley's comet passed between the Sun and Earth on May 18, although in spite of many attempted observations, no trace of the nucleus could be seen as the comet transited the solar disk. This proved that the nucleus must be tiny and the gas around it very tenuous. During this time, it was thought that the Earth may pass through the tail, although there is no evidence that this actually occurred. The pronounced curve of the tail seems to have taken it away from the Earth, preventing a passage of the planet through it. The closest approach of the comet to Earth was on May 20, when the distance between the two bodies was 21 million kilometers. For a time afterward, the comet became a prominent evening object for American observers, and many useful results were obtained by astronomers at Lick Observatory and Mount Wilson Observatory in California. A number of changes in the comet's structure were seen, and many spectroscopic observations were taken. These showed the presence of a large number of different molecules in the comet, and helped astronomers to understand more clearly its chemical constitution.

As the comet started on its journey back to the outer regions of the solar system, it grew steadily fainter. It was last seen when beyond the orbit of Jupiter in a photograph taken on June 15, 1915, on its way toward aphelion in 1948. The next return would be accompanied by an unprecedented campaign by astronomers and space scientists to expand their understanding of comets in general, and Halley's comet in particular.

The return of 1985-1986, the most recent to date, provided astronomers with their best chance yet of exploring a comet. Unlike other bright comets, many of which appear suddenly, the orbital path of Halley's comet is known with both great precision and great accuracy. Therefore, it was possible to plan missions by robotic space probes to rendezvous with the comet during its last return. For a comet rendezvous mission, the position of the comet at time of interception must be known well in advance, as was the case with Halley's comet. In all, five space probes were sent to examine the comet.

Two of these were the Soviet Vega probes, launched in December, 1984, to release balloons into the Venusian atmosphere. Along their way, the probes encountered Halley's comet on March 6 and March 9, 1986, at distances of 8,890 kilometers and 8,030 kilometers, respectively. Among the equipment they carried were cameras, infrared spectrometers, and dust-impact detectors.

The two Japanese probes carried out their investigations from greater distances. Sakigake, launched in January, 1985, flew by the comet on March 11, 1986, at a distance of 6.9 million kilometers. Its primary purpose was to investigate the interaction between the solar wind and the comet at a large distance from the comet. One of the main aims of Suisei, launched in August, 1985, was to investigate the growth and decay of the hydrogen corona. Suisei flew past the comet on March 8, 1986, at a distance of 151,000 kilometers.

By far the most ambitious, and most successful, of the probes dispatched to Halley's comet was the European Giotto, named in honor of the Italian painter Giotto di Bondone. It launched toward the comet on July 2, 1985. Giotto was cylindrical in shape, with a length of 2.85 meters and a diameter of 1.86 meters. Its payload included numerous dust-impact detectors, a camera for imaging the nucleus and inner coma of Halley's comet, and a photopolarimeter for measuring the brightness of the coma. Giotto flew within 610 kilometers of the nucleus on March 14, 1986, at a speed of more than 65 kilometers per second. Data collected by Giotto were immediately transmitted back to Earth via a special high-gain antenna mounted on the end of the space probe facing away from the comet. Information was received back on Earth by the 64-meter antenna at the Parkes ground station in Australia. At the opposite end, Giotto was equipped with a special shield to protect it from impacts by dust particles during its passage through the comet's halo.

Exploration of Halley's comet by space probes was a truly international effort, the images and measurements obtained by the Soviet Vega craft helping scientists to target Giotto precisely. From Earth, the nucleus of a comet is hidden from view by the material surrounding it. Not until the Vega images were received was its position established and the subsequent trajectory of Giotto determined. During the close encounter, all instruments performed well, although disaster struck immediately before closest approach to the nucleus. A dust particle weighing merely one gram impacted Giotto. This temporarily knocked the spacecraft and its antenna out of alignment with Earth, and for thirty tense minutes contact was lost. The problem was rectified, and contact was reestablished. After the encounter, it was found that

Launched in 1985, the Giotto space probe passed by Comet Halley's nucleus in 1986. (European Space Agency)

approximately half of the scientific equipment had suffered damage, although scientists were able to redirect the craft and put it on a course back to Earth. Tests carried out by the European Space Agency in 1989 paved the way for reactivation of Giotto, which set up a pass within 22,000 kilometers of the Earth and placed Giotto in a new orbit that allow it to intercept another comet. On July 10, 1992, Giotto flew close to Comet Grigg-Skjellerup, at a point just twelve days in advance of the comet's closest passage to the Sun, a time when its activity was approaching maximum.

CONTEXT

Although study of Halley's comet has taught scientists much about comets in general, there still remains much to learn about these ghostly visitors. Halley's comet provides a chance to investigate the origins of the solar system. Cometary explorations by space probes could include rendezvous missions during which a probe would position itself close to a cometary nucleus for a prolonged period and perhaps send a lander to the surface of the nucleus. The possibilities of such a mission were being examined by the National Aeronautics and Space Administration (NASA) at the time of Halley's 1986 visit. Known as Comet Rendezvous and Asteroid Flyby (CRAF), this mission would have enabled scientists to undertake close-up exploration of both asteroids and comets. Unfortunately, budget cuts led to the cancellation of CRAF. Sample return missions, by which scientists can examine at first hand material plucked from the heart of a comet, also remain a possibility. Some astronomers and scientists hope for a mission that will carry a human crew to Halley's comet during its next apparition, in 2061.

More realistically in the meantime, NASA was able to launch its Deep Space 1 probe and demonstrate the capability of an ion propulsion system to drive a spacecraft to effect rendezvous with an asteroid and a comet. On September 22, 2001, Deep Space 1 flew within 2,200 kilometers of Comet Borrelly, performing measurements and taking high-resolution images. The Deep Impact mission slammed a copper impactor into the Comet Tempel 1 on July 4, 2005, to expel surface material and excavate a crater on the comet's nucleus. The flyby portion of the Deep Impact spacecraft observed the collision of its impactor and analyzed material thrown up from the formation of an impact crater on the cometary nucleus. Then in January, 2006, the Stardust mission returned samples to Earth released from Comet Wild 2; those samples were collected at a distance of 240 kilometers from the comet's nucleus. ESA launched the Rosetta spacecraft in 2004 and set it on a trajectory toward an encounter with the comet 67P/Churyumov-Gerasimenko in May, 2014. Rosetta is designed to orbit the comet and later release a small lander named Philae to touch down on the comet's nucleus and perform in situ analyses of surface materials.

Brian Jones

FURTHER READING

Beatty, J. Kelly, Carolyn Collins Petersen, and Andrew Chaikin, eds. *The New Solar System*. 4th ed. Cambridge, Mass.: Sky, 1999. Filled with color diagrams and photographs, this popular work covers solar system astronomy and planetary exploration through the Mars Pathfinder and Galileo missions. Accessible to the astronomy enthusiast. Provokes excitement in the general reader, who gains an explanation of the need for greater understanding of the universe.

Faure, Gunter, and Teresa M. Mensing. *Introduction to Planetary Science: The Geological Perspective*. New York: Springer, 2007. Designed for college students majoring in Earth sciences, this textbook provides an application of general principles and subject material to bodies throughout the solar system. Excellent for learning comparative planetology.

Gingerich, Owen. "Newton, Halley, and the Comet." *Sky and Telescope* 71 (March, 1986): 230-232. Provides background information on Sir Isaac Newton and Edmond Halley. Describes how Halley used Newton's work with gravitation to draw comparisons between the orbital motions of comets seen in 1531, 1607, and 1682, which led to his conclusion that they were all sightings of the same object and his predict of its return in 1758. Suitable for the general reader.

Grewing, M., F. Praderie, and R. Reinhard, eds. *Exploration of Halley's Comet*. New York: Springer, 1989. A technical review of the information garnered by the return of Halley's comet in 1985-1986.

Harpur, Brian, and Laurence Anslow. *The Official Halley's Comet Project Book*. London: Hodder and Stoughton, 1985. A comprehensive guide to knowledge of Halley's comet prior to its exploration by space probe. As well as a general description of comets, the book contains details of Edmond Halley and his work, many facts relating to Halley's comet and its previous appearances, and a detailed description of the 1910 apparition of the comet. Includes a discussion on the pronunciation of Halley's name and a collection of poems written about the comet in 1910. A useful book for the general reader, containing many items not printed elsewhere.

McBride, Neil, and Iain Gilmour, eds. *An Introduction to the Solar System*. Cambridge, England: Cambridge University Press, 2004. A complete description of solar system astronomy suitable for an introductory college course, filled with supplemental learning aids and solved student exercises. Accessible to nonscientists as well. A Web site is available for educator support.

Sekanina, Zdenek, ed. *The Comet Halley Archive Summary Volume*. Pasadena, Calif.: Jet Propulsion Laboratory (International Halley Watch), California Institute of Technology, 1991. A collection of observations made by various observers of Halley's comet's return in 1985-1986.

Whipple, Fred L. "The Black Heart of Comet Halley." *Sky and Telescope* 73 (March, 1987): 242-245. An examination of the information received regarding the nucleus of Halley's comet and what it tells scientists. Comparisons are drawn between previous models of the structure of cometary nuclei and current knowledge. Suitable for the general reader.

_____. *The Mystery of Comets*. Washington, D.C.: Smithsonian Institution Press, 1985. Chapter 4, "Halley and His Comet," outlines the life and work of Edmond Halley and his involvement with cometary orbits. Chapter 5, "The Returns of Halley's Comet," describes the apparitions of Halley's comet from the earliest sightings to 1910; chapter 24, "Space Missions to Comets," is a description of the various space probes that intercepted Halley's comet during its return in 1985-1986. Suitable for the general reader.

See also: Ceres; Comet Shoemaker-Levy 9; Comets; Dwarf Planets; Earth's Oceans; Eris and Dysnomia; Extraterrestrial Life in the Solar System; Infrared Astronomy; Kuiper Belt; Lunar Craters; Meteorites: Achondrites; Meteorites: Carbonaceous Chondrites; Meteorites: Chondrites; Meteorites: Nickel-Irons; Meteorites: Stony Irons; Meteoroids from the Moon and Mars; Meteors and Meteor Showers; Oort Cloud; Planetary Classifications; Planetary Formation; Pluto and Charon; Solar System: Element Distribution; Solar System: Origins; Solar Wind; Ultraviolet Astronomy.

Comet Shoemaker-Levy 9

Category: Small Bodies

The spectacular collision of comet Shoemaker-Levy 9 with Jupiter in July, 1994, provided valuable information about comets, Jupiter's atmosphere, and the Jovian role in diminishing potentially cataclysmic Earth-damaging debris in the inner solar system.

OVERVIEW

On the night of March 24, 1993, the husband-and-wife team of Eugene ("Gene") and Carolyn Shoemaker together with their colleague David Levy were using the Schmidt telescope at Mount Palomar Observatory in California to take photographs in connection with a project designed to discover near-Earth celestial objects. Gene Shoemaker, who had recently retired from the U.S. Geological Survey's Astrogeology Research Program, which he had established, was an expert in Earth-orbit-crossing asteroids and comets. In their five years together, the Shoemaker-Levy team had already discovered eight comets, and they were pleased when one of their photographs of the sky near Jupiter revealed what Levy called "the strangest comet" he had ever seen. Its several tails and bat-shaped wings of dust reminded him of the American Stealth Bomber. They quickly realized that this fragmented comet was an important discovery, and, three days later, they made comet Shoemaker-Levy 9 (SL 9) public in a circular published by the International Astronomical Union (IAU). Following tradition, the comet was named after its discoverers, with the number indicating it was the ninth comet that this team had found. Its formal IAU name became D/1993F2, in which the prefix "D" indicated that it was a periodic comet that later "disappeared," 1993 was the year of discovery, the suffix "F" represented the half-month of discovery (F = March 16-31),

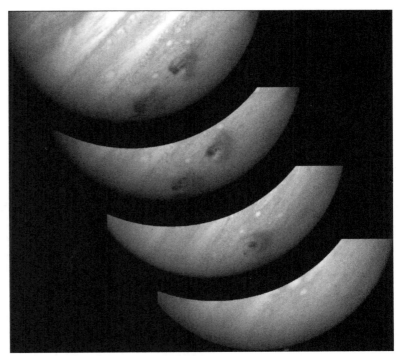

A photo mosaic showing the comet Shoemaker-Levy 9 as it impacted Jupiter in 1994. (JPL/NASA/STScI)

and "2" meant it was the second discovery in that half-month.

Other observers, stimulated by this announcement, returned to photographs that they had taken before March 24 and confirmed the discovery of the Palomar team. Using these and other data, astronomers calculated the orbit of the comet, which, unlike all earlier comets, orbited Jupiter rather than the Sun. Its highly elliptical orbit had an apojove (the point farthest away from Jupiter) of nearly 50 million kilometers and a period of nearly two years. Later data helped refine this orbit and provided clues about the comet's early history. Like most comets, it had orbited the Sun, but several decades earlier it had been captured by Jupiter's gravity. In early July, 1992, it traveled so close to Jupiter, about 20,000 kilometers from the Jovian cloud tops, that the giant planet's powerful gravity broke the comet into twenty-one separate fragments, each of which collected a coma of dust.

Following the convention established for previous fragmented comets, the twenty-one discernible pieces were labeled with letters of the alphabet (excluding I and O), and so SL 9's fragments, which averaged a few kilometers in diameter, ran from A to W, with the brightest (and presumably the largest) piece called Q. With the orbit and fragments identified, astronomers soon realized that this piecemeal comet was on a collision course with Jupiter. For SL 9 there would be no escape from Jovian orbit to return to the Kuiper Belt, now believed to be the source of Jupiter Family (JF) comets; instead, SL 9 faced extinction within sixteen months of its discovery.

Because astronomers knew when the comet would collide with Jupiter, they had time to organize observatories around the world, including in such remote locations as Antarctica, to make preparations to collect data on this unprecedented event. Because the initial impact would take place on the side of Jupiter hidden from the Earth, several spacecraft would play important roles, particularly in the earliest observations. These spacecraft included Galileo, well on its way to study Jupiter; the Hubble

Space Telescope (HST); Ulysses, which had been studying solar poles; the Roentgen satellite (ROSAT), which had been surveying the sky for X-ray sources; and Voyager 2, which had been exploring the outer planets of the solar system.

As predicted, the comet's first fragments slammed into the Jovian atmosphere on July 16, 1994, at a speed of about 60 kilometers/second, or fast enough to traverse the United States in about a minute. When Jupiter's rotation made the crash site visible to terrestrial observers, thousands of telescopes could see the dark spots that had been created. Fragments of SL 9 continued to collide with Jupiter over the next 5.6 days. Because of the great excitement created by this astronomical event, and because the Earth had been interconnected by various computer networks, images, observations, scientific data, and personal impressions were rapidly transmitted all over the planet. Many

others experienced the event through television or through the many stories in magazines and newspapers. By the time the final fragment, W, struck Jupiter on July 22, many millions of Earthlings had shared, in some way or other, this unique interplanetary event.

KNOWLEDGE GAINED

The prodigious wealth of information created by the SL 9 event had important implications for the understanding of comets, the Jovian atmosphere, and the future history of Earth. Astronomers knew, of course, that comets could be destroyed by collisions with the Sun, planets, and satellites, but the data from the fragmentation and collision of SL 9 with Jupiter revealed that its nucleus had been neither a solid body nor a loose agglomeration of materials but something in between. When the pieces hit Jupiter, spectroscopic analysis detected the presence of several elements absent from the Jovian

The remains of Shoemaker-Levy 9 emerge after the comet's impact with Jupiter caused it to break into twenty-one pieces. (JPL/D. Seal, edited by CXC/M. Weiss)

atmosphere. These elements, which came from the comet, included such nonmetals as sulfur and silicon and such metals as iron, aluminum, magnesium, and even lithium, hitherto undetected in comets.

As expected, when the high-speed fragments of SL 9 penetrated the Jovian atmosphere, gigantic explosions and massive seismic waves resulted. Fireballs created temperatures in excess of 10,000 kelvins, which rapidly diminished to 2,000 kelvins. Collision-zone temperatures remained elevated for two weeks, but, astonishingly, smaller impact sites had higher temperatures than larger ones. A typical fireball spread from 15 to 100 kilometers in about 40 seconds, and some plumes extended to an altitude of 3,000 kilometers. These explosions also produced waves that sped across the planet at about 450 kilometers/second. These waves, which weakened in about two hours, posed a problem for astronomers. Disagreements developed about where they occurred (in the Jovian stratosphere or troposphere) and how they traveled (guided by a stable layer or generated by interlayer complexities). Just as spectroscopic analysis revealed some surprises about SL 9's chemical composition, so, too, certain elements and compounds were discovered for the first time on Jupiter: for example, diatomic sulfur and carbon disulfide. By contrast, astronomers had expected to find sulfur dioxide, but they did not.

Astronomers also used other parts of the electromagnetic spectrum to gather data on the collision. For example, radio emissions at a specific wavelength (21 centimeters) were indicative of synchrotron radiation, most likely caused by the collision's injecting very-high-speed electrons into the Jovian magnetosphere, which also experienced other changes after the impact. Since both Jupiter and comets were known to have water in their makeup, astronomers were surprised by the very small amounts of water that were detected. Perhaps the comet's fragments lost most of their water before the collision, or perhaps the comet's fragments were destroyed before they reached the planet's water layer.

For many people the highlight of the event was the creation of a series of dark spots that scarred Jupiter's southern hemisphere for several weeks. Similar to the Great Red Spot, if smaller in scale, these dark spots were the most enduring transient features ever seen on the planet, although some historians of science pointed out that, in 1790, Gian Domenico Cassini had reported unusual temporary marks on Jupiter's disk. If these had been due to a cometary collision, then SL 9's crash onto Jupiter would not have been the unique event that many touted it to be.

CONTEXT

Throughout its long history, the Earth has experienced steady and numerous collisions with interplanetary objects. The collision of the Shoemaker-Levy 9 comet with Jupiter provided astronomers with valuable insights into how the collision affected Jupiter and, by analogy, how such comets may have affected other planets, including Earth. Gene Shoemaker estimated that comets had most likely caused about a fifth of the large impact craters on Earth. Linear crater chains have been photographed on Ganymede and Callisto, two of Jupiter's satellites, and these were probably due to cometary collisions. Some scientists have speculated that if SL 9 had collided with Earth instead of Jupiter, a cataclysmic destruction of life would have occurred. According to many scientists, 65 million years ago an asteroid or comet smashed into Central America, creating massive amounts of atmospheric pollutants that helped to bring about the extinction of the dinosaurs and many other forms of life.

In the history of life on Earth other mass extinctions have occurred, and some scientists associate these with periodic comet showers. Various theories have been put forward to explain these periodicities—for example, Nemesis, a companion star of our Sun, may create perturbations in the Oort Cloud that lead to these recurrent invasions of comets into the solar system. Jan Oort was the first to suggest that this large reservoir of icy bodies might be the source of very-long-period comets. However, SL 9's collision with Jupiter revealed something very significant. Because of Jupiter's powerful gravitational field, it attracts many asteroids, comets, and other interplanetary debris, resulting in fewer collisions of these objects with the inner planets, especially Earth. Some have even called Jupiter a "cosmic vacuum cleaner." On the other hand, estimates indicate that small comets collide with Jupiter about once a century, and comets comparable in size to SL 9 hit it about once per millennium. Comet Shoemaker-Levy 9 certainly expanded knowledge about the nature and properties of comets, as well as their interactions with other members of the solar system, but astronomers also realize that they need to learn much more before they will be able to make reliable predictions about some future comet's possibly devastating collision with Earth.

Robert J. Paradowski

FURTHER READING

Fernández, Julio Angel. *Comets: Nature, Dynamics, Origin, and Their Cosmological Relevance*. Dordrecht, Netherlands: Springer, 2005. Using advanced mathematics and celestial mechanics, the author analyzes the history, structure, and behavior of comets, including SL 9. Includes 483 references and an index.

Gehrels, Tom. "Collisions with Comets and Asteroids." *Scientific American* 274, no. 3 (March, 1996): 54-59. A discussion of the likelihood of cometary and asteroid impacts on Earth, designed for the scientifically inclined general audience.

Levy, David H. *Impact Jupiter: The Crash of Comet Shoemaker-Levy 9*. New York: Basic Books, 2003. Called the definitive memoir of SL 9, this book by the comet's codiscoverer provides for the general reader a lively account of the comet, its collision with Jupiter, and the knowledge gained from this event. Illustrated with color and black-and-white photographs. Chapter references and an index.

Noll, Keith S., Harold A. Weaver, and Paul D. Feldman, eds. *The Collision of Comet Shoemaker-Levy 9 and Jupiter*. Cambridge, England: Cambridge University Press, 1996. Contains fifteen reviews by experts involved with the SL 9 event, as well as many references to the primary literature. Index.

Shoemaker, Eugene M., P. R. Weissman, and C. S. Shoemaker. "The Flux of Periodic Comets Near Earth." In *Hazards Due to Comets and Asteroids*, edited by Tom Gehrels. Tucson: University of Arizona Press, 1994. This article explores the probability of a comet like SL 9 colliding with Earth.

Spencer, John R., and Jacqueline Mitton, eds. *The Great Comet Crash: The Collision of Comet Shoemaker-Levy 9 and Jupiter*. Cambridge, England: Cambridge University Press, 1995. A collection of articles by various scientists, including the Shoemakers. Analyzes, from various perspectives, the discovery, tracking, and crash of this comet into Jupiter, as well as discussions of the many discoveries this event stimulated. In the final selections, astronomers explore the possible effects of a comet like SL 9 colliding with Earth.

See also: Callisto; Ceres; Comet Halley; Comets; Dwarf Planets; Eris and Dysnomia; Kuiper Belt; Meteorites: Achondrites; Meteorites: Carbonaceous Chondrites; Meteorites: Chondrites; Meteorites: Nickel-Irons; Meteorites: Stony Irons; Meteoroids from the Moon and Mars; Meteors and Meteor Showers; Oort Cloud; Pluto and Charon.

Comets

Category: Small Bodies

A comet is a minor body composed mostly of frozen ices typically embedded with solids. Comets revolve about the Sun in highly elliptical orbits. The Oort Cloud is a vast cloud of cometary bodies that extends billions of kilometers out from the Sun.

OVERVIEW

Comets are familiar to nearly everyone as majestic, starlike objects with long tails stretching across a wide band of the sky. The most famous comet, Halley's comet, makes its periodic return to the night skies every seventy-five years. The word "comet" is derived from a Greek word meaning "long-haired." Comets were greatly feared before the twentieth century as

A comet streaks across the night sky. (NASA)

bad omens. Since then, they have been identified and cataloged as objects that come into the inner solar system from deep space. Most of them occupy orbits that carry them far away from the Sun. Many comets make only a single approach to the Sun and then never return again, while others exist in stable, but highly elliptical, orbits that allow them to return after an extended period of time.

One of the first theories advanced to explain the makeup of comets was proposed by astronomer Fred L. Whipple. Whipple suggested that comets were dirty snowballs, essentially bodies of water ice that incorporate dust and perhaps volatives other than water. This remained the primary theory through the first four decades of the space age. Only when spacecraft began visiting comets could it be put to the test.

Until the space age, comets were studied only in visible light through optical telescopic images. The first comet to be studied using Earth-orbital instruments, which permitted observations in the ultraviolet as well as the visible, was the much-heralded Comet Kohoutek late in 1973 and early in 1974. Comet Kohoutek turned out to be something of a disappointment visually from Earth, but images and data collected by the orbiting Skylab 4 astronauts advanced the understanding of comets.

In 1986, the European space probe Giotto passed about 600 kilometers from Halley's comet as the comet made its close approach to the Sun. The probe verified existing theories that comets are made up of ices covered by black dust or soil. In other words, the spacecraft confirmed the dirty snowball model, at least for this comet. Using data taken by the spacecraft, scientists determined that the dust is composed of carbon, hydrogen, oxygen, and nitrogen. Other metals have also been discovered in comets, such as iron, calcium, nickel, potassium, copper, and silicon. Halley's comet was one of the darkest objects ever seen in the solar system; it has virtually no albedo. Only one other major body in the solar system, Saturn's satellite Iapetus, is known to be this low in albedo.

As a comet approaches the Sun, it absorbs solar radiation and becomes warmer. The main body of the comet is called the nucleus. As the nucleus warms, ices beneath the comet's soil evaporate. Because the comet has no atmosphere, evaporated substances, also called volatiles, escape into the vacuum of space. This gaseous envelope that surrounds the comet is called the coma. As the coma grows, it forms a plume of vapor that carries away some of the comet's surface dust as well. This mixture of evaporated volatiles and dust is carried away from the comet by the solar wind, is ionized by high-energy particles, and creates the spectacular tail of the comet. The comet's tail, glowing in the solar wind, can stream behind the comet for millions of kilometers. Cometary nuclei consist mostly of volatile ices and dust. That ice is nearly all water ice, but there is also evidence of ices composed of carbon dioxide and methane. More elementary compounds of nitrogen, oxygen, and carbon monoxide may exist as volatile ices.

Comets are typically small bodies. Halley's comet is an irregular potato-shaped object, 14 by 17 kilometers. In fact, some noted that images of Halley's comet captured during the Giotto mission suggested that the famous comet resembled the cartoon character Felix the Cat. The largest known comet is Chiron, which is estimated to be approximately 200 kilometers in diameter. Comets are thought to have formed as the solar system evolved. Comets were accreted out of material at the outer edge of the solar nebula that ultimately condensed to become the Sun and planets. Because cometary material was fashioned at the outer edge of the solar system, the Sun did not evaporate comets' volatiles. At the same time, the giant planets of the solar system formed at what would become the outer orbits of the solar system. These massive planets encountered the newly formed comets, and the comets that were not engulfed by the giant planets were, over the first billion years, ejected into interstellar space by the planets' massive gravitational fields. Not all comets met that fate, however. Some were gently nudged into stable orbits closer to the Sun. Others were flung into the inner solar system, eventually impacting the inner planets. There are strong reasons to believe that Earth's oceans came from cometary ices delivered to the planet during the early era of bombardment, but that is not universally accepted.

What remained after billions of years of planetary encounters was an extraordinarily large cloud of comets extending outward from orbits beyond Pluto in all directions. A virtual spherically shaped cloud of comets surrounds the Sun at a distance from 1,000 to 100,000 astronomical units (AU). This cloud, which may contain as many as two trillion comets of all shapes and sizes, is called the Oort Cloud. It is named for the Dutch astronomer Jan Hendrik Oort, who first proposed its existence in 1950. The spherically shaped Oort Cloud is not the only source of comets in the solar system. There is a disk-shaped source of comets that extends from about 35 to 40 AU out from the Sun to about 1,000 AU. This source, the Kuiper Belt, was named for the astronomer Gerard Peter Kuiper, who theorized its possible existence in 1951. The disk-shaped Kuiper Belt blends with the spherical Oort Cloud at about 1,000 AU.

The Oort Cloud is the source for long-period comets, with orbital periods of greater than two hundred years. The Kuiper Belt is most likely the primary source for short-period comets, with orbital periods of less than two hundred years, such as Halley's comet. Comets have definite life spans, unlike planets. Each time a comet streaks in toward the Sun, volatile gases stream off the comet and form a beautiful cometary tail, while also depleting the comet's total mass. The comet melts away with each pass toward the Sun. When Halley's comet streamed past the Sun in 1986, the Giotto spacecraft measured a loss of 40 tons of mass per second from the comet. If the supply of comets were not steadily replenished from deep space, they would have all been lost long ago.

The Sun is one among billions of stars in the Milky Way galaxy. In the relatively nearby region of the galaxy, there are hundreds of local stars, which are all revolving around the galactic center and are moving relative to one another. Because stars are so far apart on the average, the chance of one star colliding with another is quite low. However, the possibility of a local star passing near to or through the Oort

Cloud (which extends up to 100,000 AU away from the Sun) is very high over millions of years. It is estimated that since the solar system formed, about five thousand stars have passed within 100,000 AU of the Sun. If an object as massive as another star passed close to the Oort Cloud, it could easily cause enough gravitational perturbations to direct comets in toward the Sun.

Since the Oort Cloud is spherical, long-period comets can appear to approach the Sun from any point in space. Short-period comets, originating from the Kuiper Belt, always appear to emanate from a band along the ecliptic plane (the plane that contains the planetary orbits). After careful study of where comets actually originate, an analysis was made of their orbits. It has been discovered that there are areas of the sky that are richer in comets than others, and other areas that appear to be practically devoid of comets. Four different theories have been advanced to explain the source of these newly appearing comets. The first theory postulates that the passage of stars in or near the Oort Cloud may so affect the gravitational balance of comets that they are sent falling in toward the Sun. The second theory involves brown dwarfs, which are massive objects—about thirty times the mass of Jupiter—that are not quite planets and not quite stars. They do not have enough mass to create the conditions for thermonuclear ignition at their core. They predominantly radiate infrared energy, and they cannot be readily seen from Earth's surface. Current estimates approximate the number of brown dwarfs near the Sun to be sixty times greater than that of ordinary stars. A brown dwarf should pass through the Oort Cloud every 7 million years. Such an object would travel very slowly with respect to the Sun and would gravitationally release large swarms of comets into the solar system. These two stellar mechanisms, the action of either a passing star or a brown dwarf, are estimated to have been the source of about one-third of the observed comets.

Facts About Selected Comets

Name (no)	Period (yrs)	Perihelion Date	Perihelion Distance	Distance from Sun (AU)
Borrelly (19P)	6.86	2001-09-14	1.358	3.59
Chiron (95P)	50.7	1996-02-14	8.460	13.7
Crommelin (27P)	27.89	1984-09-01	0.743	9.20
d'Arrest (6P)	6.51	2008-08-01	1.346	3.49
Encke (2P)	3.30	2003-12-28	0.340	2.21
Giacobini-Zinner (21P)	6.52	1998-11-21	0.996	3.52
Grigg-Skjellerup (26P)	5.09	1992-07-22	0.989	2.96
Hale-Bopp	4,000	1997-03-31	0.914	250
Halley (1P)	76.1	1986-02-09	0.587	17.94
Honda-Mrkos-Pajdusakova (45P)	5.29	1995-12-25	0.581	3.02
Hyakutak	~40,000	1996-05-01	0.230	~1,165
Kohoutek (75P)	6.24	1973-12-28	1.571	3.4
Schwassmann-Wachmann 3 (73P)	5.35	2006-06-02	0.933	3.06
Tempel 1 (9P)	5.51	2005-07-07	1.497	3.12
Tempel-Tuttle (55P)	32.92	1998-02-28	0.982	10.33
West-Kohoutek-Ikemura (76P)	6.46	2000-06-01	1.596	3.45
Wild 2 (81P)	6.39	2003-09-25	1.583	3.44
Wilson-Harrington (107P)	4.30	2001-03-24	1.000	2.64
Wirtanen (46P)	5.46	2013-10-21	1.063	3.12

Source: Data are from the National Aeronautics and Space Administration/Goddard Space Flight Center, National Space Science Data Center.

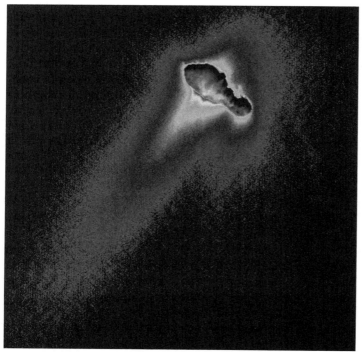

A composite of images captured by Deep Space 1, about 4,800 kilometers from Comet Borrelly, shows the comet's nucleus during the spacecraft's September, 2001, flyby. (NASA/JPL)

In the aftermath of the American decision to be the only spacefaring nation not to dispatch a spacecraft to investigate Halley's comet on its most recent appearance in the inner solar system, the National Aeronautics and Space Administration (NASA) proposed the Comet Rendezvous and Asteroid Flyby (CRAF). CRAF was a sister ship to the Cassini spacecraft. Because of budget cuts, NASA was able to save the Cassini mission, but CRAF was canceled. If it had been adopted, CRAF would have rendezvoused with Comet Kopff and remained in its vicinity for thirty-two months to observe variations in that comet's activity during different portions of its orbit. CRAF would also have dropped penetrometers into the comet to ascertain information about internal structure and make chemical analyses of surface materials.

Aspects of the ambitious CRAF concept were recycled into cheaper comet missions such as Deep Space 1, Stardust, and Deep Impact. Also, the European Space Agency (ESA) developed the Rosetta spacecraft to visit a comet.

Launched on October 24, 1998, the Deep Space 1 spacecraft began tests of an ion propulsion system and autonomous navigation system. Deep Space 1's targets were the asteroid Braille and Comet Borrelly. Flying by the comet at a relatively close distance, Deep Space 1 captured images of a comet's nucleus that had higher resolutions than any of those captured by the probes that had visited Halley's comet. Comet Borrelly was shaped much like a bowling pin and displayed emission jets not distributed uniformly across its irregular nucleus. Deep Space 1 was not outfitted with debris shields, but it survived the close encounter nevertheless. In time Deep Space 1 ran out of propellant, but the mission showed that ion propulsion could be used on a spacecraft designed to visit multiple targets such as comets.

Stardust was launched on Feburary 7, 1999, and directed toward Comet Wild 2, where it opened up special sample collectors incorporat-

According to the third theory, huge molecular clouds in interstellar space (much more massive than a single star) may pass at very large distances (tens of light-years) and may cause a release of comets through gentle perturbations of their orbits. The final theory for the source of newly appearing comets is galactic tidal action. Each galaxy has a gravitational field, which causes an attraction toward the midplane of the galaxy of all bodies (comets and stars). As these bodies orbit the galaxy, they are gravitationally influenced by one another. The galactic tide is the difference between the galactic forces acting on the Sun and the comet. Because the force of the galactic tide is very specific with respect to direction, it cannot act toward the poles of the Sun or toward the equator. Observations of cometary tracks confirm that comets from deep space do not seem to approach the Sun from these segments of the celestial sphere. This mechanism appears to explain the approach of the majority of all long-period comets entering the solar system from the Oort Cloud.

ing aerogel to capture both interplanetary and cometary dust. After the spacecraft's encounter with the comet on January 2, 2004, its sample collectors were sealed for a two-year journey back to Earth. On January 15, 2006, Stardust's sample collection unit safely reentered Earth's atmosphere and was recovered intact in Utah.

In June, 2008, researchers studying comet 26P/Grigg-Skjellerup material collected by the Stardust spacecraft announced that they had discovered new mineral grains. This mineral, named brownleeite after Donald Brownlee of the University of Washington, was a variety of manganese silicide not previously predicted by models of comets or the condensation of material from the early proto-Sun's nebula.

Deep Impact was designed as its name clearly suggests to fly to a comet and strike it, excavating material from deep below the surface. The spacecraft launched on January 12, 2005, and its onboard navigation steered the spacecraft toward the comet Tempel 1, released an impactor payload made largely of copper, and then veered out of the way to observe the resulting impact of the payload on the comet. The impactor was composed of copper, since that was an element not expected to be found naturally within the comet. The impact was observed by Deep Impact itself, as well as by the Hubble Space Telescope, the Chandra X-Ray Orbservatory, the Spitzer Space Telescope, the Swift spacecraft, and ESA's XMM-Newton observatory and Rosetta spacecraft. This coordinated effort permitted time-evolution studies of the plume and debris cloud created by the high-speed impact of the copper payload on Tempel 1.

Rosetta was launched on March 2, 2004. This was the European Space Agency's second attempt at a comet study and incorporated both a flyby craft and a lander named Philae. The mission was designed to rendezvous with the comet 67P/Churyumov-Gerasimenko in May, 2014, orbit it for many months while mapping the surface, and observe changes in the comet's activity as its distance to the Sun changed. Then the lander was scheduled to touch down on the comet on or about November, 2014, where it would secure itself to the surface in the comet's weak gravity field and then begin studies of chemical composition and physical characteristics of the comet's surface. The Rosetta mission was planned to continue through December, 2015.

APPLICATIONS

The study of comets requires detailed knowledge of the composition of the outer regions of the solar system and the space between the last planet and 100,000 AU outward from the Sun. Comet studies also seek to understand complex gravitational interactions between bodies separated by wide distances and even gravitational interactions between tiny comets and the entire galaxy. Astronomers who study comets want to learn more about their makeup, their behavior when approaching the Sun, and the makeup and evolution of the early solar system.

New comets approaching the Sun for the first time have been held in deep freeze within the Oort Cloud and are thought to be composed of primordial material of the newly forming solar

The Spitzer Space Telescope captured this infrared image of the breakup of Comet 73P/Schwassman-Wachmann 3, which began to split into pieces in 1995. (NASA/JPL-Caltech)

system. They have been tied up in the Oort Cloud for billions of years at temperatures barely above absolute zero. As they approach the Sun, their internal gases begin to stream away. Detailed study of an approaching comet's outgassing can inform planetary scientists about the composition of the early solar system. Comets and their approach have also hinted at the existence of the elusive brown dwarfs, thought to be one of the most common bodies of interstellar space. Because they are so dim, they are all but invisible from Earth. On the other hand, because brown dwarfs are thought to be so plentiful, the study of comets and their orbits may give the first real clues to the former's reality and abundance. The first serious studies of brown dwarfs came from observations made by the Spitzer Space Telescope, the final member of NASA's Great Observatory program. Spitzer detected brown dwarfs from their infrared emissions.

In the early 1980's the existence of galactic tidal action was merely speculation. Since then, careful study of comet orbits and their approaches to the inner solar system has favorably supported the theory of galactic tides. In the close approach of Halley's comet by robotic

spacecraft in 1986, a wealth of information was recovered on the shape, behavior, and composition of comets. The existence of the Oort Cloud and the concept of gravitational interactions by passing objects in space have led to the theory of periodic comet showers. Such comet showers, separated by periods of tens of millions of years, may be responsible for certain mass extinctions on Earth. These extinctions might be the result of a shower of comets originating from within the Oort Cloud, sent on their close approach to the Sun by the close passage of a star or brown dwarf to or through the Oort Cloud.

Samples of Comet Wild 2 were treated to many of the contamination safeguards used with the Apollo lunar rocks. Analyses of Stardust's captured comet material revealed some surprises. Tracks in the aerogel suggested solid materials larger than interstellar dust grains. Silicate crystals and other mineral crystals were found which required more than just mild heating, as would have been the case if the comet was largely composed of interstellar dust grains. This suggested that the theory of comet formation may need alteration. Inclusions of vanadium nitride, titanium, molybdenum, osmium, ruthenium, and tungsten were found, components that would have required high heating. Samples also contained organic materials more primitive than found in asteroidal material, compounds such as polycyclic aromatic hydrocarbons.

The collision of Deep Impact's copper payload was equivalent to five tons of TNT. Comet Tempel 1 increased in brightness sixfold as a result of the event. As much as between 10 and 25 million kilograms of comet material was ejected as a crater formed as a result of the impact. Tempel 1 material was much finer than had been expected, being more akin to talcum powder than a sandy grain. Data ruled out a loose aggregate or highly porous model of the comet's structure. Indeed,

Debris from comet 73P/Schwassman-Wachmann 3, taken in the infrared by Spitzer. The debris passes near Earth every year and is expected to cause a significant meteor-shower display in the year 2022. (NASA/JPL-Caltech)

rather unlike a dirty snowball as proposed by Whipple, Comet Tempel 1 was more like an icy dirt ball. Seen in the ejected materials in addition to volatiles were clays, carbonates, sodium, and crystalline silicates. After this encounter the flyby portion of the Deep Impact spacecraft was redirected to a planned encounter with Comet Hartley 2 in 2010.

Rosetta holds the potential for in situ analyses of comet material if the Philae lander successfully touches down safely, and can perform its chemical and physical tests. This information will greatly assist in determining how comets form and how their comas and tails develop as they travel into the inner solar system.

CONTEXT

Humankind has always looked to the heavens in awe and wonder, and sometimes in fear. Perhaps no other astronomical phenomenon except a total solar eclipse has historically evoked as much fear as comets. When the specter of fear is removed, however, they emerge as strikingly beautiful objects in the sky. It was once believed that if Earth passed through the tail of a comet, its inhabitants would die; this theory has been discredited. Comets are messengers from a time long past. Most are chunks of dirty ice, locked away in the Oort Cloud for billions of years.

Comets have been used to judge vast distances, evaluate the early composition of the solar system, and even test the idea that the gravity of the entire Galaxy can make a difference to the smallest of objects in space. Comets have been used as yardsticks to evaluate what may be the most common type of star in the galaxy, the brown dwarf—which ironically is one that is difficult to observe, even in the infrared. Comets have been called dirty snowballs. Halley's comet was so black that it was the darkest object ever seen in space. Yet, from these dirty specks of ice, planetary scientists have witnessed some

Comet Tempel 1 in an image taken in 1998 by the Deep Impact flyby craft. (NASA/JPL-Caltech)

of the most spectacular light shows. Ultimately, comets may also generate clues to some of the most fundamental secrets about the solar system and planets. From these tiny messengers, planetary scientists may unlock and examine pristine elements from the formation of the solar system.

Debris from comets provides the material that Earth passes through when annual meteor showers occur. For example, the Orionid meteor shower is leftover material from Halley's comet, the Leonids meteor shower is associated with Comet Tempel-Tuttle, and the Perseid meteor shower is material from Comet Swift-Tuttle.

Historically comets have come full circle from being seen as omens in the heavens to be feared, to celestial objects evoking a sense of wonder, and to again being objects that should be feared if they come too close and perhaps even impact Earth. Comets represent a more troubling threat than do asteroids, as comets are usually discovered only when they come in past the orbit of Jupiter. As such there is insufficient time to mount any mitigating effort if a new comet is determined to make a close pass or actually impact the Earth. Have comets impacted the Earth in the past? The answer is believed to be almost certainly yes, and indeed many believe that the majority of Earth's water came from comets encountering the early Earth. The Siberian Tunguska event of 1908, itself a curiosity that has been explained by some (without any legitimate supporting evidence) as a nuclear explosion or even the impact of an unidentified flying object, is now believed to have been the result of a comet or asteroid impact, most likely an air burst explosion of the body. Although this theory is not yet confirmed, it points out the potential for devastation that an impacting comet represents.

A comet collision was observed in 1994 when the nearly two dozen pieces of the shattered Comet Shoemaker-Levy 9 smacked into Jupiter's upper atmosphere. These pieces created temporary changes in the gas giant's appearance, many of which were the size of the Earth, indicating that a tremendous amount of energy was involved in this series of collisions. The impacts were recorded by the Hubble Space Telescope and Galileo spacecraft; the incredible magnitude of the disruption came as a surprise to the scientific community.

It was believed since roughly 1950 that gravitational disruption of the Sun's Oort Cloud by a close passage of another star was responsible for swarms of comets heading into the inner solar system, resulting in bombardment of the planets. However, in late 2008 Hans Rickman of Sweden's Uppsala Astronomical Observatory reported in *Celestial Mechanics and Dynamical Astronomy* the results of an updated computer simulation of the Oort Cloud investigated by his research group. If correct, their model indicates that sporadic stellar encounters, while indeed important in generating fresh comets that head toward the inner solar system, is not the only mechanism for sending comets toward the planets. This model accounted for galactic gravitational tidal influences on the Oort Cloud and found that the threat from comets may be more constant than previously believed. If correct, the model reinforces the need to monitor the

skies for incoming comets that might be headed our way.

Dennis Chamberland and David G. Fisher

FURTHER READING

Arny, Thomas T. *Explorations: An Introduction to Astronomy*. 3d ed. New York: McGraw-Hill, 2003. A general astronomy text for the nonscience reader. Includes an interactive CD-ROM and is updated with a Web site.

Benningfield, Damond. "Where Do Comets Come From?" *Astronomy* 18 (September, 1990): 28-36. A fine summary of comets, superbly illustrated and written for the general public. Addresses the question of the Oort Cloud and Kuiper Belt in detailed, scaled illustrations. Discusses possible linkage to the extinction of the dinosaurs and the latest satellite discoveries.

Brandt, John C., and Robert D. Chapman. *Introduction to Comets*. New York: Cambridge University Press, 2004. A text suitable for a planetary science course, this comprehensive work covers our knowledge of comets from early observations to telescopic investigations through spacecraft encounters of these mysterious and alluring bodies.

Hartmann, William K. *Moons and Planets*. 5th ed. Belmont, Calif.: Thomson Brooks/Cole, 2005. An updated version of a classic text that covers all aspects of planetary science. Well-explained material on asteroids and comets. Takes a comparative planetology approach.

Levy, David H. *The Quest for Comets: An Explosive Trail of Beauty and Danger*. New York: Plenum Press, 1994. Written by one of the codiscoverers of Comet Shoemaker-Levy 9, this book is for the general reader. It highlights the author's comet discovery program and the comet catastrophe theory.

Newburn, R. L., M. Neugebauer, and Jurgen H. Rahe, eds. *Comets in the Post-Halley Era*. New York: Springer, 2007. A collection of fifty papers compiled into a volume in Springer's Astrophysics and Space Science Library. Covers observational techniques, origin of comets, evolution of comets, and spacecraft data. Special attention is given to Comet Halley and other comets encountered by spacecraft.

Russell, Christopher T. *Deep Impact Mission: Looking Beneath the Surface of a Cometary Nucleus.* New York: Springer, 2005. A complete description of the Deep Impact mission to excavate and analyze material from Comet 9P/Tempel 1.

Sagan, Carl, and Ann Druyan. *Comet.* New York: Random House, 1985. This coffee-table book is a classic work of art, written by the most popular astronomer in the United States. Filled with beautiful color and historical black-and-white photographs and illustrations, it was the basis for the popular movie by the same name. For general audiences.

Schaaf, Fred. *Comet of the Century: From Halley to Hale-Bopp.* New York: Copernicus Books, 1997. Comet Hale-Bopp was a popular comet to observe, one that provided a better show than the most recent appearance of Halley's comet. For a general audience.

Thomas, Paul J., Roland D. Hicks, Christopher F. Chyba, and Christopher P. McKay. *Comets and the Origin and Evolution of Life.* 2d ed. New York: Springer, 2006. A collection of chapters written by experts in the field. This update of the first edition covers new understandings of Halley's comet and more recent spacecraft data. Provides insights into organic compounds found in comets, protostars, and interstellar clouds.

Verschuur, Gerrit L. *Impact! The Threat of Comets and Asteroids.* New York: Oxford University Press, 1997. Verschuur explains the change in thinking from uniformitarianism to catastrophism. Identifies the Chicxulub Crater with an impact event that led to the extinction of the dinosaurs. Warns of the potential for devastation that a comet impact on Earth would cause.

See also: Ceres; Comet Halley; Comet Shoemaker-Levy 9; Dwarf Planets; Earth's Oceans; Eris and Dysnomia; Extraterrestrial Life in the Solar System; Infrared Astronomy; Kuiper Belt; Lunar Craters; Meteorites: Achondrites; Meteorites: Carbonaceous Chondrites; Meteorites: Chondrites; Meteorites: Nickel-Irons; Meteorites: Stony Irons; Meteoroids from the Moon and Mars; Meteors and Meteor Showers; Oort Cloud; Planetary Classifications; Planetary Formation; Pluto and Charon; Solar System: Element Distribution; Solar System: Origins; Solar Wind; Ultraviolet Astronomy.

Coordinate Systems

Category: Scientific Methods

There are several astronomical coordinate systems that are in common usage. In each system, the position of an object in the sky, or on the celestial sphere, is specified by two angles, similar to latitude and longitude on Earth.

OVERVIEW

An astronomical coordinate system is a way for locating an object in the sky, or on the celestial sphere, using two angles. (The celestial sphere is an imaginary sphere of large size surrounding the Earth, representing the sky as seen from Earth.) There are several astronomical coordinate systems in common usage, and each system is based on a reference plane and a reference direction in that plane. In each system, the intersection of the reference plane and the celestial sphere is a great circle on the celestial sphere defining the "equator" of the coordinate system. The two "poles" of the system are the two points on the celestial sphere each 90° from the system's equator. Great circles passing through these poles intersect the equator of the system at right angles. One of the two angular coordinates of each coordinate system is measured from the equator of the system to the object along the great circle passing through it and the poles. Angles on one side of the equator are considered positive; those on the opposite side are negative. The other angular coordinate is measured along the equator from the reference direction to the intersection of the equator, with the great circle passing through the object and the poles. (For comparison, in the system of latitude and longitude on Earth, the reference plane is the Earth's equatorial plane, and the reference direction is the intersection of the prime meridian passing through Greenwich,

England, with the equator. Latitude is measured as the angle north or south of the equator, and longitude is measured as the angle east or west of the prime meridian).

Four astronomical coordinate systems are commonly used: alt-azimuth (or horizon), equatorial, ecliptic, and galactic.

The alt-azimuth, or horizon, system has as its reference plane the plane of the horizon, which is a great circle on the celestial sphere 90° from the zenith (the point directly over the observer). Its reference direction is the north point (the point on the horizon due north). Its latitude-like coordinate, called altitude (h), is the angle above or below the horizon (positive above and negative below). Altitude ranges from +90° at the zenith to –90° at the nadir (the point directly underneath the observer). The longitude-like coordinate, called azimuth (A), is the angle measured to the east along the horizon from the north point. Azimuth varies from 0° due north, to 90° due east, to 180° due south, to 270° due west, and to 360° (equivalent to 0°) as north is approached from the west. This system is convenient for giving approximate directions to objects in the sky at any one moment from any one location on Earth, but the altitude and azimuth for any object are different as seen from different locations on Earth, and they constantly change as objects appear to move across the sky due to the Earth's rotation.

In the equatorial system, the reference plane is the plane of the Earth's equator; its intersection with the celestial sphere defines the celestial equator. The extension of the Earth's rotational axis through the north and south terrestrial (or geographic) poles intersects the celestial sphere at the north and south celestial poles. The reference direction is given by the Sun's apparent position at the moment of the vernal equinox, when the Sun is directly above some point on the Earth's equator in March. The latitude-like coordinate, called declination, is measured as the angle north or south of the celestial equator (positive to the north and negative to the south). Declination ranges from +90° at the north celestial pole (the point on the celestial sphere directly over the Earth's north pole) to –90° at the south celestial pole (the point on the celestial sphere directly over the

Earth's south pole). The longitude-like coordinate, called right ascension, is measured to the east along the celestial equator from the vernal equinox point. Instead of specifying this angle in degrees, right ascension is given traditionally in hours, minutes, and seconds. Because of the Earth's rotation on its axis, the sky appears to turn through 360° in approximately 24 hours: exactly 24 hours of sidereal time, but 23 hours, 56 minutes, and 4 seconds of mean solar time (the difference in these two time systems is due to the Earth's revolution around the Sun). A full circle of 360° around the celestial equator is defined as 24 hours of right ascension. Thus 1 hour of right ascension corresponds to 15° of arc, 1 minute of right ascension corresponds to 15 arc minutes, and 1 second of right ascension corresponds to 15 arc seconds.

This equatorial system rotates with the apparent daily motion of the sky, so declination and right ascension of most astronomical objects remain nearly constant for reasonably short times of up to a few years. However, over the long run, declination and right ascension change as a result of two main factors: precession and proper motion. (Note that the Sun, Moon, and planets appear to move through the sky with respect to the stars, so their right ascension and declination noticeably change much more rapidly, over timescales of hours to months.) Precession is a slow change in the direction of the Earth's rotational axis, which traces out a double-cone figure in space over approximately 25,800 years, due to the gravitational pull of the Moon and Sun on the Earth's equatorial bulge. As a result, the positions of the celestial poles and celestial equator do not remain fixed on the celestial sphere, but they shift in a predictable way. Each of the two celestial poles traces out a circle with an angular diameter of 47° on the celestial sphere over a period of 25,800 years. Because of precession, the equinoxes shift westward relative to the stars so that gradually the seasonal constellations change. The vernal equinox (occurring in March), now is located in Pisces, but 2,000 years ago it was in Aries, and in about 600 years it will shift into Aquarius. Polaris currently is the north pole star, but at the time of the building of the Great Pyramid in Egypt, about 5,000 years ago, the star Thuban in Draco was the north pole star. In about 12,000 years, the star Vega in Lyra will approximately mark the north celestial pole. Because of precession, catalogs listing the declination and right ascension of stars, nebulae, galaxies, and other objects must specify the epoch (year) for which the coordinates are rigorously correct. Corrections must be calculated using standard precessional formulas to convert the listed coordinates to other years.

The declination and right ascension of individual stars also change due to their proper motion, which is the change in direction to a star as seen from Earth due to the star's actual motion through space relative to our solar system. For most stars, the proper motion is small enough so that its practical effect on declination and right ascension can be ignored, but some stars have proper motions of several arc seconds per year (the largest proper motion known is that of Barnards star, 10.4 arc seconds per year), so over timescales of decades the declination and right ascension of these stars can change by arc minutes. (A very long-term effect of proper motions is that over timescales of thousands of years or more, the familiar shapes of constellations slowly change.)

In the ecliptic system, the reference plane is the plane of the Earth's orbit around the Sun. The term "ecliptic" is used both for the Earth's orbital plane and for the great circle that marks its intersection with the celestial sphere. The reference direction is the same as in the equatorial system: the apparent position of the Sun at the moment of the vernal equinox. The term "vernal equinox" is used both for the moment in time in March when the Sun is directly above some point on the Earth's equator and for one of the two intersections of the ecliptic and the celestial equator on the celestial sphere. (The other intersection is called the autumnal equinox and marks the Sun's apparent position when it is directly above some point on the Earth's equator in September.) The latitude-like coordinate, called ecliptic latitude, is measured as the angle north (positive) or south (negative) of the ecliptic; it ranges from +90° to –90°. The longitude-like coordinate, called ecliptic longitude, is measured to the east along the ecliptic from the vernal equinox point; it ranges from 0° to 360°.

This coordinate system is especially useful in giving the position of solar-system objects. Because many solar-system objects have orbits around the Sun that are not inclined very greatly to the plane of the Earth's orbit around the Sun, such solar-system objects will be seen near the ecliptic and thus have small ecliptic latitudes.

The galactic coordinate system has a reference plane that is the mean plane of our galaxy, the Milky Way (defined primarily by 21-centimeter, 1,400-megahertz radio observations of neutral hydrogen, which is concentrated in the galactic plane), and a reference direction that points to the galactic center. The latitude-like coordinate, called galactic latitude, is measured as the angle north (positive) or south (negative) of the galactic equator. The longitude-like coordinate, called galactic longitude, is measured along the galactic equator from the direction to the galactic center (as viewed from the poition of our solar system) toward the direction of galactic rotation. It varies from 0° in the direction of the galactic center, to 90° in the direction of general galactic rotation, to 180° in the direction opposite to the galactic center (the galactic "anticenter"), to 270° opposite to the direction of general galactic rotation, and up to 360° as the direction to the galactic center is approached again. This system is useful for indicating the location of objects relative to the Milky Way.

APPLICATIONS

The primary use of any astronomical coordinate system is to specify the location of celestial objects for observation. Most sources list the equatorial coordinates of right ascension and declination, together with the epoch (year) for which the right ascension and declination are rigorously correct. In the past, astronomers had to calculate precessional corrections to adjust the right ascensions and declinations to the current date, and then calculate the angles to set the telescope based on the time of observation. However, now the settings of most major telescopes have been computerized so that the right ascension, declination, and listing epoch can be input directly, and the correct settings will be calculated automatically.

Another use for old catalogs listing right ascension and declination is to determine proper motions. The old coordinates, corrected for precession, are compared to new coordinates. Any differences are due to the proper motions of the objects.

In the case of an object in our solar system (such as a planet, asteroid, or comet), often the elements of its orbit around the Sun (such as semimajor axis, eccentricity, inclination, or time of perihelion passage) will be given. These can be used to calculate the object's ecliptic latitude and longitude for a specific time and date, and in turn these can be converted into its right ascension and declination for that same time and date. A list of these coordinates for a series of dates provides an ephemeris of where to observe the object.

In the study of the structure of the Milky Way, coordinates of galactic latitude and galactic longitude often are most useful for visualizing where objects are located. Of course, galactic coordinates can be converted to equatorial coordinates of right ascension and declination for setting telescopes to observe the objects.

CONTEXT

The earliest references to locations and motions of stars and planets can be traced back about twenty-five hundred years, when Babylonian observer-priests recorded the movement of planets relative to the stars. In the hands of the Greeks, the Babylonian results for periodic and irregular celestial motions became the basis for geometric models trying to explain the structure and motions of the universe. Ptolemy, between 296 and 272 B.C.E., measured positions for a number of stars in terms of their angular distances above or below the celestial equator (today this would be called declination), as well as differences in their angular positions parallel to the celestial equator (today this would be called differences in right ascension). Around 150 B.C.E., the Greek astronomer Hipparchus compiled one of the first systematic star catalogs, giving the coordinates and apparent magnitudes for about 850 stars.

The development of astronomical coordinate systems was important for several practical reasons. The most significant was that it allowed for the construction of an accurate calendar, which was essential for weather prediction and agriculture. The length of the year could be fixed, months and days could be intercalated, and the passing of the solstices and equinoxes could be established. In addition, astronomical coordinate systems led to the development of celestial navigation, enabling commerce to expand to new trade areas and allowing more of the world to be explored.

Earl G. Hoover and Richard R. Erickson

FURTHER READING

Boucher, C., ed. *Earth Rotation and Coordinate Reference Frames*. International Association of Geodesy Symposia 105. New York: Springer, 1990. Discusses techniques for determining Earth's rotation, including laser ranging and very long baseline interferometry (VLBI).

Fraknoi, Andrew, David Morrison, and Sidney Wolff. *Voyages to the Stars and Galaxies*. Belmont, Calif.: Brooks/Cole-Thomson Learning, 2006. A well-written, thorough college textbook for introductory astronomy courses. Includes a section on coordinates.

Freedman, Roger A., and William J. Kaufmann III. *Universe*. 8th ed. New York: W. H. Freeman, 2008. College-level introductory astronomy textbook, thorough and well written. Includes a section on celestial coordinates.

Kovalevsky, Jean. *Modern Astrometry*. New York: Springer, 1995. An introduction to astrometry. Covers how to find motions, positions, dimensions, and other data about astronomical objects. Also discusses observational techniques and equipment.

Lankford, John, ed. *History of Astronomy: An Encyclopedia*. New York: Garland, 1997. An encyclopedia focused on the history of astronomy, women in astronomy, social aspects, and national contexts.

Maran, Stephen P., ed. *The Astronomy and Astrophysics Encyclopedia*. Foreword by Carl Sagan. New York: Van Nostrand Reinhold, 1992. More than four hundred articles arranged alphabetically. Full of diagrams and photographs. Easy to use and read.

Moche, Dinah L. *Astronomy: A Self-Teaching Guide*. 6th ed. New York: John Wiley & Sons, 2004. A self-instructional text designed so that students with no formal astronomy background can easily learn basic principles and concepts. The material in each chapter is presented in short, numbered sections. The chapter on understanding the starry night is especially recommended for its coverage of the coordinate systems. An excellent book for the upper-level high school and lower-level college students.

Pannekoek, A. *A History of Astronomy*. London: Barnes & Noble Books, 1969. As the title denotes, this is a history and as such is written in a nonscientific style. Of special interest are the early chapters covering the Babylonians, Assyrians, and Chaldean contributions. Recommended for the college-level student and interested general reader.

Pasachoff, Jay M., and Will Tirion. *Field Guide to the Stars and Planets*. 5th ed. Boston: Houghton Mifflin, 1999. An excellent handy reference. Suitable for all high school upper-level students, college students, and hobbyists. Useful tables give star names and coordinates. In addition, one chapter has an easy-to-read discussion on coordinates, time, and calendars.

Rey, H. A. *The Stars: A New Way to See Them*. 1952. Reprint. Boston: Houghton Mifflin, 1988. A simple, clearly written, and charming book. Includes a good introduction to celestial coordinates and time. Written by the author of the "Curious George" children's stories.

Schneider, Stephen E., and Thomas T. Arny. *Pathways to Astronomy*. 2d ed. New York: McGraw-Hill, 2008. Very thorough college textbook for introductory astronomy courses, divided into many short sections on specific topics. Contains sections on coordinate systems. However, being the first edition of a revised text, there are several typographical errors throughout.

See also: Earth System Science; Earth's Rotation; Gravity Measurement; Hertzsprung-Russell Diagram; Infrared Astronomy; Neutrino Astronomy; Optical Astronomy; Radio Astronomy; Telescopes: Ground-Based; Telescopes: Space-Based; Ultraviolet Astronomy; X-Ray and Gamma-Ray Astronomy.

Coronal Holes and Coronal Mass Ejections

Category: The Sun

The corona is the outermost layer of the Sun. It is extremely hot but so tenuous that it is visible only when a solar eclipse blocks the brighter photosphere. Coronal holes are less dense regions of the corona where coronal matter streams outward into interplanetary space. Coronal mass ejections occur when magnetic field lines in the solar corona snap and eject large clumps of solar material into interplanetary space.

OVERVIEW

The outermost layer in the Sun's atmosphere is the corona (which means "crown"). Gas in the corona can reach temperatures of a few million kelvins. This gas is very thin, however, with a density on the order of 10^{-12} kilograms/meter^3. Thus, the corona is faint, so faint that it cannot normally be seen because its feeble light is overwhelmed by the much brighter photosphere. The corona must be observed optically either during a total solar eclipse or by using a coronagraph. The latter is a disk, blocking the photosphere, in the focal plane of the telescope.

The chromosphere is the layer of the Sun's atmosphere between the photosphere and corona. The chromosphere is only a few thousand kilometers thick. The temperature of the gas in the chromosphere is slightly higher than that in the photosphere. In the approximately 100-kilometer-thick transitional region between the chromosphere and corona, the temperature rapidly increases from about 6,000 kelvins to a few hundred thousand kelvins.

The Sun is a ball of gas without anything that could be considered a solid surface. The Sun's photosphere, however, is the closest the Sun has to a surface. The photosphere is relatively opaque and blocks our view of the solar interior, so most photographs or observations of the solar disk show the Sun's photosphere. It is also the coolest layer of the Sun. The bottom layer of the photosphere is at a temperature of 5,800 kelvins. The photospheric temperature drops with increasing height to a temperature of about 4,500 kelvins at the top of the photosphere, begins to increase in the chromosphere, and is extremely high in the corona.

Relatively cool stars are reddish in color, while hot stars are bluish. The Sun's corona, at a few million kelvins, is much hotter than most stars, so it emits most of its energy in the extreme ultraviolet (the shortest ultraviolet wavelengths) to X-ray region of the electromagnetic spectrum. The photosphere is not hot enough to emit significant amounts of energy in this spectral region. Fortunately for human beings, Earth's atmosphere blocks most extreme ultraviolet and X radiation, so astronomers study the Sun at these wavelengths from satellites.

Extreme ultraviolet and X-ray pictures of the Sun show a bright corona and dark photosphere, which is the reverse of optical pictures showing a bright photosphere and much fainter corona. Solar astronomers therefore study the solar corona using extreme ultraviolet or X-ray images. At these wavelengths the corona shows structures that are not visible at optical wavelengths.

One common structure in the X-ray corona is the coronal hole. On X-ray images of the Sun's corona, coronal holes show up as dark areas because they are regions where the corona is much more tenuous than normal. In the coronal holes, the corona does not show up in X rays and the photosphere below is very dark at X-ray wavelengths. Gas density in the coronal holes is typically about one-tenth the density of the normal portions of the corona. Near the north and south poles of the Sun, coronal holes tend to be relatively stable. Near the equatorial and midlatitude regions of the Sun, coronal holes are less stable. Coronal hole activity varies with the Sun's magnetic activity cycle. Coronal holes are therefore in some way related to the Sun's magnetic field. The largest coronal holes, which are a few hundred thousand kilometers in diameter, can last for months. More typical coronal holes are tens of thousands kilometers in diameter. These smaller coronal holes typically last only for hours rather than months.

Most of the corona contains coronal loops, which are magnetic field structures. The solar magnetic field lines come up from the lower layers of the Sun, loop into the corona, then flow back down into the solar interior. Charged particles, such as protons and electrons, in strong magnetic fields generally travel in spiral paths around the magnetic field lines. The magnetic forces do not allow these particles to travel across the magnetic field lines. Solar material is a plasma in which electrons are separated from the atomic nuclei; it is composed of charged particles and flows along these coronal loops.

Coronal loops do not exist in coronal holes. In coronal holes the magnetic field lines from the Sun's interior do not loop back into the Sun. They extend outward into interplanetary space. In the coronal holes, solar material moves upward from the Sun's lower layers along these magnetic field lines. Rather than falling back down into the Sun, this material—which is mostly protons (hydrogen nuclei) and electrons with occa-

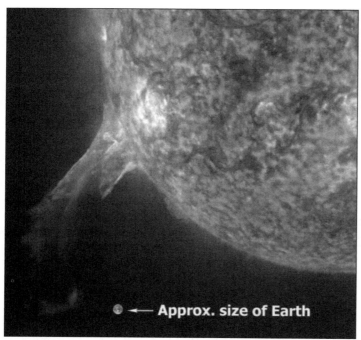

This huge CME was imaged in 2003 by the SOHO spacecraft. (NASA/European Space Agency)

sional heavier, ionized atoms—streams out into interplanetary space and leaves a low-density coronal hole. Coronal holes therefore contain solar matter flowing into interplanetary space and are a major source of the solar wind.

The solar wind consists of charged particles from the Sun flowing outward into space. Were it not continually replenished from the lower layers of the Sun, the solar wind would evaporate the corona in about a day. A few billion kilograms of solar material flow outward in the solar wind every second. The Sun permanently loses this mass. It would seem that the Sun might quickly evaporate from the cumulative effect of the coronal holes and solar wind, but the Sun is very much more massive than the Earth. Hence, in the nearly five billion years of its existence, the Sun has lost less than one-tenth of 1 percent of its mass to the outflow of the solar wind.

Coronal mass ejections occur when coronal loops break. Normally solar material flows along the coronal loops from the solar interior, into the corona, and back into the interior. However, occasionally the magnetic field lines in a coronal loop break. When this happens, the loop more or less explodes. The solar material in the loop is no longer confined by the magnetic field; it shoots outward into space. These events are called coronal mass ejections, or CMEs.

A typical CME flings about 10^{12} to 10^{13} kilograms of solar material into space. Typically a CME releases 10^{24} to 10^{25} joules of energy. Because CMEs are related to the Sun's magnetic field, they occur more frequently during the maxima of the solar activity (sunspot) cycle. CMEs happen as often as a few times a day. When a CME is pointed toward Earth, the resulting magnetic storm can severely disrupt long-distance communications on Earth and cause increased auroral activity.

KNOWLEDGE GAINED

In contrast to the Sun's photosphere, the spectrum of the Sun's corona contains emission lines, which form from a hot, thin gas. Many of the emission lines in the Sun's coronal spectrum

These huge loops of solar plasma above the Sun's photosphere extend many times Earth's diameter. (NASA)

are lines not visible from Earth. Astronomers originally thought these lines might be a new element, but they turned out to be what scientists call forbidden lines. The existence of these forbidden lines was an early clue to the extremely low density of the corona.

Although the corona can be studied optically from the ground during eclipses, much of our knowledge of the corona, coronal holes, and coronal mass ejections comes from satellites, particularly those equipped to observe the Sun at extreme ultraviolet and X-ray wavelengths as well as optically. Such studies started in earnest in the 1970's using X-ray telescopes on the Skylab mission. During this crewed mission, solar astronomers first noticed the connection between coronal holes and the solar wind.

Other satellites have been important to understanding coronal phenomena. The Japanese Yohkoh satellite was launched in 1991 and fell back to Earth in 2005. Yohkoh for the first time allowed daily images of the corona allowing solar astronomers to study rapid changes in the coronal structure. The joint European and National Aeronautics and Space Administration (NASA) Solar and Heliospheric Observatory,

SOHO, mission launched in 1995, and the NASA Transition Region and Coronal Explorer (TRACE) mission, launched in 1998, also made many contributions to our understanding of coronal and other solar phenomena. SOHO discovered a magnetic carpet on the Sun's surface that plays a major role in providing the energy needed to heat the corona. TRACE takes very high-resolution, extreme ultraviolet images of the corona. Although TRACE can image only a small region of the corona at one time, it does allow very detailed studies of coronal phenomena.

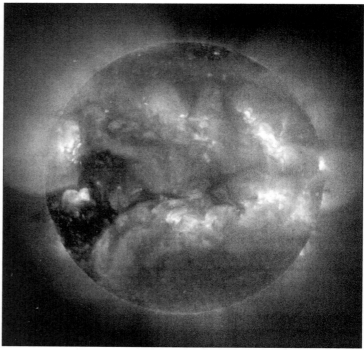

The dark areas on the left of this image are coronal holes. (SOHO/NASA)

CONTEXT

Coronal mass ejections as well as variations in the solar wind can affect Earth, causing geomagnetic storms. These geomagnetic storms are often referred to as space weather. The aurora borealis and aurora australis, also known as the northern and southern lights, are caused by these geomagnetic storms. Hence they are more likely to be visible when a coronal mass ejection reaches Earth. Other geomagnetic effects are less benign. Earth's upper atmosphere expands a little and disrupts long-distance radio communications that depend on radio waves reflecting off the ionosphere or communication satellites. The expanded upper atmosphere can cause some friction on low-Earth-orbit satellites leading to orbital decay and eventually falling back to Earth. Geomagnetic storms caused by coronal mass ejections can also disrupt the electrical power grid.

Coronal holes and coronal mass ejections are part of the complex magnetic phenomena of the Sun's corona. Coronal holes seem to play a still poorly understood role in the Sun's magnetic activity cycle. These phenomena do not exist in isolation. The corona and its magnetic fields interact with the Sun's chromosphere, photosphere, interior, and their magnetic fields. Via the solar wind, originating in coronal holes, and coronal mass ejections, the Sun's corona also interacts with Earth. To understand any facet of this complex Sun-Earth system completely, astronomers need to understand all the other facets.

Paul A. Heckert

FURTHER READING

Chaisson, Eric, and Steve McMillan. *Astronomy Today*. 6th ed. New York: Addison-Wesley, 2008. Chapter 16 of this readable introductory astronomy textbook covers the Sun. It has a good discussion of coronal holes, coronal mass ejections, and their role in producing the solar wind.

Frazier, Kendrick. *Our Turbulent Sun*. Englewood Cliffs, N.J.: Prentice-Hall, 1980. Provides a good, readable account of our knowledge of the Sun through 1979.

Freedman, Roger A., and William J. Kaufmann III. *Universe*. 8th ed. New York: W. H. Freeman, 2008. Chapter 16 of this introductory astronomy textbook is a complete, readable overview of our knowledge of the Sun, including the corona.

Golub, Leon, and Jay M. Pasachoff. *Nearest Star: The Surprising Science of Our Sun.* Cambridge, Mass.: Harvard University Press, 2001. This well-written book gives a detailed summary of our knowledge of the Sun.

Heckert, Paul A. "Solar and Heliospheric Observatory." In *USA in Space.* 3d ed. Edited by Russell Tobias and David G. Fisher. Pasadena, Calif.: Salem Press, 2006. This article describes the SOHO solar observatory mission, which was used to study the Sun's outer layers, including the corona. This mission helped us understand the Sun's magnetic activity and why the temperature increases in the chromosphere and corona. It also observed and returned data on many coronal mass ejections.

Hester, Jeff, et al. *Twenty-first Century Astronomy.* New York: W. W. Norton, 2007. Chapter 13 of this readable astronomy textbook is about the Sun.

Morrison, David, Sidney Wolf, and Andrew Fraknoi. *Abell's Exploration of the Universe.* Philadelphia: Saunders College Publishing, 1995. The Sun is covered in chapter 26 of this classic astronomy textbook.

Schrijver, Carolus J. "The Science Behind the Solar Corona." *Sky and Telescope* 111, no. 4 (April, 2006): 28-33. A good article discussing the latest understanding of the solar corona.

Zeilik, Michael. *Astronomy: The Evolving Universe.* 9th ed. Cambridge, England: Cambridge University Press, 2002. A well-written introductory astronomy textbook. Chapter 12 is an overview of the Sun.

Zeilik, Michael, and Stephen A. Gregory. *Introductory Astronomy and Astrophysics.* 4th ed. Fort Worth, Tex.: Saunders College Publishing, 1998. A textbook designed for undergraduate physics or astronomy majors, it goes into more mathematical depth than most introductory astronomy textbooks. Chapter 10 covers the Sun, including the corona.

See also: Earth-Sun Relations; Earth's Magnetic Field: Origins; Earth's Magnetic Field at Present; Red Giant Stars; Solar Chromosphere; Solar Corona; Solar Evolution; Solar Flares; Solar Geodesy; Solar Infrared Emissions; Solar Interior; Solar Magnetic Field; Solar Photosphere; Solar Radiation; Solar Radio Emissions; Solar Seismology; Solar Structure and Energy; Solar Ultraviolet Emissions; Solar Variability; Solar Wind; Solar X-Ray Emissions; Sunspots.

Cosmic Rays

Category: The Cosmological Context

High-energy cosmic rays are samples of material from outside the solar system. Elemental and isotopic compositions of the cosmic rays constrain models for element production in a variety of astrophysical sources.

OVERVIEW

Cosmic rays are charged particles, electrons, and positively charged ions ranging from protons to the heaviest elements, which arrive at Earth from space. About 98 percent of cosmic rays are positively charged nuclei, with most of the remainder being negatively charged electrons. Although some of the lowest-energy cosmic rays are particles emitted by the Sun, most cosmic rays are too energetic to be confined to the solar system and are samples of material from other parts of the galaxy. Because cosmic rays are charged particles, their paths from their sources to Earth are bent by magnetic fields in the galaxy. As a result, traditional astronomy in which electromagnetic radiation intercepted by a detector, such as a telescope, is traced back in a straight line to its source is not possible with cosmic rays. Nevertheless, cosmic rays provide important clues to the processes that occur in stars, supernovae, and other astrophysical sources.

Determination of the composition of cosmic rays permits comparison to the composition of the Earth, lunar samples returned by the Apollo missions, meteorites, and the Sun. This allows processes by which elements are produced within stars to be examined and compared to theoretical models for nucleosynthesis.

The nucleus of each element has a unique

charge, so methods of determining the composition of cosmic rays require a measurement of charge on each individual cosmic-ray particle. Generally, these techniques require two independent measurements. The first measurement might determine the rate at which cosmic rays lose energy in traversing the detector. This rate of energy loss is proportional to the square of the charge-to-speed ratio of the particle. A second measurement might then determine the velocity, or some other property that depends on velocity, in a manner different from that for determining the rate of energy loss. From these two measurements, the charge can be determined.

A number of innovative charge-measurement detectors have been developed. These detectors can be divided into three general categories: recording detectors, such as photographic emulsions; visual detectors, such as cloud chambers; and electronic detectors, such as Geiger-Müller counters.

In the late 1940's, groups of cosmic-ray investigators from the University of Minnesota and the University of Rochester employed photographic emulsions carried to high altitudes by balloons to determine the charge and energy of the cosmic rays. High altitudes, frequently above 27,000 meters, were required because collisions between incoming cosmic rays and air molecules can cause cosmic rays to fragment into several lighter nuclei, thus altering their composition. At high altitudes, the probability of such a collision is low. Therefore, balloon detectors measure the primary composition of the particles in space. These early experiments demonstrated that, of the nuclei in cosmic rays, about 87 percent are hydrogen, or protons; 12 percent are helium; and the remaining 1 percent are heavier nuclei. It is the composition of these heavier nuclei that contains the clues to the nucleosynthetic processes.

Following the initial discovery of heavy nuclei among cosmic rays, the emphasis in cosmic-ray research shifted to the determination of the charge spectrum, or relative abundances, of each element. Early experiments made use of the Earth's magnetic field as a velocity selector. Paths of charged particles are bent when they encounter a magnetic field, so only particles exceeding a given cutoff energy can penetrate through a region of given magnetic field intensity. The magnetic field of the Earth is so strong near the equator that only cosmic-ray particles with velocities very close to the speed of light can penetrate. Thus, for those cosmic rays detected near the equator, the magnetic cutoff identifies the velocity to be approximately (just under) the speed of light. A single measurement of the rate of energy loss for these particles provides a measurement of their charge.

These early experiments indicated the difficulty of de-

Showers of cosmic rays fall to Earth after striking the atmosphere in this artist's conception. (Simon Swordy/NASA)

tection of the heavy nuclei among cosmic rays. A one-square-meter detector placed in space, above the Earth's atmosphere and outside the Earth's magnetic field, would register several hundred non-solar protons per second and about one-seventh that number of helium nuclei. However, only one or two nuclei heavier than carbon would be measured every second, and the detector would register a single iron nucleus every fifteen seconds. To observe a single lead nucleus would require several months of detector operation. Cosmic-ray astrophysicists recognized that large detectors with long exposure times would be required to determine accurately the composition of the heavy cosmic rays.

In 1956, Frank McDonald, a physicist at Iowa State University, developed a combination of two electronic detectors—a scintillation counter and a Cherenkov counter—to determine the charge and velocity of the cosmic rays. This combination of detectors provided good measurements of the elemental abundances for elements up to iron. Elements heavier than iron were so rare that their identification required a new technique. In the mid-1960's, Robert Fleischer, Buford Price, and Robert Walker, researchers at the General Electric Research and Development Center, found that trails of ionizing particles were recorded within certain types of plastics and that these trails could be revealed later by etching the plastic in an appropriate chemical agent. They demonstrated that if the rate at which the trail was etching as well as the total etchable length were both measured, the charge and energy of the particle could be determined. Balloon flights with these plastic detectors provided information on the composition of the heavier elements in cosmic rays.

In the 1970's, cosmic-ray researchers employed orbiting Earth satellites to increase the duration of their measurements. Large plastic detectors were flown for several months on the National Aeronautics and Space Administration's (NASA's) Skylab space station. The IMP-7 and IMP-8 satellites, launched in 1972 and 1973, respectively, provided good measurements on the isotopic composition of lighter cosmic rays. In 1978, the third High-Energy Astronomical Observatory (HEAO 3), carrying a 6-square-meter electronic detector, provided high-quality measurements of the abundances of nuclei up to bismuth. Cosmic-ray studies continue at polar stations, with small university-based instruments examining the space environment surrounding Earth and with national and international spacecraft journeying throughout the solar system.

APPLICATIONS

Astrophysicists generally believe that the only elements present in the early universe were hydrogen, helium, and perhaps small amounts of lithium, beryllium, and boron. Most of the elements now present were produced in stellar explosions, or by nucleosynthesis in stars. Theoretical calculations show that the elemental and isotopic abundances produced depend on the particular conditions of the nucleosynthetic event. Thus, abundances of the elements and isotopes of material from outside Earth's solar system might be different from those of solar-system material, and those differences would provide clues to the differing nucleosynthetic conditions at those sites. This comparison requires a knowledge of the cosmic-ray composition at the source.

The composition of the cosmic rays can be altered during their journey through space to Earth. Radioactive decay will remove those radioactive elements with short half-lives, compared to the time it took for cosmic rays to reach Earth. Collisions between cosmic rays and interstellar gas will cause fragmentation of some of cosmic rays.

Measurements of cosmic-ray composition provide clues to the "age" of cosmic rays, that is, the duration of their journey through space. The light nuclei—lithium, beryllium, and boron—are much more abundant in cosmic rays than in solar-system materials. These excess light nuclei are believed to have been produced by spallation, or collisional fragmentation with interstellar gas atoms. Since the abundance of interstellar gas atoms is known from other astronomical measurements, the amount of excess light elements provides a measure of the duration of the cosmic-ray journey. Astrophysicists indicate that cosmic rays currently arriving at Earth began their journey about 10 million

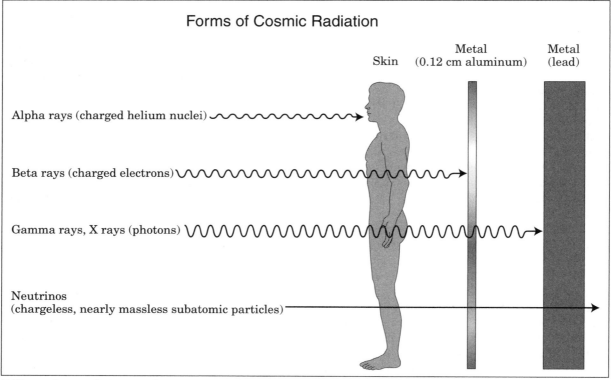

Forms of Cosmic Radiation

Skin Metal (0.12 cm aluminum) Metal (lead)

Alpha rays (charged helium nuclei)

Beta rays (charged electrons)

Gamma rays, X rays (photons)

Neutrinos
(chargeless, nearly massless subatomic particles)

Different forms of cosmic radiation can penetrate different forms of matter: Alpha rays cannot penetrate skin; beta rays can penetrate skin but not metal; gamma rays can penetrate both but are stopped by lead; and neutrinos, chargeless, nearly massless particles, can penetrate even lead, making them extremely difficult to detect. Although they interact very little with matter, neutrinos are believed to be produced in the nuclear reactions at the core of the Sun and other stars and may constitute a large portion of the "missing mass" of the universe.

years ago. Since the solar system formed about 4.5 billion years ago, the cosmic rays may be sampling a much younger type of material than the solar system.

Once the age of cosmic rays is known, the abundances of heavier elements detected at Earth can be corrected back to the source by removing the spallation contribution. Generally, elemental composition of cosmic rays is similar to that of the solar system, but the differences provide clues to differences in the nucleosynthetic processes.

The largest difference in the heavy element composition is for the isotope neon 22, which is four times more abundant relative to the other neon isotopes in cosmic rays than in solar-system matter. Isotopic measurements also show excesses of magnesium 25, magnesium 26, silicon 29, and silicon 30 in cosmic rays when

compared to solar-system matter. These latter discrepancies could be explained if the cosmic-ray sources were stars with initial abundances of carbon, nitrogen, and oxygen about twice that seen in Earth's Sun. Nevertheless, even this alteration of the stellar composition cannot explain the unusually high abundance of neon 22.

Abundances of the heavier elements may provide clues to the site of the production of cosmic rays. Astrophysicists have identified several different nucleosynthetic processes. Two major ones both proceed by the addition of neutrons to light target nuclei. In the s-process, the time between successive neutron capture events is long enough that the new nucleus can be transformed (beta decay) to a stable nucleus before the next capture. This process occurs in the interiors of stars. In the r-process, neutron capture proceeds so rapidly that beta decay is not possible between individual capture events, leading to production of more neutron-rich elements. This process is believed to occur in explosive processes such as supernovae. Only the r-process can produce elements heavier than bismuth.

Since astrophysicists have suggested that supernovae are a likely source of cosmic rays, they would be expected to contain the r-process elemental and isotopic abundance signatures. The presence of elements heavier than bismuth in cosmic rays would suggest an r-process origin. Thus far, experimental results are ambiguous. Rare events, possibly attributable to elements heavier than bismuth, were reported from balloon flights carrying photographic emulsions and plastic detectors. However, the HEAO 3 detected no such events. Because of the scarcity of these heavy elements, longer-duration, large-area cosmic-ray detectors will be required to resolve the question.

Elemental and isotopic measurements on cosmic rays indicate that their sources differ in significant ways from the source of solar-system material. Because of these differences, more precise measurements of elemental and isotopic compositions of the heavier elements are required for detailed comparisons to the nucleosynthetic models.

CONTEXT

Formulation of a detailed model of the nucleosynthesis of the heavy elements by Geoffrey Burbidge, Margaret Burbidge, William A. Fowler, and Fred Hoyle in 1957 provided predictions of the elemental and isotopic abundances expected from the r-process and s-process. This information, coupled with the discovery of heavy elements in cosmic rays in the late 1940's, suggested comparison of cosmic-ray composition with the predictions of the nucleosynthetic models. Rapid advances in electronic detectors in the 1950's made such comparisons possible, but the limited flight duration of high-altitude balloons restricted the number of elements that could be measured because of the low abundance of heavy elements. The use of Earth satellites in the 1970's significantly increased the duration of cosmic-ray composition experiments. Nevertheless, even these long-duration satellite experiments were inadequate to answer the question of the abundance of heavy elements in the cosmic rays.

Development of high-resolution electronic detectors, permitting high-quality determinations of isotopic composition, showed significant differences between the neon, magnesium, and silicon isotopic abundances in cosmic rays and solar-system matter. Advances in modeling of the nuclear processes in stellar interiors allowed astrophysicists to calculate that most of these discrepancies were consistent with stellar nucleosynthesis, with carbon, nitrogen, and oxygen abundances approximately double that of Earth. Long-duration, large-area cosmic-ray detectors, possibly on a space station, will be required to determine the abundances of elements heavier than bismuth, allowing direct comparison of the cosmic-ray composition with that expected for r-process nucleosynthesis in supernovae, which are suggested as the cosmic-ray source.

George J. Flynn

FURTHER READING

Cronin, James W., Thomas K. Gaisser, and Simon P. Swordy. "Cosmic Rays at the Energy Frontier." *Scientific American* 276, no. 1 (January, 1997): 44-50. A technical but thorough examination of our understanding of high-energy cosmic rays.

Foerstner, Abigail. *James Van Allen: The First Eight Billion Miles*. Iowa City: University of Iowa Press, 2007. An engaging portrait of the legendary physicist James Van Allen. Discusses his contributions to the World War II effort as well as the advancement of studies of Earth's geomagnetic environment, his early efforts to study cosmic rays using balloon-launched rockets, the Explorer 1 story, and Van Allen's continuing participation in studying space physics until his death in 2006.

Friedlander, Michael W. *Cosmic Rays*. Cambridge, Mass.: Harvard University Press, 1989. Well-illustrated account of the history of cosmic-ray astronomy. Deals with methods of detection, elemental and isotopic composition, and implications for the cosmic-ray sources.

Ginzburg, V. L., and S. I. Syrovatskii. *The Origin of Cosmic Rays*. New York: Macmillan, 1964. A technical account of cosmic-ray astrophysics. Includes a good discussion of how the light element abundances provide an age for the cosmic rays. Suitable for college physics students.

Pomerantz, Martin A. *Cosmic Rays*. New York: Van Nostrand Reinhold, 1971. A classic text suitable for readers with only an introductory physics background. Describes the interactions of cosmic rays with matter and how these interactions are used to detect and determine the properties of the cosmic rays. Well illustrated.

Rossi, Bruno. *Cosmic Rays*. New York: McGraw-Hill, 1964. A firsthand account by one of the pioneers of cosmic-ray physics. Describes how cosmic rays are detected and discusses ideas about their origins.

Van Allen, James A. *The Magnetospheres of Eight Planets and the Moon*. Oslo, Norway: Norwegian Academy of Science and Letters, 1990. A technical summary of all major magnetic structures in the solar system, written by the famous and prolific researcher after whom the Van Allen belts are named.

Wefel, John P. "Matter from Outside Our Solar System: New Insights. Part 1, The Astrophysical Framework." *The Physics Teacher* 20 (April, 1982): 222-229. Presentation of cosmic rays at a level useful for both high school and undergraduate physics.

_____. "Matter from Outside Our Solar System: New Insights. Part 2, Experimental Measurements and Interpretation." *The Physics Teacher* 20 (May, 1982): 289-297. Discusses the history of cosmic-ray physics, the mechanisms of nucleosynthesis, the construction of cosmic-ray detectors, and the implications of the composition on the sources. Well illustrated. Suitable for high school science students.

See also: Big Bang; Cosmology; Earth's Magnetosphere; Electromagnetic Radiation: Nonthermal Emissions; Electromagnetic Radiation: Thermal Emissions; Extraterrestrial Life in the Solar System; General Relativity; Interplanetary Environment; Interstellar Clouds and the Interstellar Medium; Lunar Rocks; Milky Way; Radio Astronomy; Solar Flares; Solar Magnetic Field; Space-Time: Distortion by Gravity; Space-Time: Mathematical Models; Universe: Evolution; Universe: Expansion; Universe: Structure; X-Ray and Gamma-Ray Astronomy.

Cosmology

Category: The Cosmological Context

Cosmology is the study of the structure and evolution of the universe, including the eventual development of our solar system. It brings together many branches of astronomy and physics, including general relativity and high-energy particle physics.

OVERVIEW

Originally, cosmology was a branch of philosophy devoted to understanding the nature of reality and the origin and structure of everything that exists. With the growth of astrophysics during the nineteenth and twentieth centuries, cosmology rapidly became a major area of research in astronomy and physics, and its focus narrowed to the origin and evolution of energy and matter in the universe as a whole. Cosmology today is concerned with the large-scale structure of the universe, including the distribution of billions of galaxies and galaxy clusters throughout space and time. Cosmologists use physical laws to derive mathematical models of the early universe to within 10^{-43} second of its beginning. They also extrapolate physical processes into the distant future to predict the future of the universe and its contents. Nevertheless, modern cosmology still retains some philosophical qualities.

Modern cosmology has its basis in Albert Einstein's general theory of relativity, which he published in 1915. In 1922, the Russian mathematical physicist Alexander Alexandrovich Friedmann derived two types of solutions to the field equations of general relativity in which the universe initially expands with time. In one type (called "open"), the universe continues to expand forever. In the other type (called "closed"), the universe expands to some maximum size, after which it contracts. In 1927, the Belgian priest and cosmologist Abbé Georges Lemaître independently derived the same two solutions to the general relativistic field equations and went on to speculate about the cause of the expansion; he suggested the universe began from a compact, dense initial state—the "primeval atom"—which disintegrated and dispersed into all the atoms in the universe today.

Edwin P. Hubble presented observational evidence that the universe actually is (or at least appears to be) expanding, when in 1929 he published his research that showed a linear correlation between a galaxy's distance from Earth and the redshift of its spectrum. The cosmological explanation for the redshift is that, as the universe expands, the wavelengths of all electromagnetic radiation are stretched. The color red has the longest wavelength of all components of visible light. When wavelengths of visible light are stretched and shifted toward the red, they are said to be redshifted. The use of the term "redshift" has also been extended to refer to a shift toward longer wavelengths in any region of the electromagnetic spectrum.

The generally accepted cosmological theory describing the origin and expansion of the universe is the big bang theory, which can be traced back to the proposal by Lemaître of a compact, dense initial state that somehow exploded. The physicist George Gamow expanded on Lemaître's idea in the 1930's and 1940's, making it physically more rigorous and incorporating new work in nuclear physics. The name "big bang" was coined by Fred Hoyle, one of the developers of the rival steady state, continuous creation theory as a derogatory term for an explosive ori-

An image from the Hubble Space Telescope's Ultra Deep Field shows galaxies going back billions of light-years, to within a billion years of the big bang. (NASA/ESA/S. Beckwith, STScI)

gin, but it rapidly caught on and was adopted by proponents and opponents alike. By the mid- to late 1960's, observational evidence had eliminated competing theories, and the big bang became generally accepted. Throughout the remainder of the twentieth century and into the twenty-first century, the big bang has been further modified and refined, and now it incorporates the latest work in high-energy particle physics.

The big bang theory maintains that space, time, energy, and matter were all created from a primordial explosion, now dated at about 13 to 14 billion years ago. Space expanded isotropically, and as it expanded, the universe cooled. Modern physical theories can be used to trace the development of the universe back to 10^{-43} second after the big bang, when the temperature was 10^{32} kelvins, but before that time the theories break down. At the even higher temperatures prior to that time, scientists think all four fundamental forces of nature—gravity, strong nuclear, weak nuclear, and electromagnetism—were all unified as one force, indistinguishable from one another. As the universe expanded and cooled, the forces gradually separated from each other one by one—first gravity, then strong nuclear, and finally weak nuclear from electromagnetism at about 10^{-10} second and 10^{15} kelvins.

The early universe was a hot, dense "soup" of interacting high-energy photons and subatomic particles, with energy and mass being transformed back and forth. When two photons with enough energy collided, their energy could be converted to mass, producing a matter-antimatter pair of particles in a process called pair production. When a particle and its antiparticle collided, they mutually annihilated each other with their mass being converted into two high-energy gamma-ray photons.

As the universe expanded and cooled, eventually photons did not have enough energy to produce any more particle-antiparticle pairs. Thereafter, the particles and their antiparticles collided and mutually annihilated. Since equal numbers of particles and antiparticles had been created, they all should have annihilated each other and the universe today would be devoid of matter and antimatter. Alternatively, some

segregation process might have separated the matter and antimatter into distinct regions, and the universe today would consist of equal but separate concentrations of matter and antimatter. However, the universe today appears to be composed almost entirely of matter throughout. An asymmetry in the weak nuclear force seems to provide a way for antimatter but not matter to decay, so there was a slight excess of matter particles by about one in a billion. By the end of a few seconds, the temperature had dropped to several billion kelvins, and particle creation and annihilation had ceased. The small excess of matter that survived is the matter of the universe today.

Among the particles to survive were quarks. They combined to form protons and neutrons; a proton is two "up" quarks and one "down" quark, while a neutron is one "up" quark and two "down" quarks. Protons and neutrons are the particles that make up atomic nuclei. Single protons are hydrogen nuclei, and they combined with neutrons to form nuclei of other light elements such as deuterium (also called heavy hydrogen, with one proton and one neutron), helium (two protons and one or two neutrons), and small amounts of lithium (three protons and three or four neutrons) and beryllium (four protons and three neutrons). However, after about fifteen minutes, the temperature had dropped to a few hundred million kelvins, too cool for further nucleosynthesis. Heavier nuclei would be formed much later, in nuclear fusion reactions in stars.

The early universe was dominated by electromagnetic radiation; that is, the spatial density of electromagnetic energy was greater than the spatial density of matter. Both densities decreased as the universe expanded, but the electromagnetic energy density decreased faster than the matter density. Several thousand years after the big bang, the electromagnetic energy density dropped below the matter density at a time referred to as the crossover time. After that, the universe was dominated by matter, since its density was greater.

Around 300,000 to 500,000 years after the big bang, the temperature had dropped to about 3,000 kelvins. Electrons could then join with protons (bare hydrogen nuclei) to form electrically neutral hydrogen atoms. Free electrons are very effective at scattering photons, but electrons in atoms are not able to do so. As a result, the universe changed from being very opaque to becoming transparent to electromagnetic radiation. Photons could now travel freely through the universe. This was the source of the cosmic microwave background radiation currently observed at a temperature of about 3 kelvins, because the wavelength of the electromagnetic radiation has been greatly stretched (redshifted) by the expansion of the universe.

Within the first few hundred million years, matter clumped together by gravitational attraction to form protogalaxies or pregalactic fragments, and within them further gravitational clumping formed the first stars. These protogalaxies were relatively small, but through mergers they developed into larger systems, the galaxies of the universe today. Galaxies range in size from dwarfs, containing a few tens of millions of stars, to giants, with more than ten trillion stars. Galaxies are not distributed randomly through space but are grouped into galaxy clusters; poor clusters contain only a few tens of galaxies, while rich clusters have thousands of members. Galaxy clusters in turn are grouped into superclusters. Between them are large, nearly empty regions called voids.

Within galaxies, stars form from clouds of gas and dust called nebulae. Stars initially heat up and begin to shine by gravitational contraction, but this is a relatively brief stage in the life cycle of a star. During most of their energy-producing lives, stars generate energy by nuclear fusion reactions in which lighter atomic nuclei are fused into heavier nuclei with the release of energy. Stars with many times the Sun's mass can synthesize nuclei as heavy as iron in their interiors. When they explode as Type II supernovae at the end of their energy-producing lives, the tremendous energy released synthesizes nuclei heavier than iron. The explosion disperses the elements the star formed during its life out into space, there to enrich the nebulae from which new generations of stars form.

It had been assumed that in the future, the expansion of the universe will slow down, due to the gravitational attraction between galaxies. The only question has been how rapidly the expansion is decelerating. If the deceleration were small, the universe would expand forever, at a gradually decreasing rate; if the deceleration were large enough, someday the universe would stop expanding and begin to contract at ever-increasing speed. Beginning in the 1990's, astronomers tried to determine how much the universe is slowing down by measuring the expansion rate at great distances (and hence at great times in the past) and comparing it to the expansion rate at smaller distances (and more recent times). Contrary to expectations, they found the expansion rate in the past was slower than it is now, indicating that the expansion of the universe is accelerating. The cause for this acceleration is unknown, but it is called "dark energy." If the acceleration continues, the distances between galaxy clusters will grow at an ever-increasing rate. Eventually, all the matter in galaxies will be processed into stars, all the stars will use up their sources of energy and go out, and the universe will grow cold and dark.

METHODS OF STUDY

Cosmology is studied both theoretically and observationally, the two complementing each other. New observations need to be interpreted by theories, and theories need to be confirmed by further observations. Cosmological observations are now made over the entire electromagnetic spectrum, from high-energy, short-wavelength gamma rays, through X rays, ultraviolet, visible light, infrared, microwaves, to long-wavelength radio waves. Not all electromagnetic radiation penetrates Earth's atmosphere, so parts of the electromagnetic spectrum must be observed from satellites above Earth orbit, above the atmosphere.

The speed at which all forms of electromagnetic radiation travel through a vacuum is exceedingly fast but finite, being very close to 300,000 kilometers per second. Therefore, looking out to greater distances means looking further back in time. When astronomers look at Earth's moon, at a distance of about 400,000 kilometers, they see it as it was about 1.3 seconds earlier. When one observes the Sun, at a distance of about 150 million kilometers, it appears as it was 8 minutes and 20 seconds earlier. The nearest star system outside our solar system, the Alpha Centauri system, is at a distance of 4.3 light-years, meaning that we see that system as it was 4.3 years ago. Distant galaxies are billions of light-years away, so we see them as they were billions of years ago. In this manner it is possible to observe the early universe and its contents by observing at larger and larger distances.

There are two primary observational anchors in modern cosmology. The first is the Hubble law, the relationship between galaxy redshifts and distances, first discovered by Edwin Hubble. This provides the basic observational evidence that the universe is expanding. Other explanations for the redshifts of galaxy spectra have been proposed. For example, the "tired light" hypothesis posits that photons lose energy and hence are shifted to lower frequencies and longer wavelengths as they travel immense distances. However, none of these alternative explanations fits the observed data, with a minimum of extra assumptions, as well as the expanding universe does—which explains the redshifts of distant objects as due to the stretching of wavelengths of electromagnetic radiation as the universe expands.

The second observational anchor in cosmology is the cosmic microwave background (CMB) radiation, which is the firmest observational evidence supporting a big bang origin to the expanding universe. It was first detected accidentally by Arno A. Penzias and Robert W. Wilson in 1965 as part of their work on a communication satellite project at AT&T's Bell Laboratories in Holmdel, New Jersey. Using a large radio horn antenna, they found a uniform microwave background signal coming from all directions. It was identified by Robert Dicke and his colleagues at Princeton University as greatly redshifted radiation from a few hundred thousand years after the big bang, the time when the universe became transparent. Subsequent observations of it by Earth-orbiting spacecraft—the Cosmic Background Explorer (COBE) and the Wilkinson Microwave Anisotropy Probe (WMAP)—and high-altitude balloons launched in Antarctica, including Balloon Observations of Millimetric Extragalactic Radiation and Geomagnetics (BOOMERANG), showed that the radiation exactly fits a blackbody spectral curve for a temperature of 2.73 kelvins. However, it is not precisely uniform. There are temperature variations of up to about 0.00001 kelvin over areas of the sky with an angular size of about 1° of arc (about twice the apparent angular size of the Sun and full Moon as seen in our sky). These temperature variations are thought to represent slight differences in density in the early universe that ultimately produced the "lumpy" universe of clusters of galaxies that observed today.

On the theoretical side, modern cosmology draws primarily upon general relativity and high-energy particle physics. Solutions to the simplest form of the field equations of general relativity predicted an expanding universe before it was confirmed observationally. High-energy particle physics provides insights into the processes that likely occurred in the high-temperature, high-energy environment of the early universe.

Grand unified theories (GUTs) seek to unify the strong nuclear, weak nuclear, and electromagnetic forces as manifestations of a single more fundamental force, and theories of everything (TOEs) try to include gravity with the other three forces. It is thought that this unification, meaning the forces are indistinguishable from each other, occurs at extremely high temperatures and energies, the conditions that existed in the very early universe. Thus the very early universe serves as a laboratory to test such theories.

CONTEXT

The value of cosmology lies in understanding the structure and organization of the universe, where it came from, and how it will develop in the future. Cosmology gives us a perspective on our place in the universe. At the same time, cosmology provides a way to test the laws of physics on a grand universal scale.

Two of the major unsolved questions in cosmology involve the nature of dark matter and dark energy. There are numerous situations in astronomy in which a study of the dynamics of a system implies gravitational forces that far exceed what the observed mass can account for. The deficit in observed mass was originally called "missing mass." However, the dynamical mass calculations seem reliable, so astronomers now generally use the term "dark matter," since the mass really is not missing, it is just not observable in any part of the electromagnetic spectrum. Dark matter probably includes nonluminous ordinary matter that has not been observed yet, such as small conglomerates of nonradiating matter, black dwarfs, and black holes. However, indications are that most dark matter is much more exotic—completely unknown forms of matter that do not interact with ordinary matter except gravitationally. Possible candidates include a class of particles called WIMPs (weakly interacting massive particles) and cosmic strings (long, thin, massive lines of unbroken symmetry left over from the early universe in which the strong, weak, and electromagnetic forces remain unified).

Dark energy, which drives the acceleration of the expansion of the universe, is even more enigmatic. Mathematically, it may take the form of Einstein's cosmological constant in the equations derived from general relativity that describe the expanding universe. Physically, its nature is completely unknown.

Various observations indicate that the geometry of the universe is almost precisely flat. That means the average density of matter and energy throughout the universe must almost exactly equal a value called the critical density, about 10^{-26} kilograms per cubic meter. Observed luminous matter accounts for about 1 percent of this. Allowing for probable nonluminous but ordinary dark matter gives about 3 percent more. It is estimated there is about 26 times more dark matter (both ordinary and exotic) than luminous matter, so exotic dark matter accounts for 23 percent of the critical density. The total for all forms of matter comes to about 27 percent; thus dark energy contributes about 73 percent of the average density of the universe. That means approximately 96 percent of the universe consists of dark matter and dark energy, about which we know virtually nothing. Only about 4 percent consists of ordinary matter, both luminous and nonluminous.

Another cosmological puzzle is that the universe seems "fine-tuned" for life. If the physical laws and constants of the universe were much different from what they are, life as we know it would be impossible. Stars and planets would not form, or they would not last long enough for life, especially intelligent life, to develop. One explanation for this fine-tuning is called the anthropic principle: the idea that the universe has to be the way it is, because otherwise we would not exist to ask about such things. However, some scientists find the odds overwhelmingly against the universe being the way it is solely by chance, proposing instead that the universe in some way may have been deliberately designed for life. To avoid the theological implications of deliberate design, other scientists suggest that the universe is just one of many alternate parallel universes, each with its own unique set of laws and constants; human beings occupy the one that allows the existence of life.

Workable GUTs and TOEs are needed to describe the very early universe. One theory for unifying all four forces requires eleven dimensions—the familiar three dimensions of space and one of time, plus seven more dimensions. The extra dimensions would be rolled up into structures too small to detect. In some versions of this theory, particles such as quarks and electrons are really multidimensional membranes wrapped around the extra dimensions. Multidimensional membranes are also called M-branes or just branes. It has even been suggested that collisions between branes lead to big bangs, with each collision and big bang creating a new universe.

Through the 1980's, most of the parameters that characterize the universe and its expansion were very poorly known, often with more than a factor of two uncertainty. Since then, new observations have dramatically narrowed the range of uncertainty. For example, current determinations of the Hubble constant—the slope of the redshift-distance relation, which is the rate at which the universe is expanding—are generally between 65 and 75 kilometers per second per megaparsec. (A megaparsec is a million parsecs or 3,260,000 light-years.) These values mean that the average speed with which other galaxies recede from us increases by 65 to 75 kilometers per second for every million parsecs (or 3,260,000 light-years) of distance from us. This small range of values for the Hubble constant, together with the matter and energy density percentages (27 percent matter, 73 percent energy), yields a time back to the big bang of 13 to 14 billion years ago.

Cosmology started as a branch of philosophy, but it has become an integral part of astronomy and physics. It has shifted from being primarily a speculative, qualitative endeavor to becoming a precise, quantitative science. Nevertheless, many aspects of cosmology remain philosophical in scope. Cosmologists speculate whether or not our universe is the only universe, whether or not there are additional dimensions to the known universe that have yet to be discovered, and how the four principal forces of nature might have been unified in the very early universe right after the big bang. Since humans are part of the universe, it can be said that "we are the universe contemplating itself."

David Wason Hollar, Jr., and
Richard R. Erickson

FURTHER READING

Bartusiak, Marcia. *Thursday's Universe*. New York: Times Books, 1986. A thorough survey of major twentieth century breakthroughs and theories in astrophysics. Discusses the development of major cosmological principles and the people behind these ideas. Big bang cosmology is described clearly from very early stages of the universe to the distant future universe.

Chaisson, Eric, and Steve McMillan. *Astronomy Today*. 6th ed. New York: Addison-Wesley, 2008. A well-written college-level textbook for introductory astronomy courses. Two chapters provide a thorough discussion of cosmology.

Fraknoi, Andrew, David Morrison, and Sidney Wolff. *Voyages to the Stars and Galaxies*. Belmont, Calif.: Brooks/Cole-Thomson Learning, 2006. A well-written, thorough college textbook for introductory astronomy courses. Several chapters contain material concerning cosmology.

Freedman, Roger A., and William J. Kaufmann III. *Universe*. 8th ed. New York: W. H. Freeman, 2008. College-level introductory astronomy textbook. Two chapters deal with cosmology.

Guth, Alan H., and Paul J. Steinhardt. "The Inflationary Universe." *Scientific American* 250 (May, 1984): 116-129. This general review article is an excellent presentation of big bang cosmology with the inclusion of inflation, in which the very early universe briefly undergoes extremely rapid expansion.

Hawking, Stephen W. *A Brief History of Time: From the Big Bang to Black Holes*. New York: Bantam Books, 1988. This enormously popular best seller presents a clear, outstanding discussion of cosmology. Hawking describes the evolution of the universe and grand unified theories, among other topics.

Kippenhahn, Rudolf. *Light from the Depths of Time*. New York: Springer, 1987. An exciting description of cosmology and the universe. Kippenhahn, an astrophysicist, covers the history of cosmological thought in the twentieth century and outlines the evolution of the universe and basic cosmological principles using humorous fictional characters.

Schneider, Stephen E., and Thomas T. Arny. *Pathways to Astronomy*. 2d ed. New York: McGraw-Hill, 2008. Very thorough college textbook for introductory astronomy courses, divided into many short units on specific topics. Several units provide a thorough discussion of cosmology.

Seielstad, George A. *At the Heart of the Web: The Inevitable Genesis of Intelligent Life*. Boston, Mass.: Harcourt Brace Jovanovich, 1989. An excellent survey of cosmological thought. Seielstad describes the order and evolution of the universe while stressing the anthropic principle.

Silk, Joseph. *The Big Bang*. Rev. ed. New York: W. H. Freeman, 1989. A comprehensive, readable discussion of cosmological views on the origin and evolution of the universe.

See also: Big Bang; Cosmic Rays; Electromagnetic Radiation: Nonthermal Emissions; Electromagnetic Radiation: Thermal Emissions; General Relativity; Interstellar Clouds and the Interstellar Medium; Milky Way; Space-Time: Distortion by Gravity; Space-Time: Mathematical Models; Universe: Evolution; Universe: Expansion; Universe: Structure.

Dwarf Planets

Categories: Planets and Planetology; Small Bodies

The discovery of many new bodies orbiting the Sun beyond the orbit of formerly outermost Neptune—including at least one larger than Pluto—created a crisis in astronomy. It became evident that a new definition was required to distinguish these objects from traditional planets. The term "dwarf planet" was introduced to include planetary objects smaller than planets but larger than asteroids, resulting in the demotion of Pluto from its status as a planet.

OVERVIEW

The concept of a planet has a long history, leading to a total of nine planets in the solar system until discoveries in the early twenty-first century led to new definitions that excluded Pluto. The word "planet" originates from a Greek word meaning "wanderer" and for centuries was applied to celestial objects that shifted positions relative to the "fixed" stars. In classical antiquity, seven such objects were identified and were associated with mythical gods: the Sun, the Moon, Mercury, Venus, Mars, Jupiter, and Saturn. The Latin names for the seven days of the week were based on these seven celestial deities. In Greek thought, the planets were believed to orbit the Earth along complex paths determined by a combination of circles.

During the scientific revolution of the sixteenth and seventeenth centuries, it was shown that five of the classical planets revolve around the Sun in elliptical orbits, along with the Earth-Moon system. Late in the eighteenth century, British astronomer William Herschel, aided by his sister Caroline, discovered Uranus, the first planet to be discovered with the aid of a telescope. Early in the nineteenth century, Sicilian astronomer Giuseppe Piazzi discovered what he thought was a new planet, smaller than Mercury and orbiting the Sun between Mars and Jupiter. He called it Ceres. However, when many smaller bodies with similar orbits were discovered in the next few decades, they were called asteroids, and Ceres was demoted from its status as a planet (though later reinstated as a "dwarf planet"). The asteroids are believed to be remnants from the formation of the solar system.

By the middle of the nineteenth century, investigations into slight deviations in the elliptical orbit of Uranus led to the discovery of Neptune by the German astronomer Johann Galle. Using Sir Isaac Newton's law of universal gravitation, astronomers were able to determine the masses of all but two of the eight known planets from the motions of their satellites, with Jupiter as the most massive, at 318 times the Earth's mass. Perceived deviations in the orbit of Neptune led to the discovery of Pluto in 1930 by American astronomer Clyde Tombaugh. Pluto's orbit differed from those of the other planets, with its large inclination from the ecliptic plane and its highly elliptical shape that brings it closer to the Sun than Neptune during some 20 years of its 248-year period. It was also found to be much smaller than the outer gas giant planets and to consist mostly of icy materials.

In 1977 Charles Kowal discovered a small, icy planetoid orbiting the Sun between Jupiter and Uranus, later named Chiron. In the 1990's several similar, cometlike objects were found between Jupiter and Neptune and are now called centaurs. Pluto's status as the ninth planet began to be suspect in 1978, when its satellite Charon was discovered and Pluto's mass was found to be only 0.2 percent of Earth's mass. That is much less than even Mercury, at 5.5 percent of the mass of the Earth. Pluto's mass was too small to have produced deviations in Neptune's orbit, which were then found to be negligible.

Then, in 1992 after a five-year search using digital cameras and computerized analysis, David Jewitt and Jane Luu of the Massachusetts

Institute of Technology (MIT) discovered the first of many similar icy objects beyond Neptune in a region called the Kuiper Belt. Existence of such a region had been predicted by Dutch American astronomer Gerard Kuiper. It is similar to the asteroid belt between Mars and Jupiter but about twenty times wider and populated by icy objects rather than the rocky and metallic bodies found in the asteroid belt. The Kuiper Belt extends from the orbit of Neptune, between 30 and about 55 astronomical units (AU), and is believed to contain thousands of objects larger than 100 kilometers in diameter.

More than 130 Kuiper Belt objects (KBOs) have been found with nearly the same 248-year period as Pluto at about 40 AU from the Sun.

These "plutinos" complete their orbits twice during three orbits of Neptune, referred to as a 2:3 gravitational resonance. KBOs with other resonances, such as 3:5 and 4:7, are called cubewanos, and a few objects are found beyond a 1:2 resonance at 55 AU and with 330-year periods. Some objects have been found beyond 55 AU but are believed to have been scattered from the Kuiper Belt into a region called the scattered disk containing scattered disk objects (SDOs). Planetesimal objects in these latter two regions (KBOs and SDOs) are called trans-Neptunian objects (TNOs).

Astronomers began to view Pluto as the largest member of the new class of plutinos, and some started to question its status as a planet.

The dwarf planets Sedna and Pluto shown in size comparisons with other bodies. (NASA/JPL-Caltech/R. Hurt, SSC-Caltech)

In 2003 a team from the California Institute of Technology (CalTech), working at Mount Palomar Observatory north of San Diego and led by Mike Brown, discovered an SDO at about 97 AU from the Sun, now called Eris. When a satellite was discovered in 2005, the mass of Eris was found to be 27 percent larger than that of Pluto, and a few astronomers began to refer to it as the tenth planet. Most astronomers, however, recognized that many TNOs might be larger than Pluto and that either they would also have to be classified as planets or Pluto would have to be reclassified to distinguish such objects from the traditional planets.

The definition of a planet was placed on the agenda for the General Assembly of the International Astronomical Union (IAU) meeting in August of 2006. An initial draft

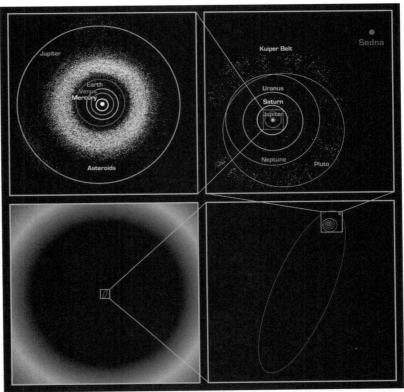

Four panels show the location of the dwarf planet Sedna. (NASA/JPL-Caltech/R. Hurt, SSC-Caltech)

proposal recommended that Pluto be retained as a planet and that Ceres, Charon, and Eris be added to the list of planets. This recommendation was made by astronomers who viewed both Pluto and its satellite Charon as planets in a double-planet system, since each body rotates about a point located between the two. After many objections, an alternate proposal was offered by the Uruguayan astronomer Julio Fernández, who suggested an intermediate category for objects like Pluto, which are large enough to be nearly round but too small to clear their orbits of other planetesimals. The IAU accepted this proposal, and by unanimous vote it was agreed to call these intermediate objects "dwarf planets," with smaller objects to be called "small solar-system bodies." By further vote, it was agreed that Pluto is a dwarf planet.

These definitions as voted in Resolution 5A by the Twenty-sixth General Assembly of the IAU are as follows:

The IAU therefore resolves that planets and other bodies in our Solar System, except satellites, be defined into three distinct categories in the following way:

(1) A "planet" is a celestial body that (a) is in orbit around the Sun (b) has sufficient mass for its self-gravity to overcome rigid body forces so that it assumes a hydrostatic equilibrium (nearly round) shape, and (c) has cleared the neighbourhood around its orbit.

(2) A "dwarf planet" is a celestial body that (a) is in orbit around the Sun (b) has sufficient mass for its self-gravity to overcome rigid body forces so that it assumes a hydrostatic equilibrium (nearly round) shape (c) has not cleared the neighbourhood around its orbit, and (d) is not a satellite.

(3) All other objects, except satellites, orbiting the Sun shall be referred to collectively as "small solar-system bodies."

In three footnotes, this IAU resolution agreed that the eight planets are Mercury, Venus,

Earth, Mars, Jupiter, Saturn, Uranus, and Neptune. It also agreed to establish a process for assigning objects to the category dwarf planet or another status. It also suggested that small solar-system bodies include most solar system asteroids, most TNOs, comets, and other small bodies. In the same meeting, the IAU announced only three members of the dwarf planet category: Ceres, Pluto, and Eris.

KNOWLEDGE GAINED

The new definitions of planets, dwarf planets, and small solar-system bodies have helped clarify both the nature of these objects and the structure of the solar system, as well as stimulating new research about them. The new definitions have led to new searches for dwarf planets and new research on criteria for hydrostatic equilibrium shape (nearly round) and orbital dominance (clearing the neighborhood).

The IAU maintains a dwarf planet watch-list of about a dozen candidates, which keeps changing as new candidates are found and as more is learned about the physics of existing candidates. Current candidates include the plutinos Orcus and Ixion, cubewanos Quaoar and Varuna, and the SDO Sedna, all of which are similar in size to or larger than Ceres (975 kilometers in diameter) but are not yet established as round. Observations indicate that icy bodies of more than about 400 kilometers reach hydrostatic equilibrium, but rocky objects with more rigid interiors might require at least 800 kilometers. The only other asteroid candidate seems to be Vesta, the second largest at 530 kilometers, which appears to be round except for a large impact crater. The Dawn space probe, scheduled to orbit Vesta by 2011, may resolve its status. Estimates range from forty to two hundred candidates in the Kuiper Belt and more beyond it.

CONTEXT

The new definitions of "planet" and "dwarf planet" highlight the increasingly complex nature of the solar system as more is discovered about it. The definitions have also, however, introduced many ambiguities and criticisms.

The new definitions do incorporate accepted theories for the evolution of the solar system and appeal to observational criteria. As planets

formed from the dust and planetesimals of the solar disk, their gravity attracted more matter and they eventually dominated their orbits. However, if planetesimals were sufficiently disturbed by gravitational forces, such as those from nearby Jupiter, they never formed planets and remained as asteroids. Although no planets have completely cleared their orbital neighborhoods, even Mars, as the least dominant planet, has collected more than five thousand times as much material as that which remained in its orbit, while Ceres is only 0.33 times larger and Pluto only 0.07 times larger than the remaining material in their orbits.

Critics complained, however, that the new definitions were arbitrary, since no planet has completely cleared its orbit, and that the round shape of hydrostatic equilibrium is ambiguous, since there are various degrees of roundness. Others voiced concerns about the demotion of Pluto from its longtime status as a planet. Although the National Aeronautics and Space Administration (NASA) decided to accept the new definitions, many respected astronomers, including the director of the New Horizon mission to Pluto, Alan Stern, remained opposed, and his team continued to refer to Pluto as a planet. The discussions and debates would continue at later meetings of the IAU and as more was learned about solar-system objects and their physics.

Joseph L. Spradley

FURTHER READING

Freedman, Roger A., and William J. Kaufmann III. *Universe*. 8th ed. New York: W. H. Freeman, 2008. College-level introductory text covering the field of astronomy. Contains descriptions of astrophysical questions and their relationships.

Hartmann, William K. *Moons and Planets*. 5th ed. Belmont, Calif.: Thomson Brooks/Cole, 2005. An excellent introductory college text on planetary science by one of the leaders in the field. It has good chapters on the formation of the solar system and on asteroids and other small solar-system bodies.

Serge, Brunier. *Solar System Voyage*. Translated by Storm Dunlop. New York: Cambridge University Press, 2000. This well-illustrated book describes the solar system and discusses issues related to the definition of planets.

Sobel, Dava. *The Planets*. New York: Viking, 2005. A very readable account by a popular science writer of the nature and history of planets and asteroids, and of the scientists who study them.

Soter, Steven. "What Is a Planet?" *Scientific American* 132, no. 6 (January, 2007): 2513-2519. A planetary scientist discusses the controversy over the revised definition of a planet, including both its flaws and the scientific advantages of the concept of a dwarf planet.

Weintraub, David A. *Is Pluto a Planet? A His-torical Journey Through the Solar System*. Princeton, N.J.: Princeton University Press, 2006. This book traces the concept of a planet from antiquity to the present day, providing the historical and astronomical context for deciding if Pluto is a planet.

See also: Ceres; Comet Halley; Comet Shoemaker-Levy 9; Comets; Eris and Dysnomia; Kuiper Belt; Meteorites: Achondrites; Meteorites: Carbonaceous Chondrites; Meteorites: Chondrites; Meteorites: Nickel-Irons; Meteorites: Stony Irons; Meteoroids from the Moon and Mars; Meteors and Meteor Showers; Oort Cloud; Pluto and Charon.

Earth-Moon Relations

Categories: Earth; Planets and Planetology

The Moon is the closest astronomical body to the Earth, with a mass approximately 1.2 percent that of Earth. This unusually large fraction gives the Moon significant influence over the orbital and rotational motion of Earth, creating tides strong enough to have important geologic and oceanographic effects, among them variations in the length of the day. The Moon, along with the Sun, causes Earth's spin axis to precess with a period of 26,000 years.

OVERVIEW

The Moon is the most prominent astronomical body after the Sun. It is the closest astronomical body to the Earth, orbiting at an average center-to-center separation of 384,000 kilometers. The Moon has a radius of 1,740 kilometers; at this distance, it appears to be 0.5° in angular width. The mass of the Moon is 3.74×10^{22} kilograms and the density of the Moon is 3.3 grams per cubic centimeter. Earth, by contrast, has a mass of 5.97×10^{24} kilograms and a radius of 6,380 kilometers, giving it a density of 5.5 grams per cubic centimeter, substantially more than that of the Moon. The lower density of the Moon, along with its lack of a magnetic field, argues that the Moon lacks a molten metallic core such as the Earth has.

Earth is close enough for material thrown off of the Moon by meteorite impact (called "ejecta") to fall onto it. A small number of meteorites discovered in desert areas or in Antarctica closely resemble lunar rocks collected by the Apollo astronauts and have been verified as of lunar origin.

The Moon and Earth are gravitationally bound. They orbit around a common point, called the barycenter, with a period of 27.3 days. This period is called the sidereal month and rep-

resents the time for the Earth-Moon system to complete one rotation with respect to the stars. The synodic month, by contrast, is 29.5 days, the time between successive full Moons.

The Earth-Moon system is gravitationally bound to the Sun. Hence, the barycenter orbits the Sun in obedience to Johannes Kepler's three laws of planetary motion: the orbit of the barycenter is an ellipse with the Sun at one focus; the line from the center of the Sun to the barycenter sweeps out equal areas in equal times; and, the cube of the radius of the barycenter orbit is proportional to the square of the period. The barycenter lies on a line joining the center of Earth to the center of the Moon, at

The Clementine spacecraft captured this image of the Moon in earthshine. The Sun peeks from behind the Moon, with Venus in the far background as the pearl of light near the top. (NASA)

a point 4,680 kilometers from the center of Earth. This distance is 73 percent of the radius of Earth. An observer on Mars would see Earth displaced from the ideal elliptical orbit by as much as ⅜ of its diameter.

The motion of Earth about the barycenter is superimposed on the elliptical motion of the barycenter about the Sun in a complicated manner. Earth oscillates back and forth across the barycenter ellipse, spending half of a synodic month inside the ellipse (toward the Sun) and the other half outside the ellipse. Simultaneously, the Earth oscillates above and below the ecliptic (the plane of Earth's orbit) spending half of a sidereal month above the plane and the other half below it. These back-and-forth and up-and-down oscillations are not necessarily synchronous. When the Earth is inside the ellipse, the Moon is outside it, and vice versa. A similar arrangement holds for the up-and-down displacements. Absent the Moon, the center of the Earth would coincide with the barycenter and the planetary motion of the Earth would be close to the elliptical ideal. The Earth-Moon system, on the other hand, has one of the most complicated motions in the solar system.

The Earth-Moon system is like an unbalanced dumbbell tumbling end over end about the barycenter. The gravitational pull is the "bar" holding the dumbbell together. The sides of the Earth and Moon facing each other can be referred to as the "inner" sides, and the opposite sides of each can be referred to as the "outer" sides. The gravitational force falls off with distance, making the gravitational pull on the inner side of each body stronger than the gravitational pull on the outer side. This inequality of forces is referred to as gravitational tidal force. Neither the Earth nor the Moon is rigid. Each is plastic enough to change shape under the influence of the tidal force. The gravitational pull of the Moon raises a bulge in the Earth more or less directly under the Moon; the bulge is

The Galileo spacecraft returned separate images of Earth and its moon that were later compiled into this montage. (NASA/JPL)

matched by a similar one at a location more or less directly opposite the Moon. The bulge in the ocean presents itself as the familiar tides. Similar but less familiar tides exist in the atmosphere and in the Earth's crust. The rotation of the Earth attempts to carry these bulges away from the point directly under the Moon, resulting in a slight sideways component to the mutual gravitational pull. This sideways pull acts as a brake on the rotational motion of the Earth, slowing it down and increasing the length of the day. The increase is approximately one-thousandth of a second per century, but it has been accumulating since the creation of the Moon billions of years ago. Growth-ring counts in fossil coral from 400 million years ago seem to indicate that the year (whose length should not change) consisted of about 400 days back then; now a year consists of about 365 days. In other words, the length of the day has increased by about 10 percent in the past 400 million years.

The sideways pull on the Moon is in the direction of its orbital motion around Earth. Extra energy imparted by the pull increases the radius of the Moon's orbit and also increases the length of the siderial month. Since the length of the day is increasing faster than the length of the sidereal month, eventually the two will become equal and the day and month will be the same. At that time, the Earth will always present the same face to the Moon, just as the Moon always presents the same face to the Earth today.

The rotation of the Earth gives it an oblate shape that is thicker at the equator than through the poles. The gravitational pull of the Sun and Moon on this equatorial bulge acts as a

The Moon viewed from Earth orbit, with the Earth's atmosphere rising above the planet's surface in the foreground. (NASA)

torque that causes the Earth to precess like a top. The spin axis of the Earth currently points toward Polaris, the pole star, but this is only an accident of history. In 13,000 years, Vega (in the constellation Lyra) will be the pole star.

KNOWLEDGE GAINED

The bulk of Earth-Moon interactions are gravitational and are known from earthbound observations. The apparent location of the Sun in the zodiac on the first day of spring (recognized as the day that the Sun rose due East and set due West) held great cultural and religious significance to ancient civilizations and was monitored closely. Over the centuries, it became clear that this location, originally in the constellation Taurus, had moved to the constellation Aries. The Greek astronomer Hipparchus discovered this fact about 130 B.C.E. and from it deduced the 26,700-year circular motion of the north celestial pole. In 1530, Nicolaus Copernicus recognized this as due to drift of the Earth's rotational axis with respect to the fixed stars, and Sir Isaac Newton in 1687 showed the phenomenon to be an effect of Moon's gravitational influence on the Earth.

Edmond Halley in 1693 and Immanuel Kant in 1754 used Newtonian gravitational theory to calculate the locations, dates, and times of total solar eclipses discussed in ancient Greek and Roman documents. Their calculations argued that the eclipses could not have taken place at the dates and places recorded. The discrepancies were eventually traced to changes in the length of the day due to tidal braking.

Starting with Apollo 11, each subsequent lunar landing mission (except the ill-fated Apollo 13) brought back significant amounts of lunar rock for scientific study. Oxygen derived from the lunar material proved to have the same ratio of isotopes as oxygen found on Earth. In contrast, oxygen retrieved from meteorites believed to be of Martian origin had substantially different isotopic ratios.

This discovery, in conjunction with the observation that the Moon lacks an iron core, led to the impact theory of lunar origin. In this theory, the young Earth and a body approximately the size of Mars collided some 4.5 billion years ago. The collision threw a substantial amount of the

The Earth viewed from the Moon, as the Apollo 8 astronauts began their orbit on December 29, 1968. (NASA)

Earth's crust into space, where some of the material coalesced into the Moon, with the remainder falling back to Earth. Since this happened after the bulk of the iron in the proto-Earth had sunk into the core, the material that formed the Moon was relatively iron-free.

CONTEXT

The combined motion of the Earth and the Moon around their common barycenter is one of the most complicated problems in celestial mechanics. Newton once referred to it as the only problem that ever gave him a headache. Several factors complicate the solution. The influence of the Sun makes the problem a three-body gravitational interaction rather than the simpler two-body problem conquered by Kepler. Unlike the two-body problem, the three-body problem cannot be solved in closed analytic form; particular approximate solutions exist for special configurations, but the Sun-Earth-Moon trio does not conform to any of them.

The Earth and Moon are also too close for either to be regarded as point masses. Further, neither is purely spherical: the Earth is ellipsoidal, with an equatorial bulge as a product of its rotation; the Moon is oval as a result of a permanent tidal bulge on the side facing the Earth. The rotational period of the Moon equals its orbital period, so that one face perpetually faces the Earth, but the orbit is not circular, so that the Moon moves along the orbit at a varying rate. This causes the side of the Moon facing the Earth to rock back and forth, a motion known as libration. The deviation from circularity (called the eccentricity) is itself variable, driven by the gravitational pull of the Sun, so that the the extent of the libration waxes and wanes. This variation in eccentricity is called evection.

Billy R. Smith, Jr.

FURTHER READING

Brusche, P., and J. Sundermann. *Tidal Friction and the Earth's Rotation*. Berlin: Springer, 1978. A scholarly work. The first paper, "Pre-Telescopic Astronomical Observations" by F. R. Stephenson covers the historical eclipse data that revealed the slow increase in the length of the day due to tidal braking.

Comins, N. *What if the Moon Didn't Exist? Voyages to Earths That Might Have Been*. New York: HarperCollins, 1993. An astronomer examines how the Earth might have evolved without the interaction of a massive nearby Moon. An unusual but entertaining and engaging exploration of Earth-Moon interactions. Index and bibliography.

Darwin, G. *The Tides and Kindred Phenomena in the Solar System*. San Francisco: W. H. Freeman, 1962. A thorough discussion of tides from all perspectives: oceanographic, hydrodynamic, geological, astronomical, and historical. Contains a chapter on the less familiar tides in the atmosphere and in the Earth's crust. The remarks on tidal coupling between the Sun and planets are unfortunately out of date.

Ferguson, Kitty. *Tycho and Kepler: The Unlikely Partnership That Forever Changed Our Understanding of the Heavens*. New York: Walker, 2002. This book engagingly describes what is probably the most fruitful and important collaboration in all of the physical sciences. Tycho Brahe's naked-eye observations were the most accurate ever obtained before the invention of the telescope. They revealed hitherto unknown variations in the motion of the Moon; Brahe's observations of Mars made it possible for Kepler to discover the laws of planetary motion that bear his name. Well illustrated (includes six pages of color plates), with notes, bibliography, and index.

Kolerstrom, Nicholas. *Newton's Forgotten Lunar Theory: His Contribution to the Quest for Longitude*. Santa Fe, N.Mex.: Green Lion Press, 2000. Before the invention of the ship's chronometer, navigators used the motion of the Moon to determine longitude. Doing so accurately requires a very accurate theory of the Moon's orbital motion. In Newton's time, this was an area of immensely important scientific research. This book outlines his efforts at solving this exceedingly difficult problem. Chapter 1 describes the motions of the Sun, Earth, and Moon.

Moore, Patrick. *On the Moon*. London: Cassell, 2001. Contains an excellent nontechnical discussion of lunar motion and tides.

See also: Auroras; Earth's Age; Earth's Differentiation; Earth's Magnetic Field: Origins; Earth's Magnetic Field: Secular Variation; Earth's Magnetic Field at Present; Earth's Magnetosphere; Earth's Origin; Earth's Rotation; Earth's Shape; Earth's Structure; Lunar History; Planetary Orbits: Couplings and Resonances; Solar System: Element Distribution; Van Allen Radiation Belts.

Earth-Sun Relations

Categories: Earth; Planets and Planetology; The Sun

The fundamental Earth-Sun relationship, from which all others derive, is that the Earth rotates on its axis as it revolves around the Sun. The relationships between the Earth and the Sun determine the Earth's "heat budget" and control life on the planet. Earth motions also produce noticeable periodic changes in the apparent path of the Sun across the sky, perhaps most obvious in the seasonal change in directions of sunrise and sunset, and the length of time the Sun is above the horizon.

OVERVIEW

Earth-Sun relations are the dominant controls of life on Earth. The Sun is a star, and its electromagnetic radiation warms the Earth and supplies the energy that supports life on the planet. Earth-Sun relations determine the amount, duration, and distribution of solar radiation that is received by Earth. The Earth's rotation on its axis produces day and night, and its revolution around the Sun and the tilt of its rotational axis result in the seasons; these processes serve to distribute solar radiation over the Earth. Earth's atmosphere and oceans influence the reflection, absorption, and transfer of solar energy. The result of these interacting phenomena is a "heat budget" on Earth that is hospitable to life.

The Sun radiates electromagnetic energy from every part of its spherical surface. Earth, 150 million kilometers away, intercepts only a minute portion of the Sun's radiation, about one two-billionth. The small amount of the Sun's energy that strikes Earth is Earth's energizer. It sustains life on Earth and drives weather systems and oceanic circulation. Solar energy from the past has been preserved in the form of fossil fuels—coal, petroleum, and natural gas.

Perhaps the most remarkable aspect of Earth—remarkable because it is rare in our solar system—is its relatively narrow range of moderate temperatures. The adjectives "hot" and "cold" are frequently used in describing our weather. In relation to the temperatures that are found elsewhere in the solar system, Earth is always moderate, and the words "hot" and "cold" better describe conditions on the other planets. The mean temperature of Earth is about 15° Celsius (59° Fahrenheit, or 288 kelvins); the absolute extremes recorded anywhere on Earth are 58° Celsius (136° Fahrenheit, or 331 kelvins) in North Africa and −89° Celsius (−128° Fahrenheit, or 184 kelvins) in Antarctica. Few inhabitants of Earth will ever experience a temperature range of much more than 60° or 70° Celsius in a lifetime. Compare these temperatures with those of Earth's nearest neighbor, the Moon, where temperatures range from about 120 to −170° Celsius (393 to 103 kelvins) between the day- and nightsides. Earth's so-called sister planet, Venus, has a surface temperature of about 450° Celsius (723 kelvins). The outer planets of the solar system experience a permanent deep freeze, below −100° Celsius (137 kelvins).

The best demonstration of the moderate nature of Earth's temperature is the presence of the world's oceans. Water can exist in the liquid state only in the narrow temperature range of 0 to 100° Celsius (273 to 373 kelvins) at Earth's surface atmospheric pressure. Yet almost 98 percent of Earth's water remains in the liquid state. The polar ice caps contain 2 percent, and a minute portion is water vapor in the atmosphere at any time. Currently 71 percent of Earth's surface is covered by oceans of liquid water, and it has had large oceans for much of its existence as a planet.

The factors that cause Earth to experience such moderate temperatures are complex and interrelated. The Sun is the source of the energy, yet being the right distance from the Sun cannot be the sole cause of Earth's moderate temperature—witness the Moon. Rather, the explanation has to do with Earth's atmosphere, its oceans, and its motions relative to the Sun.

Earth's atmosphere moderates the planet's temperature during both daylight and darkness. During daylight, the atmosphere blocks excessive amounts of solar radiation from reaching Earth's surface and thus prevents overheating. During darkness, the atmosphere retards the escape of heat in the form of longwave infrared energy back into space and thus prevents excessive overnight cooling.

The oceans, also, have a pronounced effect on the heat budget of Earth. Water has the highest specific heat of any common substance. That means more heat is needed to raise the temperature of water than to raise the temperature of most other materials. Summers and daylight periods are kept cooler by the water's ability to absorb great amounts of solar energy without the water's temperature being raised significantly. During winter and during darkness, the water slowly gives up large amounts of heat without significant cooling of the water. Thus, the oceans act as a huge temperature buffer and, along with the atmosphere, add a moderating effect to temperature extremes.

Another factor in moderating temperature variations is the rate of Earth's rotation on its axis—one rotation in twenty-four hours. Rotation causes places on Earth to be alternately turned toward and away from the Sun, as though it were on a rotisserie. The relatively rapid rotation prevents places on Earth from overheating or overcooling. If Earth rotated significantly more slowly, so one side were exposed to the Sun for a much longer time, the illuminated side would become considerably hotter, while the dark side would cool down considerably more. For example, the planet Mercury rotates on its axis once in 58.6 Earth days, and it revolves around the Sun in 88 days; in other words, it makes two orbits around the Sun in the same time it completes three rotations on its axis. As a result, a given spot on the planet's surface is exposed to sunlight for 88 Earth days, and then is in darkness for 88 days more. The resultant temperature extremes range from about 430 to −170° Celsius (703 to 103 kelvins). Similarly, a given spot on the surface of the Moon is exposed to sunlight for a little over two Earth weeks and then is in darkness for about two weeks more, resulting in temperatures that range from about 120 to minus 170° Celsius (393 to 103 kelvins) between the day and night sides.

One complete revolution, or orbit, of Earth around the Sun defines the time unit of one year. During a single revolution, Earth rotates on its axis 365.25 times; therefore, there are 365 days in most calendar years, with an extra day every fourth year (leap year). The orbit of Earth around the Sun is an ellipse, which lies in a plane called the ecliptic plane. The Sun is located at one of the two foci of the ellipse; thus the distance of Earth from the Sun varies during the year. The point on the orbit where Earth is closest to the Sun is called perihelion; it occurs on or about January 3 each year at an Earth-Sun distance of about 147 million kilometers. The point on the orbit farthest from from the Sun is called aphelion; it occurs on or about July 4 each year at an Earth-Sun distance of about 152 million kilometers.

This variation in Earth's distance from the Sun does alter the amount of solar radiation that is received by Earth, but it is not the cause of the seasons. Perihelion, when Earth is nearest to the Sun and seemingly when Earth would be the warmest, occurs during winter in the Northern Hemisphere, and aphelion occurs during the Northern Hemisphere's summer. Thus the distance variations are out of phase with the seasons in the Northern Hemisphere, but in phase with seasons in the Southern

Hemisphere. In both cases, the distance variations modify seasonal temperatures but do not cause the seasons themselves.

The cause of the seasons is the fact that Earth's axis of rotation is tilted 23.5° from the perpendicular to the ecliptic plane, which is the plane of Earth's orbit around the Sun. The orientation in space of Earth's rotational axis changes only very slowly, so that during one year (one orbit around the Sun) it remains nearly constant in position. As Earth revolves around the Sun, the axis in the Northern Hemisphere is alternately tilted toward and away from the Sun. When Earth's North Pole is tilted toward the Sun, the Northern Hemisphere receives more solar radiation than does the Southern Hemisphere, resulting in summer in the Northern hemisphere and winter in the Southern Hemisphere. When Earth's North Pole is tilted away from the Sun, the opposite occurs.

At the point in the orbit when the North Pole is tilted most directly away from the Sun, the Sun is exactly overhead at noon at the Tropic of Capricorn (23.5° south latitude) on Earth, and the entire area south of the Antarctic Circle experiences continuous daylight. This position in the orbit and the moment of time when it occurs both are referred to as the December solstice, which occurs around December 21 each year. For the Northern Hemisphere, it is the "winter solstice," but for the Southern Hemisphere, it is the "summer solstice." Six months later, when the North Pole is tilted most directly toward the Sun, the Sun is exactly overhead at noon at the Tropic of Cancer (23.5° north latitude) on Earth, and the entire area north of the Arctic Circle experiences continuous daylight. This position in the orbit and the moment in time when it occurs both are known as the June solstice, which occurs about June 21 each year. Approximately halfway in between the two solstices are the two equinoxes, occurring about March 20 or 21 and September 22 or 23 each year. On the two equinoxes, the Sun is directly overhead at noon at the Equator (0° latitude). Both Northern and Southern Hemispheres receive equal solar radiation then. On the two equinoxes, the periods of daylight and darkness are equal all over the Earth (equinox means "equal night"), and the Sun rises due east and sets due west.

Between March and September, when the Sun is overhead as seen from north of the Equator, sunrise is north of east and sunset is north of west for all locations that experience sunrise and sunset on a particular date, both Northern and Southern Hemispheres. (The places that do not experience sunrise and sunset are those areas near the poles that are experiencing either continuous daylight or continuous darkness on that date.) Sunrise is farthest north of east and sunset is farthest north of west on the June solstice, after which they both begin a southward migration. Between September and March, when the Sun is overhead as seen from south of the Equator, sunrise is south of east and sunset is south of west for all places that experience sunrise and sunset on a particular date, both Northern and Southern Hemispheres. (Again, the places that do not experience sunrise and sunset are those areas near the poles that are experiencing either continuous daylight or continuous darkness on that date.) Sunrise is farthest south of east and sunset farthest south of west on the December solstice, after which they both begin a northward migration to repeat the pattern.

METHODS OF STUDY

The seasonal variations in the Sun's apparent daily motion across the sky were noted by many ancient cultures. Various stone structures built hundreds to thousands of years ago around the world—from Stonehenge on England's Salisbury Plain to Caracol in Mexico's Yucatán peninsula to the Bighorn Medicine Wheel high in Wyoming's Bighorn Mountains to Mystery Hill in southern New Hampshire—display alignments pointing toward the rising and setting points of the Sun on the solstices and equinoxes.

More recently it has been determined that the tilt and orientation of Earth's rotational axis and the eccentricity of Earth's elliptical orbit change slowly and cyclically with time. The tilt of Earth's rotational axis relative to a perpendicular to the ecliptic plane (the plane of Earth's orbit around the Sun) now is about 23.5°, but it varies between approximately 21.5 and 24.5° over a cycle of 41,000 years. A greater tilt results in more extreme summer and winter

climates, while a smaller tilt means summers are not as hot and winters are not as cold.

Earth's rotational axis also slowly wobbles like that of a giant top, tracing out in space a double cone over a period of 26,000 years. This wobble is due to the gravitational pull of the Moon and the Sun on Earth's equatorial bulge. At the present time, we are closest to the Sun during northern winter and farthest away during northern summer, but as a result of precession, in 13,000 years we will be farthest during northern winters and closest during northern summers, making seasons more severe in the Northern Hemisphere but less severe in the Southern Hemisphere.

Finally, the shape of Earth's orbit around the Sun slowly alternates between being more nearly circular and slightly more elliptical over a period of about 100,000 years due to gravitational perturbations by the other planets. The interplay of all these changes alter slightly the solar radiation received in the Northern and Souther Hemispheres and hence their seasonal climate variations. The Serbian astrophysicist Milutin Milanković was the first to study the effects of these changes and link them to the multiple advances and retreats of large-scale continental glaciation in the Northern Hemisphere during the Pleistocene epoch (the last two million years of geologic time).

Many solar phenomena such as sunspots, prominences, and flares vary in number and frequency of occurrence over a period of about eleven years (the solar activity cycle). This is caused by changes in the Sun's magnetic field, which reverses direction with each solar cycle. Long-term studies of the Sun show that the Sun's activity level varies over timescales of hundreds of years, becoming more or less active. The changes in solar activity seem to be related to changes in Earth's climate, as recorded in old documents and preserved in the width of annual tree rings. Very little solar activity was observed during the 1600's and 1700's, a period known as the Maunder minimum. During this time, Europe and northeastern North America were colder (the so-called Little Ice Age) and western North America experienced prolonged droughts. In contrast, solar activity seems to have been unusually high from sometime in the

1000's until about 1250, a time known as the Medieval Optimum (also known as the medieval grand maximum) when the climate was warmer than it is today. This time marked the height of the Vikings' expansion, when they established colonies in Greenland and Newfoundland. The colonies were abandoned or died out when solar activity declined and the climate turned colder.

Over still longer timescales, theories of stellar evolution applied to the Sun indicate that it has slowly increased in brightness since it formed about 4.5 billion years ago and that it will continue to do so for several billion years more, until it begins to run out of hydrogen fuel in its interior. When that happens, the Sun will expand relatively rapidly to become a red giant star several hundred to more than a thousand times brighter than it is now. These changes will greatly increase Earth's temperature, eventually making it uninhabitable.

Accurate measurements of the length of the day show small, erratic changes in Earth's rotation period, but on average an Earth day is lengthening by about 0.001 second every century. This slowing of Earth's rotation is due to tidal friction. The gravitational effects of the Moon and to a smaller extent the Sun produce tides on Earth, which gradually retard Earth's rotation. To conserve angular momentum, the Moon's distance from Earth is slowly increasing. This has been confirmed by accurately measuring the out-and-back travel time of laser beams bounced off retroreflectors left on the Moon's surface by the Apollo Moon landing missions. These processes will continue until, in the distant future (at least billions of years), Earth's rotation will become tidally locked with the Moon's revolution around Earth, both taking about 47 of our present days. However, this is happening so slowly that the Sun probably will become a red giant first.

Evidence of changes in Earth's heat budget in the past (from the recent past to the ancient past) have come from many disciplines, including history, geology, paleontology, climatology, and astronomy. Satellite studies of Earth over the past several decades have opened many new ways to monitor present conditions and look for predictors of possible future changes. For exam-

ple, satellite images of the oceans at various wavelengths of the electromagnetic spectrum are analyzed to detect any slight temperature changes over time that may portend changes in Earth's climate. Sensitive satellite-borne instruments measure the intensity of sunlight in remote areas. Satellite imagery also provides an accurate record-base for changes in areas of snow cover in polar regions.

CONTEXT

Life on Earth is profoundly dependent upon the relationships between Earth and the Sun. The temperature of Earth is set by a balance between Earth's absorption of electromagnetic energy from the Sun and the subsequent reradiation of that energy from Earth as heat back into space. Life on the planet is dependent on this balance and the moderate temperatures that result.

Rotation influences Earth like a rotisserie, turning the planet so as to expose all sides to the Sun during the twenty-four-hour day for a more even heat. The atmosphere protects Earth from overheating by day and from overcooling at night. Earth's "greenhouse effect" is a result of the atmosphere's ability to trap solar radiation as heat during the day and retard its escape back into space at night, when the Sun is not above the horizon. Earth's heat budget is a product of many factors, not all of which are fully understood. Intense research continues on possible causes and effects of changes in Earth's heat budget. Being able to predict future changes is of prime importance so we can either prepare for them or try to avert them.

John H. Corbet

FURTHER READING

Ahrens, C. Donald. *Meteorology Today: An Introduction to Weather, Climate, and the Environment.* 8th ed. Florence, Ky.: Brooks/Cole, 2006. This introductory college-level text on meteorology presents a thorough treatment of weather phenomena and explains the seasons and the effects of solar energy on the atmosphere. Written for students with little background in science or mathematics. Includes many illustrations.

Chaisson, Eric, and Steve McMillan. *Astronomy*

Today. 6th ed. New York: Addison-Wesley, 2008. Very well-written college-level textbook for introductory astronomy courses. Part of one chapter deals with Earth motions and the seasons; part of another, with solar activity.

Fraknoi, Andrew, David Morrison, and Sidney Wolff. *Voyages to the Stars and Galaxies*. Belmont, Calif.: Brooks/Cole-Thomson Learning, 2006. A well-written, thorough college textbook for introductory astronomy courses. Has sections dealing with sky motions, the seasons, and solar activity.

Freedman, Roger A., and William J. Kaufmann III. *Universe*. 8th ed. New York: W. H. Freeman, 2008. College-level introductory astronomy textbook, thorough and well written. Includes sections on sky motions, the seasons, and solar activity.

Gabler, Robert E., Robert J. Sager, Sheila M. Brazier, and D. L. Wise. *Essentials of Physical Geography*. 8th ed. Florence, Ky.: Brooks/Cole, 2006. A general introductory-level text on physical geography. Covers rotation, revolution, solar energy, and the elements of weather and climate. Well illustrated. Suitable for the general reader.

Harrison, Lucia Carolyn. *Sun, Earth, Time, and Man*. Chicago: Rand McNally, 1960. This book is considered the classic reference for Earth-Sun relations. Although dated, it is an excellent source of information and offers a remarkably extensive coverage of Earth-Sun relations.

Ruddiman, William F. *Earth's Climate: Past and Future*. 2d ed. New York: W. H. Freeman, 2008. A college textbook, suitable for both introductory and upper-level undergraduate courses. Contains much material on climate change and its causes.

Schneider, Stephen E., and Thomas T. Arny. *Pathways to Astronomy*. 2d ed. New York: McGraw-Hill, 2008. Very thorough college textbook for introductory astronomy courses, divided into many short sections on specific topics. Has several sections on the motions of Earth, equinoxes and solstices, and solar activity.

Strahler, Arthur N., and Alan H. Strahler. *Modern Physical Geography*. 4th ed. New York: John Wiley & Sons, 1992. In this general college-level text on physical geography, Earth-Sun relations are discussed within the context of the study of weather and climate. Diagrams are well employed to explain Earth's orbit, rotation, revolution, and axis tilt. Easy to read.

Tarbuck, Edward J., and Frederick K. Lutgens. *Earth: An Introduction to Physical Geology*. Illustrated by Dennis Tasa. 9th ed. Upper Saddle River, N.J.: Pearson Prentice Hall, 2008. Several chapters in this introductory text deal with the solar system. The nature of solar activity and the Earth's motions are explained. Well illustrated and accessible.

See also: Auroras; Earth-Moon Relations; Earth's Magnetic Field: Origins; Earth's Magnetic Field: Secular Variation; Earth's Magnetic Field at Present; Earth's Magnetosphere; Earth's Rotation; Earth's Shape; Eclipses; Greenhouse Effect; Main Sequence Stars; Solar Corona; Solar Flares; Solar Infrared Emissions; Solar Magnetic Field; Solar Radiation; Solar Radio Emissions; Solar Ultraviolet Emissions; Van Allen Radiation Belts.

Earth System Science

Categories: Earth; Scientific Methods

Earth system science views the planet Earth as a dynamic, unified system of simultaneous, interacting forces. In particular, Earth system science focuses on achieving a better understanding of the effects of human interaction with the environment.

OVERVIEW

A new approach to studying the Earth's systems views them as a set of interacting forces all operating simultaneously rather than separate Earth science disciplines to be studied in isolation. This new and promising viewpoint came about as a result of a growing recognition of the interactive nature of Earth's forces, as exerting influences on one another, as opposed to the idea that these forces act independently. The Earth is a constantly changing world with dramatic tectonic activity, volcanism, mountain building, earthquakes, dynamic oceans, severe storms, and varying climatic patterns and atmospheric conditions. Scientists using this "systems approach" view the Earth as a unified whole, and instead of concentrating attention on one component at a time, they use total global observation methods (attempting to model Earth as a whole) together with numerical modeling.

The Earth systems science approach was first detailed by an Earth System Science Committee (ESSC) appointed by the Advisory Council of the National Aeronautics and Space Administration (NASA). In 1986, the committee completed a three-year study of research opportunities in Earth science and recommended that an integrated, global Earth observation and information system be adopted and in full operation by the mid-1990's. The committee's *Overview Report* was released on June 26 of that year. Requests for the findings of the committee from the National Oceanic and Atmospheric Administration (NOAA) and the National Science Foundation (NSF), along with other federal agencies, have drawn the agencies—especially NASA, NOAA, and NSF—into a scientific alliance. The committee's report outlined immediate needs in several wide-reaching areas: scientific understanding of the entire Earth as a system of interacting components; the ability to predict both natural and human-induced changes in the Earth system; strong, coordinated research and observational programs in NASA, NOAA, and NSF as the core of a major U.S. effort; long-term measurements, from space and from Earth's surface, to describe changes as they occur and as a basis for numerical modeling; modeling, research, and analysis programs to explain the functioning of individual Earth system processes and their interactions; a sequence of specialized space research missions focusing on Earth systems, including the Upper Atmosphere Research Satellite (UARS), the joint United States/France Ocean Topography Experiment (TOPEX/POSEIDON), the Geopotential Research Mission (GRM), and an Earth-observing system using polar-orbiting platforms planned as part of the U.S. Space Station complex combining NOAA and NASA instrumentation.

Earth system science utilizes new technologies in global observations, space science applications, computer innovations, and quantitative modeling. These new tools of advanced technology allow scientists to probe and learn about the interactions responsible for Earth evolution and global change. Examples of research made possible by new tools are the opportunity to include the effects of global atmospheric motions in models of ocean circulation; the study of volcanic activity as a link between convection in the Earth's mantle and worldwide atmospheric properties; and the tracing of the global carbon cycle through the many transformations of carbon by biological organisms, atmospheric chemical reactions, and the weathering of Earth's solid surface and soils. In addition, recent advances in these technologies have had the immediate practical effect of improving the quality of human life in areas such as weather prediction, agriculture, forestry, navigation, and ocean-resource management.

The goal of Earth system science is to obtain a scientific understanding of the entire Earth system on a global scale by describing how its component parts and their interactions have evolved, how they function, and how they may be expected to continue to evolve on all timescales. This evolution is influenced by human activities—for example, the depletion of the Earth's energy and mineral resources and the alteration of atmospheric chemical composition—that sometimes are easily identified. The overall long-range consequences of these human actions are difficult to predict; the changes do not occur quickly enough for immediate recognition and, indeed, often take decades to evolve fully. The challenge to Earth system science is to develop the capability to predict those changes that will occur in the twenty-first century, both naturally and in response to human activity. To meet this challenge, vigorous investigations are being undertaken that include global observations, information systems built to process global data, and existing numerical models that already are contributing to a detailed understanding of individual Earth components and interactions. Such programs require interdisciplinary research support and interagency cooperation.

Observations from space, the best vantage point from which to obtain the comprehensive global data required to discriminate among worldwide processes operating on both long and short timescales, are essential to the study of the Earth as a system. Rapid variations in atmospheric and ocean properties, and the global effects of volcanic eruptions, ocean circulations, and motions of the Earth's crustal plates are examples of such processes. The Space Science Board of the National Academy of Sciences recommended orbital observation as a major method of global study; the Earth System Science Committee accepted the recommendations and expanded on them. Of particular value are NASA and NOAA satellites already on station in orbit, such as the Laser Geodynamics Satellites, which employ laser ranging to measure motions and deformations of Earth's crustal plates. Weather satellites already have supplied a sizable fund of data about the atmosphere and oceans, facilitating a good start on numerical modeling of weather variations. Other programs that have yielded coordinated studies of specific Earth system processes include the Earth Radiation Budget Experiment (1984), the Laser Geodynamics Satellites (1976 and 1983), the Navy Remote Ocean Sensing System (1985), and the Upper Atmosphere Research Satellite (1982).

In order to implement the full measure of the Earth system science concept, advanced information systems are needed to process global data and allow analysis, interpretation, and quantitative modeling. Also required is the implementation of additional satellite observations that yield ocean color imaging, scanning radar altimeters for surface topography, and atmospheric monitors. In addition, vigorous programs of ground-level measurements are needed to complement, validate, and interpret the global observations from space. International cooperation is essential to the success of Earth system science; the development of management policies and mechanisms are required to encourage cooperation among agencies around the globe in order to ensure the coordination necessary for a truly worldwide study of the Earth. A number of major international research programs are already operating, such as the World Climate Research Program, sponsored by the International Council of Scientific Unions and the World Meteorological Organization. To accomplish the many objectives of these programs, the Earth System Science Committee recommends two specific goals in which the three major U.S. agencies—NASA, NOAA, and NSF—must work closely together. The first goal is to establish and develop the advanced information systems and management structures required by Earth system science as a cooperative venture, and the second is to build close cooperation in programs of basic research.

The Orbiting Carbon Observatory which was lost during launch on February 24, 2009, was designed to be the first spacecraft to study Earth's atmospheric carbon dioxide, a main cause of global warming. (NASA/JPL)

METHODS OF STUDY

The most significant tools for global observation are Earth-orbiting satellites that can precisely measure large areas of the Earth at one time. Meteorological satellites, for example, gather enormous amounts of data about temperature, weather patterns and forces, and atmospheric changes and components; instruments aboard these satellites can gather data on and monitor variations of climate and storm systems, adding to the growing fund of global information. These satellites are placed in geosynchronous orbit at an altitude of 35,000 kilometers over the equator; at that altitude their orbital period is the same as the Earth's rotation—one day—so they remain over the same spot on Earth and can continuously monitor the same region.

Earth observation satellites, working in the infrared band of the spectrum, allow scientists to gather imagery and information about volcanic activity, earthquakes, geological formations, mineral resources, and geographic changes to provide still another perspective on the Earth. Orbiting the Earth from pole to pole many times a day, they are able to make a record of large sections of the Earth in a twenty-four-hour period as the Earth rotates underneath the satellite's orbital path. Earth observation satellites also carry instruments that measure temperatures, record cloud cover, and monitor catastrophic changes.

Other satellites measure ocean dynamics such as the temperature of large sections of seas and oceans, wave action, ocean water content, and relationships between water and the land it touches. Special instruments aboard these satellites are designed to monitor ice conditions and snowfall at sea and watch for changes in polar regions.

Still other spacecraft carry radar-imaging devices to measure precise distances and relationships between land features. The International Space Station has, as one of its most important objectives, the function of a permanently orbiting platform on which both humans and unattended instruments can work over long periods of time to monitor Earth activities and topography. The space station will be able to contribute large amounts of data because it can function both as information gatherer and processor using advanced onboard automated equipment such as specialized computers.

Although much of the instrumentation for Earth system science will be space-borne, much of it also will have to be ground-based, at the sites where data need to be gathered: near volcanoes, earthquakes, hurricanes, tornadoes, and thunderstorms, for example. Such phenomena must be measured on the ground to determine their effects on other Earth-surface processes. Ground data can then be compared and synthesized with data gathered from space to offer a broader view.

One of the most valuable of tools is the computer, for the receipt, storage, retrieval, analysis, and supply of large quantities of information. Ground-based and space-borne computers work in conjunction with each other for the comparison and large-scale analysis of data, which can be networked to any place on Earth. Computers also are used to generate theoretical models of various kinds of processes. By feeding weather data from the past hundred years into a computer, for example, scientists can begin to construct long-term models of weather patterns and global changes in climate and precipitation. Another study method is the creation and management of global information systems into which is fed data from countries all over the world; all nations can retrieve data for their own research as well as input data to add to the ongoing process of worldwide data analysis.

CONTEXT

The new methodology of Earth system science offers the opportunity to study the Earth from a more integrated perspective and to raise public awareness of the human practices that are affecting the planet. It is important that citizens of the twenty-first century understand the forces and processes that can cause global changes because, individually and collectively, they are contributors to those changes. Human contributions include continued clear-cutting of vast forest areas, thus inviting massive deforestation (destruction of forests); removal of protective trees and underbrush from areas adjacent to desert areas, thus encouraging rampant desertification (the spread of desert conditions);

and pollution of the atmosphere and waterways. Over time, these practices can slowly deplete Earth's natural resources and upset the fragile balance of nature worldwide. Human activities can trigger events that could cause long-term environmental damage.

Thomas W. Becker

FURTHER READING

Asrar, Ghassem. *EOS: Science Strategy for the Earth Observing System.* New York: American Institute of Physics, 1994. Describes the Earth Observing System program: its investigations, capabilities, and educational activities. For undergraduates and science readers.

Earth System Science Committee. *An Integrated Global Earth Observation and Information System to Be in Full Operation by the Mid-1990's.* Boulder, Colo.: University Corporation for Atmospheric Research, 1986. Written by the key people who created the method, a basic, concise presentation of the Earth Observation and Information System.

Kump, Lee R., James Kasting, and Robert Crane. *The Earth System.* Upper Saddle River, N.J.: Prentice Hall, 2003. An introductory work for those new to the Earth system field. Addresses the carbon cycle and events in Earth's history that help explain current global changes. For a general audience.

MacKenzie, Fred T. *Our Changing Planet: An Introduction to Earth System Science and Global Environmental Change.* Upper Saddle River, N.J.: Prentice Hall, 2002. Aimed at nonscientists, this volume covers all areas of Earth system science, including global change associated with both natural and human sources.

Matthews, Samuel W. "This Changing Earth." *National Geographic* 143 (January, 1973): 1-37. One of the earliest articles to describe the Earth's dynamic processes in a language that the public could readily understand. Addresses plate tectonics and takes the reader on a historic tour of the development of modern Earth science. Superb diagrams and supportive photography.

National Aeronautics and Space Administration Advisory Council. *Earth System Science Overview.* Washington, D.C.: Government Printing Office, 1986. This fifty-page document details in easy-to-understand language all the intricate natural mechanisms at work on the planet. Describes the entire Earth system science concept and outlines how the discipline's tools and methods will be brought together to focus on a global data-gathering, archiving of information, and international cooperative efforts. For high school students and general readers.

Skinner, Brian J. *The Blue Planet: An Introduction to Earth System Science.* New York: John Wiley, 1995. Good introduction to the field of Earth system science. Contains a series of essays covering both methods of research and current progress, some written by leading scientists. For a general audience.

See also: Auroras; Earth-Moon Relations; Earth-Sun Relations; Earth's Age; Earth's Atmosphere; Earth's Composition; Earth's Core; Earth's Core-Mantle Boundary; Earth's Crust; Earth's Crust-Mantle Boundary: The Mohorovičić Discontinuity; Earth's Differentiation; Earth's Magnetic Field: Origins; Earth's Magnetic Field: Secular Variation; Earth's Magnetic Field at Present; Earth's Magnetosphere; Earth's Mantle; Earth's Oceans; Earth's Origin; Earth's Rotation; Earth's Shape; Earth's Structure; Eclipses; Gravity Measurement; Greenhouse Effect; Van Allen Radiation Belts.

Earth's Age

Category: Earth

Determining the age of the Earth is one of the great achievements of science. Until the eighteenth century, it was generally believed that the Earth was several thousand years old, and all its features—including mountains and valleys—had been produced by catastrophes such as great floods and earthquakes. The new science of geology gradually showed that the Earth was billions of years old and it had taking its form through slow, uniform processes operating over long periods of time.

OVERVIEW

In the middle of the seventeenth century, Joseph Barber Lightfoot of Cambridge University in England penned the following words:

> Heaven and Earth, center and circumference, were made in the same instant of time, and clouds full of water, and man was created by the Trinity on the 26th of October, 4004 B.C.E., at 9 o'clock in the morning.

At the time that Lightfoot wrote those words, this statement expressed the most informed opinion on the age of the Earth. The year 4004 B.C.E. had been calculated by James Ussher, the Anglican archbishop of Armagh, Ireland, by adding up the ages of the patriarchs recorded in the Old Testament. This was the method that most scholars used to date the Earth, and much effort was expended analyzing the first few books of the Old Testament.

A little over a century later, a Scottish physician and gentleman farmer named James Hutton (1726-1797) suggested that there was a better way to determine the past history of the Earth than by poring over biblical genealogies. Hutton believed that processes currently operating in nature could be extrapolated back in time to shed light on the historical development of the Earth. This idea—that past processes are essentially the same as present processes—is called uniformitarianism. In 1785, he presented his new views on geology in a paper entitled "Theory of the Earth: Or, An Investigation of the Laws Observable in the Composition, Dissolution, and Restoration of Land upon the Globe." Uniformitarianism became the foundation of the newly developing science of geology. Charles Lyell (1797-1875), who was born in the year of Hutton's death, extended these new ideas and helped lay the foundation for what was becoming a powerful new science. A major argument was over the age of the Earth. Was it really millions or billions of years old, as indicated by new discoveries and theories, or was it only a few thousand years old, as everyone had previously believed?

According to current theories, the matter that makes up the Earth as well as everything else in the universe was originally created in the big bang about 13 to 14 billion years ago. Hydrogen, most helium, and trace amounts of lithium and beryllium formed in the immediate aftermath of the big bang, while the atoms of all the other elements were formed by nuclear fusion reactions in massive stars as they generated energy during their "lives" and then exploded as supernovae.

About 4.5 to 4.6 billion years ago, the matter that would become our solar system was part of a nebula, an interstellar cloud of gas and dust, in the disk of the Milky Way galaxy. The portion of this nebula that would become our solar system, called the solar nebula, began to contract under the influence of gravity. Most of the material in the solar nebula collapsed to the center and formed the Sun. The planets and everything else orbiting the Sun formed from the leftover debris through condensation and accretion. All this took place comparatively rapidly, over a time period of a few tens of millions to at most one hundred million years.

Since the initial formation of the Earth, many processes have been taking place in it and on it: Unstable atomic nuclei have radioactively decayed into nuclei of other elements; the Earth's early rotation rate has slowed due to friction from tides generated by the Moon and Sun; mountains have risen under the influence of global tectonics and have been worn away by the ceaseless activities of erosion; and evolution of life

Half-Lives of Some Unstable Isotopes Used in Dating

Parent Isotope	Daughter Product	Half-Life Value
Uranium 238	Lead 206	4.5 billion years
Uranium 235	Lead 207	704 million years
Thorium 232	Lead 208	14.0 billion years
Rubidium 87	Strontium 87	48.8 billion years
Potassium 40	Argon 40	1.25 billion years
Samarium 147	Neodymium 143	106 billion years

Source: U.S. Geological Survey.

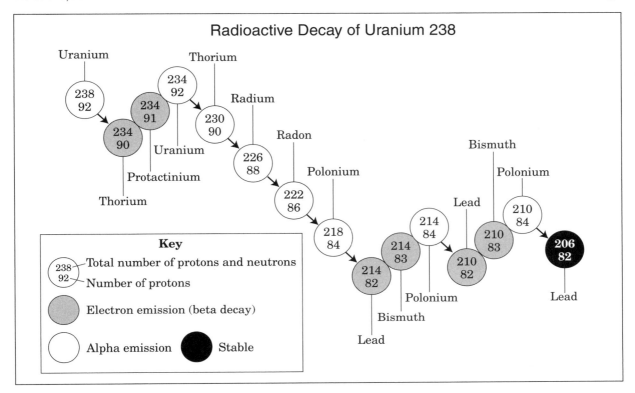

Radioactive Decay of Uranium 238

Uranium

Thorium

Radium

Radon

Polonium

Bismuth

Polonium

Lead

Uranium

Protactinium

Thorium

Key

238 — Total number of protons and neutrons
92 — Number of protons

Electron emission (beta decay)

Alpha emission   Stable

Lead

Bismuth

Polonium

Lead

has transformed the planet, changing barren wastelands into complex ecosystems teeming with diverse forms of life. These various processes have left their marks on the Earth; by studying them, scientists have begun to reconstruct the history of the Earth, in some cases all the way back to the origin of the Earth about 4.5 to 4.6 billion years ago.

METHODS OF STUDY

Many scientific attempts were made during the 1800's to try to determine the age of the Earth. Most involved some process that produced a noticeable change in something. By measuring the magnitude of the change and the rate at which it occurs, the age could be calculated. For example, suppose water is pouring into a bucket at the rate of 1 gallon per minute. If the bucket contains 10 gallons, one can calculate that the process started 10 minutes ago. However, the validity of that result hinges on several assumptions, such as (1) that the rate of water inflow has been constant, (2) that there was no water in the bucket at the start, and

(3) that there are no holes in the bucket letting some water drain out.

Similar problems beset the early attempts at dating the Earth. Some of them included estimating how long it would take the Earth to cool and harden from a blob of molten rock and metal, how long it would take to accumulate the entire thickness of sedimentary rocks exposed all around the world, and how long it would take for the oceans to become as salty as they are. Because of uncertainty in the assumptions inherent in these methods, the calculated ages were not at all consistent, ranging from tens of thousands to hundreds of millions of years.

Much more accurate and consistent ages for Earth materials and events have been obtained by radiometric dating, a technique developed during the 1900's. This method uses radioactive decay, a process in which the unstable nucleus of an atom of one element (called the parent) spontaneously transforms into a nucleus of another element (called the daughter). Although it is impossible to say precisely when a single unstable parent nucleus will decay, a large number will decay at a definite rate referred to as half-life. A half-life is the period of time during which half the parent nuclei that are present will decay. Thus after one half-life, one-half of the parent nuclei will have decayed and one half will remain; after two half-lives, three-fourths will have decayed and one-fourth will remain; after three half-lives, seven-eighths will have decayed and one-eighth will remain; and so on. The decay rate or half-life appears to be constant, since nothing seems to have any effect on it. Measuring the ratio of daughter to parent in a sample can tell how long the parent has been decaying and thus how old the specimen is.

The nucleus of an atom is a dense packing of particles called protons with positive electrical charge and neutrons with no electrical charge. All nuclei of a particular chemical element have the same number of protons, but they can have different numbers of neutrons. Nuclei of the same element but with different numbers of neutrons are called isotopes of that element and are identified by their atomic mass number (which is simply the total number of protons and neutrons). For example, all carbon nuclei have six protons. Most carbon nuclei also have six neutrons, and this isotope is called carbon 12. Some carbon nuclei have eight neutrons, and that isotope is called carbon 14.

The three nuclear decay processes underlying geologic radiometric dating are (1) alpha decay, in which the parent nucleus emits an alpha particle, a helium nucleus of two protons and two neutrons; (2) beta decay, in which the parent nucleus emits a beta particle, an electron, when a neutron in the parent nucleus turns into a proton and electron; and (3) electron capture, in which the parent nucleus captures an electron that combines with a proton in the nucleus to form a neutron. The specific isotope decay schemes most commonly used for geologic dating are potassium 40 to argon 40 (half-life 1.3 billion years, via electron capture), rubidium 87 to strontium 87 (half-life 47 billion years, via beta decay), uranium 238 to lead 206 (half-life 4.5 billion years, via eight alpha and six beta decays), uranium 235 to lead 207 (half-life 713 million years, via seven alpha and four beta decays), and thorium 232 to lead 208 (half-life 14.1 billion years, via six alpha and four beta de-

cays). Carbon 14 decays to nitrogen 14 via beta decay with a half-life of only 5,730 years; this is so short that it is limited to dating only very recent geologic events, although it has been very useful for archaeological and historical dating.

Radiometric dating has been applied to thousands of rock specimens from all over the Earth. The oldest rock formation found on Earth so far is the Acasta gneiss from near Great Slave Lake in northern Canada, dated at 4.03 billion years. Even older mineral grains—small crystals of zircon dated at 4.3 to 4.4 billion years—have been found in younger sedimentary rocks from the Jack Hills area of western Australia. It is difficult to find very ancient rocks on the surface of the Earth, because most of the Earth's surface has undergone many changes since the Earth was formed.

The currently accepted age for the Earth, 4.5 to 4.6 billion years, was obtained by radiometrically dating meteorites that fell to Earth from space. These meteorites are believed to be remnants from the time when the planets, including Earth, were forming in the early solar system. Similar dates have been obtained from a few of the rocks brought back by the Apollo landings on the Moon, which is believed to have formed at about the same time as the Earth.

CONTEXT

The problem of the age of the Earth is part of a much larger scientific question, which exists at the interface between the study of the Earth and its various processes (which often have practical benefits) and the more esoteric question of the origin and evolution of the universe as a whole. On the practical side, knowledge of the Earth is necessary to predict geologic disasters (such as earthquakes and volcanoes) and to search for geologic resources (such as oil and metallic ores). From a more esoteric point of view, the age of the Earth is important because it speaks to the most fundamental questions that are asked about our place in the universe: How old is this planet, and how was it formed? In the century or two since advances in geological science overthrew the seventeenth century notion of a much younger Earth, people have struggled with finding a new place in the universe. Proponents of "creation science" still argue that the Earth is thousands, not billions, of years old. Legal battles rage over the issue of whether schools across the United States should teach geochronology that is based on religious dogma rather than on scientific research. (In contrast, in Europe this is not a contentious issue, with little public questioning of the antiquity of the Earth based on geologic evidence.) While many questions remain about the details of the formation of the Earth, two facts seem clear: First, the Earth owes its origin to the same processes that brought the solar system into existence; second, those processes can be dated with a high degree of confidence at between 4.5 and 4.6 billion years ago.

Karl Giberson

FURTHER READING

Brush, Stephen G. *Nebulous Earth: The Origin of the Solar System and the Core of the Earth from Laplace to Jeffreys*. Cambridge, England: Cambridge University Press, 1997. Describes how thinking about the origin of the solar system changed and includes discussions of the origin of the Earth-Moon system.

Condie, Kent, and Robert Sloan. *Origin and Evolution of Earth: Principles of Historical Geology*. Upper Saddle River, N.J.: Prentice Hall, 1998. An easy-to-read text covering the complexities of the Earth's history, life, and how Earth's subsystems interact. Also discusses dating methods, planetary evolution, and ancient climates.

Dalrymple, G. Brent. *Ancient Earth, Ancient Skies: The Age of the Earth and Its Cosmic Surroundings*. Stanford, Calif.: Stanford University Press, 2004. A book designed for nonscientists who want to learn about the Earth's age. Covers the manner in which scientists collect their data and describes the conclusions to which the data have led them.

Haber, Frances C. *The Age of the World: Moses to Darwin*. Baltimore: Johns Hopkins University Press, 1959. Reprint. Westport, Conn.: Greenwood Press, 1978. Focuses not on estimates of the age of the Earth but rather on the historical controversy that emerged when nonbiblical values for the age of the Earth began to be accepted. Provides insight into the conflict between science and dogma.

Hartmann, William K. *Moons and Planets*. 5th ed. Belmont, Calif.: Thomson Brooks/Cole, 2005. A college textbook beyond the introductory level, with an approach based on comparative planetology. Offers much information on the origin of the solar system in general and the of Earth in particular.

Hurley, Patrick M. *How Old Is the Earth?* Garden City, N.Y.: Doubleday, 1959. One of the few full-length books on geochronology for the layperson. Even though published fifty years ago, it is still useful, as much of the broad overall outline has not changed appreciably since its publication, although many specific details have.

Lewis, Cherry. *The Dating Game: One Man's Search for the Age of the Earth*. Cambridge, England: Cambridge University Press, 2002. The story of Arthur Holmes and the evolution of calculating the age of rocks. Written by a geologist who makes the more technical details understandable to a general audience.

Ozima, Minoru. *The Earth: Its Birth and Growth*. Cambridge, England: Cambridge University Press, 1981. A translation of a Japanese book that was written by a scientist whose specialty is geochronology. Written at an introductory level.

Tarbuck, Edward J., and Frederick K. Lutgens. *Earth: An Introduction to Physical Geology*. Illustrated by Dennis Tasa. 9th ed. Upper Saddle River, N.J.: Pearson Prentice Hall, 2008. This college-level textbook for introductory geology courses is well written and illustrated. It has two chapters on geologic dating and the historical development of the Earth.

Thackray, John. *The Age of the Earth*. New York: Cambridge University Press, 1989. About forty pages long, and published by a British geological museum, this concise volume contains more pictures than text, but the pictures, most in color, are helpful and make this an interesting source.

Wicander, Reed, and James Monroe. *Historical Geology*. 5th ed. Florence, Ky.: Brooks/Cole, 2006. An undergraduate text covering all major areas of historical geology. Provides a history of the Earth and events that have shaped it.

Earth's Atmosphere

Category: Earth

The chemical composition of the atmosphere has changed significantly over the history of the Earth. The composition of the atmosphere has been influenced by a number of processes, including interaction with the solar wind; "outgassing" of volatiles (materials that easily vaporize to form gases) originally trapped in the Earth's interior during its formation; the geochemical cycling of carbon, nitrogen, hydrogen, and oxygen compounds between the surface, the ocean, and the atmosphere; and the origin and evolution of life.

OVERVIEW

About 4.5 to 4.6 billion years ago, the primordial solar nebula, a part of a large interstellar cloud of gas and dust, began to contract under the influence of gravity. This contraction led to the formation of the Sun and the rest of the solar system including the Earth. The primordial solar nebula was composed mostly of hydrogen gas, with a smaller amount of helium, still smaller amounts of carbon, nitrogen, and oxygen, and still smaller amounts of the rest of the elements of the periodic table.

As the solar nebula contracted, its rotational speed increased to conserve angular momen-

tum. Most of its mass contracted to its center, there becoming the proto-Sun, while the remaining matter was spun off into an equatorial disk. Within the disk, matter condensed from the gaseous state into small, solid grains. Only materials with high melting-point temperatures could condense near the developing proto-Sun; materials with lower melting points condensed farther out. The small solid grains collided with each other and stuck together in a process called accretion that led to the growth of larger bodies called planetesimals. Continued accretion resulted in fewer but larger planetesimals that eventually grew into protoplanets and finally planets, such as Earth.

About the time that the newly formed Earth was reaching its approximate present mass, it may have acquired a temporary atmosphere of hydrogen, helium, methane, ammonia, water vapor, and carbon dioxide—gases that were common in the solar nebula. However, even if such an atmosphere had surrounded the very young Earth, it would have been very short-lived. The Earth's mass was too small to have enough gravity to retain hydrogen and helium for long, and those gases would have quickly escaped into space. Ammonia and methane were chemically unstable in the early Earth's environment and were readily destroyed by ultraviolet radiation from the young Sun. Also, as the young Sun went through its T-Tauri phase of evolution, very strong solar winds (the supersonic flow of protons and electrons from the Sun) would have quickly stripped most of the rest of this primitive atmosphere away.

The early Earth was heated by the intense bombardment of remaining planetesimals and the decay of radioactive elements, to the point where it at least partly melted. The heating and melting released volatiles trapped in its interior by a process called outgassing, forming a gravitationally bound atmosphere. (It is believed that the atmospheres of Mars and Venus also originated in this manner.) The period of extensive volatile outgassing may have lasted for many tens of millions of years. The outgassed volatiles probably had roughly the same chemical composition as do present-day volcanic gaseous emissions: by volume about 80 percent water vapor, 10 percent carbon dioxide, 5 per-

This oblique photograph of Earth, taken from the space shuttle on September 4, 1997, shows the atmosphere and cloud cover over the northwestern African continent. (NASA/JPL/UCSD/JSC)

cent sulfur dioxide, 1 percent nitrogen, and smaller amounts of hydrogen, carbon monoxide, sulfur, chlorine, and argon.

The water vapor that outgassed from the interior soon reached its saturation point, which is controlled by the atmospheric temperature and pressure. Once the saturation point was reached, the atmosphere could not hold any additional gaseous water vapor. Any new outgassed water vapor that entered the atmosphere would have precipitated out of the atmosphere as rain that fell and formed the Earth's vast oceans. Only small amounts of water vapor remained in the atmosphere—ranging from a fraction of a percentage point to several percent by volume, depending on atmospheric temperature, season, and latitude.

The outgassed atmospheric carbon dioxide, being very water-soluble, readily dissolved into the newly formed oceans and formed carbonic acid. In the oceans, carbonic acid formed ions of hydrogen, bicarbonate, and carbonate. The carbonate ions reacted with ions of calcium and magnesium in the ocean water, forming carbonate rocks, which precipitated out of the ocean and accumulated as seafloor carbonate sediments. Most of the outgassed atmospheric carbon dioxide formed carbonates, leaving only trace amounts of gaseous carbon dioxide in the atmosphere (about 0.035 percent by volume).

Sulfur dioxide, the third most abundant component of volatile outgassing, was chemically transformed into other sulfur compounds and sulfates in the atmosphere. Eventually, the sulfates formed atmospheric aerosols and diffused out of the atmosphere onto the surface.

The fourth most abundant outgassed component, nitrogen, is chemically inert in the atmosphere and thus was not chemically transformed, as was sulfur dioxide. Unlike carbon dioxide, nitrogen is relatively insoluble in water and, unlike water vapor, does not condense out of the atmosphere. For these reasons, nitrogen remained in the atmosphere to become its major constituent (now 78.08 percent by volume). In this way, volatile outgassing led to the formation of the Earth's atmosphere, oceans, and carbonate rocks.

The molecules of nitrogen, carbon dioxide, and water vapor in the early atmosphere were acted upon by solar ultraviolet radiation and atmospheric lightning. In the process, molecules of formaldehyde and hydrogen cyanide could have been chemically synthesized, which would have precipitated and diffused out of the atmosphere into the oceans. In the oceans, the formaldehyde and hydrogen cyanide may have entered into polymerization reactions that eventually led to the chemical synthesis of amino acids, the building blocks of living systems. The synthesis of amino acids from nitrogen, carbon dioxide, and water vapor in the atmosphere and ocean is called chemical evolution. Chemical evolution preceded and provided the material for biological evolution.

There is chemical trace evidence for the existence of microbial living organisms on the Earth by about 3.8 billion years ago; the oldest known simple fossils are at least 3.5 billion years old. These earliest living organisms were anaerobic since there was no free oxygen in the atmosphere and oceans. Photosynthesis evolved in one or more of these early microbial groups, such as cyanobacteria. In photosynthesis, the organism utilizes water vapor and carbon dioxide in the presence of sunlight and chlorophyll to form carbohydrates, used by the organism for food. In the process of photosynthesis, oxygen is given off as a metabolic by-product. The production of oxygen by photosynthesis was a major event on the Earth and transformed the composition and chemistry of the early atmosphere. As a result of photosynthetic production, oxygen built up to become the second most abundant constituent of the atmosphere (now 20.9 percent by volume).

The evolution of atmospheric oxygen had very important implications for the evolution of life. The presence and buildup of oxygen led to the evolution of respiration, which replaced fermentation as the energy production mechanism in living systems. Accompanying and directly controlled by the buildup of atmospheric oxygen was the origin and evolution of atmospheric ozone, which is chemically formed from oxygen. The production of atmospheric ozone resulted in shielding the Earth's surface from biologically lethal solar ultraviolet wavelengths between about 200 and 300 nanometers. Prior to the evolution of the atmospheric ozone layer, early life was restricted to a depth of at least several meters below the ocean surface. At this depth, the ocean water offered shielding from solar ultraviolet radiation. The development of the atmospheric ozone

Layers of the Earth's Atmosphere

Storm clouds hover over Earth in this image taken from the International Space Station. (NASA)

layer and its consequent shielding of the Earth's surface permitted early life to leave the safety of the oceans and go ashore for the first time in the history of the planet. Theoretical computer calculations indicate that atmospheric ozone provided sufficient shielding from biologically lethal ultraviolet radiation for the colonization of the land once oxygen reached about one-tenth of its present atmospheric level.

Mercury, Venus, and Mars formed in a fashion similar to Earth, but they developed very differently because of their masses and distances from the Sun. They all experienced a period of heating, partial melting, and volatile outgassing of the same gases that led to the formation of the Earth's atmosphere. However, Mercury's distance from the Sun is so close (resulting in high temperatures) that its relatively weak gravity (due to its small mass) was unable to retain more than a thin trace of gases, and thus today it has virtually no atmosphere. In the case of Venus and Mars, the important dif-

ference is that the outgassed water vapor never existed in the form of liquid water on the surfaces of those two planets.

Because of Venus's closer distance to the Sun (108 million kilometers versus 150 million kilometers for Earth), its lower atmosphere was too hot to permit the outgassed water vapor to condense out of the atmosphere. Thus, the outgassed water vapor remained in gaseous form in the atmosphere and, over geological time, was broken apart by solar ultraviolet radiation to form hydrogen and oxygen. The very light hydrogen gas quickly escaped from the atmosphere of Venus, and the heavier oxygen combined with surface minerals to form a highly oxidized surface. In the absence of liquid water on the surface of Venus, the outgassed carbon dioxide remained in the atmosphere and built up to become the overwhelming constituent of the atmosphere of Venus (about 96 percent by volume). The outgassed nitrogen accumulated to comprise only about 4 percent by volume of the atmosphere of Venus. This carbon dioxide and nitrogen atmosphere is very massive—it produces an atmospheric pressure at the surface of the Venus about ninety times the surface pressure of Earth's atmosphere. If the outgassed carbon dioxide in the atmosphere of Earth had not left via dissolution in the oceans and resultant carbonate rock formation, the Earth's surface atmospheric pressure would be about seventy times greater than at present, with carbon dioxide comprising about 98-99 percent of the atmosphere and nitrogen about 1-2 percent. Thus, the atmosphere of Earth would closely resemble that of Venus. The thick carbon dioxide atmosphere of Venus causes a very significant greenhouse temperature enhancement, giving the lower atmosphere and surface of Venus a temperature of about 750 kelvins (about 477° Celsius), which is hot enough to melt lead. For comparison, the average surface temperature of Earth is only about 288 kelvins (about 15° Celsius).

Like Venus, Mars has an atmosphere composed primarily of carbon dioxide (about 95 percent by volume) and nitrogen (about 3 percent by volume). Because of Mars's greater distance from the Sun (228 million kilometers versus 150 million kilometers for Earth), the temperature of the surface of Mars was too low to support the presence of liquid water. There may be very large quantities of outgassed water in the form of ice or permafrost below the surface of Mars. In the absence of liquid water, the outgassed carbon dioxide remained in the atmosphere. The atmospheric pressure at the surface of Mars, however, is only about 7 millibars (the average surface atmospheric pressure on Earth is 1,013 millibars). The smaller mass of the atmosphere of Mars compared to the atmosphere of Venus and Earth may be attributable to the smaller mass of Mars and, therefore, the smaller mass of volatiles trapped in the interior of Mars during its formation. In addition, it appears that the amount of gases trapped in the interiors of Venus, Earth, and Mars during their formation decreased with increasing distance from the Sun. Venus appears to have trapped the greatest amount of gases and was the most volatile-rich planet, Earth trapped the next greatest amount, and Mars trapped the smallest amount.

The atmospheres of the outer planets—Jupiter, Saturn, Uranus, and Neptune—all contain appreciable quantities of hydrogen and helium, along with methane and ammonia. It is believed that the atmospheres of these planets, unlike the atmo-

The SMART 1 spacecraft took this image of Earth from 70,000 kilometers above the surface in May, 2004. Clouds and weather patterns are visible from Scandinavia (top) to northwestern Africa. (European Space Agency)

spheres of the terrestrial planets Venus, Earth, and Mars, are captured remnants of the primordial solar nebula. Because of the outer planets' large masses and their great distance from the Sun resulting in their very low temperatures, hydrogen, helium, methane, and ammonia are stable and long-lived constituents of their atmospheres.

METHODS OF STUDY

Information about the origin, early history, and evolution of the Earth's atmosphere comes from a variety of sources. Information on the origin of Earth and other planets is based on theoretical computer simulations. These computer models simulate the collapse of the primordial solar nebula and the formation of the planets. Astronomical observations of what appear to be equatorial disks and the possible formation of planetary systems around young stars have provided new insights into the computer modeling of this phenomenon. Information about the origin, early history, and evolution of the atmosphere is based on theoretical computer models of volatile outgassing and the geochemical cycling and photochemistry of the outgassed volatiles. The process of chemical evolution—which led to the synthesis of organic molecules of increasing complexity, the precursors of the first living systems on the early Earth—is studied in laboratory experiments. In these experiments, mixtures of gases simulating the Earth's early atmosphere are energized by ultraviolet radiation, electrical discharges, or heated rocks, simulations of energy sources available on the early Earth. The resulting products are analyzed by chemical techniques.

One of the parameters affecting atmospheric photochemical reactions, chemical evolution, and the origin of life was the flux of solar ultraviolet radiation incident on the early Earth. Astronomical measurements of the ultraviolet emissions from young, sunlike stars have provided important information about ultraviolet emissions from the young Sun during the very early history of the atmosphere.

Geological and paleontological studies of the oldest rocks and the earliest fossil records have provided important information on the evolution of the atmosphere and the transition from an oxygen-deficient to an oxygen-sufficient atmosphere. Studies of the biogeochemical cycling of the elements have provided important insights into the later evolution of the atmosphere. Thus, studies of the origin and evolution of the atmosphere are based on a broad cross-section of the sciences, involving astronomy, geology, geochemistry, geophysics, and biology as well as atmospheric chemistry.

CONTEXT

Studies of the origin and evolution of the atmosphere have provided new insights into the processes and parameters responsible for global change. Understanding the history of the atmosphere provides insight into its future. Today, atmospheric changes being studied for their possible long-term effects include the buildup of greenhouse gases like carbon dioxide and the depletion of ozone in the stratosphere. The study of the evolution of the atmosphere has provided new insights into the biogeochemical cycling of elements between the atmosphere, biosphere, land, and ocean. Understanding this cycling is a key to understanding environmental problems and possible remedies. Studies of the origin and evolution of the atmosphere have also provided new insights into the origin of life and the possibility of life outside the Earth.

Joel S. Levine

FURTHER READING

Ahrens, C. Donald. *Essentials of Meteorology: An Invitation to the Atmosphere*. 5th ed. Florence, Ky.: Brooks/Cole, 2007. An updated version of a classic meteorology textbook. Explains tricky concepts in an easy-to-understand way. Suitable for students and nonscientists.

_____. *Meteorology Today*. 8th ed. Florence, Ky.: Brooks/Cole, 2006. A common text used for introductory meteorology college courses but can also be understood by general audiences. Comes with a CD-ROM learning aid, which includes chapter tests and multimedia tutorials.

Chaisson, Eric, and Steve McMillan. *Astronomy Today*. 6th ed. New York: Addison-Wesley, 2008. Very well-written college-level textbook for introductory astronomy courses with an entire chapter on the formation of the planets.

Fraknoi, Andrew, David Morrison, and Sidney Wolff. *Voyages to the Stars and Galaxies*. Belmont, Calif.: Brooks/Cole-Thomson Learning, 2006. A textbook for introductory astronomy courses that offers several sections dealing with the origin of the solar system and planetary atmospheres.

Freedman, Roger A., and William J. Kaufmann III. *Universe*. 8th ed. New York: W. H. Freeman, 2008. College-level introductory astronomy textbook with several sections on the origin of the solar system, including coverage of planetary atmospheres.

Hartmann, William K. *Moons and Planets*. 5th ed. Belmont, Calif.: Thomson Brooks/Cole, 2005. A college textbook beyond the introductory level, its approach is based on comparative planetology. Includes information throughout the text on the origin of the solar system and its planetary atmospheres.

Henderson-Sellers, A. *The Origin and Evolution of Planetary Atmospheres*. Bristol, England: Adam Hilger, 1983. A technical treatment of the variation of the atmosphere of the Earth over geological time and the processes and parameters that controlled it. Chapters cover the mechanisms for long-term climate change, the atmospheres of the other planets, planetary climatology on shorter timescales, and the stability of planetary environments.

Holland, H. D. *The Chemical Evolution of the Atmosphere and Oceans*. Princeton, N.J.: Princeton University Press, 1984. A comprehensive and technical treatment of the geochemical cycling of elements over geologic time and the coupling between the atmosphere, ocean, and surface. Includes coverage of the origin of the solar system, the release and recycling of volatiles, the chemistry of the early atmosphere and ocean, the acid-base balance of the atmosphere-ocean-crust system, and carbonates and clays.

Levine, Joel S., ed. *The Photochemistry of Atmospheres: Earth, the Other Planets, and Comets*. Orlando, Fla.: Academic Press, 1985. A series of review papers dealing with the origin and evolution of the atmosphere, the origin of life, the atmospheres of Earth and other planets, and climate. The book contrasts the origin, evolution, composition, and chemistry of Earth's atmosphere with the atmospheres of the other planets. It contains two appendixes that summarize all atmospheric photochemical and chemical processes.

Lewis, John S., and Ronald G. Prinn. *Planets and Their Atmospheres: Origin and Evolution*. New York: Academic Press, 1983. A comprehensive treatment of the formation of the planets and their atmospheres. Begins with a detailed account of the origin and evolution of solid planets via coalescence and accretion in the primordial solar nebula; then discusses the surface geology and atmospheric composition of each planet.

Schneider, Stephen E., and Thomas T. Arny. *Pathways to Astronomy*. 2d ed. New York: McGraw-Hill, 2008. A thorough college textbook for introductory astronomy courses, divided into many short sections on specific topics. Contains a unit on the origin of the solar system and several sections on the origins of planetary atmospheres.

Schopf, J. William, ed. *Earth's Earliest Biosphere: Its Origin and Evolution*. Princeton, N.J.: Princeton University Press, 1984. A comprehensive group of papers on such subjects as the early Earth, the oldest rocks, the origin of life, early life, and microfossils. Chapters address the oldest known rock record, prebiotic organic syntheses and the origin of life, Precambrian organic geochemistry, the transition from fermentation to anoxygenic photosynthesis, the development of an aerobic environment, and early microfossils. Technical.

Tarbuck, Edward J., and Frederick K. Lutgens. *Earth: An Introduction to Physical Geology*. Illustrated by Dennis Tasa. 9th ed. Upper Saddle River, N.J.: Pearson Prentice Hall, 2008. This college-level textbook for introductory geology courses is well written and fully illustrated, with a chapter on the origin and historical development of the Earth and its atmosphere.

See also: Auroras; Earth-Sun Relations; Earth's Origin; Earth's Rotation; Earth's Shape; Earth's Structure; Eclipses; Greenhouse Effect; Planetary Atmospheres; Van Allen Radiation Belts.

Earth's Composition

Category: Earth

The Earth consists of a metallic core surrounded by a rocky mantle, which in turn is surrounded by a thin, rocky crust. Much of the crust is covered by an ocean of liquid, salty water. Surrounding it all is the Earth's atmosphere. Of the rocky and metallic material at or below the surface, only the crust, and in a few locales samples of mantle, are available for direct laboratory study. The composition of most of the Earth's interior is inferred by indirect means, primarily from the study of meteorites.

OVERVIEW

About 4.5 to 4.6 billion years ago, the solar system was formed from a cloud of gas and dust called the solar nebula. The cloud contracted gravitationally, with most of it forming the Sun. The planets and other objects that today orbit the Sun formed by condensation and accretion in an equatorial disk that developed around the early proto-Sun. Small, solid grains condensed from the gas as it cooled. As the grains in the equatorial disk orbited the proto-Sun, they collided and stuck together, accreting into small bodies called planetesimals. As the planetesimals collided and grew into protoplanets, their gravitational fields increased, so they swept up more material in the equatorial disk. The innermost planets—Mercury, Venus, Earth, and Mars—were formed mainly from dense metals and rocks, while the outer planets—Jupiter, Saturn, Uranus, and Neptune—were formed mostly of gases and volatile ices. During or shortly after Earth's accretion, differentiation occurred; the denser metals, such as iron and nickel, sank to the core of the early Earth, while the less dense rocky material rose to the outer portions of the planet.

Samples of the Earth's crust are readily available. Geologic processes have brought samples from the upper part of the mantle to the Earth's surface in certain locales. Most of the Earth's interior is inaccessible to direct study, but meteorites offer clues to its composition.

Most meteorites are remnants of the earliest period of planetary formation. They are classified into three main groups based on composition: stony meteorites, stony-iron meteorites, and iron meteorites. Stony meteorites comprise the most abundant group and are composed of silica-associated, or lithophile, elements such as those found in the Earth's crustal materials. Stony-iron meteorites are composed of roughly equal parts of rock (typically the mineral olivine) suspended in a matrix of iron. Iron meteorites are composed of iron (about 80 to 90 percent) along with siderophile elements such as nickel.

Iron meteorites are particularly suggestive to scientists when they attempt to model the composition of the Earth's core. The average density of the entire Earth is about 5.5 grams per cubic centimeter, while the average density of crustal rocks is only about 2.7 grams per cubic centimeter for continental crust and 3.0 grams per cubic centimeter for oceanic crust. This simple comparison indicates that the core must be substantially denser than the average for the entire Earth, and the only reasonably abundant element with about the right density is iron.

The core has two parts: a solid inner core, with a radius of 1,300 kilometers and a density of about 12 to 13 grams per cubic centimeter, and a molten outer core, 2,200 kilometers thick, with a density of about 10 grams per cubic centimeter. The inner core is mostly iron and nickel under high pressure (to make it solid), while the molten outer core probably contains, besides iron and nickel, lighter elements such as sulfur, silicon, oxygen, carbon, and hydrogen. As a whole, the core comprises about one-sixth of the Earth's volume and about one-third of the Earth's mass.

Almost all the remaining two-thirds of the Earth's mass is contained in the mantle, making the mass of the crust, oceans, and atmosphere insignificant in comparison. The mantle is rich in dense, ultramafic rocks such as peridotite, composed mostly of the minerals olivine and pyroxene.

Shortly after (or perhaps during) the initial condensation of grains and their accretion into planetesimals and protoplanets, the Earth's thermal history began through the process of radioactive decay. During this early thermal

Chemical Composition of Earth's Crust

Element	Weight (%)	Volume (%)
Oxygen (O)	46.59	94.24
Silicon (Si)	27.72	0.51
Aluminum (Al)	8.13	0.44
Iron (Fe)	5.01	0.37
Calcium (Ca)	3.63	1.04
Sodium (Na)	2.85	1.21
Potassium (K)	2.60	1.88
Magnesium (Mg)	2.09	0.28
Titanium (Ti)	0.62	0.03
Hydrogen (H)	0.14	—

period, radioactive nuclides (atoms of specific isotopes) decayed, producing substantial heat that led to at least partial melting. Much of the heating is attributable to the decay of potassium 40 and short half-life elements such as aluminum 26. After as little as perhaps 100,000 years, the planet separated into the iron-nickel core and magnesium-iron-silicate lower mantle. Over a longer timescale (probably more than ten million years but no more than a few hundred million), the high-volatility compounds (such as lead, mercury, thallium, bismuth, water in hydrated silicates, carbon-based organic compounds, and the noble gases) all migrated to the surface, where the material was outgassed or melted into magmas in a continuous period of crustal reprocessing that lasted for several hundred million years.

Separated into three main layers—the core, mantle, and crust—the Earth is an active body, its internal heat far from exhausted. The complexity of the chemical composition increases with each successive outward layer. This generalized model gives a framework for examining the relationships of Earth materials.

Earth's wide range of pressure and temperature regimes helps explain why several thousand distinct minerals and numerous rock types composed of different combinations of minerals have been recognized in samples of the crust and upper mantle. Sampling a variety of crustal rocks leads to a determination of elemental abundance in the crust. By mass, approximately one-half of the crust is oxygen

and approximately one-fourth is silicon. These two elements, plus aluminum, iron, calcium, sodium, potassium, and magnesium, make up more than 99 percent of the Earth's crust. Silicon and oxygen combine to form the silicon-oxygen tetrahedron, consisting of a single silicon atom surrounded by four oxygen atoms evenly spaced around it three-dimensionally at the corners of a tetrahedron. This silicon-oxygen tetrahedron joined to additional tetrahedra and/or atoms of other elements forms the class of minerals called silicates, by far and away the most common minerals in the crust.

As ultramafic magmas cool, successive minerals crystallize and settle out via reaction series. As the temperature drops in the melt zone, a discontinuous series (a set of discrete reactions) can occur. Magnetite, an oxide of iron and titanium, is the first to settle out, at about 1,400° Celsius (1,700 kelvins). Olivine, a silicate mineral with a crystal lattice structure of individual silicon-oxygen tetrahedra joined together by other ions (commonly iron and magnesium) and a density between 3.2 and 4.4 grams per cubic centimeter, is the next to crystallize out of the melt. Then comes pyroxene, a silicate mineral with its silicon-oxygen tetrahedra connected in long single chains and a density of 3.2 to 3.6 grams per cubic centimeter. As temperatures in the magma drop to near 1,000° Celsius (1,300 kelvins), the next to crystallize is amphibole, a silicate mineral with its silicon-oxygen tetrahedra joined in long double chains and a still lower density of 2.9 to 3.2 grams per cubic centimeter. As the cooling progresses, the lattice structures increase in complexity with biotite mica, with its silicon-oxygen tetrahedra joined in planar sheets. Paralleling this discontinuous series of reactions is the continuous reaction series of plagioclase feldspar. It has a full three-dimensional lattice of silicon-oxygen tetrahedra, and it varies continuously from being calcium-rich at high temperatures of crystallization to sodium-rich at lower temperatures. Finally, at still lower temperatures down to about 1,000 kelvins (700° Celsius), come potassium feldspar, muscovite mica, and quartz.

With this information, one can start to hypothesize about how the crust and its ocean basins and continents evolved. The oldest Earth materials yet identified are zircon crystals, possibly dating back 4.4 billion years, found in the Jack Hills area of Australia, while the oldest known continental rocks—the Acasta gneiss from the Northwest Territories of Canada—are about 4 billion years old, and they were metamorphosed from earlier igneous rocks. This means that within a few hundred million years after the initial formation of the Earth through condensation, accretion, and differentiation, the first crustal rocks of the Archean eon formed. They probably were composed of olivine, pyroxene, and anorthite (calcium-rich plagioclase feldspar), which crystallized out of basaltic magmas that rose to the surface and cooled and hardened. The early crust, which may have been similar to the anorthosite that makes up much of the ancient highlands on Earth's moon, formed a sheet that was fractured into pieces and subjected to heating through radioactive decay. Differentiation led to the formation of thicker granitic regions surrounded by the thinner basaltic crust. This was the beginning stage in the development of today's crust, which consists of two main types: the denser, thinner, mafic or basaltic oceanic crust and the less dense, thicker, felsic or granitic continental crust. The onset of plate tectonics moved the early continental fragments, causing them to collide and weld themselves together into continental shields in episodes of mountain-building, called orogenies.

The Earth's original inventory of gases appears to have been lost very early in its history, to be replaced with a secondary atmosphere through volcanic outgassing and perhaps impacts of volatile-rich cometary nuclei and carbonaceous chrondrite meteorites. Extensive volcanic activity and high surface temperatures gradually diminished until the hydrosphere (water cycle) was established and oceans appeared.

Life on Earth existed at least 3.5 billion years ago, as evidenced by microfossils similar to modern cyanobacteria (blue-green algae). With the oceans growing in volume and salinity and the development of oxygen-releasing life-forms, Earth's geochemistry became more complex. By the beginning of the Paleozoic era, about 540 million years ago, the oxygen content of the atmosphere had reached 1 percent of its present level. Life-forms significantly shaped the Earth's chemical composition. Multicelled

Primary Rocks and Minerals in Earth's Crust

Rocks	% Volume of Crust	Minerals	% Volume of Crust
Sedimentary		Quartz	12
Sands	1.7	Alkali feldspar	12
Clays and shales	4.2	Plagioclase	39
Carbonates (including		Micas	5
salt-bearing deposits)	2.0	Amphiboles	5
		Pyroxenes	11
Igneous		Olivines	3
Granites	10.4	Clay minerals (and	
Granodiorites, diorites	11.2	chlorites)	4.6
Syenites	0.4	Calcite (and aragonite)	1.5
Basalts, gabbros,		Dolomite	0.5
amphibolites, eclogites	42.5	Magnetite (and	
Dunites, peridotites	0.2	titanomagnetite)	1.5
		Others (garnets, kyanite,	
Metamorphic		andalusite, sillimanite,	
Gneisses	21.4	apatite, etc.)	4.9
Schists	5.1		
Marbles	0.9	**Totals**	
		Quartz and feldspars	63
Totals		Pyroxene and olivine	14
Sedimentary	7.9	Hydrated silicates	14.6
Igneous	64.7	Carbonates	2.0
Metamorphic	27.4	Others	6.4

Source: Michael H. Carr et al., *The Geology of the Terrestrial Planets*, NASA SP-469, 1984. Data are from A. B. Ronov and A. A. Yaroshevsky, "Chemical Composition of the Earth's Crust," American Geophysical Union Monograph 13.

Typical Composition of Rocks That Compose Much of the Earth's Mantle or Crust

Oxide	Unmelted Peridotite in the Mantle	Basalt Formed at Oceanic Ridges or Rises	Andesite Formed at Subduction Zones	Granite Rock Along Continental Subduction Zones	Continental Rift Basalt	Shale	Sandstone Near the Source	Sandstone Far from the Source	Limestone
SiO_2 (silicon oxide)	45.0	49.0	59.0	65.0	50.0	58.0	67.0	95.0	5.0
TiO_2 (titanium oxide)	0.4	1.8	0.7	0.6	3.0	0.7	0.6	0.2	0.1
Al_2O_3 (aluminum oxide)	8.7	15.0	17.0	16.0	14.0	16.0	14.0	1.0	0.8
Fe_2O_3 (ferric iron oxide)	1.4	2.4	3.0	1.3	2.0	4.0	1.5	0.4	0.2
FeO (ferrous iron oxide)	7.5	8.0	3.3	3.0	11.0	2.5	3.5	0.2	0.3
MnO (manganese oxide)	0.15	0.15	0.13	0.1	0.2	0.1	0.1	—	0.05
MgO (magnesium oxide)	28.0	8.0	3.5	2.0	6.0	2.5	2.0	0.1	8.0
CaO (calcium oxide)	7.0	11.0	6.4	4.0	9.0	3.0	2.5	1.5	43.0
Na_2O (sodium oxide)	0.8	2.6	3.7	3.5	2.8	1.0	2.9	0.1	0.05
K_2O (potassium oxide)	0.04	0.2	1.9	2.3	1.0	3.5	2.0	0.2	0.3
Volatiles (water or carbon dioxide)	1.0	1.0	1.0	2.0	1.0	8.0	2.0	1.0	42.0

Note: Compositions are given as weight percentages of the element oxide in the entire rock.

animals in the oceans scrubbed carbon dioxide from the atmosphere and locked it up in the carbonate rocks, forming biochemically precipitated limestones. By the latter part of Paleozoic era, about 300 million years ago, coal formed as a result of the first land forests being periodically inundated by ocean transgressions.

METHODS OF STUDY

Perhaps no other Earth science is as speculative as that of early Earth history and the geochemical evolution of the Earth. Some of the major challenges confronting Earth scientists are questions about how the Earth's crust formed and when plate tectonic movement began. It is generally accepted by most Earth scientists that heat flow was substantially greater and hence crustal formation occurred more rapidly in Archean times. Despite the problems of extrapolating back to a time when the first solid rocks were forming, the established models are based on some solid lines of evidence.

In 1873, American geologist James D. Dana made one of the initial advances in the study of the Earth's internal chemical composition when he suggested that analogies could be drawn from the study of meteorites. Geochemists studying meteorites today have derived radiometric dates of 4.4 to 4.6 billion years for many of them—corresponding to the initial epoch of condensation and accretion in the solar nebula. Because meteorite types approximate the elemental distribution in the Earth, they are valuable samples of what the Earth formed from.

Geophysicists use seismic waves from earthquakes to study the structure of the Earth's interior. Variations in speed as the waves pass through the Earth, and reflection and refraction of them at internal boundaries, have revealed a differentiated Earth with a very dense metallic core, a less dense rocky mantle, and an even less dense rocky crust "floating" on top. The well-established theory of plate tectonics holds that the crust and upper mantle together form rigid lithospheric plates that are moving, driven by slow convection currents in the mantle.

The drive to study Archean rocks was partly fueled by the United States Apollo missions to the Moon, which returned rocks of comparable age from the lunar surface. Interest in Archean crustal evolution was further aroused by the discovery of Archean lavas called komatiites around greenstone belts (which are agglomerations of Archean basaltic, andesitic, and rhyolitic volcanics, along with their sediments derived by weathering and erosion). Komatiites are ultramafic lavas that formed at temperatures greater than about 1,100° Celsius (1,400 kelvins) and may be fragments of the first crust. Work by field geologists in regions with exposed Archean rocks found successively older granitic rocks—

3.8 billion years in western Greenland, 3.9 billion years in Antarctica, and 4.0 billion years in Canada's Northwest Territories. Even older detrital zircons with radiometric ages between 3.8 and 4.4 billion years were discovered in somewhat younger sedimentary rocks in western Australia. The zircon find is significant because it sets an approximate birth date for early continental crust, as zircon is a reasonably common though minor constituent of granitic igneous rocks. The Australian zircons probably formed in early continental igneous rocks and then were eroded, transported, and deposited with other sediments in the sandstones in which they were found.

Geochemists have refined their study of these ancient rocks with more sophisticated methods to determine isotope ratios in them. Instruments common in geochemical laboratories today use X-ray diffraction and gamma-ray spectral analysis to determine which isotopes are present. Isotope ratios in rocks are of particular interest to geochemists because they provide clues as to chemical cycles in nature. The equilibria of these cycles, as indicated by the isotope ratios, offer insights into volcanic, oceanic, biological, and atmospheric cycles and conditions in the past.

CONTEXT

Perhaps no other area of scientific study is as intriguing and controversial as that of the origin and evolution of the Earth. Geochemists and geophysicists have been at the forefront of the quest to understand the Earth's present geology in terms of its past. Before the 1960's, little was known of the Earth's history during early Precambrian times. This lack is significant when one considers that the Precambrian comprises about eight-ninths of the geologic timescale.

It is likely that improved techniques used to analyze rocks and minerals in the laboratory will continue to provide a better understanding of the formation of the Earth's crustal materials and the evolution of moving lithospheric plates. Radiometric dating and isotope analysis will help unravel the relationships between the greenstone belts and granulite-gneiss associations that typify Archean formations on all continents.

Studying features and materials on other solar system bodies will also lead to a better understanding of the early Earth and its evolution. Similarities and differences in Earth's early history are expected to be revealed by future space probes to the Moon, Mars, Venus, Mercury, and asteroids.

David M. Schlom

FURTHER READING

Fyfe, W. S. *Geochemistry*. Oxford, England: Clarendon Press, 1974. Part of the Oxford Chemistry series, this work was written for lower-division college chemistry students. Although in some respects dated, it is nevertheless a brief (about one-hundred-page) and excellent introduction to the science of geochemistry. Of special interest is chapter 9, "Evolution of the Earth." Bibliography, glossary, index.

Gregor, C. Bryan, et al. *Chemical Cycles in the Evolution of the Earth*. New York: John Wiley & Sons, 1988. A systems approach to geochemistry, this book is suitable for the serious college student. Although filled with graphs, tables, and chemical equations, sections are still accessible to the layperson as well. Discussions of mineralogical, oceanic, atmospheric, and other important chemical cycles are extensive and the work is well referenced.

Hartmann, William K. *Moons and Planets*. 5th ed. Belmont, Calif.: Thomson Brooks/Cole, 2005. A college textbook beyond the introductory level, its approach is based on comparative planetology. Provides much information on condensation and accretion in the solar nebula, as well as the composition of the terrestrial planets generally and Earth in particular.

Kroner, A., G. N. Hanson, and A. M. Goodwin, eds. *Archaean Geochemistry: The Origin and Evolution of the Archaean Continental Crust*. Berlin: Springer, 1984. A collection of reports by the world's leading geochemists studying the geochemistry of the world's oldest rocks. Although many of the articles are technical in nature, the abstracts, introductions, and summaries are accessible to a college-level reader interested in the work of top international scientists.

Levin, Harold L. *The Earth Through Time*. 5th ed. Fort Worth: Saunders College Publishing, 1996. A thorough and readable college text on historical geology. Filled with illustrations, photographs, and figures, this book is also suitable for the layperson. Chapters on planetary beginnings, origin and evolution of the early Earth, and plate tectonics are of special interest. Contains an excellent glossary and index.

Salop, Lazarus J. *Geological Evolution of the Earth During the Precambrian*. New York: Springer-Verlag, 1983. A top Soviet geologist conducts an exhaustive survey of Precambrian geology. Suitable for a college-level reader with a serious interest in the subject. Contains numerous graphs and tables, with extensive references.

Tarbuck, Edward J., and Frederick K. Lutgens. *Earth: An Introduction to Physical Geology*. Illustrated by Dennis Tasa. 9th ed. Upper Saddle River, N.J.: Pearson Prentice Hall, 2008. This college-level textbook for introductory geology courses is well written and illustrated. It has a full chapter on the origin, historical development, and composition of the Earth.

Wedepohl, Karl H. *Geochemistry*. New York: Holt, Rinehart and Winston, 1971. An older but still good and accessible brief introduction to geochemistry fundamentals. Contains an excellent chapter on meteorites and cosmic abundances of the elements. Suitable for the nontechnical reader, with index and references. A good starting point for those unfamiliar with mineral formation.

See also: Comets; Earth-Moon Relations; Earth-Sun Relations; Earth System Science; Earth's Age; Earth's Atmosphere; Earth's Core; Earth's Core-Mantle Boundary; Earth's Crust; Earth's Crust-Mantle Boundary: The Mohorovičić Discontinuity; Earth's Differentiation; Earth's Mantle; Earth's Oceans; Earth's Origin; Earth's Structure; Lunar History; Lunar Interior; Planetary Interiors; Planetary Tectonics; Planetology: Venus, Earth, and Mars; Solar System: Element Distribution; Terrestrial Planets.

Earth's Core

Category: Earth

The core is the Earth's densest, hottest region. The thermal energy released by the core's continuous cooling stirs the overlying mantle into slow, convective motions that drive plate tectonics and hence are ultimately responsible for moving continents, building mountains, fueling volcanoes, and producing earthquakes.

OVERVIEW

The Earth's core extends from a depth of about 2,900 kilometers down to the center of the Earth, 6,378 kilometers below the surface at the equator. The outer part of the core is molten, while the central, inner part is solid. Ambient pressures inside the core range from about 1.4 million to 3.6 million atmospheres, temperatures range from about 4,300 to 5,800 kelvins, and densities range from about 10 to 13 grams per cubic centimeter. Being about twice as dense as the rest of the planet, the core contains one-third of the Earth's mass but occupies a mere one-sixth of its volume.

Surrounding the core is the mantle. The boundary between the solid mantle and the underlying molten outer core is the core-mantle boundary (CMB), a surface that demarcates the most fundamental compositional discontinuity in the Earth's interior. Below it, the core is mostly made of iron-nickel alloys. Above it, and all the way to the surface, the mantle and overlying crust are made mostly of silicates (the most abundant group of rock-forming minerals). The core has lower seismic-wave transmission speeds and higher densities than the mantle, a consequence of it having a different chemical composition. The molten outer core probably contains about 80 to 90 percent (by mass) iron-nickel alloys with a 20 to 10 percent mix of sulfur, silicon, oxygen, and maybe even hydrogen. The solid inner core contains less of each of the lighter elements and may be almost entirely iron-nickel alloys that solidified out from the molten outer core. The boundary between the molten outer core and the solid inner core is known as the inner core boundary (ICB);

it appears sharp to seismic waves, which easily reflect off it. The entire core must be a good electrical and thermal conductor because of its metallic composition. The mantle, in contrast, is composed mainly of crystalline silicate minerals of magnesium and iron, and therefore is a good electrical and thermal insulator.

This sharp contrast in physical properties is the end product of the way in which the Earth evolved thermally, gravitationally, and chemically. The Earth formed by the accretion of planetesimals about 4.5 billion years ago along with the rest of the solar system. As the early Earth was slowly heated by radioactivity, the iron in it suddenly melted and sank by gravity toward the center in a cataclysmic "iron catastrophe," forming the core; silicate minerals were left behind to form the mantle and crust. Calculations show that iron sinking to the core must have released great amounts of energy that would have heated and at least partially melted the entire Earth. Cooling of the outer parts proceeded rapidly by convection, but the silicate mantle created a thermal barrier for the iron-rich core, which, not being able to cool down as readily, remained molten. The inner core then began to form at the center, where the pressure was greatest and solidification was (barely) possible.

The most tangible consequence of the existence of an electrically conducting, fluid outer core is the presence of the Earth's magnetic field, which has existed for at least 3.5 billion years with a strength not very different from what it has today. There can be no permanently magnetized substances deep inside the Earth. Magnetic materials lose their magnetism as their temperature increases above the Curie temperature (around 800 kelvins for most magnetic substances), and the interior below a depth of about 30 kilometers is at temperatures well above the Curie point. The process that generates and maintains the geomagnetic field is attributed to a self-exciting dynamo mechanism—that is, an electromagnetic induction process that transforms the motions of the conducting fluid into electric currents, which in turn induce a magnetic field that strengthens the existing field. In order for the system to get started, at least a small magnetic field (perhaps a weak primordial interplanetary field) needs to be present to initiate the generation of electric currents that induce a stronger magnetic field. The increased magnetic field in turn induces stronger currents, which further strengthen the field, and so on. As the magnetic field increases beyond a certain point, it begins to affect the fluid flow through the mechanical Lorentz force, which is induced in a conductor as it moves across a magnetic field. The stronger the magnetic field, the stronger the Lorentz force becomes and the more it tends to modify the motion of the fluid so as to oppose the growth of the magnetic field. The result is a self-regulating mechanism which, over time, attains a steady state.

A source of energy is required to maintain the fluid flow in the molten outer core that produces the electrical currents that induce the magnetic field. One possibility is that the necessary energy to maintain the flow is provided by the growth of the solid inner core as it is fed by the crystallization of iron from the molten outer core. This process could provide enough gravitational energy to keep the core hot throughout, thus driving convection in the molten outer core both thermally and compositionally.

A most extraordinary feature of the core-generated magnetic field is that over the past few hundred million years at least, it has reversed its polarity with irregular frequency, the intervals between reversals varying from less than 100,000 to more than 10 million years. (For convenience, the present orientation of the field is considered normal.) Some igneous rocks and some sedimentary rocks, if they contain iron-bearing mineral grains, can acquire and preserve the magnetism that exists at the time and place where they form. Therefore, rocks formed throughout geologic time have recorded the alternating pattern of normal and reversed Earth magnetism. This sequence of magnetic reversals contains clues to the core's nature.

Geophysicists are eager to determine whether the outer core is vigorously convecting as a consequence of the inner core's growth. If that were the case, convection would be delivering a large flow of heat to the mantle, whose low thermal conductivity would create a barrier to the outward flow of the heat. As a result, the local temperature gradient at the base of the mantle would probably be very high, so that a layer there would be gravitationally unstable. From this layer, thermal inhomogeneities would rise through the mantle in the form of plumes of buoyant, hot, lower-mantle material. Several such plumes might reach the upper mantle or set the entire mantle into convection. These convection currents would be responsible for the motion of the tectonic plates on the Earth's surface and, consequently, for

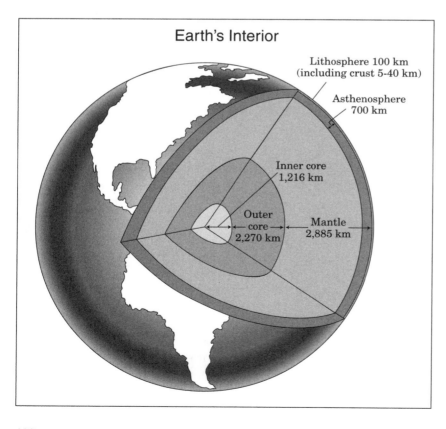

Earth's Interior

Lithosphere 100 km
(including crust 5-40 km)

Asthenosphere
700 km

Inner core
1,216 km

Outer
core
2,270 km

Mantle
2,885 km

the drifting of continents, the uplifting of mountain ranges, the formation of oceanic basins, and the occurrence of volcanic eruptions and earthquakes. These observable effects of plate tectonics thus are linked to the internal cooling of the Earth and the growth of the inner core.

METHODS OF STUDY

Knowledge of the structure, physical properties, and composition of the core is entirely based on indirect evidence gathered mostly from analyzing seismic waves, studying the Earth's gravitational and magnetic fields, and conducting laboratory experiments on the behavior of rocks and minerals at high pressures and temperatures. The first evidence for the existence of the core was presented in a paper suggestively titled "The Constitution of the Interior of the Earth, as Revealed by Earthquakes," published in 1906 by Richard D. Oldham, of the geological survey of India. Thirty years later, Inge Lehmann, from the Copenhagen seismological observatory, presented seismic evidence for the existence of the inner core. In the past few decades, with the advent of high-speed computers and technological advances in seismometry, seismologists have developed increasingly sensitive instrumentation to record seismic waves worldwide and sophisticated mathematical techniques that allow them to construct models of the core that explain the observed data.

Seismic waves provide the most important data about the core. Earthquakes and explosions generate elastic body waves that propagate throughout the Earth. These seismic waves penetrate deep into the Earth and, after being reflected or transmitted, travel back to the surface to be recorded at seismic stations around the world. The most direct information that seismic or elastic waves carry is their travel time. Knowing the time it takes for elastic waves to traverse some region of the Earth's interior allows the calculation of their speed of propagation in that region. The speed of seismic waves strongly depends on the density and rigidity of the material through which they propagate, so estimates of the mechanical properties of the Earth's interior can be derived from seismic travel time analyses. Seismic waves that propagate through the interior of the Earth are

of two types: compressional waves (also called primary or P waves) and transverse shear waves (also called secondary or S waves). Compressional waves produce volume changes in the medium they travel through; shear waves produce shape distortion without volume change. If the medium has some rigidity (that is, if it is solid), both P and S waves can propagate through it. If the medium has no rigidity (that is, if it is liquid or gaseous), it offers no resistance to a change in shape, so S waves cannot propagate through it, although P waves can.

S waves are not transmitted through the outer core. Therefore, the outer core material has no rigidity, but behaves as a fluid would. Similar observations suggest that the inner core is solid. However, the actual rigidity of the inner core is very difficult to estimate since shear waves inside the inner core are isolated from the mantle by the outer core and can only travel through it as P waves converted from S waves at the inner core boundary. Nevertheless, when the whole Earth is set into vibration by a very large earthquake, the average rigidity of the inner core can be estimated by comparing the observed frequencies of oscillation with those theoretically computed for models of the Earth that include a solid inner core. Model studies have indicated that the inner core is indeed solid, because a totally liquid core model does not satisfy the observations.

The average speed of P waves through the entire Earth is about 10 kilometers per second, whereas the P-wave speed in the rocks accessible to measurement at the Earth's surface is about 4 to 5 kilometers per second. The speed of S waves is a little more than half that of P waves in solids and is zero in perfect fluids. The P-wave speed drops abruptly across the core-mantle boundary, from 13.7 kilometers per second at the base of the mantle to 8.06 kilometers per second at the top of the core. From this point down, the speed steadily increases to 10.35 kilometers per second at the inner core boundary, where it jumps discontinuously to 11.03 kilometers per second at the top of the inner core. From there to the center of the Earth, the P-wave speed increases slowly to reach 11.3 kilometers per second. The S-wave speed increases from zero at the inner core boundary to around 3.6 kilometers per second at the Earth's center. The core's density abruptly increases from 5,500 kilograms per cubic meter at the base of the mantle to nearly 10,000 kilograms per cubic meter at the top of the core. From there, the density increases slowly to nearly 13,100 kilograms per cubic meter at the Earth's center. For comparison, the density of mercury at room temperature and ambient pressure is 13,600 kilograms per cubic meter.

The theory that the core is mostly iron is consistent with iron being cosmically more abundant than other heavy elements and with the high electrical conductivity the core needs in order to generate the Earth's magnetic field. The fluidity of the outer core has been demonstrated by measurements not only of seismic wave transmission but also of the oscillation period of gravitational waves in the core excited by the lunisolar tides. The existence of a sustained magnetic field is also consistent with a fluid outer core.

Seismic data can probe the inner core only partially from the Earth's surface, unless the source of the seismic waves and the receivers are located antipodally to each other. Such an arrangement allows scientists to measure seismic waves that have penetrated the center of the Earth.

New views of the Earth's interior are produced, sometimes unexpectedly, by the analyses of data collected by satellite missions. Data from orbiting satellites that measure tiny variations of the Earth's gravitational field, combined with computer-aided seismic tomography of the Earth's interior, have revealed large-density anomalies at the base of the mantle and a large relief of more than 2 kilometers along the core-mantle boundary.

Seismic tomography uses earthquake-generated waves that penetrate the interior in a multitude of directions to map its three-dimensional structure, just as computerized medical tomography uses multiple X-ray images to create a three-dimensional view of internal organs of the body. Essential to the success of these studies is installation of dense networks of seismic sensors all over the surface of the Earth; this installation, however, would be very expensive.

CONTEXT

Any study of the Earth provides insight into the nature and future of the planet and, consequently, the future of humankind. A more complete understanding of how the Earth's core works could result in predictions of the geomagnetic field's activity for years to come, including perhaps an upcoming magnetic reversal. According to the best estimates, a reversal does not occur instantaneously, but takes at least a few thousand years. That means that during a reversal, there is a time interval of very weak or even zero field intensity. Under such conditions, the magnetic shielding that prevents highly energetic charged particles of the solar wind from reaching the Earth's surface would disappear, leaving the surface directly exposed to intense particle radiation that could directly be potentially lethal, or that could at least lead to increased cancers and genetic mutations.

The inner core has not yet been thoroughly explored with seismic waves. One reason is that it is the most remote region of the Earth and therefore the most difficult to study; another is that it is hidden beneath the "seismic noise" created by the crust, mantle, and outer core. The inner core, however, holds the key to the understanding of the Earth's early history and its subsequent development as a planet. It would be possible in principle to investigate the inner core in more detail by deploying an array of highly sensitive seismic sensors antipodal to a seismically active region. However, despite the wealth of unique data that could be obtained from such an experiment, it would be a very expensive endeavor.

J. A. Rial

FURTHER READING

Bolt, Bruce A. *Inside the Earth: Evidence from Earthquakes*. San Francisco: W. H. Freeman, 1982. An elementary treatment of what is known about the Earth's interior, mostly through the study of seismic waves, the author's major field of research. The book contains abundant diagrams that illustrate important results of the investigation of the core and mantle. For readers with some knowledge of mathematics, the book includes brief derivations of important formulas, separated by "boxes" from the main text. Includes anecdotal descriptions of great scientific discoveries along with personal views of the history and development of seismology.

Fowler, C. M. R. *The Solid Earth: An Introduction to Global Geophysics*. 2d ed. New York: Cambridge University Press, 2004. An updated version of a widely used textbook for introductory geophysics courses. Designed for students with some knowledge of physics and calculus.

Jacobs, J. A. *The Earth's Core*. 2d ed. New York: Academic Press, 1987. A highly technical text, but perhaps the best reference for a detailed description of the most accepted core models. The tables—which give the numerical values of the density, temperature, rigidity, and wave velocity distributions within the Earth—are of interest to anyone wanting a quantitative description of the core. A long list of research articles is included.

Tarbuck, Edward J., and Frederick K. Lutgens. *Earth: An Introduction to Physical Geology*. Illustrated by Dennis Tasa. 9th ed. Upper Saddle River, N.J.: Pearson Prentice Hall, 2008. This college-level textbook for introductory geology courses is well written and illustrated. There is a very good chapter on the Earth's interior.

Van der Pluijm, Ben, and Stephen Marshak. *Earth's Structure*. 2d ed. New York: W. W. Norton, 2003. An introductory text on structural geology and tectonics. Designed for undergraduate students.

Vogel, Shawna. *Naked Earth: The New Geophysics*. New York: Plume, 1996. Covers geophysics research and theories since 1960. Includes information about Pangaea, the supercontinent cycle, and the reversals of the Earth's magnetic field. For general audiences.

See also: Earth-Moon Relations; Earth-Sun Relations; Earth's Age; Earth's Composition; Earth's Crust; Earth's Crust-Mantle Boundary: The Mohorovičić Discontinuity; Earth's Differentiation; Earth's Mantle; Earth's Oceans; Earth's Origin; Earth's Structure; Planetary Interiors; Planetology: Venus, Earth, and Mars; Terrestrial Planets.

Earth's Core-Mantle Boundary

Category: Earth

The core-mantle boundary is a pronounced discontinuity separating the outer core from the mantle of the Earth. It is a chemical and mineralogical as well as a thermal and mechanical boundary. The topography of the core-mantle boundary is believed to be controlled by the dynamic processes in the mantle and the outer core.

OVERVIEW

The core-mantle boundary (CMB) is a prominent discontinuity within the Earth. It is located at a radius of about 3,500 kilometers and a depth below the surface of about 2,900 kilometers. The mantle above the boundary is largely solid, of relatively low temperature, and primarily composed of silicate minerals rich in magnesium and iron. The outer core below the boundary is liquid, of higher temperature, and composed of dense materials such as iron-nickel oxides and iron-nickel sulfide alloys. This boundary separates two dynamic systems: one operating in the mantle as hot spots and convection cells, the other in the outer core consisting of convection currents and eddies of the core fluid. The motions of the core fluid appear to be responsible for the Earth's magnetic field.

The lowermost part of the mantle, labeled the D″ (pronounced "dee double prime") layer, is called the core-mantle transition zone. It is approximately 200-300 kilometers thick and is located just above the CMB. Seismic waves from earthquakes and explosives detonated at or near the surface show significant speed variations within the D″ layer over lateral or horizontal distances of 1,000 kilometers or more. Longitudinal (or compressional) P waves travel faster in the portions of this layer that are located below North America, China, the eastern part of the Indian Ocean, and off the Pacific coast of Chile; P waves travel more slowly in the D″ layer below the southern part of Africa, the New Hebrides Islands, the South Pacific Ocean, and the Argentine Basin. Similar variations in speed have also been observed for transverse (or shear) S waves, which travel faster in this layer under the American continents, Asia, the northern Indian and Pacific oceans, and Antarctica, and more slowly under the Central and South Pacific Ocean, the Atlantic Ocean, most of Africa, and the southern part of the Indian Ocean. These speed variations in the D″ layer appear to continue upward in the mantle and may be related to the thermally induced convection currents and hot spots in the mantle.

Improvements in instrumentation have enabled scientists to simulate in the laboratory the physical and chemical conditions of the lower mantle and the outermost core. The lower mantle is thought to be composed primarily of magnesium-iron silicates with the compact perovskite crystal structure. Some aluminum-calcium silicates and magnesium-iron oxides may also be present, but their relative abundance is not known. Laboratory measurement of the melting point of perovskite lead to estimates of the temperature of the D″ layer varying from 3,300 to 4,300 kelvins. Similar studies of outer core materials, which are primarily iron-nickel sulfides and iron-nickel oxides, indicate that the temperature of the outermost core is at least 4,300 kelvins. Thus, the temperature increases by about 1,000 kelvins through the D″ layer, resulting in partial melting of some minerals, thereby making the zone soft with anomalous characteristics.

Seismologists studying the core-mantle boundary (CMB) by means of reflected waves from the core have long been frustrated by the strong scatter of the reflected amplitudes. A major part of this scatter is believed to be the result of vertical undulations of the core-mantle boundary. The lateral extent of these undulations is of the order of thousands of kilometers. The elevation of the boundary may change by as much as 5 to 8 kilometers above or below its normal depth. Topographic highs of the CMB have been observed beneath the Indian Ocean, the Pacific Ocean, and the Atlantic Ocean (particularly in the North Atlantic). The CMB is depressed below the Tonga-Karmadec islands, the China-Japan region, Central Africa, and off the west coast of South America. Because most of these areas are associated with the subduction

of oceanic plates, the structure of the CMB is thought to be caused by the dynamic processes in the mantle, which may be related to the convection processes in the outer core. Subduction of a lithospheric plate is associated with downwelling convective flow in the mantle. When the convective flow reaches the core boundary, it depresses the CMB into the hot, liquid core. Core fluids may partially invade the "topographic low" of the CMB, altering the chemical composition of the D″ layer. Similarly, beneath an upwelling zone of mantle flow, liquid core material may be "sucked" up into the mantle, creating a topographic high of the CMB. At the topographic "lows" of the CMB, mantle material is subjected to the higher temperatures of the outer core and may melt; it then recrystallizes at the topographic "highs" of the CMB. Thus, the overall effect is to smooth the CMB, which is continually disturbed by the convective circulations in the mantle. With heat dissipation in the mantle, the outer core slowly cools and the core materials crystallize and underplate the mantle. Therefore the outer core slowly shrinks with time, and the CMB gets deeper.

The D″ core-mantle transition zone and core-mantle boundary are not only a compositional boundary but also a thermal boundary, where the temperature increases by at least 1000 kelvins. Thermal coupling between the mantle and the outer core, however, may change laterally, resulting in a variable heat flow across the boundary. Although no consensus has been reached among scientists, it is possible that the mantle dynamics are at least partially responsible for controlling the heat flow.

The Earth behaves like a large magnet. The magnetic field of the Earth—the geomagnetic field—undergoes changes known as secular variations. The origin of the geomagnetic field appears to be related to motions of the outer core fluid. Studies suggest that the deep mantle and the outer core play a significant role in shaping the secular variations. Upwellings in the outer core material are associated with the hotter and seismically slower regions of the D″ zone; downwellings are associated with the colder and seismically faster regions in D″. Cold regions in the mantle transmit greater amounts of heat from the outer core, thereby setting up

Earth's Discontinuities

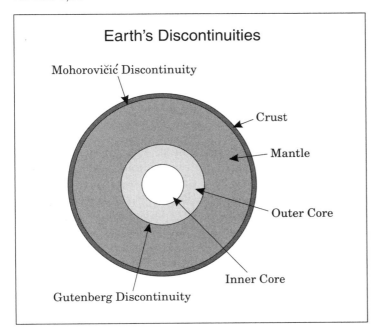

mantle circulations. The topography of the CMB is controlled by the circulations in the mantle as well as in the outer core. The topographic relief of the CMB may also set up a lateral temperature gradient which may be responsible for the secular variation of the magnetic field.

METHODS OF STUDY

Various subdisciplines of geophysics contribute to investigating the nature and the structure of the core-mantle boundary (CMB) and the D″ core-mantle transition layer. They include, among others, seismology, geodesy, geodynamics, high-temperature and high-pressure mineral physics, geothermometry, and geomagnetism. Seismology has been the most important among all these subdisciplines and has contributed most of the information about the Earth's interior.

Seismology deals with earthquakes and the propagation of earthquake waves through the Earth. Whenever an earthquake occurs, different types of seismic waves are generated. Surface waves travel along the Earth's surface, and longitudinal (or compressional) P waves and transverse (or shear) S waves travel through the interior of the Earth. It is often helpful to vi-

sualize the direction of travel of P and S waves as rays originating from an earthquake focus, or hypocenter, and radiating in all directions through the Earth. Because of the increased rigidity and incompressibility of rocks downward, the speeds of these waves increase with depth. As a result, the downgoing seismic rays (except for vertical or near-vertical rays) are curved back toward the surface. The seismographic stations that are farther away from the epicenter record seismic waves that penetrate through the deeper layers in the Earth.

The outer core has no rigidity, since it is liquid. Consequently, P waves slow down abruptly as they cross the CMB, from 13.5 to 8.5 kilometers per second, and they are sharply refracted or bent. S waves do not propagate through liquids, so they stop at the CMB. As a result, there are shadow zones beyond about 11,000 kilometers from the epicenter where neither P nor S direct waves are detected at the ground surface. The presence of liquid outer core was discovered through the existence of the shadow zones and the absence of core-transmitted S waves.

Seismic waves emerging at steep angles from the hypocenter encounter the CMB. Part of the incident energy is reflected back from the boundary, and the rest is refracted through the outer core. P waves can be reflected back as P and as S waves, designated as PcP and PcS waves (or phases) respectively. Similarly, S waves reflected back from the CMB as P and S waves are designated as ScP or ScS waves. These core-reflected phases have been important in the study of the nature, shape, and depth of the CMB. Because S waves cannot travel through liquids, the refracted energy in the outer core propagates in the form of P waves. These refracted P waves are designated as K phases. Thus PKP is a wave that travels from the hypocenter in the mantle as a P wave, propagates as a K (that is, P) in the outer core, and reemerges in the mantle as a P wave. Similarly, SKS and other combinations, such as PKS and

SKP, are often observed in the seismic records. A joint study of the core-reflected phases (for example, PcP) and the core-refracted phases (for example, PKP) is often important in resolving the depths and topography of the CMB. Seismic rays incident at a large angle on the CMB are diffracted. Study of these diffracted waves provides important information on the D″ zone above the CMB. Using the waveform modeling techniques, scientists are determining the thickness and fine structures of the D″ zone.

Another important tool is seismic tomography. It utilizes the same principle used in computed tomography (CT) scan X rays of humans. In a CT scan, the X-ray source and imager are rotated around the body and a large number of X-ray images are recorded. A computer processes these images and forms a three-dimensional image of the internal organs of the subject. The seismological data collected worldwide can similarly be processed to form a three-dimensional image of the Earth's interior. Seismic tomography provides valuable information on the CMB as well as the Earth's mantle.

A large earthquake sets the Earth vibrating like a bell. If the Earth were perfectly spherical with uniform layering, it would produce a pure tone, vibrating at a preferred frequency. Departures from the spherical shape of the Earth, as well as depth-related discontinuities, produce additional tones involving distortions of the Earth. Thus, recordings of these various modes of the Earth's vibrations, known as free oscillations, can furnish information about the shape of the CMB.

Satellite measurements of the Earth's gravity field and the geodetic observations of the geoid can also provide information on the CMB. Theoretical models of the Earth's interior, particularly the mantle, the D″ zone, and the CMB, can be constructed to match observed geoidal undulations and gravity anomalies. It appears that a 2- to 3-kilometer variation in elevation of the CMB can explain 90 percent of the observed large-scale gravity anomalies. Astronomic observations of the Earth's wobble also furnish additional constraints on the shape of the CMB. The Earth has an equatorial bulge caused by its rotation. The Moon pulls at the bulge and attempts to align it along the orbital plane of the Moon, generating a wobble, or a nutational motion, of the Earth's axis. (This motion is similar to the wobble of a spinning top or a gyroscope.) Deformation of the CMB produces certain irregularities in the nutational motion. Studies of these irregularities indicate that the undulations of the CMB are less than 1 kilometer in height.

Major developments in instrumentation have made it possible to simulate in the laboratory the temperature and pressure conditions of the deep mantle. Scientists can now study how the crystal structures of minerals change with increased temperature and pressure. Measurements of the electrical properties of rocks under high pressure, and possible alloying of iron by sulfur and oxygen that may occur in the outer core, are also being studied. These investigations are important for complete understanding of the mineral compositions, structure, temperature, and pressure environment of the Earth's deep interior.

CONTEXT

Scientists from various geophysical subdisciplines have made a concerted effort to investigate the structure and nature of the CMB and the deep interior of the Earth because it is important from several perspectives. The CMB is believed to be associated with deep mantle plumes, the mantle convection currents that drive the lithospheric plates and may be responsible for secular variations of the geomagnetic field. As the most pronounced discontinuity within the Earth, the undulations at the CMB may also cause regional gravity anomalies and can affect the transmission of seismic waves. The transmission effects of seismic waves crossing the CMB provide information about the geometry and the physical and chemical properties of materials at the CMB, as well as in the mantle above and the outer and inner core below. Furthermore, because core-reflected phases travel along vertical or near-vertical paths in the mantle, they are often utilized to study heterogeneity and seismic behavior in the mantle. Knowledge of the nature of the CMB is necessary to determine these mantle characteristics.

A committee on Studies of the Earth's Deep Interior (SEDI), under the joint auspices of the

International Union of Geodesy and Geophysics (IUGG) and the American Geophysical Union (AGU), facilitates international exchange of scientific information about the Earth's interior. In addition, special sessions on the Earth's deep interior are held at most AGU meetings.

D. K. Chowdhury

FURTHER READING

Bolt, Bruce A. *Earthquakes*. New York: W. H. Freeman, 1988. This volume presents information on the Earth's interior obtained from seismological studies. Suitable for high school and college levels.

_____. *Inside the Earth*. San Francisco: W. H. Freeman, 1982. A good introduction to seismology for the nonscientist, this well-illustrated, concise book summarizes the seismological methods and the results.

Fowler, C. M. R. *The Solid Earth: An Introduction to Global Geophysics*. 2d ed. New York: Cambridge University Press, 2004. An updated version of a widely used textbook for introductory geophysics courses. Designed for students with some knowledge of physics and calculus.

Tarbuck, Edward J., and Frederick K. Lutgens. *Earth: An Introduction to Physical Geology*. Illustrated by Dennis Tasa. 9th ed. Upper Saddle River, N.J.: Pearson Prentice Hall, 2008. This college-level textbook for introductory geology courses is well written and illustrated. There is a very good chapter on the Earth's interior, that includes sections on the CMB and the D″ layer.

Vogel, Shawna. *Naked Earth: The New Geophysics*. New York: Plume, 1996. Covers geophysics research and theories since 1960. Includes information about Pangaea, the supercontinent cycle, and the reversals of the Earth's magnetic field. For general audiences.

See also: Auroras; Earth-Moon Relations; Earth-Sun Relations; Earth System Science; Earth's Age; Earth's Atmosphere; Earth's Composition; Earth's Core; Earth's Crust; Earth's Crust-Mantle Boundary: The Mohorovičić Discontinuity; Earth's Differentiation; Earth's Magnetic Field: Origins; Earth's Magnetic Field: Secular Variation; Earth's Magnetic Field at Present; Earth's Magnetosphere; Earth's Mantle; Earth's Oceans; Earth's Origin; Earth's Rotation; Earth's Shape; Earth's Structure; Eclipses; Greenhouse Effect; Van Allen Radiation Belts.

Earth's Crust

Category: Earth

The crust is the outermost layer of the Earth. The dynamic changes involved in the creation and destruction of crustal rock also fuel volcanoes, cause earthquakes, concentrate mineral deposits, and liberate gases and water that form the atmosphere and ocean.

OVERVIEW

The crust of the Earth is the outermost layer of the Earth. It is distinct from the region of rock lying beneath it, called the mantle, in that the rocks that comprise the crust have different compositions and a lower density. Continental crust is composed mostly of granitic rocks with densities around 2.7 grams per cubic centimeter, and oceanic crust is composed mostly of basaltic rocks with densities around 3.0 grams per cubic centimeter. In contrast, rocks from the upper mantle have densities around 3.3 grams per cubic centimeter, and probably are composed mostly of peridotite.

The rocks of the Earth's crust are quite varied. They can be classified as belonging to one of three broad groups, depending on how they formed: igneous, sedimentary, and metamorphic. These three groups are parts of what is referred to as the rock cycle, which depicts the way in which rocks from each of these groups can provide the raw material to form rocks in any other group.

Igneous rocks are formed by cooling and crystallization from molten material called lava (if on the surface) or magma (if below the surface). Igneous rocks that cool and harden on the surface are said to be extrusive, and those that cool and harden below the surface are said to be intrusive. Intrusive igneous rocks cool more slowly and thus usually contain larger mineral crystals (large enough to be seen with the unaided eye). Extrusive igneous rocks cool more rapidly and thus either contain smaller, microscopic mineral crystals or are glassy, containing no crystals at all. Some common igneous rocks are granite and gabbro (intrusive) and rhyolite, basalt, and obsidian (extrusive).

Sedimentary rocks are formed from sedi-

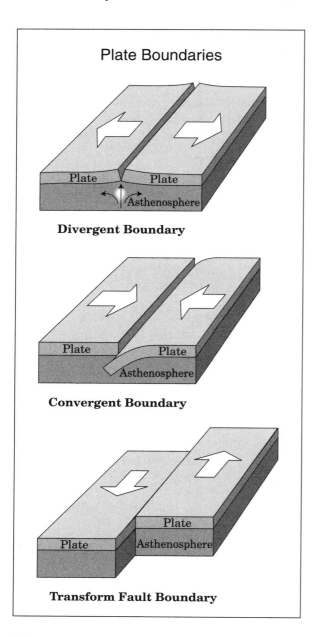

Plate Boundaries

Divergent Boundary

Convergent Boundary

Transform Fault Boundary

ment, the remnants of other rocks that were weathered and eroded when exposed at the surface. Weathering processes may be chemical (such as dissolving in water or acid, or being oxidized by oxygen from the atmosphere) or physical (such as being broken apart when water in cracks freezes and expands, or roots grow into cracks). Sediment can be transported by various agents (such as moving water or blowing wind) and is ultimately deposited in layers. The sediment may then be compacted and cemented, forming sedimentary rock. Some common sedimentary rocks include sandstone, shale, and limestone.

Metamorphic rocks are formed from other rocks that have been subjected to pressures and temperatures high enough to change the crystalline structure of the minerals in the rock (but not high enough to melt the rock). Such changes often occur in the deeper parts of the crust, where the temperature and pressure are greater (around 400 to 1,200° Celsius, and 3,000 to 15,000 atmospheres). Some common metamorphic rocks are slate, schist, gneiss, and marble.

The boundary between the crust and the mantle is known as the Mohorovičić discontinuity or simply the Moho. The depth of the Moho, and thus the thickness of the crust, varies widely. The crust is thickest under the continents, averaging about 40 kilometers and reaching a maximum of 70 kilometers beneath young high mountain chains such as the Himalayas. Under the ocean basins, the crustal thickness averages only about 7 kilometers.

The crust and uppermost mantle (down to a depth of about 100 kilometers) constitute the rigid lithosphere, which is divided into a number of separate blocks called plates. These plates are slowly moving, driven by slow convective motions in the soft, plastic part of the mantle called the asthenosphere, which is directly beneath the lithosphere.

The crust is constantly being created, deformed, and destroyed by processes at plate boundaries. There are three types of boundaries between plates: divergent, where plates are pulled apart; convergent, where plates collide; and transform or side-slip, where plates slide horizontally past each other. Divergent plate boundaries, where plates move apart, are

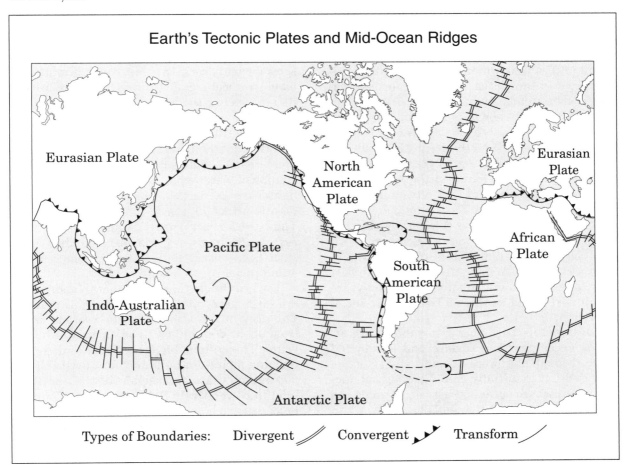

Earth's Tectonic Plates and Mid-Ocean Ridges

Eurasian Plate

North American Plate

Eurasian Plate

African Plate

Pacific Plate

South American Plate

Indo-Australian Plate

Antarctic Plate

Types of Boundaries: Divergent   Convergent   Transform

marked by the ocean ridge-rift system where new ocean basin crust is produced by hot upwelling magma. The Mid-Atlantic Ridge (located in the Atlantic Ocean about midway between North and South America to the west and Europe and Africa to the east) and the Red Sea (between Africa and Arabia) are examples of divergent boundaries. Convergent plate boundaries, where plates come together, can be divided into three subtypes, depending on the nature of the plate (oceanic or continental) on each side of the boundary: oceanic-oceanic, oceanic-continental, and continental-continental. At oceanic-oceanic and oceanic-continental convergent boundaries, subduction occurs. Subduction is a process in which an oceanic plate descends back down into the mantle. Subduction zones are marked by deep-sea trenches that are located where oceanic plates bend down into

the mantle, there to be heated and remelted, fueling chains of volcanoes parallel to the trench. The Mariana Trench in the western Pacific and the Peru-Chile Trench along the west coast of South America are examples of oceanic-oceanic and oceanic-continental convergent boundaries respectively. At continental-continental convergent boundaries, neither plate subducts, but instead crumple and buckle along the collision zone forming high mountain ranges like the Himalayas between India and the rest of Asia. At transform or side-slip plate boundaries, two plates slide horizontally past each other. The San Andreas fault in California is a classic example of a transform plate boundary.

New ocean basin crust forms at the ocean ridge-rift system, which is located along divergent plate boundaries. As the existing lithospheric plates are pulled apart, probably by very slow lateral flow in the underlying asthenosphere, upwelling magma fills the gap. It cools and hardens, producing the igneous rocks basalt and gabbro, which add new bands of oceanic crust to the plates. The newly formed oceanic crust spreads away from the ocean ridge-rift in a process known as seafloor spreading. The age of oceanic crust increases systematically with distance from the ocean ridge-rift system. The oldest known seafloor crust, about 180 million years old, is found in the western Pacific. Apparently in that sort of time frame or less, oceanic crust reaches a subduction zone and is recycled back into the mantle. In contrast, many continental rocks are several hundred million years old, and the oldest continental rocks found so far, from the Northwest Territories of Canada, are about 4 billion years old, almost as old as the 4.5 billion-year age of the Earth.

Continental crust does not subduct. It remains on or near the surface and is recycled there through the processes and stages of the rock cycle. It is initially created along subduction zones where oceanic crust is consumed. As the lithospheric plate descends back down into the mantle, it is heated. Eventually, differential melting occurs; the minerals with lower melting temperatures melt and rise as molten blobs of magma, either cooling and hardening below the surface as igneous intrusions or erupting at the surface as volcanoes. These igneous intrusions and volcanoes are composed of granite, rhyolite, and similar igneous rocks that are less dense than the basalts and gabbros of oceanic crust. They form continental crust, which, because of its lower density, does not subduct but remains at the surface. When two continental plates collide, they crumple and weld themselves together, forming a high mountain range with roots that extend downward, increasing crustal thickness there. Thus, continents grow with time by two processes occurring along their edges: volcanism near subduction zones and accretion. Continents also can be broken apart when rifts develop in them, like the East African Rift. If the rift continues to grow, eventually it becomes a long narrow arm of the ocean, called a linear sea, like the Red Sea between Africa and Arabia.

The crust is thickest under young mountain belts, called orogenic belts, piling upward and sinking downward simultaneously to form a thick wedge of rock. In this sense, it is much like a buoyant iceberg, floating with the majority of its mass below the water. The buoyancy of the less dense crustal rocks floating on the more dense mantle rocks is known as isostasy or flotational equilibrium. Just as the iceberg must reach a flotational level by displacing a volume of water equal to its mass, so must the lighter crustal rocks displace a volume of denser mantle rocks to reach their buoyancy level. Thus the thickest crust is found under the highest mountainous regions because they have such deep roots, while the thinnest crust is found under ocean basins.

Toward the center of continental landmasses are large areas of old rocks known as cratons. The ages of rocks found in the cratons range from about 600 million to as much as 4 billion years. The cratons have been free of deformation and mountain-building for at least the last 600 million years. Consequently, their surfaces tend to be relatively flat as a result of surface processes such as weathering and erosion acting on the exposed rocks over geologically long periods of time. Cratons have long and complex histories. Over large areas of them, highly deformed rocks from the deep roots of ancient mountain regions are exposed, and in other regions the ancient mountain roots are covered by more recent layers of sedimentary rocks.

METHODS OF STUDY

Seismic waves created by both earthquakes and artificial explosions are used to probe the interior of the Earth, including its top layer, the crust. One type of seismic wave that travels through the Earth is a longitudinal compressional wave called a P wave (for primary wave); this type of wave is analogous to an acoustic or sound wave, alternately compressing and stretching the material it travels through. The density, rigidity, and compressibility of the material determine the wave speed in it. P waves travel at speeds between about 2 and 6 kilometers per second near the surface, because of the wide range of compositions of surface rocks as well as the presence of open space and fluids

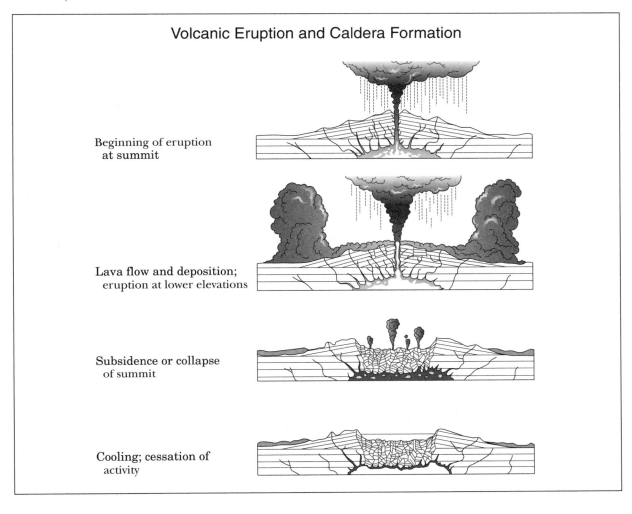

Volcanic Eruption and Caldera Formation

Beginning of eruption
at summit

Lava flow and deposition;
eruption at lower elevations

Subsidence or collapse
of summit

Cooling; cessation of
activity

within them. (For comparison, sound waves travel at about 0.3 kilometers per second through the lower atmosphere.) P-wave speed reaches about 6 to 7 kilometers per second in the lower crust just above the Moho. It has been found in the laboratory that metamorphic rocks known as granulites, which can form when basalts and gabbros are subjected to the pressures and temperatures of the lower crust, have P-wave speeds in the proper range. Furthermore, granulites are similar to some rock samples brought to the surface in volcanic pipes that are thought to have originated in the lower crust.

The thickness of continental crust has been determined by the study of seismic waves that are reflected off the Moho, as well as those that

refracted by it. The increase in density and the resulting increase in speed of seismic waves when they cross the Moho from crust to mantle cause their wave path to bend or refract. Waves that cross the Moho at what is termed the critical angle will travel along right beneath the Moho and be refracted back to the surface at the same angle. By Snell's law of refraction, the sine of the critical angle equals the ratio between the crust and mantle speeds. The travel time of these refracted seismic waves from their source till they are recorded on a seismograph is determined by the crust and mantle speeds, and the thickness of the crust. Using critically refracted P waves, thicknesses have been estimated for much of the crust. It is possible to check the crustal thickness determined from critically refracted waves by using reflected waves. This has been applied with particular success in continental areas with artificial acoustic wave sources such as explosives and vibrator trucks.

CONTEXT

The Earth's crust is in a state of dynamic evolution, with rock materials being created, deformed, and destroyed. This dynamic evolution of the crust is brought about by the movement of large lithospheric plates composed of the crust and upper mantle, up to 100 kilometers thick. Processes at plate boundaries can produce volcanism and earthquakes.

Volcanic activity over the billions of years of the Earth's existence has provided the water vapor and other gases necessary to form the oceans and atmosphere by the release of gases trapped in lavas that reach the surface. An understanding of volcanoes, including how, why, and where they occur, requires an understanding of the Earth's crust and crustal dynamics. It is especially important in the areas with active volcanoes to be able to assess the hazards they pose.

Plate motion also produces earthquakes where two plates rub against each other. Usually the strongest occur at transform plate boundaries and at subduction zones along convergent boundaries. Eventually, knowledge of how crustal rocks change and respond before impending earthquakes may allow their prediction.

Exploration for important economic minerals is guided by knowledge about the evolution and composition of the crust. The concentration of valuable metal deposits, such as gold and copper, occurs during volcanic activity at ocean ridge sites where new oceanic crustal rocks are being created. Consequently, exploration efforts for such metallic ores can be directed toward identifying ancient ridge site locations. The formation of continental sedimentary rocks in the Gulf of Mexico traps organic materials that will be turned into oil and natural gas. Looking for similar types of sedimentary rocks in ancient crustal environments aids in the search for petroleum and natural gas.

David S. Brumbaugh,
revised by Richard R. Erickson

FURTHER READING

Bally, A. W. *Seismic Expression of Structural Styles.* Tulsa, Okla.: American Association of Petroleum Geologists, 1983. An excellent visual treatment of the structure and layering primarily of the upper crust throughout the world. Sections into the crust of offshore Scotland and northwest Germany show the Moho. Suitable for a broad audience from general readers to scientific specialists.

Bott, M. H. P. *The Interior of the Earth.* New York: Elsevier, 1982. This book was intended for undergraduate and graduate students of geology and geophysics as well as for other scientists interested in the topic. The plate tectonic framework of the outer part of the Earth is strongly emphasized.

Brown, G. C. *The Inaccessible Earth: An Integrated View to Its Structure and Composition.* 2d ed. New York: Chapman and Hall, 1993. A good general introduction geared toward the undergraduate college student. The primary topics are the internal state and composition of the Earth. Included is background material on seismology and three chapters discussing the Earth's crust.

Fowler, C. M. R. *The Solid Earth: An Introduction to Global Geophysics.* 2d ed. New York: Cambridge University Press, 2004. An updated version of a widely used textbook for introductory geophysics courses. Designed for students with some knowledge of physics and calculus.

Knapp, Ralph E. *Geophysics.* Exeter, England: Pergamon Press, 1995. Thorough treatment of the physics of Earth. For the serious student.

Smith, David G., ed. *The Cambridge Encyclopedia of Earth Sciences.* New York: Cambridge University Press, 1982. This general reference provides an excellent overview of the Earth sciences. Chapter 10 is an extensive discussion of the Earth's crust, including useful illustrations and diagrams. Contains a glossary, an index, and recommendations for further reading.

Tarbuck, Edward J., and Frederick K. Lutgens. *Earth: An Introduction to Physical Geology.* Illustrated by Dennis Tasa. 9th ed. Upper Saddle River, N.J.: Pearson Prentice Hall,

2008. This college-level textbook for introductory geology courses is well written and illustrated. There are very good chapters on the Earth's interior and plate tectonics.

Taylor, Stuart R., and Scott M. McLennan. *The Continental Crust: Its Composition and Evolution*. Boston: Blackwell Scientific, 1985. A text aimed at undergraduate and graduate geology and geophysics students as well as general Earth scientists. It is clearly written and has excellent, well-rounded scientific references.

Vogel, Shawna. *Naked Earth: The New Geophysics*. New York: Plume, 1996. Rigorously covers geophysics research and theories since 1960.

See also: Earth's Age; Earth's Composition; Earth's Core; Earth's Core-Mantle Boundary; Earth's Differentiation; Earth's Mantle; Earth's Oceans; Earth's Origin; Earth's Structure; Terrestrial Planets.

Earth's Crust-Mantle Boundary: The Mohorovičić Discontinuity

Category: Earth

The Mohorovičić discontinuity, or Moho (the boundary between the crust and mantle), was discovered in 1909 through the observation of an abrupt change in the speed of seismic waves traveling below the surface of the Earth. Its discovery was among the first evidence for the now-famous "onion" model of the layers of the earth, but many of its fundamental properties are still not well understood.

OVERVIEW

On October 8, 1909, Central Europe was struck by a large earthquake centered in Croatia, near the village of Papuspsko. Andrija Mohorovičić, director of the Meteorological Observatory at the nearby University of Zagreb, collected data on the quake and its aftershocks.

In the seismic records from stations at intermediate distances from the epicenter of the quake, he identified two sets of seismic P (longitudinal, or compressional) and S (transverse, or shear) waves, one set arriving at the recording stations sooner than the second. He correctly interpreted the first set of seismic waves to arrive as having traveled deeper and hence faster through material that had a higher density (and possibly different composition) than the crust material through which the shallower and slower second set of waves had passed. Just as light waves moving from water or glass into air bend or refract because of a sudden increase in speed, so also these seismic waves changed speed and refracted at some boundary, then traveled just under the boundary between these layers for some distance before being refracted again back into the lower-density layer and traveling through it to the surface. This boundary, now known to divide the crust from the mantle, was later named after its discoverer, but the somewhat unwieldy title Mohorovičić discontinuity commonly is shortened to Moho. Similar seismic boundaries have since been discovered between the mantle and outer core, and between the inner and outer cores.

The thickness of the crust, and hence the depth of the Moho, varies from 3 kilometers at mid-ocean ridges to 70 kilometers under young, high mountain ranges created by continental collisions, such as the Himalayas. It is important to note that the Moho does not mark the bottom of the tectonic plates; each plate is made of the crust and the uppermost mantle welded together to form the rigid lithosphere, which moves over the mantle's plastic layer known as the asthenosphere.

Since its discovery, the Moho has been the subject of intense scrutiny by geophysicists because it is the closest boundary to the Earth's surface and hence the easiest to study. One of the first debates was whether the Moho represented a transition between rock layers of different chemical composition or between rock layers of similar composition but different mineral and crystalline structure (phase changes caused by changes in temperature and pressure). Since the 1960's, the former model has been that favored by most geologists, although

The Mohorovičić Discontinuity

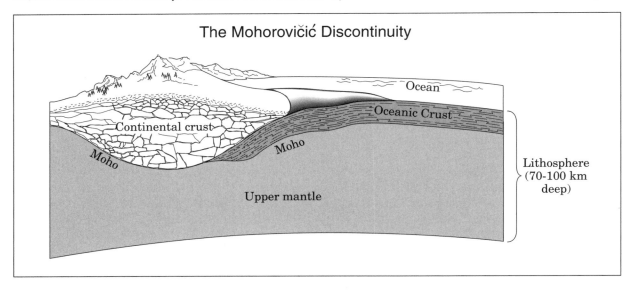

some alternate models based on phase transitions still exist. The crust appears to be composed mainly of basaltic rocks in ocean floors and granitic rocks in continents, while the upper mantle is largely made of peridotite. This model is supported by the study and comparison of continental rocks, ophiolite complexes (which are slices of former ocean floor thrust above sea level), and xenoliths (which are pieces of the lower crust and upper mantle brought up to the surface by volcanic activity).

Worldwide studies of seismic data have added considerably to our understanding of the Moho. Seismic refraction data have been gathered not only from earthquakes but also from nuclear weapons testing and other underground explosions. Seismic reflection methods—which generate seismic waves by means of a near-surface explosion or truck-mounted vibrator—have also been widely used. These waves travel down through the crust, reflect off the Moho, and travel back to the surface, like light waves bouncing off a mirror. The out-and-back travel time can be used to determine the depth of the boundary, similar to the way a bat emits sounds that echo off obstacles in its path. Studies at some locations have also detected a change in the electrical conductivity of rocks near the Moho.

While this broad array of experimental data would suggest that the Moho is well defined, the opposite is actually the case. The problem is that the transition depths determined by these different methods often disagree. Therefore, the classic or seismic Moho (defined as the refraction transition, where the velocity of P waves jumps to 7.6 kilometers per second) may differ from the reflection boundary, the electrical conductivity boundary, and most important, from the rock-type or petrologic boundary, which is usually regarded as the true crust-mantle boundary. Compounding this problem is the fact that the seismic Moho is absent in some locations, with the velocity of the P waves only gradually increasing with depth. Taken together, these studies led to a questioning of the early model of the Moho as a thin, sharp, well-defined boundary in favor of a transition layer of definite thickness (perhaps several kilometers), over which the composition of the rocks changes. The Moho also has a complex, multilayered structure in regions of complex tectonic history. These observations have led geologists to view the Moho as a dynamic entity that evolves over time, in terms of both its petrology and its physical structure and substructures.

KNOWLEDGE GAINED

Although these generalizations about the nature of the Moho are fairly well established, the details of the structure and creation of the Moho are less well understood. Due to the difference in ocean crust and continental crust in terms of composition, thickness, and tectonic interactions, detailed studies of the Moho are usually divided into either oceanic or continental. In addition, the problem of understanding these fundamental processes is so complex that many researchers in this area utilize seismic and petrological data to model the Moho in a single geographic region rather than attempt to make a unified model of the entire Moho.

More is known about the oceanic Moho than its continental counterpart, largely thanks to studies of ophiolite complexes. Distributed worldwide, these samples of former (ancient) oceanic crust and upper mantle allow for direct study of the chemical and mineral composition of the rocks. Coupled with seismic refraction and reflection data, they paint a reasonably clear picture of the oceanic Moho and its history. Its structure is complex, with interlocking layers ranging in composition from mafic (basaltic) at the top to ultramafic at the bottom, finally merging with the peridotite of the mantle. The total thickness of the oceanic Moho layer ranges from 0 to 3 kilometers. Oceanic Moho is created from the same source as oceanic crust, namely the upwelling of magma at mid-ocean ridges such as the Mid-Atlantic Ridge (which is currently increasing the width of the Atlantic

Ocean). It forms rather quickly in geologic terms, within a few thousand years, and after initial creation it is not significantly modified.

The continental Moho is much less well understood. Petrologic information can be gathered from xenoliths and exposed samples of the crust-mantle boundary uplifted by tectonic forces, although the latter must be used with caution, since such rocks have been changed by the same forces that lifted them to the surface. These tectonic forces, such as continental collisions, along with the much longer time frame covered by continental crust samples, lead to a greater complexity in the data. Because of the greater age of the continental rocks, the problem of how the Moho forms in continental regions is central to our understanding of how the early Earth differentiated into layers in the first place.

At least four hypotheses have been suggested for how the continental Moho forms. In the relic Moho model, the continental Moho is the relic of the oceanic Moho, surviving the assembly processes (such as continental collisions and accretion of terranes along the coast) that build the continents. How this might occur without severely disrupting the Moho is unclear, especially given the fact that the crust itself is certainly changed in this process. The magmatic

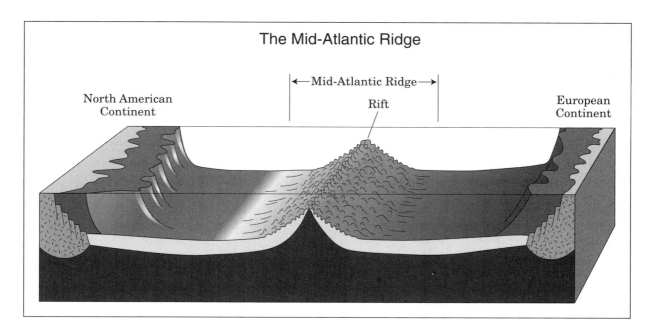

The Mid-Atlantic Ridge

North American Continent

←— Mid-Atlantic Ridge —→

Rift

European Continent

underplating hypothesis suggests that as continents assemble, new Moho material is added beneath them in a process similar to that which creates sills, horizontal intrusions of magma sandwiched between rock layers. In this model, the Moho would be younger than the continental crust, since it is created after the crustal material is set into place.

These first two hypotheses rely on igneous processes, while the final two hypotheses suggest the continental Moho consists of "reworked" rocks. In the metamorphic/metasomatic front model, the Moho is created by additional metamorphism of rocks of the lower crust and/or the upper mantle. Finally, the regional décollement hypothesis posits that the Moho forms as the crust and mantle physically decouple at structurally weak zones, especially under high temperatures.

It should be noted that this list is not suggested as being complete, and all four processes may occur in different geographical areas, depending on the particular geological conditions present. This conclusion has been suggested by researchers working on the Canadian LITHOPROBE program as a result of their study of the subsurface geology of North America. Therefore, it may be that there is no single explanation for the creation and evolution of the continental Moho, and the geologic history of each geographical region must be interpreted to develop a sensible model of its Moho structure and evolution.

CONTEXT

All methods described so far to study the Moho have relied on indirect testing, with the exception of samples of ophiolites and xenoliths. However, as previously mentioned, relying on these samples as truly representative of the Moho in general is unwise, since they may have been severely modified by tectonic forces. The most reliable rock sample would obviously be one obtained directly from the current Moho through drilling. However, given the great depths involved and the resulting technological difficulties, no direct Moho sample has yet been obtained, despite a number of programs designed to do so.

The most infamous such project was dubbed Project Mohole, proposed to the National Science Foundation in 1957 and funded until 1966. This three-stage project completed only its first phase, experimental drilling off the coasts of California and Mexico, before it was canceled by Congress (after a series of bureaucratic and financial problems). Despite the failure of the project, geologists remained committed to large-scale drilling projects in the ocean floor, leading to a series of ongoing projects such as the Deep Sea Drilling Project and Ocean Drilling Project. Although individual projects have reached depths of more than a kilometer, identifiable Moho samples have yet to be obtained, in keeping with the problems of identifying the seismic Moho with the petrological crust-mantle boundary and the complex nature of the Moho in general.

Therefore, nearly a century after the discovery of the Moho, its secrets continue to intrigue geologists. Despite the gaps in our understanding of the terrestrial crust-mantle boundary, planetary geologists are currently applying terrestrial methods and models to its neighbors in space. For example, lunar data suggest that the Moon's Moho lies about 30 to 70 kilometers beneath its surface, being shallowest beneath the maria. Models of Martian structure suggest that its Moho may lie between 6 and 100 kilometers beneath its surface. Such studies add to our understanding of the geologic evolution of these rocky worlds and how their evolution is similar to, yet differs from, that of Earth.

Kristine Larsen

FURTHER READING

Bascom, Willard. *A Hole in the Bottom of the Sea*. Garden City, N.Y.: Doubleday, 1961. An enthusiastic and detailed (but one-sided) popular-level survey of the scientific motivation behind Project Mohole, written by the project's original director.

Cook, Frederick A. "Fine Structure of the Continental Reflection Moho." *Geological Society of America Bulletin* 114, no. 1 (2002): 64-79. This technical review of seismic reflection data focuses on Canada yet contains a valuable overview of the topic and a lengthy bibliography.

Eaton, David W. "Multi-genetic Origin of the Continental Moho: Insights from LITHO-

PROBE." *Terra Nova* 18, no. 1 (2008): 34-43. This technical paper reviews the four hypotheses for the formation of the continental Moho described above. The bibliography is extensive.

Fowler, C. M. R. *The Solid Earth: An Introduction to Global Geophysics*. 2d ed. New York: Cambridge University Press, 2004. An updated version of a widely used textbook for introductory geophysics courses. Designed for students with some knowledge of physics and calculus.

Griffin, W. L., and Suzanne Y. O'Reilly. "Is the Continental Moho the Crust-Mantle Boundary?" *Geology* 15 (1987): 241-244. This seminal technical paper was among the first to succinctly and precisely challenge the assumption that the seismic and geochemical boundaries between the crust and mantle are identical.

Jarchow, Craig M., and George A. Thompson. "The Nature of the Mohorovičić Discontinuity." *Annual Review of Earth and Planetary Sciences* 17 (May, 1989): 475-506. A detailed overview of the general structure of the Moho and the main difficulties with the simplistic standard model.

Tarbuck, Edward J., and Frederick K. Lutgens. *Earth: An Introduction to Physical Geology*. Illustrated by Dennis Tasa. 9th ed. Upper Saddle River, N.J.: Pearson Prentice Hall, 2008. This college-level textbook for introductory geology courses is well written and illustrated. There is a very good chapter on the Earth's interior.

Van der Pluijm, Ben, and Stephen Marshak. *Earth's Structure*. 2d ed. New York: W. W. Norton, 2003. An introductory text on structural geology and tectonics. Designed for undergraduate students.

Vogel, Shawna. *Naked Earth: The New Geophysics*. New York: Plume, 1996. Covers geophysics research and theories since 1960. Includes information about Pangaea, the supercontinent cycle, and the reversals of the Earth's magnetic field. For general audiences.

See also: Earth-Moon Relations; Earth-Sun Relations; Earth System Science; Earth's Age; Earth's Atmosphere; Earth's Composition; Earth's Core; Earth's Core-Mantle Boundary; Earth's Crust; Earth's Differentiation; Earth's Magnetic Field: Origins; Earth's Magnetic Field: Secular Variation; Earth's Magnetic Field at Present; Earth's Magnetosphere; Earth's Mantle; Earth's Oceans; Earth's Origin; Earth's Rotation; Earth's Shape; Earth's Structure; Planetary Interiors.

Earth's Differentiation

Category: Earth

Earth's differentiation refers to the separation of matter in the early Earth into "layers" or zones with different chemical compositions and physical properties, during or shortly after its formation about 4.5 to 4.6 billion years ago. The structure that developed consists of a metallic core surrounded by a rocky mantle overlain by a thin, rocky crust.

OVERVIEW

The Earth consists of nearly spherical concentric layers arranged according to density. The core, composed mainly of metals like iron and nickel, is the central part of the Earth. Its radius is about 3,400 kilometers, and its density ranges between about 10 to 13 grams per cubic centimeter. The crust is the outermost layer at the surface of the Earth. This layer is very thin, its thickness ranging from as little 5 kilometers under ocean basins to as much as 70 kilometers under young, high mountain ranges. The average density of continental crust is about 2.7 grams per cubic centimeter, while ocean basin crust is somewhat denser, at about 3.0 grams per cubic centimeter.

The most common minerals in the crust are the silicates, which have as their basic crystal lattice structure the silicon-oxygen tetrahedron (one silicon atom surrounded three-dimensionally by four evenly spaced oxygen atoms). These tetrahedra combine with atoms of many other elements, especially aluminum, iron, calcium, sodium, potassium, and magnesium, to form the large class of minerals called silicates. Between the crust and the core is the mantle, which is rocky like the crust, but composed of denser silicate minerals with larger percentages of iron and magnesium. The mantle is about 2,900 kilometers thick, and ranges in density from about 3.3 grams per cubic centimeter at its top, just below the crust, to about 5.6 grams per cubic centimeter at its base, just above the core.

Earth and the other planets in the solar system formed about the same time the Sun did, approximately 4.5 to 4.6 billion years ago. Planet formation occurred by condensation of solid dust grains in the solar nebula, followed by accretion of those solid grains to form planetesi-

mals that grew in size through continued accretion. There are two major models of how this condensation and accretion occurred: homogeneous accretion and inhomogeneous or heterogenous accretion.

If cooling of the solar nebula and the resulting condensation occurred faster than accretion, the material swept up in the accretion process would have been well mixed in its composition. The early Earth would have been homogeneous, or undifferentiated, composed of random grains of metal alloys, oxides, silicates, and carbonaceous materials throughout. During or shortly after accretion, the early Earth became hot enough to at least partially melt.

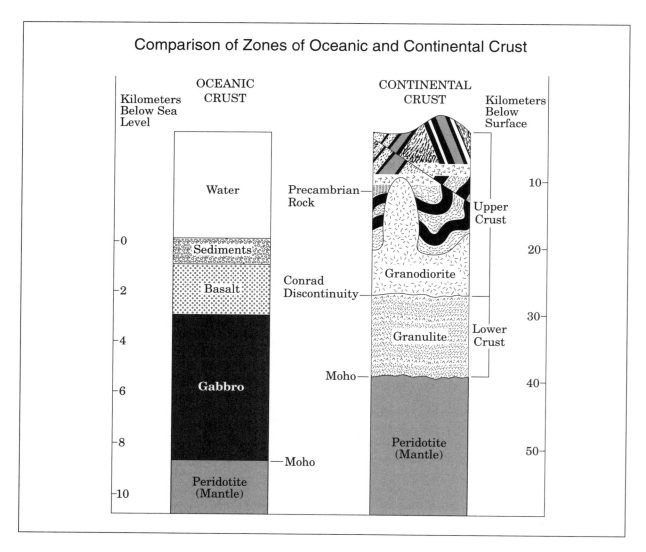

Dense blobs of molten metals like iron and nickel sank to the center to form the core, while less dense silicate and oxide minerals remained to form the mantle. The material near the surface also melted, becoming a magma ocean a few hundred kilometers deep. When it cooled and hardened, it formed a thin crust of basaltic silicate minerals probably similar to today's ocean basin crust. Radioactive elements such as uranium and thorium are relatively abundant in the crust because of their tendency to combine with low-density crustal minerals.

In order for this homogeneous accretion model to be correct, there must have been some processes that heated the early Earth enough to at least partially melt. Several likely processes have been proposed. First, radioactive isotopes were much more common in the early Earth because they had not yet had time to decay; radioactive decay of these isotopes would have produced much heat. This heat source has greatly decreased through time as the isotopes with short half-lives have decayed, but it is still probably the major source maintaining Earth's internal heat today. Second, as planetesimals continued to collide with the proto-Earth, their kinetic energy (energy of motion) was converted on impact into heat energy. Third, heat is produced by gravitational compression. As more and more material was added to Earth, progressively greater temperatures were generated in the interior. Furthermore, as Earth grew larger in size, it tended to insulate itself, raising the internal temperature by trapping interior heat and impeding its outward flow. The homogeneous accretion model requires that the combined effects of all these process raised the interior temperature of Earth enough to cause at least partial melting. Once that occurred, dense lumps of metal began to sink toward the center. In doing so, they converted gravitational potential energy into more heat, leading to more melting, more sinking, even more heating, and so on, quickly becoming a

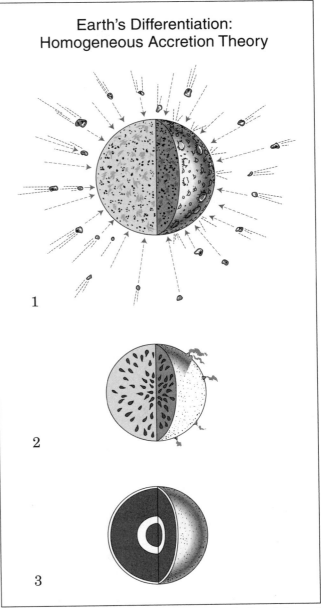

Earth's Differentiation: Homogeneous Accretion Theory

1

2

3

Homogeneous accretion theory states that the Earth formed by randomly sweeping up debris (meteors, dust) in its orbit around the Sun. (1) This process, which occurred about 4.6 billion years ago, caused the debris to be added on (accreted) to the early Earth in a sort of "snowball" effect that made the Earth progressively larger and still undifferentiated, or homogeneous. (2) Gravitational compression, increasing temperatures in the interior, initiated differentiation, as iron, nickel, and other denser elements sank to the core. (3) The result was that lighter elements rose to the surface to form the crust, and elements of intermediate density stayed between, forming today's core, mantle, and crust.

runaway collapse of a metallic core. So much heat would have been produced by the collapse that the entire core and probably much of the mantle would have been molten.

However, some computer models of the early Earth indicate that not enough heat would have been generated internally to result in partial melting. An alternative to this possible heating problem of homogeneous accretion is provided by the heterogeneous or inhomogeneous accretion model. If accretion occurred faster than cooling of the solar nebula and the resulting condensation, then the material swept up in the accretion process would have varied with time. The first to condense and accrete would be the materials with higher melting points, like metal alloys and oxides, forming metal-rich cores. Later as the temperature in the solar nebula dropped further, silicate minerals would condense and be swept up, forming rocky mantles around the metallic cores, and eventually even less dense rocky crusts at the surface.

In this heterogeneous or inhomogeneous accretion model, differentiation occurred during accretion as a result of changes in what had condensed and was ready to be picked up. Instead of accreting as an originally homogeneous body, which then became layered as denser material sank and less dense material rose through melting, Earth would have accreted its layers in sequence.

In spite of their differences, the two models agree on a number of major points about Earth's formation, the most basic being that the Earth originated from condensation and accretion in a large cloud of dust and debris around the early Sun about 4.5 to 4.6 billion years ago. The models also agree that differentiation began very early in Earth's history, either as part of the accretion process (in the inhomogeneous accretion model) or within the first few tens of millions of years (in the homogeneous accretion model). A primitive crust must have developed within the first few hundred million years. Zircon crystals possibly dating back to 4.4 billion years ago have been found in somewhat younger rocks in Australia's Jack Hills region, and the oldest known definitely continental rocks—the Acasta gneiss from Canada's Northwest Territories—date back to 4 billion years ago.

Whichever of the two models is correct, the differentiation of the Earth did not end with the formation of the core, mantle, and crust. The early crust almost certainly consisted mainly of mafic and ultramafic rocks, which are composed of relatively dense silicate minerals rich in iron and magnesium. Today, there are two distinct types of crust: continental crust, composed mainly of granitic or felsic rocks with an average density of about 2.7 grams per cubic centimeter, and oceanic crust, composed mostly of basaltic or mafic rocks with an average density of about 3.0 grams per cubic centimeter. The differentiation of crustal material is thought to be the result of plate tectonic processes that presumably have been occurring ever since the first primitive crust was formed. The motion of the tectonic plates and the crustal differentiation that accompanies it are thought to be driven by slow convection currents in the mantle.

Rocks from oceanic crust underlying the ocean basins are mainly basaltic or mafic, consisting mostly of minerals that are not quite as rich in iron and magnesium and thus not as dense as ultramafic rocks, although they are denser than granitic rocks. The separation of mafic from ultramafic material presumably takes place at the ocean ridge-rift system located along divergent plate boundaries. As the tectonic plates spread apart, mafic material rises to produce new basaltic oceanic crust while the ultramafic material remains in the mantle.

Rocks from continental crust are mainly granitic or felsic, consisting primarily of less dense silicate minerals rich in potassium and sodium. The separation of granitic continental rocks from basaltic oceanic crust is thought to be the result of subduction, which occurs at convergent plate boundaries, where an oceanic plate descends (or subducts) under another plate, either oceanic or continental. As the subducting oceanic plate descends back into the mantle, it is heated and differentially melted. The minerals that preferentially melt are those with lower melting temperatures. These are mostly the lower density silicate minerals common in granitic rocks. They rise as molten blobs of magma and produce the granitic rocks of continental crust, either erupting through the surface as volcanoes, or cooling and hardening below the surface in magma chambers. The bulk of the oceanic crust, consisting of mafic rocks, returns into the mantle.

The early Earth along with the other planets was heavily bombarded by planetesimals in the latter stages of accretion, and this bombardment may have contributed to the formation of early continents. Large impacts would have melted the early crust and upper mantle. Denser minerals would have sunk while less dense minerals would have risen during cooling and crystallization, possibly forming a granitic continental "nucleus" around which additional continental material was added later by subduction.

Whether large impacts contributed to early continent formation or not, the amount of continental crust has continued to increase with time, as new continental crust is produced at subduction zones, and once formed, generally remains continental crust (although it can change from one rock type to another). In contrast, oceanic crust is continually recycled, being produced at the ocean ridge/rift system and consumed at subduction zones.

METHODS OF STUDY

It is not possible to study directly the differentiation of the Earth because it occurred at least 4 billion years ago. Most of our current theories and models are based on inferences from indirect methods of study. Some tentative clues about the first stages of condensation and accretion are provided by the protoplanetary disks (or proplyds) that have been discovered around very young stars.

Most meteorites are remnants from the early solar system and thus yield more evidence about the condensation and accretion processes that occurred then. Meteorites are divided into three main groups based on composition: stony meteorites, iron meteorites, and stony-iron meteorites. A subgroup of stony meteorites are the chondrites, which consist of glassy spherules called chondrules embedded in a mixture of small grains of various silicate minerals and metal alloys and sometimes carbonaceous compounds. These meteorites are thought to be relatively unaltered samples of the accretion pro-

cess. Another subgroup of stony meteorites are the achondrites, which lack chondrules and instead show evidence of heating and melting followed by cooling and crystallization. The iron meteorites, composed mostly of iron and nickel, appear to have formed under extremely high temperatures and pressures such as exist in planetary cores. The achondrites and the iron meteorites are thought to have originated from early planetesimals or protoplanets that underwent differentiation into rocky mantles and metallic cores before they were shattered by collisions in the early solar system. The stony-iron meteorites, consisting of intermixed rocky and metallic parts, may be pieces of planetesimals or protoplanets that only partially differentiated before they were broken up by collisions.

Laboratory studies used to model differentiation show that iron and nickel will separate from silicate minerals at an early stage, and thus would be available to form a metallic core.

The next major group to separate are the silicate minerals olivine and pyroxene; rich in iron and magnesium, they form the ultramafic rock peridotite, which is thought to be a major constituent of the mantle. Laboratory studies of peridotite also indicate how oceanic and continental crust would differentiate from it. As peridotite rises up from the mantle at divergent plate boundaries, the pressure and temperature of the magma's environment begin to decrease dramatically. Ultramafic material remains in the mantle, while mafic material rises to become new basaltic oceanic crust. At convergent plate boundaries where subduction is occurring, the basaltic oceanic crust is pushed underneath the other plate back into the mantle, where it heats and partially melts. Laboratory studies show that its partial melting creates magmas that rise, cool, and harden into minerals found in granitic continental crust.

The structure and physical properties of Earth's interior are revealed by seismic waves from earthquakes that pass through Earth. Two types of body waves that travel through the body or interior of Earth are generated. P waves (or primary waves, so named because they travel faster and arrive first) oscillate back and forth in the direction of travel and will go through solid or liquid material. S waves (or secondary waves, so named because they travel more slowly and arrive second) oscillate at right angles to the direction of travel and will go only through solids but not through liquids. S waves are not received at locations on the opposite side of Earth from an earthquake. This "S-wave shadow zone" indicates that Earth has a core that is composed of molten material. On the other hand, P waves are received on the opposite side of the Earth, but they arrive sooner than they would if the entire core were molten, indicating that the inner part of the core is solid, while only the outer core is molten. Using records of the arrival times of P and S waves at seismic stations around the globe, scientists can determine the speeds of P and S waves as they travel through different parts of Earth's interior, and the speeds (together with knowledge of whether the material is solid or molten) reveal that the density increases with depth. Studying how and where the seismic waves are reflected or refracted indicates the depth of the boundaries between the crust and mantle, the mantle and outer core, and the outer core and inner core.

CONTEXT

The events and processes involved in Earth's differentiation shaped the Earth we now inhabit. The better those events and processes are understood, the better we will understand the planet as it is today.

Ores are rocks that are naturally enriched in some desirable mineral resource through differentiation. Many common rocks contain trace amounts of usable minerals, but they are present in such small quantities that it is not economical to try to extract them. Knowledge of differentiation processes can aid in locating ore bodies that are economical to mine.

Most earthquakes are produced by plate motion. The processes that formed the mantle and crust established the pattern of plate tectonics. Knowledge of plate tectonics and the differentiation occurring as part of it adds to our knowledge of how and where earthquakes occur.

Studies of Earth's differentiation can be extended to better understand the formation and someday possibly to utilize the resources of other solar-system objects. The other terrestrial planets—Mercury, Venus, and Mars, and also Earth's moon—underwent planetary differentiation at about the same time as did Earth. The asteroids are remnants of the planetesimals that never formed large planets. All these objects probably are rich in useful materials that we may someday have the technology to retrieve.

Earth's differentiation is not yet complete. The crust continues to evolve due to plate tectonics. Physiochemical changes continue to occur in the core and mantle as the interior slowly cools. Extrapolating these processes and comparing Earth to planets that have or have not undergone similar processes makes it possible to draw conclusions regarding the long-term future of Earth.

Michael L. McKinney

FURTHER READING

Encrenaz, Thérèse, et al. *The Solar System.* New York: Springer, 2004. A thorough exploration of the solar system from early telescopic observations through the space missions that have investigated all planets with the exception of Pluto. Compares Earth to the other planets.

Hartmann, William K. *Moons and Planets.* 5th ed. Belmont, Calif.: Thomson Brooks/Cole, 2005. A college textbook beyond the introductory level, its approach is based on comparative planetology. Provides much information on condensation and accretion in the solar nebula, as well as differentiation in the terrestrial planets generally and Earth in particular.

Levin, Harold L. *The Earth Through Time.* 5th ed. Fort Worth, Tex.: Saunders College Publishing, 1996. A highly respected and widely used college textbook for introductory historical geology. Very well illustrated and clearly written; an excellent introduction to the subject. Technical references for further study. Contains a summary of Earth's growth and differentiation.

Ozima, Minoru. *The Earth: Its Birth and Growth.* Translated by J. F. Wakabayashi. New York: Cambridge University Press, 1981. An excellent overview of Earth's differentiation from the beginning of planetary condensation to the present. Suitable for interested laypersons and advanced high school students. Technical in parts, but covers many basic concepts.

Tarbuck, Edward J., and Frederick K. Lutgens. *Earth: An Introduction to Physical Geology.* Illustrated by Dennis Tasa. 9th ed. Upper Saddle River, N.J.: Pearson Prentice Hall, 2008. This college-level textbook for introductory geology courses is well written and illustrated. It has a full chapter on the origin and historical development of Earth that includes differentiation processes.

Wicander, R., and J. Monroe. *Historical Geology.* St. Paul, Minn.: West, 2006. A survey of Earth history, with a good summary discussion of Earth's differentiation. A basic college-level text, but readable for the layperson and the advanced high school student.

See also: Earth-Moon Relations; Earth-Sun Relations; Earth's Age; Earth's Atmosphere; Earth's Composition; Earth's Core; Earth's Core-Mantle Boundary; Earth's Crust; Earth's Magnetic Field: Origins; Earth's Magnetic Field: Secular Variation; Earth's Magnetic Field at Present; Earth's Magnetosphere; Earth's Mantle; Earth's Oceans; Earth's Origin; Earth's Rotation; Earth's Shape; Earth's Structure; Lunar History; Solar System: Element Distribution.

Earth's Magnetic Field: Origins

Category: Earth

The Earth has a dipole magnetic field that is roughly aligned with its rotational axis. A dynamo effect in the Earth's molten outer core is the most likely source of most of the magnetic field.

OVERVIEW

The Earth's magnetic field is primarily a dipole field, meaning it has two well-defined magnetic poles, called "north" and "south" (the prefix "di" comes from a Greek term for "two"). This is the type of field produced by a bar magnet or an electric current flowing in a wire loop. The Earth's iron core once was thought to act like a giant bar magnet, possibly due to a remanent field frozen in place from some primordial magnetic field that existed when our solar system was forming. Now, however, electrical currents produced by fluid motions in the molten outer part of the Earth's iron core are theorized to be the source of the magnetic field. This conclusion is based on models of the Earth's interior structure and variations in the magnetic field over both historic and geologic timescales.

The ultimate source of any magnetic field is the movement of electric charges. Wires carrying electric currents, for example, have magnetic fields around them because of the electric charges (electrons) moving through the wires. The electrons surrounding the nucleus of an atom are moving, and this produces a minute magnetic field. In a magnet, the atoms are aligned in such a way that these small fields add together to produce the larger field of the magnet. The conclusion, therefore, is that electric currents within the Earth produce its magnetic field through a process referred to as the geodynamo.

To determine the Earth's interior structure, seismic waves from earthquakes act as probes as they pass through the Earth. They reveal that the Earth's interior consists of three major zones or layers: the crust, mantle, and core. The crust and underlying mantle are composed mainly of rocky material, which is a good electrical insulator. The innermost region, the core, is composed of metals, most probably iron with a small percentage of nickel and an even smaller percentage of other elements, and thus it is a good electrical conductor. The inner core (out to a radius of about 1,300 kilometers) is solid, but the outer core (from 1,300 to about 3,500 kilometers radius) is molten.

The temperature of the solid inner core is estimated to be about 5,800 kelvins. Heat flowing outward through the outer molten core sets up convection currents, in which hotter, less dense fluid rises. When it transfers its heat to the overlying mantle, the fluid cools, becomes denser, and sinks. The convection may also be partly driven chemically. If iron crystallizes at the bottom of the molten outer core to add to the solid inner core, the remaining fluid contains less iron and so is less dense, augmenting the thermally driven upward motion. Simple convection currents are deflected by the Earth's rotation in a process called the Coriolis effect. Computer models of the outer molten core that incorporate both convection and rotation show that the fluid moves in a number of spiraling columns aligned roughly parallel with the Earth's rotational axis.

All that is needed to "jump start" the geodynamo is a weak background magnetic field, perhaps provided by the solar wind or a remanent primordial field. As the metallic fluid moves through the background field, electrical currents are induced that in turn generate their own magnetic fields. This produces a positive feedback that reinforces the electrical currents and the overall magnetic field. The approximate alignment of the fluid's spiral motion with the rotational axis produces a dipole field with the magnetic poles near the rotational poles. The energy to produce the stronger magnetic field comes from the motion of convection and rotation. This process does not continue generating an ever-increasing field; it levels off since it becomes harder to generate an even stronger field as the field strength increases.

The geodynamo process explains many features of the Earth's magnetic field. Although the magnetic poles apparently remain close to the rotational poles, there is a shift in the position of the magnetic poles and changes in the

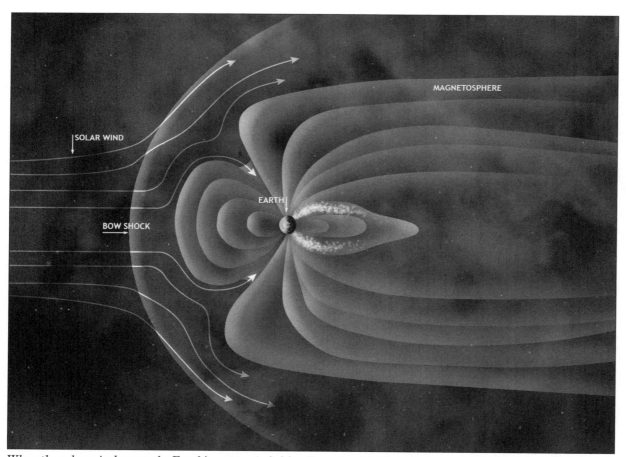

When the solar wind meets the Earth's magnetic field, a "bow shock" effect results, and the hot gases and radiation of the solar wind are deflected. Electrons are channeled by the magnetic field to create auroras during periods of high solar activity. (NASA/CXC/M. Weiss)

field strength over periods of years to centuries; this could be caused by changes in the convection currents within the molten outer core.

Furthermore, paleomagnetic studies indicate that the Earth's magnetic field has reversed polarity at irregular intervals many times in the geologic past, the last reversal occurring about 700,000 years ago. The geodynamo process is unstable over long periods of time and can decay and regrow with changed polarity. Geologists have constructed models of dynamos that are simple versions of the geodynamo, and when set in operation, these models display changes in the field's intensity and polarity. The geodynamo explains the origin of about 90 to 95 percent of the Earth's magnetic field; the rest probably comes from fields associated with magnetic minerals in the Earth's crust, more complicated irregularities in the convective motions of the molten outer core, and external sources such as the solar wind interacting with the Earth's ionosphere.

METHODS OF STUDY

The Earth's magnetic field is generated in its interior. A clue to the composition of the interior is provided by the Earth's average density. Dividing the Earth's mass by its volume shows the average density to be about 5.5 times the density of water. Common rocks from the surface are about 3 times denser than water. Therefore a portion of the Earth's interior must be much denser than surface rocks in order to yield the average value. Only metals have the required density, but some metals, such as aluminum, are too low in density, and others, such as uranium, are too high. Still others, such as gold and silver, are close to the required density but are too rare. Iron is a good candidate, since it has the right density and is fairly abundant.

The Earth's interior structure can be probed using seismic waves produced by earthquakes. Body waves from the earthquake travel through the Earth's interior. Their speed and direction of travel are determined by the density and elastic properties of the material through which they are traveling. The waves also are reflected off boundaries between different layers. When the transmitted and reflected waves reach the surface, they are recorded on seismographs. Analysis of the seismograms obtained at seismic stations all around the globe reveals that the Earth has three main zones or layers: the surface crust, the mantle, and the central core. One type of body waves, called S waves, are transverse waves that can travel only through solids, not liquids. Their presence or absence recorded on seismograms reveals which parts of the interior are solid or liquid. The crust and mantle are solid (except for isolated pockets of molten material called magma). The inner part of the core also is solid, but the outer core is molten.

Models of the Earth's interior that combine chemical composition estimates with seismic-wave data indicate that the crust and mantle are composed of rocky material, while the core (both molten outer and solid inner parts) is mostly iron, with some nickel and other elements. The composition and pressure from these models can be used to calculate the melting-point temperature as a function of depth. The mechanical properties of the layers (whether solid or liquid, as indicated by seismic waves) can then be used to determine whether the actual temperature is below or above the calculated melting-point temperature. Anchored by measurements of heat flow at the surface and the increase of temperature with depth recorded in mines and wells, the geotherm—a graph of actual temperature versus depth—can be drawn. This is how the temperature of the Earth's core, about 5,800 kelvins, is determined, showing that there is enough heat energy to drive the convection necessary for the geodynamo process.

Remanent magnetism, the evidence of past magnetic fields preserved in some igneous and sedimentary rocks, can be measured with various types of magnetometers. These data, together with records of changes in the Earth's magnetic field during historic times, show how the field varies in strength and orientation and has even reversed polarity many times in the past.

CONTEXT

The geodynamo mechanism that generates the Earth's magnetic field also can explain the presence or absence of magnetic fields for other planets, their moons, and the Sun. Mars and Earth's moon have extremely weak magnetic fields, probably because their iron cores are so small, and they may have cooled to the point that their iron cores are no longer molten so convection cannot occur. Venus also has an extremely weak field; although it is nearly the same size and mass as Earth and probably has a similar internal structure with a molten outer core, it rotates very slowly.

Mercury has a magnetic field about an order of magnitude stronger than the fields of Venus, Mars, and the Moon, but about two orders of magnitude weaker than the field of the Earth; although Mercury rotates slowly, its relatively large iron core might still have a molten convective zone. Jupiter's magnetic field is more than ten times stronger than Earth's, and Saturn's is about two-thirds as strong as Earth's; their fields are probably due to convection occurring in the liquid metallic hydrogen in their interiors (Jupiter has much more than Saturn) and their rapid rotation. Jupiter's moons Europa and Ganymede also have small magnetic fields, probably due to convection in electrically conductive salty oceans beneath their icy crusts. The Sun has a strong magnetic field that reverses polarity every eleven years; it is produced by the movement of ionized gas in the convection zone of the Sun's interior.

Stephen J. Shulik

FURTHER READING

Busse, F. H. "Recent Developments in the Dynamo Theory of Planetary Magnetism." In *Annual Review of Earth and Planetary Sciences* 11 (May, 1983): 241-268. An outline of the dynamo theory is given, with models of the dynamo for various planets. Observational evidence is discussed, along with the paleomagnetic data, geomagnetic reversals, and secular variation. Includes references and figures.

Fowler, C. M. R. *The Solid Earth: An Introduction to Global Geophysics.* 2d ed. New York: Cambridge University Press, 2004. An updated version of a widely used textbook for introductory geophysics courses. Designed for students with some knowledge of physics and calculus.

Garland, G. D. *Introduction to Geophysics.* 2d ed. Philadelphia: W. B. Saunders, 1979. Used as a text for introductory geophysics, this book covers, in sections 17.4 and 17.5, the cause of the main field and the dynamo theory. A few equations, but many figures and graphs that are of interest to the less technically informed reader. At the end of the chapter is a listing of thirty-two references.

Gubbins, D., and T. G. Masters. "Driving Mechanisms for the Earth's Dynamo." In *Advances in Geophysics.* Vol. 21, edited by B. Saltzman. New York: Academic Press, 1979. This article looks at such topics as the physical and chemical properties of the core and energy sources for the magnetic field. References are located at the end of the article. Mathematics and numerous figures and tables are included.

Hartmann, William K. *Moons and Planets.* 5th ed. Belmont, Calif.: Thomson Brooks/Cole, 2005. A college textbook beyond the introductory level, its approach is based on comparative planetology. Has a section on the generation of planetary magnetic fields by the dynamo mechanism.

Lapedes, D. N., ed. *McGraw-Hill Encyclopedia of Geological Sciences.* New York: McGraw-Hill, 1978. Pages 704-708, under the heading "Rock Magnetism," provide a concise description of many aspects associated with rock magnetism: how rock magnetization occurs, the present field, magnetic reversals, field generation, secular variation, and apparent polar wandering, among other subjects. Very readable, with no mathematics and a fair number of graphs, tables, and figures.

Merrill, R. T., and M. W. McElhinney. *The Earth's Magnetic Field.* New York: Academic Press, 1983. The authors cover much of the material associated with the Earth's field. Chapters 7 and 8 deal with the origin of the field, and chapter 9 covers the origin of secular variation and field reversals. Mathematical equations and thirty-eight pages of references. Numerous tables and figures.

Motz, L., ed. *Rediscovery of the Earth.* New York: Van Nostrand Reinhold, 1979. As a collection of articles for the nonscientist by scientists renowned in their respective fields, the text makes very interesting reading, augmented with many colorful illustrations. The chapter "The Earth's Magnetic Field and Its

Variations" is written by Dr. Takesi Nagata, who has authored hundreds of articles on diverse aspects of geophysics besides the Earth's magnetic field, and covers a wide range of magnetic field topics. Two pages are devoted to the origin of the field. Includes a small amount of mathematics and only a few references.

Smith, D. G., ed. *The Cambridge Encyclopedia of Earth Sciences*. New York: Crown, 1981. Chapter 7, "The Earth as a Magnet," contains a discussion of the field's origin. The text is well written at a nontechnical level, with many colorful diagrams and figures.

Stacey, F. D. *Physics of the Earth*. New York: John Wiley & Sons, 1977. Under section 8.4, "Generation of the Main Field," the author provides a short, technical description of the origin of the field, which will be of interest to the more advanced student. Equations are rather formidable, but several figures illustrating the dynamo effect are included. A large number of references at the end of the text. Many other areas of geophysics are covered at a technical level.

Tarbuck, Edward J., and Frederick K. Lutgens. *Earth: An Introduction to Physical Geology*. Illustrated by Dennis Tasa. 9th ed. Upper Saddle River, N.J.: Pearson Prentice Hall, 2008. This college-level textbook for introductory geology courses is well written and illustrated. The chapter on the Earth's interior has a section on the generation of the magnetic field by the geodynamo process. The chapter on plate tectonics has a section on geomagnetic reversals.

Vogel, Shawna. *Naked Earth: The New Geophysics*. New York: Plume, 1996. Covers geophysics research and theories since 1960. Includes information about Pangaea, the supercontinent cycle, and the reversals of the Earth's magnetic field. For general audiences.

See also: Auroras; Earth-Moon Relations; Earth-Sun Relations; Earth's Magnetic Field: Secular Variation; Earth's Magnetic Field at Present; Earth's Magnetosphere; Planetary Magnetospheres; Planetary Orbits: Couplings and Resonances; Van Allen Radiation Belts.

Earth's Magnetic Field: Secular Variation

Category: Earth

At every point on the Earth, its magnetic field has a direction, as indicated by a compass needle free to pivot three-dimensionally, and an intensity or strength. The direction and intensity of the magnetic field change over timescales of years to millennia, a phenomenon known as secular variation. Over longer geologic timescales of tens of thousands to millions of years, the Earth's field reverses polarity, a phenomenon called geomagnetic reversal.

OVERVIEW

Secular variation of the Earth's magnetic field refers to changes in the field's direction and intensity, a phenomenon manifested everywhere on the Earth's surface. Its most obvious effect is a gradual shift in the direction which an ordinary compass needle points. It also is seen in changes of inclination, the angle at which a magnetic needle suspended by its center tilts below the horizontal, as well as variations in the field intensity or strength. These changes appear to be noncyclic and occur over timescales of years to millennia.

The direction of the Earth's magnetic field at any point is specified by two angles called declination and inclination. The north end of a magnetic compass needle points approximately to the north, but not exactly. The angle between geographic or true north (defined by the Earth's north rotational pole) and magnetic north originally was called magnetic variation, but now is called declination. It was first noticed around the twelfth century, when it was thought to be caused by abnormalities in the compass needle's magnetization or suspension. However, by the early sixteenth century, Europeans had accepted declination as a phenomenon of the Earth's magnetism. Inclination, the downward tilt of a compass needle free to pivot three-dimensionally, was also discovered during that century. Hence William Gilbert could write in 1600 of both declination and inclination as natural features of the Earth's magnetism.

Earth's Magnetic Field: Secular Variation

The Solar System

By the early sixteenth century, Europeans had noticed that the declination varies from place to place (and this helped convince them that declination was a feature of the Earth's field and not due to flaws in their compasses). The discovery arose in the practices of navigation, chart making, and crafting of magnetic compasses—all activities connected with exploration. Perhaps Christopher Columbus, and certainly Sebastian Cabot, noted that while compass needles pointed east of north near Europe, they pointed west of north in the New World.

All three measures of the magnetic field—declination, inclination, and intensity—vary over the entire planet. These variations are most easily depicted with maps on which curved lines connect points that have the same value of one of these three parameters. For example, maps that display curved lines of equal magnetic declination are called isogonic maps. The first such printed map was produced in about 1701 by Edmond Halley of comet fame. Initially it was hoped that isogonic maps could be used to determine longitude by compass. Maps similar to isogonic maps but displaying lines of equal inclination or equal intensity also can be drawn.

Meanwhile, Henry Gellibrand announced in 1635 that declination changes over time as well as space. He found that the magnetic declination for London had shifted from 11.3° east of north in 1580 to 4.1° east of north in 1634. Later investigators discovered that the inclination and the intensity of the magnetic field also gradually change. For example, between 1700 and 1900, the inclination at London decreased from about 75° to 67°. Currently the overall intensity of the dipole field is decreasing at the rate of about 6 percent per century. If the field were to continue to decrease at this rate, it would drop to zero in about sixteen hundred years.

The agonic line is the "line of zero declination," along which compass needles point exactly toward geographic or true north. One aspect of secular variation has been the westward drift of the agonic line. This drift can be depicted on maps; just as one can map the magnetic parameters, one can also map how these parameters change, by drawing curved lines connecting points that change at the same rate. For example, all points where the declination is shifting westward at 10° per century would be connected together. These charts, known as isoporic charts, came into wide use in the mid-twentieth century. Areas of most rapid change are called isoporic foci. These isoporic foci are drifting westward, just as the agonic line is. While this westward drift has been a persistent feature of secular variation since Gellibrand, first pointed it out, some evidence exists for eastward drifts during prehistoric times.

Geologic evidence of ancient magnetic fields preserved in some igneous and sedimentary rocks shows that the geomagnetic field has reversed polarity many times over intervals of tens of thousands to millions of years. These geomagnetic reversals have played a major role in plate tectonics. They provide evidence of seafloor spreading and can be used to determine its rate. They indicate the locations of continents in the geologic past and thus can be used to trace continental drift.

The geomagnetic field and its secular changes traditionally have been attributed to the Earth's interior. Thus theories of the source of the field and the causes of its secular variation are necessarily indirect and extremely diverse, given the inaccessibility of the interior for direct study. About four hundred years ago, Gilbert suggested that the Earth behaves as if it had a bar magnet or magnetic dipole of extraordinary intensity at its center. In 1674, Robert Hooke asserted that the magnetic dipole axis of the Earth is tilted about 10° from the axis of rotation and that the dipole axis rotates westward around the rotational axis every 370 years. In 1683, Halley proposed that a double dipole pattern with four magnetic poles provided a better fit to worldwide declination data than just two magnetic poles, and in 1692, he suggested that his four poles could explain secular variation. Two of these poles he assigned to the Earth's outer crust and the other two to a central nucleus, which rotated slightly more slowly than the crust, on the same axis. The crustal magnetic poles were fixed in place, and as the nucleus, rotating a bit more slowly, drifted slowly westward relative to the crust, so did its magnetic poles. This explained, he thought, the drift of the agonic line. Theories that the core is per-

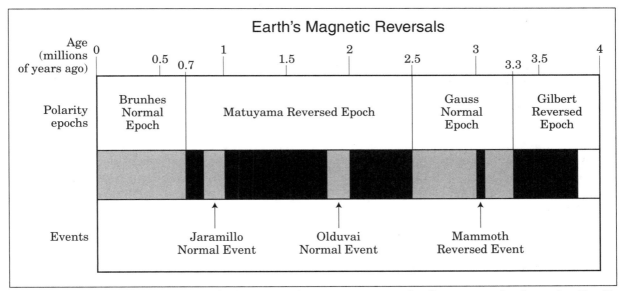

Geologic evidence of ancient magnetic fields preserved in some igneous and sedimentary rocks shows that Earth's geomagnetic field has reversed polarity many times over intervals of tens of thousands to millions of years.

manently magnetized were later ruled out when models of the Earth's interior showed that it is too hot; its temperature is above the Curie temperature of all known permanently magnetized materials. The Curie temperature (or Curie point) is the temperature above which a material is no longer permanently magnetic.

It is known that moving electrical charges generate magnetic fields. In particular, an electric current flowing around a wire loop produces a magnetic dipole field through the center of the loop. During the nineteenth and twentieth centuries, theories were developed that attributed the magnetic field and its secular variations to electric currents in the Earth's interior. One hypothesis was that the rotation of the Earth's iron core carried the charges with it, and this motion generated the field. This theory reached its highest state of development around 1950 in work by Patrick M. S. Blackett, but since then it has gradually lost favor. An alternative hypothesis is that the flow of molten metal in the interior carries the charge that generates the field. First introduced in rudimentary form in the nineteenth century, this theory became increasingly sophisticated with the investigations of Walter Elsasser beginning in 1939 and Sir Edward Crisp Bullard starting in 1948. Elsasser

proposed that the combination of the movement of molten metal and the simultaneous flow of electricity in it produced both the Earth's main dipole field and its secular variations. This dynamo was driven, he suggested, by the heat generated by the decay of radioactive materials in the interior. Convection of hotter, less dense materials upward and of colder, denser materials downward, he said, produced the dynamo.

There now is general agreement that the geomagnetic field and its secular variation are the result of fluid motions in the molten outer part of the Earth's metallic core. Various models have been developed to show that convective motion in the molten outer core, modified by the Coriolis effect due to the Earth's rotation, can produce the observed field and its secular variations. Metals are good electrical conductors because electrons can move easily through metals. The molten metal of the outer core, as it moves through the magnetic field, makes electrons in the metal move, inducing electric currents in the metal that in turn generate the magnetic field. Thus the geodynamo is self-sustaining, but it is not a perpetual motion machine; it needs an energy source to drive the motion. The geodynamo does not create its magnetic field from nothing; rather, it converts some other form of energy into magnetic energy. The two most probable energy sources to drive the convection and generate the field are heat from the decay of radioactive materials and the crystallization and settling of iron (and other dense metals) to the solid inner core.

In the end, one must remember that models of the geodynamo and its energy sources are tentative. Many debates are still waged over the details of the geodynamo and how it produces secular variation. This area of geophysical theory is a most active and challenging one, and it is in rapid flux.

METHODS OF STUDY

The simplest way to detect secular variation is to observe the changing declination of a magnetic compass over some decades; until the twentieth century, that was the only way. All the instruments employed by famous investigators of geomagnetism, from Gilbert in 1600 to Carl Friedrich Gauss in the 1830's, used adaptations of the compass to measure the magnetic parameters and their changes. Among other goals, these scientists aimed to measure these elements more accurately, so as to reveal secular change in shorter time intervals. During the twentieth century, however, there was a sustained trend to replace traditional magnetic needle instruments with ones based on other applications of electromagnetic principles.

Around 1900, research-quality Earth inductors were developed to replace the dip needle and circle in measuring inclination. The idea behind this first of the new electrically based geomagnetic instruments is simple. Rotate a coil of wire about its own diameter in a magnetic field. If the rotational axis differs from the direction of that field, an electric current is induced in the coil, but if the axis coincides with the field, the current will cease. This "null" method now is used to measure inclination more accurately and easily.

The Earth inductor was followed in the 1930's by the flux-gate magnetometer. This instrument is based on a high-permeability alloy, that is, one that magnetizes readily. Around two cores of such material are wound two coils of wire, in opposite directions, that carry the same alternating current, so that the same reversing magnetic field is produced in both cores, but 180° out of phase. When placed in the Earth's field, the changes in the magnetic fields of these two cores do not cancel out, and this changes the current flowing in the coil around each core by different amounts. The net current is related to the component of the Earth's magnetic element in the direction the magnetometer is pointing. When, however, this magnetometer is oriented parallel to the Earth's field, no current is produced. The flux-gate magnetometer has seen wide use in aerial geomagnetic surveys.

Other generations of magnetic instruments have appeared since the flux-gate. Some of the most useful are proton precession magnetometers, rubidium vapor magnetometers, and superconducting magnetometers. These devices take advantage of principles of quantum physics. Some of them, like the proton precession instrument, measure only the total intensity of the Earth's field. Others, like the superconducting magnetometer, are directional. Both types are many times more sensitive than older instruments and also perform much faster.

Magnetic surveys have been an essential part of the method of studying secular variation. All over the world, teams of observers have established "repeat stations," or places for careful observation of the magnetic field parameters at various time intervals. Magnetic surveys have been greatly facilitated not only by the new instruments mentioned above but also by the way those instruments are used. Surveys are now often conducted very quickly with instruments carried by airplanes and satellites (such as MAGSAT). Data that once took decades to gather are now collected in months. Moreover, the extensive calculations needed to analyze global data have been greatly accelerated by computers. Worldwide magnetic charts are produced much more frequently now than in 1900, and the study of secular variation is thus much more detailed.

Equally impressive changes have been wrought by the use of geomagnetic methods to study the magnetic properties of rocks. Igneous and sedimentary rocks that contain iron grains can record the Earth's magnetic field at the time they formed. The phenomenon is called remanent magnetism or paleomagnetism. Until the development in the middle of the twentieth century of techniques for measuring remanent magnetism in rocks, secular variation studies were limited to data obtained by direct measurement of the Earth's field during historic times. Little was known of the magnetic field before 1600. Past phenomena that have been revealed by these methods include reversals of the magnetic field polarity and geomagnetic excursions.

The study of geomagnetism has come a long way with the rapid development of new instruments and methods. No longer is the purpose restricted to just a description of the main field and its variations. With the new sensitivity and portability made possible by electronics, geomagnetic secular variation has become a useful tool in many diverse scientific endeavors, such as archaeological dating of artifacts, magnetostratigraphic dating of sediments, determining rates of seafloor spreading, and tracing continental drift, in addition to the traditional effort to understand processes occurring in the Earth's core.

CONTEXT

Most people are familiar with the magnetic compass, and many know at least roughly how to use it. Two activities which demand close attention to magnetic declination and its secular variation are reading topographic maps and navigating at sea.

In the margin of topographic maps, there usually are arrows which point to true north and magnetic north. With this declination information, one can relate directions on the map to compass readings in the field. In areas where the secular variation of declination occurs rapidly, it is also necessary to know when declination readings were last measured and the rate of their change. For example, near Tay River in the Canadian Yukon, the declination was listed as 33°, 25 arc minutes, east of north in 1979 and decreasing at 3.3 arc minutes per year. Thus, if the secular variation there continued at that rate, in a century the declination would change by 5°, 30 arc minutes, to 27°, 55 arc minutes east of north. Secular variations cannot be predicted reliably over so long a period, however, and maps are therefore updated regularly in magnetic surveys.

Information regarding declination at sea and especially near the coast is of even greater importance. Every ship is sometimes beset by fog, and thus an essential bit of navigational data is the present declination. Up-to-date charts are, again, the best means to avoid dealing with secular variation. As the date of the magnetic declination recedes into the past, however, reliable information concerning its secular change becomes more important.

The deep interior of the Earth is inaccessible to direct study. Thus scientists must watch closely for clues received at the Earth's surface about the conditions and processes in the interior. Magnetic secular variation is one of the ways information can be obtained about the geodynamo. The geomagnetic changes occurring over geologic time have been preserved in some igneous rocks as they cooled and some sedimentary as they settled polarity reversals of the Earth's main magnetic field. This remanent magnetism or paleomagnetism records what the field was like in the past. Such data gathered around the world from rocks of various ages reveal that the geomagnetic field has reversed its polarity many times in the geologic past. These geomagnetic reversals typically occur at intervals of tens of thousands to millions of years; the last one happened about 700,000 years ago. The geomagnetic reversals can be tied in to the chronology of the Earth and are an important element in plate tectonics.

Gregory A. Good

FURTHER READING

Backus, George. *Foundations of Geomagnetism*. New York: Cambridge University Press, 1996. Describes in detail the mathematical and physical foundations of geomagnetism. Technical and more advanced than introductory texts.

De Bremaecker, Jean-Claude. *Geophysics: The Earth's Interior*. New York: John Wiley & Sons, 1985. This well-written text is intended for college-level students with some calculus and some physics background. Nevertheless, the author is careful to explain difficult concepts or mathematical statements. Chapter 10, "Magnetostatics," and chapter 11, "The Earth's Magnetic Field," can be read separately from the rest of the book to provide an in-depth survey of geomagnetism, its measurement, and its secular variation. Especially useful are the technical appendixes on mechanical quantities, magnetic quantities, data about the Earth, notation, and some relevant mathematics. One of the best treatments available.

McConnell, Anita. *Geomagnetic Instruments Before 1900*. London: Harriet Wynter, 1980. This short book provides one of the clearest expositions of the basics of geomagnetism for the lay reader. Includes illustrations of many of the basic early forms of instrumentation, especially European.

Tarbuck, Edward J., and Frederick K. Lutgens. *Earth: An Introduction to Physical Geology*. Illustrated by Dennis Tasa. 9th ed. Upper Saddle River, N.J.: Pearson Prentice Hall, 2008. This college-level textbook for introductory geology courses is well written and illustrated. The chapters on the Earth's interior and plate tectonics have sections on the Earth's magnetic field and its reversals in the geologic past.

Thompson, Roy, and Frank Oldfield. *Environmental Magnetism*. London: Allen & Unwin, 1986. This book captures the broad range of possible applications of knowledge of magnetism in the study of the Earth that have appeared since the 1950's, from the study of magnetic minerals to biomagnetism. This is an introductory, nonmathematical, college-level text. Although its chapters on basic magnetic principles are valuable, the most unusual feature of the book is the many application chapters. Especially relevant to secular variation are chapters: "The Earth's Magnetic Field"; "Techniques of Magnetic Measurements"; "Reversal Magnetostratigraphy"; and "Secular Variation Magnetostratigraphy."

See also: Auroras; Earth-Moon Relations; Earth-Sun Relations; Earth's Magnetic Field: Origins; Earth's Magnetic Field at Present; Earth's Magnetosphere; Planetary Magnetospheres; Planetary Orbits: Couplings and Resonances; Van Allen Radiation Belts.

Earth's Magnetic Field at Present

Category: Earth

The study of the Earth's magnetic field is important from both academic and practical perspectives. The Earth is the only planet of the inner solar system with a strong magnetic field, and this provides clues about the formation of the Earth and the other inner planets. The Earth's magnetic field deflects and traps high-energy charged particles, providing a shield to protect life. It can disrupt modern communication and electrical systems, but it also can point to the location of ore deposits.

OVERVIEW

The study of the Earth's magnetic field is a branch of geophysics, which combines geology and physics to investigate various physical characteristics of the Earth. The ultimate source of any magnetic field is moving electrical charge, such as an electric current flowing in a wire. Approximately 90 to 95 percent of the Earth's magnetic field is thought to be produced by electrical currents in the Earth's molten metallic outer core, a mechanism referred to as the geodynamo.

The Earth's field is predominantly a dipole field, meaning it has two magnetic poles; the prefix "di" is derived from the Greek word meaning "two." This is the type of field produced by a bar magnet or an electric current flowing in a wire loop. By definition, the north pole of a bar magnet is the end that points northward on Earth at the present time. Since like magnetic poles repel and unlike magnetic poles attract, that means the Earth's magnetic south pole is located in the Northern Hemisphere, and the Earth's magnetic north pole is located in the Southern Hemisphere. Magnetic field lines are a way of visualizing the direction of a magnetic field, and by convention they point in the direction the north pole of a bar magnet would point. Magnetic field lines leave the Earth's surface in the Southern Hemisphere, arc over the Earth, and reenter the Earth in the Northern Hemisphere. The magnetic poles are the two places where the field lines leave and enter the Earth's surface precisely vertically.

The pole in the Northern Hemisphere is called the north magnetic pole (but remember it is a magnetic south pole), and the pole in the Southern Hemisphere is called the south magnetic pole (although it is a magnetic north pole). At first this may seem confusing, but note the distinction in terminology: the geographic hemisphere that the pole is in comes before the words "Magnetic Pole," while the type of pole comes between words "magnetic" and "pole." The field's strength is about 0.6 gauss (a unit of magnetic induction) at the magnetic poles and about 0.3 gauss at the magnetic equator, where the field lines are horizontal. (For comparison, a small bar magnet has a field strength of about 1 gauss.) The difference in strength is due to the field lines bunching together at the magnetic poles and spreading apart at the magnetic equator.

The magnetic poles are not located at the geographic or rotational poles of the Earth, which are the two points where the rotational axis of the Earth intersects the surface. Currently, near the beginning of the twenty-first century, the north magnetic pole is located in the Arctic Ocean north of Canada and west of Greenland, approximately 900 kilometers from the geographic North Pole. The south magnetic pole is located in the ocean between Antarctica and Australia, approximately 2,900 kilometers from the geographic South Pole. Notice that the magnetic poles are not symmetrically located in relation to the rotational poles.

The magnetic poles also are not stationary; rather, they wander around the polar regions at varying speeds of up to tens of kilometers per year. Since the early 1900's, the north magnetic pole has moved roughly north-northwest about 1,300 kilometers, while the south magnetic pole has moved from Antarctica northward out into the ocean toward Australia. Because the geodynamo, the theoretical source of the Earth's magnetic field, is driven partly by the Earth's rotation, it is presumed that over long periods of time, the positions of the two magnetic poles average out to roughly the locations of the geographic poles. In addition, measurements of the field's strength since the mid-nineteenth century indicate that it is decreasing at a rate of about 6 percent per century. Archaeomagnetic evidence indicates the field was approximately twice as strong two millennia ago, and before that, around 3,500 B.C.E., it was only about one-half the present strength.

These changes in direction and strength over timescales of years to millennia are called secular variations. They are thought to be due to changes in the geodynamo operating in the Earth's molten outer core. Considering its past behavior, scientists cannot predict what the magnetic field will do in the future. It may continue to decrease, or it may increase. If it were to

An artist's conception of the outflowing solar wind meeting Earth's magnetic field. (NASA/ESA)

continue decreasing at the present rate, the field would drop to zero in about 1,600 years. This might lead to a magnetic reversal, in which the field re-forms but with its polarity reversed. Paleomagnetic measurements of past magnetic fields preserved in some rocks indicate this has occurred many times in the geologic past, the last time about 700,000 years ago.

The Earth's magnetic field also exhibits small, rapid changes in direction and strength over periods of hours to days, due to a variety of external effects. For example, the gravitational fields of the Sun and Moon distort the atmosphere of the Earth, in the same manner as ocean tides. Movement of electrically charged particles in the atmosphere produces a weak contribution to the magnetic field that changes with the relative positions of the Sun and Moon.

The Sun continually blows electrons, protons, and other electrically charged particles outward from its surface at speeds of hundreds of kilometers per second, a phenomenon known as the solar wind. When these charged particles encounter the Earth's magnetic field, they interact with it, producing a boundary called the magnetopause. Inside the magnetopause is the magnetosphere, the region in which the Earth's magnetic field is dominant. The solar wind changes the shape of the Earth's field. The side facing the Sun is pushed in toward the Earth by the solar wind so that the magnetopause is about 60,000 kilometers, or 10 Earth radii, from the Earth, while the field pointing away from the Sun is elongated into a magnetic tail that can extend farther than the orbit of the Moon.

Some of the solar wind particles, particularly electrons and protons, are trapped by the Earth's magnetic field. These form the Van Allen belts, which were discovered in 1958 by Dr. James Van Allen while analyzing data from a charged particle detector he had placed aboard Explorer 1, the first successful U.S. satellite. The inner belt is a torus about 3,000 kilometers above the magnetic equator; the outer belt is a larger torus about 14,000 kilometers above the magnetic equator.

The number of sunspots increases and decreases over a cycle of eleven years. Sunspots are just one of the more obvious manifestations of solar magnetic activity, and the Sun reverses magnetic polarity with each eleven-year cycle. During times of maximum solar activity, solar flares are most likely to erupt from its surface. These flares eject large numbers of highly energetic, electrically charged particles out into the solar system. If they encounter the Earth's magnetic field, they can produce magnetic storms that cause wild variations in the Earth's field. This in turn can disrupt modern communication and electrical distribution networks. It is at these times, when the Sun is most active, that auroras (the northern and southern lights) are most common. Increased numbers of charged particles from the Sun are deflected by the Earth's magnetic field and enter the Earth's upper atmosphere near the magnetic poles, where they excite air molecules, causing them to glow.

Lightning is a very rapid electrical discharge in the atmosphere; electrical charges can flow from the ground to clouds, from clouds to the ground, or from cloud to cloud. Locally, this strong but brief electrical current produces a very large increase and then decrease in the background field strength.

Magnetic anomalies distort the dipole shape of the main field. Some of these anomalies probably result from more complicated flow patterns in the molten outer core, while others probably are associated with rock units that are rich in iron. Two of the strongest known are located near Kursk, Russia, and in northern Manitoba, Canada. Running parallel to the ocean ridge-rift system are bands or strips of seafloor with alternate normal and reversed magnetic polarity that enhance or weaken the present field over them. The strips preserve a record of the Earth's past magnetic field, frozen into the igneous rocks (mainly basalt and gabbro) that cooled from lava that oozed out along the ridge-rift, and then was pushed away from the ridge-rift as new lava oozed out. This provides evidence of magnetic field reversals in the geologic past and support for the concept of seafloor spreading (one of the key parts of plate tectonics). Small anomalies can even result from man-made iron objects.

METHODS OF STUDY

The orientation of the magnetic field at any point on Earth is specified by two angles called declination and inclination. Declination is the angle between true north (the direction of the geographic or rotational north pole) and the horizontal component of the magnetic field line at that point. Thus declination is the angle between true north and the direction an ordinary compass needle points. Inclination is the angle between a horizontal line and the downward tilt of the magnetic field line at that point. Inclinations are downward (positive) in the northern hemisphere and upward (negative) in the Southern Hemisphere. The magnetic poles are located where the inclination is 90°, specifically 90° down (positive) for the pole in the Northern Hemisphere and 90° up (negative) for the pole in the Southern Hemisphere. The magnetic equator is located where the inclination is 0°.

Around the world, 130 permanent magnetic observatories have been established to record any changes in the magnetic field. It was at observatories in London and Paris that secular variations of the field were first recognized in the 1600's. Early observatories could measure only the declination and inclination of the field. Declination was measured with a compasslike device and inclination with a magnetized rod balanced so that it could pivot freely in a vertical plane.

Magnetometers for the measurement of magnetic field intensity were first developed in the mid-1800's, and a number of different types are in use today. In conjunction with the magnetic observatories on the ground, some satellites carry magnetometers for the measurement of the field from orbit, and they provide readings for virtually the entire globe.

Portable magnetometers can detect local field anomalies due to things under the surface. Geologists use them to prospect for magnetic iron ore deposits, and archaeologists use them to search for buried iron artifacts.

CONTEXT

When magnetic storms occur, modern communication and electrical distribution networks can be disrupted. Also on these occasions, auroras are more likely to occur and be seen over larger areas. The magnetic field interacts with electrically charged particles and prevents many of them from reaching the Earth's surface. It is possible that a decrease in the field

would lead to more particles reaching the surface, perhaps producing greater numbers of genetic mutations or cancers. Changes in the field strength have been suggested as a cause of some of the mass extinctions that have occurred in the geologic past.

Stephen J. Shulik

FURTHER READING

Courtillot, V., and J. L. Le Mouel. "Time Variations of the Earth's Magnetic Field: From Daily to Secular." In *Annual Review of Earth and Planetary Sciences* 16 (May, 1988): 389-486. This source covers the geomagnetic field in general and then provides an in-depth study of its variations, from short-term to very long-term. Very little mathematics; many figures.

Fowler, C. M. R. *The Solid Earth: An Introduction to Global Geophysics*. 2d ed. New York: Cambridge University Press, 2004. An updated version of a widely used textbook for introductory geophysics courses. Designed for students with some knowledge of physics and calculus.

Garland, G. D. *Introduction to Geophysics*. 2d ed. London: W. B. Saunders, 1979. Used as a text for introductory geophysics, this book contains in chapter 17, "The Main Field," readable material on the main field and its generation. Time variations are discussed as well as the external field and methods of measurement. Some equations; many figures and graphs of interest to the less specialized reader.

Jacobs, J. A., R. D. Russell, and J. T. Wilson. *Physics and Geology*. 2d ed. New York: McGraw-Hill, 1974. This introductory geophysics textbook is formidable for the average student because it uses considerable mathematics in some chapters, but chapter 8, "Geomagnetism," has sections on the present field and contains a minimum of equations and many figures and graphs. Auroras and the magnetosphere are also discussed.

Knecht, D. J., and B. M. Shuman. "The Geomagnetic Field." In *Handbook of Geophysics and Space Environment*, edited by A. S. Jursa. Springfield, Va.: National Technical Information Service, 1985. This source covers the geomagnetic field and various aspects of it: terminology, sources of the field, measurements, the main field, and sources of geomagnetic data. Some sections have no mathematics, but others have a small amount. Many figures help the reader to understand the authors' narratives. A number of references are listed at the end of the chapter.

Motz, Lloyd, ed. *Rediscovery of the Earth*. New York: Van Nostrand Reinhold, 1979. As a collection of articles for the nonscientist by scientists renowned in their respective fields, the text makes interesting reading, augmented with many colorful illustrations. The chapter "The Earth's Magnetic Field and Its Variations" is written by Dr. Takesi Nagata, who has written hundreds of articles on diverse aspects of geophysics besides the Earth's magnetic field.

Smith, David G., ed. *The Cambridge Encyclopedia of Earth Sciences*. New York: Cambridge University Press, 1982. Chapter 7, "The Earth as a Magnet," contains information about the Earth's present-day magnetic field, geomagnetic field changes, and magnetic anomalies. The text is well written at a nontechnical level, with many colorful diagrams and figures.

Stacey, F. D. *Physics of the Earth*. New York: John Wiley & Sons, 1977. In section 8.1, "The Main Field," the author provides a short, technical description of the main field that is of interest to the more advanced student. As a textbook for geophysics, it covers many other areas on a technical level.

Tarbuck, Edward J., and Frederick K. Lutgens. *Earth: An Introduction to Physical Geology*. Illustrated by Dennis Tasa. 9th ed. Upper Saddle River, N.J.: Pearson Prentice Hall, 2008. This college-level textbook for introductory geology courses is well written and illustrated. The chapter on the Earth's interior has a section on the Earth's magnetic field.

Vogel, Shawna. *Naked Earth: The New Geophysics*. New York: Plume, 1996. Covers geophysics research and theories since 1960. Includes information about Pangaea, the supercontinent cycle, and the reversals of the Earth's magnetic field. For general audiences.

See also: Auroras; Earth-Moon Relations; Earth-Sun Relations; Earth's Magnetic Field: Origins; Earth's Magnetic Field: Secular Variation; Earth's Magnetosphere; Planetary Magnetospheres; Planetary Orbits: Couplings and Resonances; Van Allen Radiation Belts.

Earth's Magnetosphere

Category: Earth

The Earth's magnetosphere is the region around the Earth in which the geomagnetic field is stronger than the interplanetary magnetic field. The outer boundary of the magnetosphere is termed the magnetopause. The flow of charged particles from the Sun, called the solar wind, pushes in the magnetopause on the side of the Earth facing the Sun in a bow shock effect and drags it out into a long magnetotail on the side facing away from the Sun. The Earth's magnetic field interacts with the charged particles of cosmic rays as well as those emanating from the Sun, protecting life from their harmful effects.

OVERVIEW

The magnetic field at Earth's surface has been used in navigation for centuries. However, studying the extent and behavior of the geomagnetic field out in space around Earth became possible only with the dawn of the space age. The launch on January 31, 1958, of Explorer 1, the first successful U.S. satellite, brought a completely unexpected discovery. Belts of electrically charged particles circling Earth and trapped by Earth's magnetic field were detected by James Van Allen when he analyzed the signals sent back from a Geiger counter he had placed on board the satellite. These belts were named the Van Allen radiation belts in his honor. Since Explorer 1, myriad spacecraft have explored Earth's magnetosphere, measuring the field's strength and direction, the numbers and energies of the charged particles in it, and its interaction with the solar wind and interplanetary magnetic field.

The magnetic field near Earth is basically a dipole field, the type of field produced by a bar magnet or an electric current flowing in a wire loop. It is as if a bar magnet were inside Earth, tilted about 10° to 15° with respect to the axis of rotation and offset from Earth's center several hundred kilometers toward the Pacific Ocean. The north pole of this hypothetical magnet actually is located in the Southern Hemisphere and is called the south magnetic pole; the south pole of the magnet is located in the Northern Hemisphere and is called the north magnetic pole. The north magnetic pole (actually a magnetic south pole) is located north of Canada and west of Greenland about 8° away from the geographic or rotational north pole, and the south magnetic pole (actually a magnetic north pole) is located off the coast of Antarctica in the direction of Australia about 26° from the geographic or rotational south pole. This somewhat confusing situation and terminology is because opposite poles attract, and by definition a magnet's north and south poles are the north-seeking and south-seeking ends of the magnet used as a compass.

Although imagining a bar magnet inside Earth is an easy way to visualize the geomagnetic field, this cannot be the actual situation, since Earth's interior is much too hot for any material to be permanently magnetic. Instead, the source of the geomagnetic field and mag-netosphere is thought to be fluid motions and electric currents in Earth's molten metallic outer core, in what is termed the geodynamo process.

With increasing distance from Earth, the geomagnetic field lines become distorted from those of a simple bar magnet by the solar wind—the flow of electrically charged particles blown from the Sun's surface out into interplanetary space. The flow of charged particles carries the solar magnetic field, which becomes the interplanetary magnetic field. The solar wind pushes back the geomagnetic field lines on the side of Earth facing the Sun and stretches them out into a magnetotail of nearly parallel field lines on the side away from the Sun. The boundary where the geomagnetic field equals the solar/interplanetary magnetic field is called the magnetopause; inside it is Earth's magnetosphere, where the geomagnetic field dominates. The magnetosphere typically extends out to about 65,000 kilometers on the side of Earth facing the Sun and up to millions of kilometers in the magnetotail pointing away from the Sun.

Where the rapidly moving charged particles of the solar wind encounter the magnetosphere, their speed abruptly decreases and their density abruptly increases, forming a shock wave called a bow shock, analogous to the water-wave pattern around the bow of a rapidly moving boat. Most of the particles in the solar wind flow

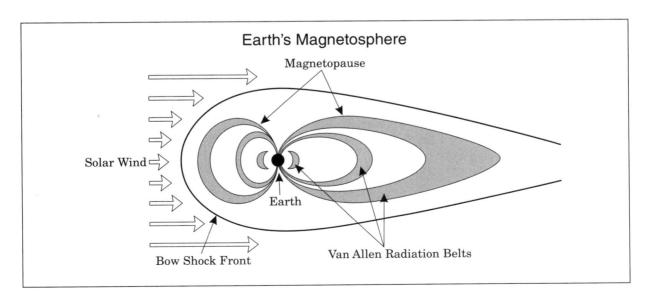

159

past Earth around this bow shock and drag the magnetosphere out into its long magnetotail. The magnetopause and its accompanying bow shock fit around Earth like a sock with a golf ball in its toe.

Some of Earth's field lines originating near a magnetic pole extend through the magnetopause into the magnetosheath, the region between the magnetopause and the bow shock. These field lines form the polar cusps or clefts where they pass through the magnetopause. It is along these field lines that some of the charged particles in the solar wind can enter Earth's magnetosphere through the polar cusps.

The magnetic field lines that exit the surface in one hemisphere and bend around Earth and reenter the surface in the other hemisphere are called closed field lines. Those that do not return but extend through a polar cusp into the magnetosheath or continue out into the magnetotail are called open field lines. At the magnetic poles, the boundary between the open field lines and the closed field lines is called the auroral oval. Because the auroral oval is the dividing line between closed field lines where particles may be trapped and open field lines where trapping is impossible, it is where the energetic charged particles of the solar wind can enter Earth's atmosphere and produce auroras. Spacecraft images have shown that the auroras are fairly continuous phenomena around the auroral ovals. Charged particles also can escape through the auroral ovals; in particular, streams consisting mainly of helium ions—the polar wind— flow out to distant regions of the magnetosphere.

Along the central plane of the magnetotail runs a flattened plasma sheet of ions (atoms that have lost one or more electrons) and electrons, separating the magnetic field lines originating from Earth's north and south magnetic poles. Currents flow crosswise in the plasma sheet and maintain the magnetic field separation. At great distances from Earth, the magnetotail may sometimes undergo a magnetic field line reconnection or collapse. This releases a large amount of stored magnetic field energy, which in turn accelerates the charged particles in the magnetosphere. Such an event is called a geomagnetic substorm. During a substorm,

high-energy electrons enter the atmosphere near the auroral oval, producing particularly strong auroras and disrupting the ionosphere and long-range radio communications, which are reflected by the ionosphere. At these times, ring currents flow around the magnetic equator at a distance from Earth of about 12,500 kilometers. Magnetometers on the ground can measure changes in the magnetosphere produced by these currents.

At times of increased solar activity, near the maximum of the solar sunspot cycle, solar flares send out strong bursts of charged particles in the solar wind, which can strongly compress the magnetosphere on the Sun side of Earth and collapse the magnetotail on the far side. Then the magnetopause may shrink to within 12,000 kilometers of Earth. When that occurs, geosynchronous satellites orbiting at a distance of about 40,000 kilometers (where the satellite revolution period equals Earth's rotation period) enter the magnetosheath and become subject to interactions with solar wind particles, causing the satellites to charge suddenly to high voltages. Some communications satellites have been disabled or have had their signals disrupted by internal sparks at such times.

When charged particles such as electrons or ions move in a magnetic field, they follow corkscrew paths around the magnetic field lines. As a charged particle corkscrews around a magnetic field line, the angle its path makes with the field line (its "pitch angle") increases as it moves into regions of greater magnetic field strengths, until it is moving in a circle perpendicular to the field line and can go no farther in the direction of the field. It has reached its "mirror point" and must now return in the direction from which it came.

The field lines of a magnetic dipole reach out from regions of high field strength near one pole to regions of weaker intensity at the magnetic equator, and then return to regions of high field strength near the opposite pole. Consequently, a particle corkscrewing around one of these field lines will be mirrored near one magnetic pole, travel back to the vicinity of the other pole and be mirrored again, and so on. High-energy particles can reach farther in toward the pole than low-energy particles, and they may even enter the atmosphere, producing the beautiful auroras, or northern and southern lights. Less energetic particles are trapped on the magnetic field lines outside the atmosphere, and they can only escape by colliding with other particles or by slowly moving across field lines in response to electric fields or changing magnetic fields. As Earth rotates, these trapped particles swing around with the rotating magnetic field.

KNOWLEDGE GAINED

Earth's magnetosphere is made up of Earth's magnetic field and the charged particles controlled by it. The magnetosphere and magnetic field can change over hours, days, or years in response to currents of charged particles and the surrounding interplanetary magnetic field.

Earth has two invisible belts of trapped energetic particles, the Van Allen belts. They were discovered by James Van Allen in 1958 using data from the first successful U.S. satellite, Explorer 1. The inner belt, which extends from 1,000 to 5,000 kilometers above the equator, is kidney-shaped in cross section and contains mainly high-energy protons. The outer belt, 15,000 to 25,000 kilometers from Earth, is crescent-shaped in cross section and contains mainly high-energy electrons. Because the charged particles are trapped in the magnetic field, they cannot easily leave the belts. They pose a danger to astronauts or sensitive electronic equipment orbiting for long periods within the Van Allen belts.

CONTEXT

It is important to understand the processes that occur in Earth's magnetosphere because they strongly affect conditions on Earth. Only since the space age began have the structure and processes of Earth's magnetosphere been subject to study.

Earth's magnetosphere performs the vital function of shielding living things from the possibly harmful effects of high-energy charged particles, such as cosmic rays and those that come from the Sun. When space travelers leave the protection of the magnetosphere, they must be shielded against charged particles, especially the intense bursts of them emitted in solar flares. Furthermore, strong solar flares may disrupt the magnetosphere, damaging or disabling Earth-orbiting satellites, long-distance radio communications, and electricity distribution networks.

At times in Earth's history when its magnetic field became very weak and reversed direction, living things may have been subjected to an intense flux of high energy charged particles, from cosmic rays and solar flare ejections, which could have led to increased rates of cancer and genetic mutation. Genetic mutations can be both good and bad. Although most mutations are deleterious (and even deadly), some are beneficial and convey an evolutionary advantage. Therefore, those times when the geomagnetic field was weak or absent may have contributed to mass extinctions or rapid evolution.

Some theorists propose that changes in the number of charged particles entering Earth's atmosphere, or changes in the solar/interplanetary magnetic field as it interacts with Earth's magnetic field, may influence the weather on Earth. Perhaps, they suggest, auroral currents modify polar air currents, leading to periods of drought or increased precipitation in different regions on Earth.

Dale C. Ferguson

FURTHER READING

Akasofu, S. I., ed. *Dynamics of the Magnetosphere*. Dordrecht, Netherlands: D. Reidel, 1980. A collection of contributions to a 1979 meeting of magnetospheric scientists in Los Alamos, New Mexico. Includes results of experiments and readable, condensed, technical summaries. This volume is mostly concerned with the disturbed magnetosphere, as during geomagnetic substorms. Well illustrated.

Akasofu, S. I., and Y. Kamide, eds. *The Solar Wind and the Earth*. Dordrecht, Netherlands: D. Reidel, 1987. Written by experts, this is a collection of chapters about the Sun and Earth and their interactions. Includes sections on Earth's ionosphere and thermosphere. The book is somewhat technical, but nevertheless clear; the history of each subtopic is well treated. Contains lists for further reading.

Allen, Oliver E. *Atmosphere*. Alexandria, Va.:

Time-Life Books, 1983. A popular book that covers all aspects of the atmosphere. A good treatment of the history of atmospheric studies. The interested layperson will find the relationship of the magnetosphere to Earth's atmosphere well explained. Contains photographs, illustrations, and a bibliography.

Chaisson, Eric, and Steve McMillan. *Astronomy Today*. 6th ed. New York: Addison-Wesley, 2008. Very well-written college-level textbook for introductory astronomy courses. Has a section on Earth's magnetosphere.

Fraknoi, Andrew, David Morrison, and Sidney Wolff. *Voyages to the Stars and Galaxies*. Belmont, Calif.: Brooks/Cole-Thomson Learning, 2006. A well-written, thorough college textbook for introductory astronomy courses. Has a section on Earth's magnetosphere.

Friedman, Herbert. *Sun and Earth*. San Francisco: W. H. Freeman, 1986. A volume that lucidly describes the Sun's effects on Earth's magnetosphere and ionosphere. Written for the layperson by a pioneer in spacecraft exploration. Of moderate length, the book contains photographs, drawings, and an appendix of references to specific topics.

Hargreaves, John K. *The Upper Atmosphere and Solar-Terrestrial Relations*. New York: Van Nostrand Reinhold, 1979. This textbook delves into the physics of the magnetosphere. It may profitably be read by specialists in the field, other physical scientists, or mathematically inclined students. Includes useful line drawings, numerous equations and references, and questions for study.

Hartmann, William K. *Moons and Planets*. 5th ed. Belmont, Calif.: Thomson Brooks/Cole, 2005. A college textbook beyond the introductory level, its approach is based on comparative planetology. Provides a succinct but easily understood description of planetary magnetic fields and magnetospheres.

Johnson, Francis S., ed. *Satellite Environment Handbook*. 2d ed. Stanford, Calif.: Stanford University Press, 1965. An excellent technical reference that has lost little value with age. It covers the near-Earth environment, from the magnetic field to micrometeoroids, and has a good section on the magnetosphere. Includes graphs, tables, line drawings, and

references. Written by many knowledgeable contributors.

Vogel, Shawna. *Naked Earth: The New Geophysics*. New York: Plume, 1996. Covers geophysics research and theories since 1960. Includes information about Pangaea, the supercontinent cycle, and the reversals of the Earth's magnetic field. For general audiences.

See also: Earth-Moon Relations; Earth-Sun Relations; Earth's Magnetic Field: Origins; Earth's Magnetic Field: Secular Variation; Earth's Magnetic Field at Present; Planetary Magnetospheres; Planetary Orbits: Couplings and Resonances; Van Allen Radiation Belts.

Earth's Mantle

Category: Earth

The mantle is the part of the Earth's interior that lies between the crust and the outer core. It is composed of rocks that are of greater density than those of the crust. The mantle generally is solid, except for pockets of molten magma, but it contains a layer called the asthenosphere where the actual temperature is so close to the melting point that the rock there is in a plastic state. It is upon this layer that the Earth's lithospheric plates ride.

OVERVIEW

The mantle is the portion of the interior of the Earth that extends from the base of the crust to the boundary with the outer core. This distance is approximately 2,900 kilometers, roughly 45 percent of the radius of the Earth. Since the thickness of the Earth's crust is not uniform, the distance from the ground surface to the upper boundary of the mantle, called the Mohorovičić discontinuity or simply Moho for short, varies significantly. The thickness of the crust in continental areas averages approximately 40 kilometers and reaches as much as 70 kilometers under young high mountain ranges, while in the ocean basins the crust is only about 5 to 7 kilometers thick.

Evidence for layers of different density within the Earth was first found in 1909 by the Croatian seismologist Andrija Mohorovičić, using seismic waves from earthquakes. Studying an earthquake that had occurred in Yugoslavia, he discovered that seismographs located a few hundred kilometers from the epicenter of the quake had recorded two sets of seismic waves. He correctly deduced that one set had traveled directly from the quake, while the other set had been refracted at a boundary somewhere below the surface where there was an abrupt change in seismic-wave speed. At stations nearer the quake, the direct set of waves arrived first, but at more distant stations, the refracted set of waves arrived first. This meant the speeds below the boundary were faster than those above; from the arrival times of the refracted waves, Mohorovičić was able to determine their speeds below the boundary. From the speeds above and below the boundary and the arrival times of both sets of waves, he could calculate the distance from the surface to the boundary, which has become known as the Mohorovičić discontinuity (or just Moho for short) in his honor. This discontinuity is taken to be the boundary between the crust above and the mantle below.

During the mid-1960's, a project to drill down through the Earth's crust to sample the mantle below the Moho was begun. The undertaking was appropriately named Project Mohole. Because of technical difficulties and cost overruns, the project was abandoned. However, we know of locations where mantle rocks have been brought to the surface by various geologic processes, so, in effect, we have been provided with free samples. The rock peridotite, composed mostly of the silicate minerals olivine and pyroxene, seems to represent the general composition of the mantle. Another likely mantle rock is eclogite, which might transform into basalt in the crust as pressure drops with reduced depth. A third type of mantle rock is kimberlite, which occurs in pipe-shaped deposits and is mined extensively for diamonds. Diamonds are a high-pressure form of carbon that originate in the mantle at depths between 100 and 200 kilometers and then are carried to the surface in kimberlite pipes.

The greatest source of information on the physical nature of the mantle comes from the study of seismic waves. Seismic wave speeds gradually increase through the upper mantle down to a certain depth and then decrease a bit in what is called the low-velocity zone, first identified by Beno Gutenberg in 1926. This low-velocity zone begins about 100 kilometers below continents, but as little as 20 kilometers below ocean ridge-rifts. It is thought this decrease in wave speed is due to a decrease in the rigidity of the rock. From the top of the low-velocity zone down to depths of about 400 kilometers, the rock is likely to be close to (but not quite at) its melting point, so the rock exhibits plastic behavior. This means it is solid, but deformable, and it is capable of flowing or oozing very slowly. This plastic zone is called the asthenosphere.

Above the asthenosphere is the rigid lithosphere, made up of the crust and very top part of the mantle. According to plate tectonic theory, the lithosphere is divided into a number of blocks called tectonic plates. These plates float on the underlying plastic asthenosphere.

Beneath the asthenosphere, wave speed begins to increase again. At depths of about 410 and 660 kilometers, sharp increases in seismic wave speed occur, representing abrupt increases in density. The pressure at 410 kilometers depth is sufficient to collapse the crystal structure of the mineral olivine into the more compact and denser mineral spinel. The even greater pressure at 660 kilometers depth converts the crystal structures of both spinel and pyroxene into the even denser mineral perovskite. Below 660 kilometers, the wave speeds gradually increase down to the boundary with the outer core. (Below this boundary, down in the outer core, primary, or P, waves slow down considerably, and secondary, or S, waves disappear, indicating the outer core is molten.) As a result of the increase in pressure with depth, the density increases from about 3.3 grams per cubic centimeter at the top of the mantle to about 5.6 grams per cubic centimeter at its base.

A critical component of modern plate tectonic theory is the heat flow that drives the motion of the plates. The Earth's interior was heated during its formation about 4.5 billion years ago, but it remains hot today mainly due to the decay of radioactive elements. It is thought that this

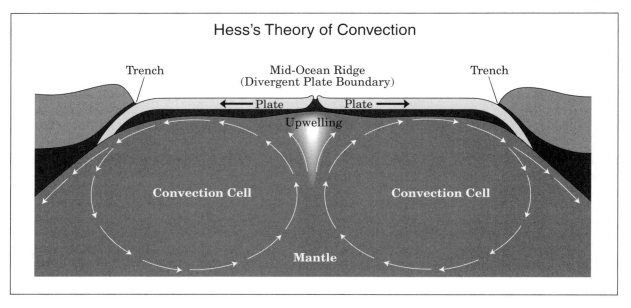

Hess's Theory of Convection

Trench

Mid-Ocean Ridge
(Divergent Plate Boundary)

Trench

← Plate

Plate →

Upwelling

Convection Cell

Convection Cell

Mantle

heat causes rock in the mantle to slowly flow, rising where it is hotter and less dense, and sinking where it is cooler and denser. It is not known whether this is due to an organized system of convection cells or to an irregular pattern of rising mantle plumes and descending lithospheric plates. The lithospheric plates are carried on the asthenosphere as it flows laterally from areas of upward movement to areas of downward movement.

It has been known since the nineteenth century that the age of the Hawaiian islands increases from southeast to northwest. In the 1960's, when plate tectonic theory was being developed, it was postulated that these volcanic islands recorded the movement of the seafloor over a hot spot or rising mantle plume. Lava from the hot spot is extruded out onto the seafloor, eventually building up to break the surface of the ocean and form a volcanic island. As the Pacific plate moves slowly toward the northwest, the newly created island is carried away from the hot spot, its volcano becomes extinct, and another volcanic island starts to form, creating the Hawaiian island chain. Another such hot spot is located under Yellowstone National Park and fuels all of its geothermal activity.

METHODS OF STUDY

Although the Earth's mantle cannot be directly observed, samples from it have been brought to the surface by geologic processes, and they can be studied in the laboratory to determine chemical composition and physical properties. However, the primary method for studying the mantle is by analyzing seismic waves generated by an earthquake or an explosion. Some seismic waves travel along the surface of the Earth. They are called L waves (L standing for "long" because of their long wavelengths); they do not travel far, but they cause most of the destructive shaking in an earthquake. Other seismic waves are called body waves because they travel through the body of the Earth, and these are the ones that are used as probes of the Earth's interior. There are two types of body waves: P (for primary) waves travel faster and arrive at seismic stations first, while S (for secondary) waves travel more slowly and arrive second. P waves are longitudinal compression waves, similar to sound waves in that they oscillate in the direction of propagation, alternately compressing and extending the material through which they travel. Like sound waves, P waves can travel through any kind of material—solid, liquid, or gas. S waves are transverse shear waves that oscillate from side to side, perpendicular to the direction of propagation. They will travel only through rigid materials, meaning they will travel only through solids, not liquids or gases.

S waves travel through the entire mantle (except for isolated pockets of molten rock called magma chambers), and this indicates the entire mantle is solid (except for the magma chambers). The speeds of both P and S waves generally increase with depth throughout the mantle, due to the increase in density with depth. The exception is in the low-velocity zone that begins at a depth of between about 20 and 100 kilometers. There both P and S waves slow down, not because of a change in density or composition, but rather because of a change in rigidity. In the low-velocity zone or asthenosphere, the temperature is close to the melting point, so the rocks, although solid and not molten, are soft and deformable, and capable of slow flow at rates of up to a few centimeters per year.

Seismic tomography uses seismic waves to construct images of the Earth's interior the same way computerized axial tomography (CAT) scans use X rays to construct images of internal organs for medical diagnoses. A CAT scan is a composite image of X rays of the body taken from a number of different directions, computer-assembled to show slices through the body; these slices can be stacked to produce a three-dimensional view of the internal organs. In a similar fashion, seismic stations all around the world provide data on waves that have traveled through the Earth in many different directions. When such seismic records are computer analyzed and combined, slices through the Earth and three-dimensional representations of the Earth's interior structure can be produced.

CONTEXT

Processes occurring in the Earth's mantle affect the Earth's surface. The role of the mantle is essential in understanding plate tectonics and the phenomena associated with it, such as earthquakes, volcanic activity, seafloor spreading, the movement of continents, and the location of mineral resources. The tectonic plates of the rigid lithosphere ride on the plastic rock of the asthenosphere in the mantle. Most earthquakes and volcanism occur along the boundaries between the plates, where the plates diverge, converge, or move sideways past each other.

For example, Southern California is prone to earthquakes because it is located along the boundary between the North American plate and the Pacific plate; as the Pacific plate slides to the northwest past the North American plate at the rate of several centimeters per year, sudden slipping of the rocks along the boundary produces earthquakes. At convergent boundaries where one plate slides back down into the mantle in a process called subduction, earthquakes and volcanism are common. The subducting plate rubs against the other plate as it descends under it, producing earthquakes. As it descends to greater depths with hotter temperatures, the minerals in it with lower melting temperatures differentially melt and rise as molten blobs of magma; if they erupt at the surface, they fuel volcanic activity. The volcanoes and earthquakes of Japan, the Philippines, the Andes Mountains along the west coast of South America, and the Cascade Mountains of the northwestern United States all are due to the subduction of one plate under another.

The mechanism that causes the movement of the lithosphere upon the asthenosphere is heat. Hot rock slowly rises from deep in the mantle, gives up its heat while moving laterally, and then sinks back down. This "conveyor belt" motion in the mantle moves the lithospheric plates above.

Processes occurring along plate boundaries as well as at isolated hot spots or mantle plumes can concentrate mineral resources into economically important deposits. Understanding these processes and being able to recognize where they occurred in the past are important in locating such valuable deposits.

David W. Maguire

FURTHER READING

Fowler, C. M. R. *The Solid Earth: An Introduction to Global Geophysics*. 2d ed. New York: Cambridge University Press, 2004. An updated version of a widely used textbook for introductory geophysics courses. Designed for students with some knowledge of physics and calculus.

Jacobs, John A., Richard D. Russell, and J. T. Wilson. *Physics and Geology*. 2d ed. New York: McGraw-Hill, 1974. A technical volume covering such topics as composition of the Earth, geochronology, isotope geology, thermal history of the Earth, magnetism, and seismic studies. The text is intended for college-level students of geology or physics. Some differential equations are used in the book.

Skinner, B. J., and S. C. Porter. *The Dynamic Earth*. 5th ed. New York: John Wiley & Sons, 2006. A well-written, well-illustrated, colorful volume on the geology of the Earth. It would be suitable for the college student beginning geology.

Tarbuck, Edward J., and Frederick K. Lutgens. *Earth: An Introduction to Physical Geology*. Illustrated by Dennis Tasa. 9th ed. Upper Saddle River, N.J.: Pearson Prentice Hall, 2008. This college-level textbook for introductory geology courses is well written and illustrated. There is a good chapter on the Earth's interior, as well as several chapters on volcanoes, earthquakes, and plate tectonics that describe their relationships with the structure and processes of the mantle.

Vogel, Shawna. *Naked Earth: The New Geophysics*. New York: Plume, 1996. Covers geophysics research and theories since 1960. Includes information about Pangaea, the supercontinent cycle, and the reversals of the Earth's magnetic field. For general audiences.

Weiner, Jonathan. *Planet Earth*. New York: Bantam Books, 1986. A colorful, well-illustrated, well-written book describing the Earth and how it is studied. This volume is the companion to the PBS television series of the same name. It is suitable for general readers.

See also: Earth's Composition; Earth's Core; Earth's Core-Mantle Boundary; Earth's Crust; Earth's Crust-Mantle Boundary: The Mohorovičić Discontinuity; Earth's Differentiation; Earth's Oceans; Earth's Origin; Earth's Structure; Planetary Tectonics.

Earth's Oceans

Category: Earth

Earth's ocean water was derived by outgassing from hydrated minerals bound up during the formation of the Earth. Subsequent evolution of the water primarily involved ions from continental and oceanic bottom sediments dissolving in the fluid medium to yield the basic saltiness characteristic of Earth's oceans.

OVERVIEW

Of all the planets in the solar system, Earth stands out as basically a watery world, distinguished from the other planets by large quantities of liquid water. In all, Earth has about 1.36 billion cubic kilometers of water, and 97.2 percent is stored in the oceans. The remaining 2.8 percent of Earth's water not in the oceans is apportioned among ice (77 percent of the total remaining water) and continental and atmospheric waters. The ice itself, now principally in the Arctic-Greenland area (1.72 million square kilometers, up to 3,200 meters thick) and the Antarctic area (12 million square kilometers, up to 4,000 meters thick), has effects ranging from climate control to providing habitats for living organisms to being a reservoir for water that, when added to or removed from the oceans in the past, has caused sea level to rise or fall more than 100 meters.

Ocean water is salty because it contains dissolved minerals; with a salinity of 35,000 parts per million, there is enough dissolved salt to cover the entire surface of the Earth to a depth of about 50 meters. This salty solution is composed primarily of sodium and chlorine ions (together constituting about 86 percent of the ions by weight), along with ions of magnesium, calcium, potassium, and sulfate and carbonate groups. Seawater is slightly basic, with a pH of 8 for the hydrogen-ion concentration.

The problem of the oceans' origin is twofold: (1) the primordial origin of the water itself and (2) the origin and rate of addition of the ions that make the oceans salty. The database for solving these problems includes the chemistry of water, the amounts and types of runoff delivered by rivers into the sea, and the composition of volcanic gases, geysers, and other vents opening to the surface, since the oceans and atmosphere are linked in origin.

Numerous sources for the Earth's water have been proposed, and the problem has not been resolved completely. Possible sources include the primordial solar nebula, the solar wind acting over time, delivery or impact degassing by bodies colliding with the Earth, and outgassing from the planetary interior. Processes controlling water on Earth involve rates, amounts, and types of outgassing, modes of planetary formation, possible chemical reactions providing water, loss rates of gases to space, and, finally, internal feedback mechanisms such as changes in Earth's albedo (reflectivity), temperature, alteration of materials, and other factors not clearly understood.

The solar wind as a primary source can be eliminated for several reasons. The basic constituents, protons, may help form water in the atmosphere by reactions with oxygen, but all evidence points to no free oxygen in the primordial atmosphere. The geologic record shows the presence of liquid water at least 4 billion years ago, but Earth was substantially devoid of free oxygen then.

Colliding bodies would be from two primary sources: meteorites and comets. The basic composition of cometary nuclei, consisting of water ice, various ions, metals, organic molecules, and dust grains, would supply enough water, provided that gigantic numbers of cometary objects struck the Earth during the first half-billion years of history. No conclusive evidence for such happenings is available at present, although a theory that Earth is still being bombarded incessantly by small comets containing large quantities of water is supported by the detection of diffuse ice balls entering the atmosphere in the 1990's. Meteoritic impact, particularly during the early stages after final planetary accretion, would definitely add water to the crust via two mechanisms. Carbonaceous chondrites, the oldest and most primitive meteorites, contain abundant volatiles, such as water, chemically bound in various minerals. Additional water, trapped in crustal and mantle rocks since Earth's accretion and differentiation, would have been released during impacts, especially by large impacting objects. It has been calculated that such impact degassing could have released 10^{22} kilograms of volatiles, quite close to the currently estimated value of 4×10^{21} kilograms for the Earth as a whole. Remnants of such ancient astroblemes are lacking, however, because of subsequent erosion, filling in by molten lava, or shifting of the continental masses over 4 billion years.

The most widely accepted origin for the

This image of Earth's Western Hemisphere, compiled from images gathered by NASA's SeaWiFS instrument, shows our planet as a world dominated by water. (NASA/GSFC/SeaWiFS Project/GeoEye, Scientific Visualization Studio)

oceans and atmosphere combines the features of the primordial solar nebula and slow outgassing from within the solidifying Earth. Original water would have been combined with silicate minerals and metallic materials during the planetary accretion process, the hydration assisted by the heating of the Earth due to infalling bodies and the decay of radioactive elements. Such wet silicates appear to be able to hold large quantities of bound water for indefinitely long periods of time. The primordial Earth, believed to have accreted cold, trapped water molecules. If Earth had started too hot, all the minerals would have been dehydrated, and if too cold, no water would have been released; a delicate balance of temperature must have been achieved. Further, the volatiles forming the atmosphere must have outgassed first, in order to provide an insulating blanket under which water could form a liquid phase.

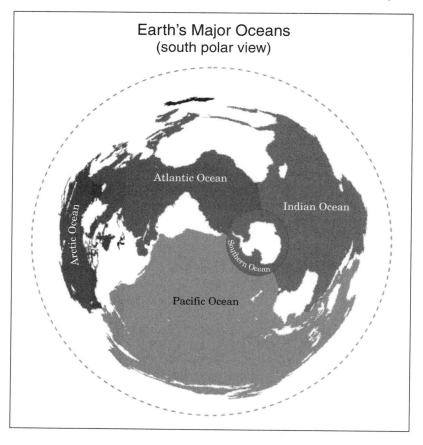

Earth's Major Oceans
(south polar view)

Atlantic Ocean

Indian Ocean

Arctic Ocean

Southern Ocean

Pacific Ocean

A secondary problem deals with how swiftly the fluids outgassed, either all at once, as many individual events, or in a continuous fashion. Most studies suggest the continuous mode of emission, with greatest reliance on data from currently active sources, such as volcanoes, undersea vents, and associated features. Fumaroles, at temperatures of 500-600° Celsius (800-900 kelvins), emit copious quantities of water, sulfur gases, and other molecules. These structures grade gradually into geysers and hot spots, areas where water is moved crustward from great depths. Magmatic melts rising in volcanoes release water and other gases directly to the surface. In Hawaii, for example, the Halemaumau Pit, the most active vent on the volcano Kilauea, emits 68 percent water vapor, 13 percent carbon dioxide, and 8 percent nitrogen, with the rest mostly sulfurous gases. Similar values are found for ocean-ridge-axis black and white smokers, where hydrothermal accretions result in spectacular deposits of minerals falling out of solution from the emerging hot mantle waters. Detailed studies show that water is trapped in the altered minerals within the basaltic crust of the oceanic plates, 5 percent of the rocks (by weight) in the upper 2 to 3 kilometers being water and hydroxide ions. Free water is known to be extremely buoyant, rising in the crust along shallow dipping faults. Bound water, subducted to great depths, would be expected to cook, moving upward as the rock density lessens and then acting as a further catalyst for melting the surrounding rocks.

In the Earth's earliest stage, the primordial atmosphere was released, only to vanish from the Earth because of overheating. In the second stage, gases were released from molten rocks, with a surface temperature of 300° Celsius (600 kelvins), providing 70 percent water and large quantities of carbon dioxide and nitrogen. In stage three, the atmosphere and oceans gradually changed as a result of volcanoes and weathering action; more and more water was deposited as liquid as the temperature fell. Then the atmosphere added oxygen either by thermal dissociation of water molecules, photochemical breakdown of high-altitude water, or photosynthetic conversion of carbon dioxide to oxygen in plants.

The saltiness of the oceans can be accounted for by the extreme dielectric constant of water, essentially ensuring that ocean water does not remain chemically pure. Geologic evidence shows the general composition of seawater to be similar over time, the content stability attributable to the continuous seawater-sediment interface. John Verhoogen has shown that only 0.7 percent of the ocean has been added since the Paleozoic era, primarily from lava materials. The saltiness is a product of acidic gases from the volcanoes (hydrochloric, sulfuric, and carbonic acids) acting to leach ions out of the common silicate rocks. Paleontological studies indicate the change in ions must have been extremely slow, as demonstrated by the narrow tolerance of organisms then alive, such as corals, echinoderms, brachiopods, and radiolarians. Present river ion concentrations differ drastically from the ocean's values, however, indicating a different atmospheric environment in the past. Robert M. Garrels and Fred T. Mackenzie have divided ocean history into three periods. In the earliest, until 3.5 billion years ago, water and volcanic acid emissions actively attacked the crust, leaching out ions and leaving residues of alumina and silicates. The next period, from 3.5 to 1.5 billion years ago, saw slow continuous chemical action attacking the sedimentary rocks, adding silica and ferrous ions. Period three, from 1.5 billion years ago onward, added ions until seawater composition reached apparent equilibrium with a mixture of calcite, potassium-feldspars, illite-montmorillonite clays, and chlorite.

Because the composition of ocean water has remained similar over much of geologic time, generally output must equal input of ions, so geochemical "sinks" must balance geochemical sources. Calcite (calcium carbonate) and silica (silicon dioxide) are removed by marine organisms to form skeletons and shells. Metals are dropped from seawater as newly formed mineral clays, oxides, sulfides, zeolites, and as alteration products at the hot-water basaltic ridges. Sulfur is removed as heavy-metal sulfides precipitating in anaerobic environments, while salts are moved in pore waters trapped in sediments. Residence times for many of the ions have been determined: for example, sodium cycles in 210 million years, magnesium in 22 million, calcium in 1 million, and silicon in 40,000. With such effective removal systems, it is truly a measure of the geochemical resistivity of the Earth's oceans to change that allows the composition to remain so stable for 4 billion years.

METHODS OF STUDY

Numerous avenues of approach have been used to investigate the ocean and its ions, including geological, chemical, and physical means. Geology has supplied basic data on the types and makeup of rocks from the earliest solidified materials to present depositional formations. Use of the petrographic microscope, involving thin sections of rocks seen under polarized light, allows the identification of minerals, providing quantity measurements of water attached to the minerals themselves. Paleontological studies of fossil organisms and paleosoils indicate the range of ions in the sea at diverse geologic periods, both by the ions themselves left in the deposited soils and rocks and through studies of the tolerance ranges for similar, twentieth century organisms. Such studies—along with sedimentological investigations of rates and types of river depositions, dissolved ion concentrations, and runoff rates for falling rain—provide determinants for comparing ion concentrations with those in the past for continentally derived materials.

Chemical analysis reveals the various ions present in seawater and rocks via two principal methods. The mass spectrometer identifies types and quantities of ions present by use of a magnetic field to accelerate the charged ions along curved paths, the curvature of the paths based on the weight and charge of the ions. Collection at the end of the paths provides pure samples of the different ions present. For solid samples, electron beam probes analyze an area only one micron in diameter. The electrons, fired at the sample, cause characteristic X rays to be emitted. The energies of the X rays identify the elements or compounds present in the sample.

Solubility studies provide residence times. Similar laboratory projects, testing the ability of water to dissolve and hold ions in solution, argue for a primordial Earth atmosphere that was essentially neutral or mildly reducing in nature. Such reduction characteristics are based on studies of the composition of Earth, supplemented by the composition of Venus and Mars as revealed by various "lander missions."

Missions in interplanetary space have also provided chemical compositions for meteoritic gases, cometary tails and nuclei, and the mixing ratios for noble gases, important for determining the origin of the solar system. Analysis of radioactive isotopes such as helium 3, an isotope of mantle origin, has allowed geophysicists to treat the Earth's mantle as a major elemental source and sink for the various geochemical cycles.

Laboratory analysis reaches two other areas. Petrographic studies of returned lunar rocks reveal that the Moon is devoid of water, lacking even hydroxyl ions. This discovery helps eliminate the solar wind and meteoritic impact as major factors in forming Earth's oceans. Furthermore, high-temperature, high-pressure metallurgical and chemical studies indicate that molten granite, at temperatures of 900° Celsius (1200 kelvins) and under 1,000 atmo-

Earth's Indian Ocean, from the space shuttle Discovery *during the STS-96 mission in 1999.* (NASA)

spheres of pressure, will hold 6 percent water by weight, while basalt holds 4 percent. Based on geochemical calculations of the amounts of magma in the planet and lavas extruded over the first billion years, all the ocean's waters can be accounted for, particularly if parts of the fluid, as steam under pressure, are a result of oxidation of deep-seated hydrogen deposits trapped within or combined with mantle rocks.

CONTEXT

Water is a ubiquitous and by far the most important molecule on Earth. All living organisms require it as a basic component of cellular structure and for numerous functions inside the body. The origin of Earth's water is highly significant, because the very presence of water may have set the scheme for all subsequent evolution, both geological and biological, on the planet. During the formation of the solar system, the accretion of various materials trapped water by hydration. Tied to the minerals, the water molecules were released through outgassing by volcanoes and other vents acting as pressure escape valves for the molten interior of the Earth. The water and other volatile gases that were released formed the atmosphere and subsequent oceans. A vital interchange was established between the ground and the atmosphere, one replenishing elements and compounds as they were lost through geochemical sinks in the normal course of history. Water, at first in the atmosphere, then as liquid seas, apparently helped to mediate the greenhouse effect, a mechanism which, if allowed to act unhindered, would have trapped infrared radiation from the Sun and overheated the early Earth. Such actions would have given the Earth the characteristics of the planet Venus: enormously hot and totally inhospitable for life's occurrence.

The outgassed water, settling as rain, also played the dominant role in shaping the landforms of Earth. As a mechanism for fluidization of rocks, it controls to a large extent the motions of magmas, helping them rise to the surface. As a weathering agent, water, in the forms of rain, snow, and ice, carves away the landscape, removing elements, as ions, to the sea. In that location, these elements became usable by early

organisms for fulfilling their biological needs, such as home building or metabolism. Water acts as a transport mechanism, a mixing agent, and ultimately a removal tool for maintaining a delicate ionic concentration range within the ocean itself. Evaporating seawater, falling as rain, breaks up rocks and forms soils with nutrients available for land-based plant life, and it provides the freshwater so necessary to non-ocean-dwelling organisms. Without the initial interplay of water on Earth, our planet, instead of being the home of countless billions of creatures, would undoubtedly be a desolate ball, revolving forever around the Sun as an improbable abode of life.

Arthur L. Alt

FURTHER READING

Brancazio, Peter J., ed. *The Origin and Evolution of Atmospheres and Oceans*. New York: John Wiley & Sons, 1964. This work is a collection of papers dealing with the chemical problems relevant to the early formation of the fluid parts of the Earth. Tracing all the basic arguments, the criteria for water formation is clearly explained and its relationship to minerals and rocks elucidated. Some heavy reading, charts, extra references.

Chamberlain, Joseph W. *Theory of Planetary Atmospheres: An Introduction to Their Physics and Chemistry*. New York: Academic Press, 1978. A detailed analysis of the characteristics of diverse atmospheres in the solar system, including water contents. By comparisons of chemical compositions and meteorological observations, criteria are established for examining the possible origins for atmospheric gases and oceans. Some mathematics, heavy reading, numerous charts, comprehensive references.

Consolmagno, Guy. *Worlds Apart: A Textbook in Planetary Sciences*. Englewood Cliffs, N.J.: Prentice-Hall, 1994. An introduction to planetary science for beginners and undergraduates. Covers all planets in the solar system.

Frakes, L. A. *Climates Throughout Geologic Time*. New York: Elsevier, 1980. A well-written explanation of how the interaction of the Earth's atmosphere and oceans has caused the climate of the Earth to change over the history of the planet. Beginning with the possible origin of ocean and atmosphere, changes are traced as revealed through the geological and paleontological records.

Garrison, Tom. *Oceanography: An Invitation to Marine Science*. Florence, Ky.: Brooks/Cole, 2007. Explores all aspects of Earth's oceans. Looks at how humans use and study the oceans. Covers events such as the December, 2004, earthquake and resulting tsunami and Hurricane Katrina.

Henderson-Sellers, A. *The Origin and Evolution of Planetary Atmospheres*. Bristol, England: Adam Hilger, 1983. This work details the theories of where the volatiles for the Earth came from and how the oceans and other planetary atmospheres came into existence from the creation of the solar system. Advanced reading.

Holland, Heinrich D. *The Chemistry of the Atmosphere and Oceans*. New York: John Wiley & Sons, 1978. A detailed reference on the basic chemical elements present in the two media and the wide variety of reactions occurring in each area. The interactions of the two areas are stressed, as are their common origin from materials outgassed from within the Earth. The action of their chemicals on the terrestrial areas is described in detail. Contains references and numerous charts of data. Advanced reading.

Knauss, John. *Introduction to Physical Oceanography*. 2d ed. Long Grove, Ill.: Waveland Press, 2005. Designed for undergraduates, covers a broad range of topics. Also includes material helpful for physicists, geologists, biologists, and chemists. Contains both descriptive and mathematical explanations.

McElhinny, M. W., ed. *The Earth: Its Origin, Structure, and Evolution*. New York: Academic Press, 1979. A readable work dealing with all the basic elements of Earth science. Starting with the theory of planetary formation, the volume covers the origin of the oceans, atmosphere, land, and life-forms. Offers an excellent description of the changes occurring on the planet throughout geologic time.

Pickard, George L. *Descriptive Physical Oceanography: An Introduction*. 5th ed. New York: Pergamon Press, 1990. An introductory work geared toward undergraduates. Covers all aspects of oceanography, including sea-ice physics, thermohaline circulation, and coral reefs.

Ponnamperuma, C., ed. *Cosmochemistry and the Origins of Life*. Dordrecht, the Netherlands: Reidel, 1982. A collection of works dealing with the distribution of elements in the universe, particularly those necessary for life. Provides information on the formation of the planetary system, showing how the chemicals combined at various temperatures to make the planets as different as they are. Discusses origins of oceans, atmospheres, and life; detailed reading with many charts and an extensive bibliography.

Seibold, E., and W. Berger. *The Sea Floor*. New York: Springer, 1982. A delightful book covering the chemistry, geology, and biology of the bottom of the Earth's oceans. That the oceans are a result of outgassing is emphasized. Well written, with a very interesting section on the black and white smokers and their relation to the origin of waters and life.

Trujillo, Alan, and Harold Thurman. *Essentials of Oceanography*. 9th ed. Upper Saddle River, N.J.: Prentice Hall, 2007. An introductory work designed to explain the complexities of oceanography to the average reader. Covers tsunamis and Hurricanes Katrina, Rita, and Wilma.

See also: Auroras; Comets; Earth-Moon Relations; Earth-Sun Relations; Earth System Science; Earth's Age; Earth's Atmosphere; Earth's Composition; Earth's Core; Earth's Core-Mantle Boundary; Earth's Crust; Earth's Crust-Mantle Boundary: The Mohorovičić Discontinuity; Earth's Differentiation; Earth's Magnetic Field: Origins; Earth's Magnetic Field: Secular Variation; Earth's Magnetic Field at Present; Earth's Magnetosphere; Earth's Mantle; Earth's Origin; Earth's Rotation; Earth's Shape; Earth's Structure; Eclipses; Greenhouse Effect; Van Allen Radiation Belts.

Earth's Origin

Categories: Earth; The Solar System as a Whole

The Earth's early formation, its subsequent internal differentiation, its active plate tectonics, and its external weathering have left little substantive evidence of its origin intact for direct study. Much about the materials and formative processes involved in the planet's origin can be deduced, however, from seismology, geomagnetics, and the study of meteorites and comets.

OVERVIEW

In order to understand the origins of the Earth, it is necessary to be aware of the sources of the materials from which it is made. The matter from which the Earth and the entire universe is made was created in the big bang, about 13 to 14 billion years ago. The processes that occurred in the first few minutes after the big bang produced all the hydrogen and most of the helium in the universe today; trace amounts of lithium and beryllium also were formed.

Later, after stars formed, they produced all the other chemical elements through nuclear fusion reactions, also called nucleosynthesis. For most of their lives, stars generate the energy to shine by fusing lighter atomic nuclei together to make heavier atomic nuclei. This requires high temperatures and densities, the conditions that exist in the interiors of stars. The first step is the fusion of four hydrogen nuclei into one helium nucleus. The next step is the fusion of three helium nuclei into one carbon nucleus, maybe adding a fourth helium to produce oxygen. This is the end of energy generation and nucleosynthesis for a Sun-like star before it puffs off its outer layers as a planetary nebula, and the exposed core cools and fades to end its life as a white dwarf and ultimately as a black dwarf. (Note that planetary nebulae have nothing to do with planets or the formation of planets. The name dates back to the 1700's, when, viewed through telescopes of that time, they looked fuzzy, like nebulae, and round, like planets).

More massive stars have a more spectacular demise. After carbon and oxygen have formed, further fusion reactions continue to form heavier elements up to iron. The production of elements even heavier than iron does not generate energy but requires the input of energy. The iron core collapses, and the outer layers collapse on top of it and then rebound, tearing the star apart in a supernova explosion. The tremendous energy released in a supernova permits the formation of the rest of the chemical elements. The chemical elements produced during the star's life and death are dispersed by the supernova explosion into interstellar space, there to enrich clouds of gas called nebulae (containing mostly hydrogen and some helium) in the heavier chemical elements.

The Sun, planets, and other bodies of the solar system formed as the result of the gravitational contraction of part of such a nebula about 4.5 to 4.6 billion years ago. The portion that would become the solar system, called the solar nebula, initially was perhaps about a light-year (about 9.5 trillion kilometers) across and was composed of about 74 percent hydrogen, about 24 percent helium, and about 2 percent all the other chemical elements. The exact mechanism responsible for the initiation of the solar nebula's contraction is still speculative. It may have involved the compression of the nebula as it passed through a spiral arm density wave as the nebula orbited the center of our galaxy, the Milky Way. It may have been triggered by a shock wave propagating through the nebula when a nearby massive star went supernova. It is generally agreed that once started, gravitational effects within the solar nebula kept the process going.

Any initial slow rotation of the solar nebula increased its speed with the contraction of the nebula to conserve angular momentum. (The same effect is seen on spinning figure skaters, whose rotational speed increases as they bring their arms close to their bodies.) The increase in rotational speed caused the nebula first to become oblate and eventually to form a flattened equatorial disk. Most of the solar nebula's mass concentrated at the center of the disk, forming the proto-Sun, which grew hotter by gravitational contraction.

As the proto-Sun formed at the center, fractional condensation began—a process in which gaseous matter solidifies into small, sand-sized grains only in regions where the ambient temperature is below the material's melting point. Only metallic grains of iron and nickel condensed close to the proto-Sun. Farther out, where the temperature was lower, they were joined by grains of silicate minerals. Still farther out, various ices of water, carbon dioxide, methane, and ammonia could condense. These solid grains collided with one another and stuck together in a process called accretion, forming planetesimals that grew in size. Within a time span of a few tens of millions to perhaps one hundred million years, the largest planetesimals grew into protoplanets, while the smaller ones became the many satellites and other minor members of the solar system.

As the protoplanets continued to grow, their gravitational influence grew as well. They could attract greater amounts of disk material, thus accelerating their growth while at the same time sweeping the surrounding interplanetary space clean. Solar radiation could then penetrate the space between the Sun and the planets, bringing light and heat to their still-evolving surfaces. It was during this time that the planets started to evolve in different ways. The third planet from the Sun, Earth, and its inner solar system neighbors (Mercury, Venus, and Mars) had relatively weak gravitational fields, which, coupled with their now high surface temperatures and exposure to the solar wind, caused them to lose significant amounts of the lighter gases. This first atmosphere, probably consisting of hydrogen, helium, methane, ammonia, carbon dioxide, and water vapor (gases common in the solar nebula), escaped from the inner planets and was blown away into the outer solar system. The outer planets, Jupiter and beyond—because of their colder temperatures, their greater masses, and consequently the lessened influence of the solar wind—were not so affected. As a result, they became the low-density gas/liquid/ice "giant" planets with small, rocky/metallic cores.

As the accretion process drew to a close, the Earth (along with the other inner planets) was subjected to a final intense bombardment of impacting planetesimals. As each colliding object struck the Earth's surface, its energy of motion was converted into heat energy. Furthermore, radioactivity levels were much higher in the very early Earth, since many of the radioactive elements with shorter half-lives had not yet decayed. As the Earth grew larger in size, it tended to insulate itself, making it more difficult for the energy released by radioactive decay in its interior to reach its surface and escape. All of these effects served to increase the early Earth's temperature to the point that it at least partially melted, allowing chemical differentiation and the development of the Earth's layered internal structure. Molten blobs of heavy metals like iron and nickel sank to the center, forming the iron-rich core. Less dense silicate and oxide minerals remained behind, forming the mantle. The surface also melted, forming a magma ocean perhaps a few hundred kilometers deep. Eventually the surface cooled and hardened into a thin, primitive basaltic crust, probably similar to present-day ocean-floor crust.

At the same time, outgassing released gases trapped in the interior at a prodigious pace, producing the Earth's second atmosphere, probably consisting mostly of water vapor, carbon dioxide, and sulfur dioxide, with smaller amounts of nitrogen, hydrogen sulfide, and other gases, but no free oxygen. (Outgassing continues today through volcanoes, fissures, and fumaroles, but at a much reduced rate.)

As the Earth cooled, water vapor in the atmosphere condensed, formed clouds, and fell as rain, forming the first streams and oceans. Due to the abundant carbon dioxide along with sulfur dioxide and hydrogen sulfide in the atmosphere, this early rain was highly acidic, resulting in rapid chemical weathering of surface rocks. The weathering products were carried by streams into the oceans, rapidly increasing their salinity. By about 4 billion years ago, the oceans had reached nearly their present volume and degree of saltiness. Large amounts of carbon dioxide from the atmosphere dissolved in the oceans, combined with other dissolved materials, and precipitated out as sediment (mostly as the mineral calcite, as calcium carbonate). As other gases were removed from the atmosphere, nitrogen remained, eventually becoming the major atmospheric constituent.

Probably by about 3.8 billion years ago, the first life appeared. Organic molecules, including amino acids, may have formed in the Earth's early atmosphere and oceans with solar ultraviolet light, lightning, or deep-sea hydrothermal vents providing the needed energy input, or they may have been delivered to the Earth's surface by impacts of comets, asteroids, and meteoroids. The earliest living organisms were anaerobic (able to survive without oxygen), since there was no free oxygen in the atmosphere and oceans. Then cyanobacteria and possibly other early organisms developed photosynthesis, using sunlight to turn water and carbon dioxide into sugars for food. This reaction released free oxygen into the oceans and atmosphere, and life evolved to utilize it to extract energy from food.

Thus the Earth was transformed from its formative stages to what it is today, a place where life thrives, where rocks are formed and weathered on the surface, and where a dynamic interior drives tectonic processes.

METHODS OF STUDY

Much of what is known about the origins of the Earth is derived by studying meteorites and comets as well as from seismology and geomagnetics. Meteorites and comets are unaltered or little-altered examples of early solar-system materials, providing information on the composition of and processes that occurred in the early solar nebula and the various types of bodies that formed from it. Seismology and geomagnetics give researchers clues about the internal structure of the planet.

Meteorites are extraterrestrial pieces of rock or metal that survived their fall through the Earth's atmosphere. The combined total composition of all meteorites is probably representative of the rocky and metallic material from which the inner planets—Mercury, Venus, Earth, and Mars—formed. Comets provide evidence of the more volatile components of the early solar system. They are composed of various "dirty" ices, indicative of materials blown out from the inner solar system by the solar wind but not before a portion was incorporated into the accreting planetesimals.

Data obtained from seismology (the study of the transmission of earthquake shock waves through the Earth) led to the discovery that the Earth's interior is divided into several distinct layers or zones. Observations of a change in speed of seismic waves near the Earth's surface led to the discovery of the Mohorovičić (Moho) discontinuity, the boundary between the crust and mantle. A low seismic velocity zone is now recognized in the upper mantle below the Moho and is used to define the lower boundary of the Earth's rigid lithosphere and the top of the "plastic," deformable asthenosphere. Another seismic discontinuity 2,900 kilometers below the surface delineates where the solid mantle is separated from the molten outer core. Later seismic work revealed the existence of a solid inner core.

Studies of the magnetic field of the Earth also give support to the zonal nature of the planet's interior. Hypotheses concerning the generation of the magnetic field within the Earth assume an iron-nickel-rich core (not unlike the iron-nickel meteorites) with a solid interior surrounded by a molten outer part. This combination could produce an electric dynamo that could sustain a magnetic field.

All these disciplines provide evidence of the Earth's formation in the early solar nebula. As more data are obtained, the picture of the Earth's origin becomes clearer and more refined.

CONTEXT

An understanding of the Earth's origin has many practical benefits. For example, the genesis of ore bodies is invaluable to prospecting for new resources. By knowing the products of various processes in the past, one is better able to predict human impact on present environments. Planetary engineering can use such information for the modification or preservation of conditions on the Earth, and maybe someday on other planets such as Mars. Meteorite size and shape studies were employed by space engineers in designing reentry vehicles and in studying their aerodynamic properties. Theories about material behavior in zero-gravity conditions similar to those in the solar nebula have led to experiments on manufacturing techniques in Earth orbit that are impossible to conduct on the Earth's surface.

Bruce D. Dod

FURTHER READING

Beatty, J. Kelly, Carolyn Collins Petersen, and Andrew Chaikin, eds. *The New Solar System*. 4th ed. Cambridge, Mass.: Sky, 1999. A general overview of the solar system and its components, this book has been organized around comparative planetology. A discussion of the various aspects of the genesis of the planets is scattered throughout. Draws heavily on the results of space exploration and on the interdisciplinary use of science to illustrate many concepts. Contains abundant illustrations, such as full-color photographs, artwork, graphs, and charts.

Brush, Stephen G. *Nebulous Earth: The Origin of the Solar System and the Core of the Earth from Laplace to Jeffreys*. New York: Cambridge University Press, 1996. An in-depth reference work detailing twentieth century theories on solar-system formation. Also discusses lunar origin theories.

Chaisson, Eric, and Steve McMillan. *Astronomy Today*. 6th ed. New York: Addison-Wesley, 2008. A well-written college-level textbook for introductory astronomy courses. Offers an entire chapter on the formation of planets.

Fraknoi, Andrew, David Morrison, and Sidney Wolff. *Voyages to the Stars and Galaxies*. Belmont, Calif.: Brooks/Cole-Thomson Learning, 2006. A well-written, thorough college textbook for introductory astronomy courses. Several sections deal with the origin of the solar system.

Freedman, Roger A., and William J. Kaufmann III. *Universe*. 8th ed. New York: W. H. Freeman, 2008. College-level introductory astronomy textbook, thorough and well written. Part of one chapter deals with the origin of the solar system, and part of another specifically addresses the development of the Earth.

Hartmann, William K. *Moons and Planets*. 5th ed. Belmont, Calif.: Thomson Brooks/Cole, 2005. A college textbook beyond the introductory level, with an approach based on comparative planetology. Offers much informa-

tion on the origin of the solar system in general and the Earth in particular.

Horton, E., and John H. Jones, eds. *Origin of the Earth*. New York: Oxford University Press, 1990. A collection of articles written by experts in the field. Focuses on the study of the Earth and origins of the Earth-Moon system.

Hutchison, Robert. *The Search for Our Beginning*. New York: Oxford University Press, 1983. The author addresses the problem of determining the processes involved in the formation of the Earth and other solar-system bodies through the analysis of meteorites. Links astrophysics, geology, cosmochemistry, organic chemistry, and astronomy using meteoritics as the common ground. Provides historical perspectives along with space exploration results. Contains some fine illustrations, both in color and in black and white.

Morrison, David, and Tobias Owen. *The Planetary System*. 3d ed. San Francisco: Pearson/Addison-Wesley, 2003. Designed as a text for a college course in planetology, this book contains many references to the origins of the solar system and its individual components. Comparative planetology based on space exploration results, meteoritics, and other sources are utilized throughout the text to illustrate some of the evolutionary phases in the development of the planets and other solar-system objects. Extensively illustrated.

Ozima, Minoru. *The Earth: Its Birth and Growth*. Translated by J. F. Wakabayashi. New York: Cambridge University Press, 1981. Traces the genesis of the Earth and its growth while highlighting problems addressed by isotope geochemistry. The past 4.5 billion years are sketched in terms that are easy to comprehend. Alternative hypotheses and explanations are considered.

Schneider, Stephen E., and Thomas T. Arny. *Pathways to Astronomy*. 2d ed. New York: McGraw-Hill, 2008. A thorough college textbook for introductory astronomy courses, divided into many short sections on specific topics. An entire unit is devoted to the origin of the solar system.

Smart, William M. *The Origin of the Earth*. 2d ed. New York: Cambridge University Press, 1953. An older work useful for those interested in earlier hypotheses and theories of the Earth's origin.

Tarbuck, Edward J., and Frederick K. Lutgens. *Earth: An Introduction to Physical Geology.* Illustrated by Dennis Tasa. 9th ed. Upper Saddle River, N.J.: Pearson Prentice Hall, 2008. This college-level textbook for introductory geology courses is well written and fully illustrated. Contains a chapter on the origin and historical development of the Earth.

Wasson, John T. *Meteorites: Their Record of Early Solar-System History.* New York: W. H. Freeman, 1985. Written as a text for a course on solar-system genesis, this book includes topics on meteorite classification, properties, formation, and compositional evidence linking meteorite groups with individual planets. Describes how researchers use meteorites to determine conditions in the formative periods of Earth and other planets. Includes many graphs, charts, and illustrations.

See also: Comets; Earth-Moon Relations; Earth-Sun Relations; Earth System Science; Earth's Age; Earth's Atmosphere; Earth's Composition; Earth's Core; Earth's Core-Mantle Boundary; Earth's Crust; Earth's Crust-Mantle Boundary: The Mohorovičić Discontinuity; Earth's Differentiation; Earth's Magnetic Field: Origins; Earth's Magnetic Field: Secular Variation; Earth's Magnetic Field at Present; Earth's Magnetosphere; Earth's Mantle; Earth's Oceans; Earth's Rotation; Earth's Shape; Earth's Structure; Lunar History; Solar System: Element Distribution.

Earth's Rotation

Category: Earth

The rotation of the Earth results in the days and nights that provide the daily rhythm of life. Rotation causes the Earth to be flattened at the poles and to bulge at the equator. It also produces the Coriolis force, which influences the circulation of the atmosphere and oceans.

OVERVIEW

The spinning of the Earth on its polar axis is called rotation. The ancient Greeks considered the Earth to be a motionless body in the center of a geocentric (Earth-centered) universe. An exception was Heracleides (fourth century B.C.E.), who thought that the Earth did rotate. In general, the Greeks reasoned that, in their experience, if the Earth moved, they would feel some effects of it. Later, the work of Nicolaus Copernicus, Galileo Galilei, Johannes Kepler, and Sir Isaac Newton resulted in the paradigm shift to a heliocentric (Sun-centered) system in which the Earth was a planet simultaneously rotating on an axis while revolving around the Sun. The Earth rotates from west to east, thus making the Sun, Moon, planets, and stars appear to move from east to west across the sky. We commonly refer to the Sun, Moon, planets, and stars as rising and setting because it appears that they all are moving around the Earth, when in fact the Earth is rotating on its axis while revolving around the Sun.

The Earth's axis of rotation is inclined 23.5° from the perpendicular to the ecliptic (the plane of the Earth's orbit around the Sun); thus the axis of the Earth makes an angle of 66.5° to the plane of the ecliptic. The inclination causes the seasons (and the seasonal variation in the length of day and night) as the Earth orbits the Sun during the course of a year. When either the Northern or Southern Hemisphere of Earth is tilted toward the Sun, the period of daylight is longer and night is shorter in that hemisphere. The longer duration of daylight, coupled with the Sun's rays striking that hemisphere more nearly head-on, results in summer in that hemisphere. Conversely, when either hemisphere is tilted away from the Sun, the duration of daylight is shorter and night is longer, the Sun's rays strike that hemisphere more obliquely, and that hemisphere experiences winter.

Because of rotation, a point on the Earth's equator moves 1,674 kilometers per hour; the speed decreases to 1,450 kilometers per hour at 30° north or south latitude, and to 837 kilometers per hour at 60° north or south latitude. The Earth's rotation defines the unit of time called the "day." A day is defined as the interval of time between successive passages of a meridian, or line of longitude from the North to South Pole, under a reference object (for example, the Sun or a star). A day with reference to the Sun is called a solar day, and a day with reference to the stars is called a sidereal day. Because of the Earth's orbital motion around the Sun in one year, the Sun appears to move eastward relative to the stars approximately one degree per day. This makes the solar day approximately four minutes longer than the sidereal day, since the Earth must rotate a little bit farther to complete one rotation relative to the Sun as compared to the stars.

The Earth's orbit around the Sun is slightly elliptical, and the Earth's orbital speed varies with its distance from the Sun, being fastest when Earth is closest to the Sun (perihelion), around January 3, and slowest when Earth is farthest from the Sun (aphelion), around July 4. This means the Sun's apparent motion relative to the stars varies during the year, being greatest when the Earth's orbital speed is fastest at perihelion and smallest when the Earth's orbital speed is slowest at aphelion. Thus the length of the solar day as measured by a sundial (called the apparent solar day) varies slightly during the year. The length of the apparent solar day averaged over a year is called the mean solar day, and mean solar time is the basis for the 24-hour day (of 86,400 seconds) kept by clocks. (The sidereal day is 23 hours, 56 minutes, and 4.091 seconds long.)

Because of rotation, the Earth is flattened in the polar regions and bulges at the equator, thus making it slightly ellipsoidal, an oblate spheroid. The equatorial radius is 6,378 kilometers, while the polar radius is 6,357 kilometers. Thus a point on the equator is 21 kilometers farther from the center of the Earth than either pole is. Because the Earth is slightly flattened, the length of a degree of latitude changes from 109.92 kilometers at the equator to 111.04 kilometers at the poles. Also, because a point on the equator is farther from the center of the Earth and its rotational speed is faster, the effect of gravity is reduced there compared to other points on Earth. Thus an object at the equator weighs less, about 1 pound in 200, compared to the same object at either pole.

The Coriolis force or effect, named after a

nineteenth century French engineer who studied this phenomenon, is caused by the Earth's rotation. It is an apparent force that affects free-moving bodies (such as wind, water, or missiles). For example, a fired projectile will veer to the right relative to the Earth's surface in the Northern Hemisphere, and to the left in the Southern Hemisphere. Precisely on the equator itself, free-moving objects are not deflected, but as they move north or south of the equator, the deflection becomes more pronounced.

The Coriolis effect is responsible for the global prevailing wind belts. Warm air near the equator rises and flows toward the poles. Cooled at higher altitude, the air descends around 30° north and south latitudes and spreads out both toward the equator and the poles. Air flowing toward the equator is deflected westward in both hemispheres, producing the easterly trade winds of the tropics. Air flowing toward the poles is deflected eastward in both hemispheres, producing the westerly winds of temperate latitudes. The Coriolis effect determines the direction wind blows around local high and low pressure systems in the atmosphere. Air moves outward from high pressure systems and inward toward low pressure systems. In the northern hemisphere, the moving air veers toward the right, setting up clockwise rotation around atmospheric highs and counterclockwise rotation around atmospheric lows. In the southern hemisphere, the moving air veers left, setting up counterclockwise rotation around highs and clockwise rotation around lows. This effect is especially noticeable in hurricanes (also called typhoons), which are regions of extremely low atmospheric pressure. The Coriolis force also influences ocean currents, which in turn affect the climate of coasts they flow along. The ocean currents of the Northern Hemisphere tend to flow clockwise, while those of the Southern Hemisphere flow counterclockwise. Witness the Gulf Stream of the north Atlantic and the Japanese (Alaskan) current of the north Pacific, always turning to the right, while the south Atlantic flow is to the left.

Overall, the rotation of the Earth is slowing down, and the length of the day is increasing by milliseconds per century. The decrease in rotational speed is primarily a result of the tidal fric-

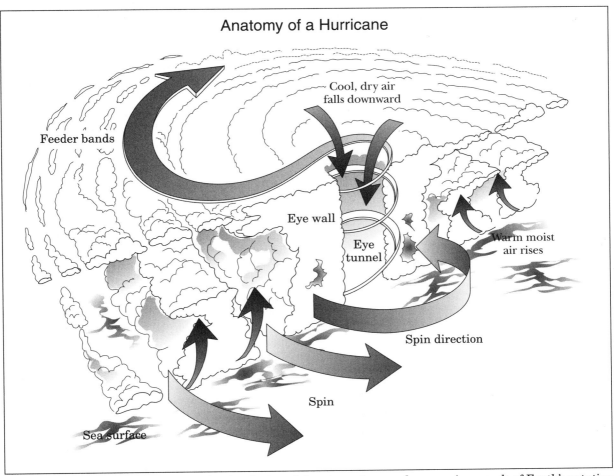

Anatomy of a Hurricane

Cool, dry air
falls downward

Feeder bands

Eye wall

Eye
tunnel

Warm moist
air rises

Spin direction

Spin

Sea surface

The Coriolis force, which influences the circulation of the atmosphere and oceans, is a result of Earth's rotation and determines the direction wind blows around local high and low pressure systems in the atmosphere. The effect is especially noticeable in hurricanes, regions of extremely low atmospheric pressure.

tion caused by the gravitational pull of the Moon and to a lesser extent the Sun. Evidence for a lengthening day during geologic history comes from the study of fossils. Clams, corals, and some other marine invertebrates add a microscopically thin layer of new shell material each day, and the thickness varies seasonally throughout the year. Counting the daily growth lines in an annual set in well-preserved fossils yields the number of days in a year. Since the length of a year (the period of the Earth's orbit around the Sun) presumably has not changed, the length of a day in past geologic times can be determined. During the early Cambrian period (540 million years ago), there were 424 days in a

year, and thus each day was about 20 hours, 40 minutes long. In the late Devonian period (365 million years ago), a year consisted of 410 days, each about 21 hours, 23 minutes long. At the beginning of the Permian period (290 million years ago), a year was down to 390 days, each about 22 hours, 29 minutes long.

As Earth's rotation slows in response to the Moon's tidal drag, the Earth's rotational angular momentum is transferred to the Moon, which increases the Moon's orbital angular momentum around the Earth, causing the Moon to move outward, away from Earth. This in turn increases the Moon's orbital period around Earth. This will continue until Earth's rotation on its axis is tidally synchronized with the Moon's revolution around Earth, both Earth and Moon keeping the same side facing each other. To conserve angular momentum, the Moon's distance and orbital period will increase to 549,000 kilometers (341,000 miles) and 46.7 of Earth's present days. Thus Earth's sidereal rotation period will be 46.7 days, and the mean solar day (time from noon to noon) will be 53.5 of Earth's present days. Since the length of the year will be unaffected, there will be only 6.8 solar days in a year.

Currently Earth's day is lengthening by an average of about 2×10^{-5} seconds each year, but the rate is quite erratic and sometimes even speeds up a bit. If this current average slowdown rate is extrapolated into the future, it will take 2×10^{11} years for Earth and Moon to become tidally locked. Alternatively, laser ranging (using retroreflectors left on the Moon's surface by the Apollo Moon landings) shows that the Moon currently is moving away from Earth at 3.8 centimeters per year. Extrapolating this rate into the future, it will take 4×10^9 years for the Moon to reach its final distance. The disagreement of these two time estimates indicates that the rates of slowdown of Earth's rotation and the increase in the Moon's distance probably will not remain constant, and the time to achieve Earth-Moon tidal lock is probably at least billions of years. However, before this can occur, the Sun probably will expand and become a red giant first.

Numerous detailed studies show that, superimposed on the systematic long-term slowdown, Earth's rotation has numerous small random changes. The reasons for these variations include transfer of angular momentum between different parts of the Earth's interior; transfer of angular momentum between the atmosphere, the oceans, and the Earth's surface; movement of air masses and changes in wind patterns; growth or shrinkage of polar ice caps; volcanic activity; earthquakes; and plate tectonic movements. The timescales for the various effects on the Earth's rotation vary, from short-term to long-term, and from systematic seasonal to erratic. It is evident from many studies that the Earth's rotational speed has varied throughout geologic time and continues

to change on a daily, weekly, monthly, seasonal, yearly, and even longer-term basis.

Currently the Earth's axis of rotation points toward Polaris, Earth's present North Star. However, the Earth's axis slowly changes direction in space in a process called precession. Recall the Earth has an equatorial bulge, and it is tilted about 23.5° from the ecliptic, the plane of the Earth's orbit around the Sun. The torque exerted by the gravitational pull of the Moon and the Sun on the Earth's equatorial bulge trying to make it line up with the ecliptic plane causes the Earth's axis to slowly precess, like the axis of a tilted spinning toy top. This precession causes the Earth's axis to trace out in space a double cone (two cones joined at their vertices) with a vertex angle of 47° (twice the 23.5° axial tilt). The Earth's axis slowly shifts direction about 50 arc seconds per year, and it takes about 26,000 years for a complete precession cycle. As the axis points to different parts of the sky in response to precession, stars other than Polaris have served and will serve as north stars. Also, in about 13,000 years (half the precessional cycle), the constellations seen during specific months on Earth will be "shifted" by six months, so that those constellations now seen during June (for example) will be seen during December, and so on. Consequently, the astronomical coordinate system of right ascension and declination slowly and systematically changes during the precessional cycle. Catalogs listing those coordinates for stars, nebulae, galaxies, and other celestial bodies must specify the epoch (year) for which the listed coordinates are rigorously correct. To point a telescope at some desired object some other year requires calculating precessional corrections to the listed coordinates.

Superimposed upon the precessional motion are two other motions. One of these is a small oscillating motion called nutation, which has a semiamplitude of 9.2 seconds of arc and a period of 18.6 years. This motion is associated with the periodic variation in the orientation of the Moon's orbital plane around the Earth with the Earth's orbital plane around the Sun. The other motion, called Chandler's wobble, has two oscillations. One of the oscillations, the Chandler component, with a period of twelve months, is a result of meteorological effects associated with seasonal changes in air masses. The second oscillation of the Chandler wobble, the 14.2-month component, is caused by shifts in the Earth's interior mass. Thus the changing direction of the Earth's rotational axis is not smooth but "wiggly" or "wobbly."

METHODS OF STUDY

Sundials were first used to mark the passage of the apparent solar day, as the shadow of the gnomon (stick) moved across the face of the dial. In 1671, the French astronomer Jean Richer made time measurements with a pendulum clock both in Paris (49° north) and in Cayenne, French Guiana (5° north) and compared the two. In French Guiana, the clock "lost" 2.5 minutes per day compared to Paris. He attributed this loss to a decrease in effective gravitational pull toward the equator due to the Earth's rotation; the practical consequence was that pendulum clocks needed to have the length of their pendula adjusted according to latitude to be able to keep accurate time.

In 1851, the French physicist Jean-Bernard-Léon Foucault hung a 25-kilogram iron ball with a 60-meter-long wire from the dome of the Panthéon in Paris, with a pin at the bottom of the ball to make marks in a smooth layer of sand underneath. After only a few minutes, the tracings in the sand showed that the plane of the ball's swing slowly rotated clockwise as seen from above. Foucault explained this as a demonstration of the Earth's rotation, which moved the attachment point on the dome and the sand on the floor, while the pendulum tried to maintain the plane of its swing in the same direction. In the 1950's, atomic clocks began to be used to measure time accurately over long periods. When time kept by these clocks was compared to time determined by the rotation of the Earth, small variations in the Earth's rotation were found.

Newer techniques used to determine length of day and polar motion involve the use of satellites and lasers. One method, called lunar laser ranging (LLR), involves the emission of light pulses from a laser on Earth to reflectors left on the Moon by Apollo and Soviet spacecraft. The returning pulses of light are received by a telescope. The total travel time is calculated to determine the Earth-to-Moon distance. By observing the time the Moon takes to cross a meridian during successive passages, this method has provided very good length-of-day measurements. Another technique involves the use of the Laser Geodynamics Satellite (Lageos). This satellite is covered by prisms that reflect light from pulsed lasers on Earth. Again, the returned beam is received by a telescope and the round-trip travel time is used to infer the one-way distance from the Earth to the satellite. This method, which includes a network of stations on Earth, can provide insight into yearly movement of crustal plates, which is believed to cause variations in the Earth's rotation. A very accurate technique known as very-long baseline interferometry (VLBI) is also being used to plot continental drift as well as variations in Earth's rotation and the position of the poles. In this method, radio signals from space (typically from quasars) are received by two radio antennas and are tape-recorded. The tapes are compared, and the difference between the arrival times of the signals at the two radio antennas is used to calculate the distance between the two. If the distance between the two antennas has changed, the crustal plates have moved.

CONTEXT

The spinning of the Earth on its polar axis once every twenty-four hours is very much a part of the daily rhythm of life. Among the primary ways Earth's rotation is felt by and governs life are its impact on our day, on gravitational pull, and on the atmosphere. Earth's rotation gives us a daily time reference by the passage of days and nights. The spinning of the Earth on its polar axis causes the Earth to bulge at the equator and to be flattened in the polar areas. Because of this phenomenon, the distance to the center of the Earth varies with latitude, and as a result the effective gravitational pull on objects on the Earth's surface also varies—objects weigh slightly less at the equator than in the polar regions of the world. The Coriolis effect, an apparent force caused by the rotation of the Earth, causes free-moving bodies to be deflected to the right in the Northern Hemisphere and to the left in the Southern Hemisphere. It governs the direction of winds as they flow in or out of pressure systems, establishing the wind belts of the world. The Coriolis force also affects the flow of ocean currents, and these patterns help to alter climates along the coasts of continents.

Roberto Garza

FURTHER READING

Bostrom, Robert C. *Tectonic Consequences of the Earth's Rotation*. New York: Oxford University Press, 2000. Reviews scientific data looking for a link between geotectonics and the Earth's rotation. The author presents a better theory to explain tidal Earth. For the more advanced reader or undergraduate.

Gould, S. G. "Time's Vastness." *Natural History* 88 (April, 1979): 18. This article summarizes the reasons for the slowing down of the Earth's rotation. It discusses the use of corals as a proof that the length of the day is increasing and that the number of days in a year is decreasing. Suitable for high-school-level readers.

Lambeck, Kurt. *The Earth's Variable Rotation: Geophysical Causes and Consequences*. New York: Cambridge University Press, 2005. Focuses on the irregular rotation of the Earth and gives detailed analysis of the various reasons for it. Covers the interdisciplinary fields of solid Earth physics, oceanography, meteorology, and magnetohydrodynamics. Technical.

McDonald, G. E. "The Coriolis Effect." *Scientific American* 186 (1952): 72. The article takes a nontechnical approach to the study of how objects move on the Earth as a result of the Coriolis effect. Suitable for high school readers.

Markowitz, W. "Polar Motion: History and Recent Results." *Sky and Telescope* 52 (August, 1976): 99. This article reviews studies of polar motion, looking at how the Earth's rotation and precessional motions are affected by various forces.

Mulholland, J. D. "The Chandler Wobble." *Natural History* 89 (April, 1980): 134. Discusses how small movements affecting the Earth's axis may be associated with other terrestrial phenomena. Suitable for high school readers.

Munk, W. H., and G. J. F. MacDonald. *The Rotation of Earth: A Geophysical Discussion*. New York: Cambridge University Press, 1960. Dated, but valuable for its detailed analytical treatment of the physics of Earth's rotation. Includes discussion of the small fluctuations in rotation as a result of redistribution of angular momentum, thought to be caused by dynamics in the fluid outer core. Designed for professional geophysicists.

Rosenburg, G. D., and S. K. Runcorn, eds. *Growth Rhythms and the History of the Earth's Rotation*. New York: John Wiley & Sons, 1975. A compilation of studies that can serve as an introduction to the methods of determining the history of the Earth's rotation. The text is suitable for college-level readers not intimidated by technical language. Each study includes a bibliography, and the book is carefully indexed by author, taxonomy, and subject.

Smylie, D. E., and L. Mansinha. "The Rotation of the Earth." *Scientific American* 225 (December, 1971): 80. This article analyzes measurements indicating that the Earth's wobble may be due to earthquakes. It is a well-illustrated article that can be understood by high school readers.

Stephenson, F. Richard. *Historical Eclipses and Earth's Rotation*. New York: Cambridge University Press, 2008. Investigates the history of the Earth's rotation by studying eclipses throughout ancient and medieval times. Shows how tides cannot be solely responsible for the lengthening of the day.

Tarbuck, Edward J., and Frederick K. Lutgens. *Earth: An Introduction to Physical Geology*. Illustrated by Dennis Tasa. 9th ed. Upper Saddle River, N.J.: Pearson Prentice Hall, 2008. This college-level textbook for introductory geology courses gives a very clear explanation of the tidal slowdown of the Earth's rotation and specific details on the number of days in the year in past geologic times based on fossil evidence.

See also: Earth-Moon Relations; Earth-Sun Relations; Earth's Magnetic Field: Origins; Earth's Magnetic Field: Secular Variation; Earth's Magnetic Field at Present; Earth's Magnetosphere; Earth's Shape; Planetary Rotation.

Earth's Shape

Category: Earth

It has been known for centuries that Earth is not a perfect sphere. The diameter of the planet is greater at the equator than it is from pole to pole. This oblateness is the result of Earth's daily rotation on its axis. Even smaller irregularities in shape have been measured by Earth-orbiting satellites.

OVERVIEW

The discovery that Earth is not a perfect sphere dates to the seventeenth century, when measurements of the distance corresponding to one degree of latitude were found to increase systematically from the equator toward both poles. Because of Earth's oblateness, one degree of latitude has a length of 110.6 kilometers at the equator and 111.7 kilometers at the poles.

Besides rotation, other forces imposed upon Earth affect the shape of the planet, a prime example being Earth tides. The oceans of the world generally have two tidal bulges, or regions where the ocean surface is relatively high, caused primarily by the gravitational attraction of the Moon and to a lesser extent by the Sun. When the Sun, Moon, and Earth all are on a straight line (the Sun and Moon either on the same side or opposite sides of the Earth), the oceans display the highest high tides and lowest low tides (spring tides), as the tidal effects of the Sun and Moon reinforce each other. When the line from the Earth to the Moon makes a right angle to the line from the Earth to the Sun, the tidal effects of the Sun and Moon partially cancel out, and the high tides are not very high and the low tides are not very low (neap tides). Not only does the water of the oceans rise and fall because of tides, but so too the surface of the "solid" Earth rises and falls very slightly due to the same tidal forces imposed by the Moon and the Sun. This periodically varying distortion is so slight as to render accurate measurements of it quite difficult.

A view of Earth from satellite distance in space would, to the naked eye, suggest that the planet is a perfect sphere, yet measurements reveal that it is not. All planets (including Earth), along with the larger satellites and asteroids, are essentially spheroidal, while smaller objects are not. Sufficiently small solid objects (such as books, boulders, and bones) can maintain any arbitrary shape because of the strength of the material of which they are composed. However, sufficiently large objects (even if solid) have internal forces due to their self-gravity that are strong enough to overwhelm the strength of whatever material composes them, and they pull themselves into spherical shapes. The critical "threshold" size depends on the density and strength of the solid material, but for the three types of solids (ices, silicate minerals, and metals, mainly iron) most common in the solar system, the threshold is approximately the same: on the order of a few hundred kilometers.

Rotation (spinning on an axis) makes a large object depart from being spherical. The Earth's rotational period, measured in a quasi-inertial frame based on distant stars and galaxies, is 23 hours, 56 minutes, and 4 seconds. As a result of this rotation, a point on the Earth's equator moves with a speed of 1,674 kilometers per hour, a point at 30° north or south latitude moves 1,450 kilometers per hour, and a point at 60° north or south latitude moves 837 kilometers per hour. The increase in rotational speed toward the equator makes the equator bulge outward, transforming a spherical shape into an oblate spheroid. The Earth's equatorial diameter is 12,756 kilometers, while its polar diameter is 12,714 kilometers. Its oblateness is defined as the difference in diameters divided by the equatorial diameter, which gives a value of 0.00336 or about one part in 298.

A comparison of the planets in the solar system shows that rotation plays a dominant role in determining oblateness. Mercury and Venus, each with very slow rotation, have no discernible oblateness and are essentially spherical. Jupiter and Saturn rotate the fastest, and both are noticeably oblate as seen in telescopes and spacecraft images. However, there are other factors. For example, Saturn rotates slightly more slowly than Jupiter, and Mars rotates slightly more slowly than Earth; yet Saturn is more oblate than Jupiter (about one part in 10 compared to one part in 15), and Mars is more

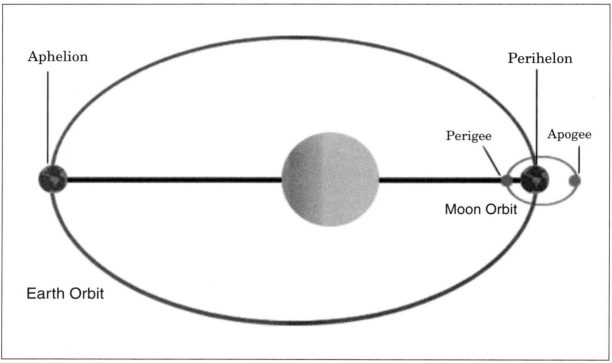

The Earth's and Moon's elliptical orbits—Earth's around the Sun, and the Moon around Earth—affect their tides and hence their shapes. (NOAA)

oblate than Earth (about one part in 200 compared to one part in 300). These "discrepancies" probably are due to the distribution of mass throughout the planet and the rigidity of the material composing the different parts of the planet.

METHODS OF STUDY

Measurements of small departures from an oblate spheroid shape became possible with the advent of the space age and the development of twentieth century instrumentation. Perturbations in the orbits of Earth-orbiting satellites show that the Earth's mass is not distributed as it would be were it a simple oblate spheroid. Some of the earliest data indicated that the Earth is slightly pear-shaped, with small bulges of up to about 100 meters near the North Pole and in a band south of the equator. More recently, satellite-mounted radar altimeters have been able to map continuously the topography of the ocean surface to an accuracy of a few centimeters by bouncing radar signals off the

water. With wave crests and troughs averaged out, ocean surfaces show small but significant deviations from a smooth oblate spheroid that reflect the topography of the seafloor underneath. Major seamounts and suboceanic ridges are clearly marked by regions of higher ocean surface above. Likewise, the major deep-sea trenches, as are common around the rim of the Pacific Ocean, are marked by troughs in the ocean surface above. This phenomenon is due to small variations in the acceleration of gravity. In the case of a seamount or ridge, there is a concentration of mass (the rock composing the feature), so gravity there is a bit stronger and attracts more water over it. In contrast, there is a deficit of mass in a trench, so gravity there is slightly weaker and attracts less water over it.

CONTEXT

The passing of geologic time has brought changes in the phenomena that control Earth's shape. The distance from the Earth to the Moon was less than at present, and Earth rotated faster in the past. These two differences would have produced larger and stronger tides, and a larger equatorial bulge. It is interesting to speculate as to what effects those changes might have had on ancient dynamic processes. Today, Earth's inhabitants suffer little effect from the planet's distortion. It cannot be observed with the naked eye, and it does not appear to play a role in weather patterns and climate.

However, one practical, though small, consequence of the Earth's shape is the variation of the effective gravitational acceleration with latitude. Due to Earth's equatorial bulge, which is the result of Earth's daily rotation on its axis, the effective gravitational acceleration, and hence the weight of objects, is slightly less at the equator than at either pole. Because of the equatorial bulge, objects at sea level at the equator are about 21 kilometers farther from the center of Earth than if they were at sea level at either pole. This alone reduces the gravitational acceleration at the equator by about 0.15 percent compared to the value at either pole. The effective gravitational acceleration at the equator is further reduced directly by Earth's rotation, since some of the gravitational acceleration that would otherwise exist is used to provide the centripetal acceleration needed to make objects follow the curved paths that keep them in contact with Earth's surface as Earth rotates. This reduces the effective gravity at the equator by about 0.35 percent. Combining the two effects, the gravitational acceleration varies from 9.832 meters per second squared at either pole to 9.780 meters per second squared at the equator, or about 0.5 percent. As a result, weight at the equator is reduced by about 0.5 percent compared to weight at either pole, so an object weighing 200 pounds at either pole weighs about 1 pound less at the equator.

John W. Foster

FURTHER READING

Greenberg, John L. *The Problem of the Earth's Shape from Newton to Clairaut.* New York: Cambridge University Press, 1995. Covers the early studies to determine the shape of the Earth by various scientists, including Isaac Newton. Explains their influence on Alexis Claude Clairaut, who confirmed Newton's belief that the Earth is flattened at the poles.

Ince, Martin. *The Rough Guide to the Earth 1.* New York: Rough Guides, 2007. A handy reference on most aspects of Earth science. Includes several diagrams and pictures to help explain material. For beginners: nontechnical and easy to read.

James, David E., ed. *The Encyclopedia of Solid Earth Geophysics.* New York: Van Nostrand Reinhold, 1989. A complete reference work on solid-Earth geophysics. Includes more than 150 articles by top scientists. Also covers topics such as geology, seismology, and gravimetry, among others. For advanced readers: detailed and technical.

King-Hele, D. "The Shape of the Earth." *Scientific American* 217 (October, 1967): 17. This article begins with the historical views of the shape of the Earth. It then discusses, with a good set of illustrations, how satellites have helped scientists to learn more about the shape of the Earth. The nontechnical approach makes the article suitable for high school and general readers.

Melchior, Paul. *The Earth Tides.* Oxford, England: Pergamon Press, 1966. A sophisticated treatment of the physical phenomenon of small distortions of Earth resulting from gravitational forces imposed by the Moon and the Sun.

Stacey, Frank D. *Physics of Earth.* 2d ed. New York: John Wiley & Sons, 1977. A reference volume on solid-Earth geophysics, including radioactivity, rotation, gravity, seismicity, geothermics, magnetics, and tectonics. Provides detailed numerical tabulations on dimensions, properties, and unit conversions.

See also: Earth-Moon Relations; Earth-Sun Relations; Earth's Age; Earth's Core; Earth's Differentiation; Earth's Magnetic Field: Origins; Earth's Magnetic Field: Secular Variation; Earth's Magnetic Field at Present; Earth's Magnetosphere; Earth's Rotation; Earth's Structure; Planetary Orbits: Couplings and Resonances; Planetary Rotation.

Earth's Structure

Category: Earth

Processes that occur in the interior of the Earth have profound effects upon the surface of the Earth and its human population. The results of such processes include earthquakes, volcanic activity, and the shielding of life-forms from solar radiation.

OVERVIEW

A simple demonstration that the Earth's interior is different from its surface is to compare the Earth's average density, calculated by dividing its mass by its volume, with the density of typical rocks from the surface. The average density is about 5.5 grams per cubic centimeter, while the density of typical surface rocks is between about 2.7 and 3.0 grams per cubic centimeter. This means that part of the interior must be composed of much denser material than surface rocks.

Evidence that the interior is differentiated into "layers" of various thicknesses, compositions, and mechanical properties comes primarily from analyzing the seismic waves produced by earthquakes that travel through the Earth. The thinnest layer is the outermost one known as the crust. The crust varies in thickness from about 5 kilometers under parts of the ocean basins up to about 70 kilometers under the highest mountain ranges of the continents. The crust is composed of a number of different rock types, but there are systematic differences between the crust of continents and that of ocean basins; continental crust is generally granitic (similar to granite), while oceanic crust is generally basaltic (similar to basalt). Both granite and basalt are igneous rocks, meaning that they cooled and hardened from hot molten material, and both are composed of silicate minerals.

However, the silicate minerals in basalt (such as pyroxene, olivine, and calcium-rich plagioclase feldspar) are comparatively rich in iron, magnesium, and calcium, giving them a generally dark color and slightly greater density (about 3.0 grams per cubic centimeter), while the silicate minerals in granite (such as quartz, potassium feldspar, and various micas) are poorer in iron, magnesium, and calcium, making them generally lighter in color and slightly lower in density (about 2.7 grams per cubic centimeter). Note that this distinction does not hold completely: Basalt and similar rocks can be found on continents, and sediments weathered from granite and similar rocks can be found in ocean basins.

The base of the crust is marked by a boundary known as the Mohorovičić discontinuity, or Moho. It represents a change in density of the rock above and below it. Rocks just below the Moho are slightly denser, about 3.3 grams per cubic centimeter, than either continental or oceanic crustal rocks. The rocks below the Moho probably are peridotite. Peridotite, composed mostly of olivine and pyroxene, is similar to basalt, but it is richer still in iron and magnesium. Peridotite is thought to represent the general composition of the layer underlying the crust, called the mantle. The mantle comprises the bulk of the Earth, representing about 80 percent by volume.

In the upper mantle, at depths starting about 100 kilometers beneath the surface and extending down to about 410 kilometers, is a zone of less rigid and more plastic, perhaps even partially melted, material called the asthenosphere. The crust and the part of the mantle above the asthenosphere, acting as a rigid unit, are known collectively as the lithosphere. The change to plastic behavior in the asthenosphere occurs because temperatures there are close to the melting point of peridotite. Although temperature continues to increase below the asthenosphere, the greater pressures at greater depths are high enough to keep the rock from melting.

The asthenosphere is thought to play an important role in movements of the lithosphere above. According to the theory of plate tectonics, the lithosphere is divided into a number of plates about 100 kilometers thick that are in constant motion at speeds of up to several centimeters per year, driven by hot convection currents of material moving slowly in the plastic asthenosphere. The hot material rises along divergent plate boundaries marked at the surface by the volcanic ridge-rift system that extends through the ocean basins around the globe. The slowly moving convection currents in the asthenosphere then move laterally away from the ridge-rifts, carrying the lithospheric plates above away from the ridge-rifts. As it moves laterally, the asthenosphere cools, becoming denser and sinking back downward. The sites where the convection currents sink are places where lithospheric plates with ocean crust on top dive into the mantle in a process called subduction. At these sites, marked at the surface by trenches in the ocean basin floor, crustal rocks may be carried into the upper mantle to depths as great as 700 kilometers. Below this level, the rock may simply be too dense for the lithospheric plates to penetrate.

There are two lower boundaries within the mantle. At 410 and 660 kilometers below the surface, abrupt increases in density occur. Although one might suspect a change in composition to account for the jump in density, laboratory studies of rocks under pressure suggest an alternative explanation. The primary mineral in peridotite is olivine. The pressures at 410 kilometers and again at 660 kilometers collapse the crystalline structure and produce denser minerals with the same iron and magnesium silicate composition. At the pressure existing at 410 kilometers, olivine converts to the denser mineral called spinel, and at the even higher pressure at 660 kilometers, both spinel and pyroxene collapse to yet a denser mineral known as perovskite. Thus the changes occurring in the mantle to produce the asthenosphere and the discontinuities at 410 and 660 kilometers are not changes in composition but instead changes in physical properties caused by temperature and pressure. The density increases from about 3.3 grams per cubic centimeter at the top of the mantle to about 5.6 grams per cubic centimeter at its base.

The next layer beneath the mantle is the outer core. This layer begins at a depth of about 2,900 kilometers beneath the surface and continues to a depth of 5,100 kilometers. There is a large density increase across the core-mantle boundary, from 5.6 grams per cubic centimeter at the base of the mantle to about 10 grams per cubic centimeter at the top of the core. Iron is the only reasonably abundant element that would have the required density at the tremendous pressure of millions of atmospheres at these depths. However, pure iron would give too high a density, so iron mixed with about 15 percent nickel, sulfur, silicon, and possibly oxygen and even hydrogen has been suggested. At the pressures and temperatures that must exist in the outer core, iron alloys would be in a molten state. Complex currents of metallic iron alloy, generated in the fluid outer core by convection and the Earth's rotation, give rise to the Earth's main magnetic field through a geodynamo process.

The core-mantle boundary represents a composition change from the silicate minerals of the lower mantle to the metals of the core. The boundary is a sharp one, but whether it is smooth and spherical in shape or irregular with "hills" or "peaks" on its surface is not known. There is some evidence from seismology that the lower mantle within 100 kilometers of the core boundary is a transition zone with a change of properties. It may consist of a mix of mantle and core material that is less rigid than the mantle rocks above it.

The innermost layer of the Earth's interior is the inner core. This region has a radius of about 1,300 kilometers where there is a boundary with the outer core. Increasing pressure at these depths requires that the iron of the inner core is solid. It is thought the solid inner core continues to grow in size as iron in the molten outer core crystallizes as Earth slowly cools. Because the solid inner core is separated from the mantle by the molten outer core, it can rotate independently. Seismic studies suggest that the inner core rotates slightly faster than the mantle and crust.

METHODS OF STUDY

Much of what is known about the structure of the Earth's interior comes from the analysis of seismic waves generated by earthquakes or by explosives detonated at or just below the surface. After passing through the Earth, the wave vibrations are recorded on seismographs located all around the world, revealing information about the part of the interior they traveled through.

The seismic waves that pass through the interior are called body waves, because they propagate through the body of the Earth and not along the surface. Body waves are of two varieties: primary (or P) waves, and secondary (or S waves). P waves are the same as acoustic or sound waves. They cause the material they traverse to move back and forth in the direction of wave travel, alternately stretching and compressing it. Like ordinary sound waves, P waves can travel through any sort of material—solid, liquid, or gas. S waves are transverse waves, which move material along the wave path from side to side. Consequently they can travel only through rigid, that is, solid, material; S waves cannot travel through liquids or gases.

Both P waves and S waves cross the asthenosphere of the upper mantle, but with reduced speed (that is why the upper mantle is sometimes called the low-velocity zone), suggesting lower rigidity but not a liquid state, since S waves do propagate through it. Therefore, it seems that the asthenosphere is a solid but plastic region, able to ooze and flow very slowly.

At the core-mantle boundary, P waves abruptly slow down by almost a factor of 2 as they enter the outer core, and S waves disappear, indicating the material of the outer core has no rigidity. Since gases cannot exist at the conditions of the outer core, the material of this region must be a liquid.

Other locations in the interior are marked by increased speeds for both P waves and S waves.

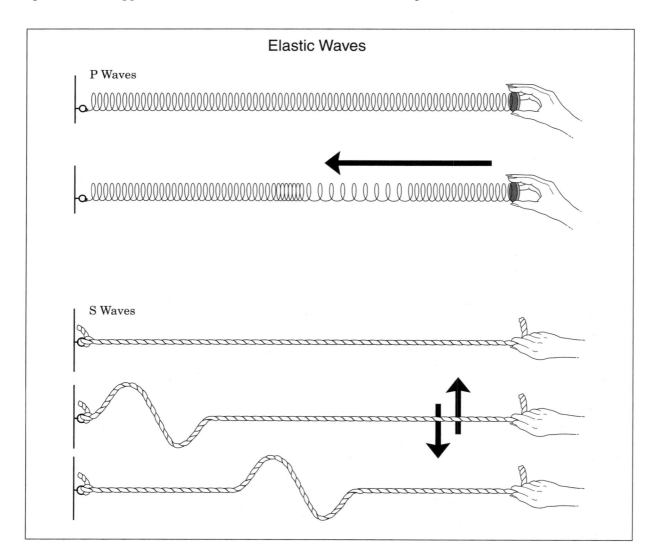

Elastic Waves

P Waves

S Waves

There is a sharp increase in speed at the Mohorovičić discontinuity at the base of the crust. The P-wave speed jumps from about 6 kilometers per second above the Moho to around 8 kilometers per second below it, while the S-wave speed jumps from about 4 to 5 kilometers per second.

Below the asthenosphere, both P-wave and S-wave speeds gradually increase to a depth of about 410 kilometers, at which point both sharply increase. Laboratory studies indicate that the increase in density when olivine collapses to form spinel accounts for the increased speeds at this depth. At a depth of around 660 kilometers, a second abrupt increase in both wave speeds occurs, here caused by a second collapse to produce the yet denser mineral perovskite. The speeds for this part of the mantle match those obtained in the laboratories from waves passing through perovskite samples placed under the kinds of pressures found at 660 kilometers. A final increase in speed is observed when P waves pass the outer to inner core boundary. This is due to a phase transition from liquid iron in the outer core to solid iron in the inner core. Such a phase transition is supported by the probable reappearance of S waves in the inner core.

Another way in which the existence of structural boundaries within the Earth can be shown from an analysis of seismic waves is to examine the way the waves reflect and/or refract when they encounter the boundaries. What the waves do depends on the angle at which they approach the boundary, as well as on the properties of materials on both sides of the boundary. P waves are reflected off the Moho, the core-mantle boundary, and the inner-outer core boundary, providing clear evidence that there are sharp boundaries between the crust and mantle, the mantle and outer core, and the outer and inner cores. Waves have also been detected reflecting off the 660-kilometer discontinuity.

Refraction or bending of waves yields further evidence. As P waves cross the mantle-core boundary, they are refracted toward the center of the Earth because their speed is less in the molten outer core. This deflection of P waves passing through the outer core leaves a gap stretching around the Earth in the form of a band extending from 100° to 140° from the epicenter of the earthquake. This gap is known as the P-wave shadow zone because no P waves reach the surface in this band.

Advances in computer science have allowed the identification of even subtler details about the Earth's interior. Computerized tomography is a technique used in medicine, in which X rays from all directions are analyzed in a computer to give a three-dimensional picture of the internal organs of the human body. Seismic tomography is an analogous approach that uses seismic waves that travel from earthquakes to seismographs around the world to map the Earth's interior. This technique includes both P and S body waves traveling through the interior as well as surface waves traveling along the surface of the Earth. By looking at the travel times of the different waves, scientists are able to compare speeds along different paths. Such an approach has resulted in maps of slow and fast regions of the mantle that probably represent less rigid (warmer) and more rigid (cooler) regions.

CONTEXT

The interior of the Earth has profound effects on the surface. The interior acts as a complex heat engine that is the driving force behind plate tectonics, resulting in the formation and evolution of oceanic and continental crust. In the process, earthquakes and volcanic activity occur that create hazards for the human population on the Earth's surface. Complete acceptance of the plate tectonic theory could not have occurred without the discovery of the asthenosphere, which makes the movement of the lithospheric plates plausible.

It is fortunate that the Earth has a magnetic field. Without it, the age of discovery and exploration would not have been possible, for navigation by magnetic compasses allowed voyages across uncharted oceans. It is now theorized that the Earth's magnetic field is generated in the molten metallic outer core by the geodynamo process. An important effect of the core-generated magnetic field is the changes it has undergone through time. In particular, at rather irregular intervals of geologic time, the magnetic poles reverse polarity; during such reversals, the magnetic field decreases in strength. Since the magnetic field shields life-forms on the Earth's surface from charged particles emitted by the Sun, there is some concern that such field weakenings during polar reversals could result in more cancers and genetic mutations. Some scientists suspect that polar reversals might be at least partly responsible for some of the mass extinctions that have occurred in the geologic past, as well as periods of rapid evolution. Thus, the surface of the Earth as well as the life-forms on it depend upon and are strongly affected by processes occurring within the Earth's interior.

David S. Brumbaugh

FURTHER READING

Bolt, Bruce A. "Fine Structure of the Earth's Interior." In *Planet Earth*. San Francisco: W. H. Freeman, 1974. An extremely well illustrated review of how seismic waves have been used to discover and define the various layers of the Earth's interior. Written at a general-interest college level. Little background or expertise in mathematics is required.

_____. *Inside the Earth: Evidence from Earthquakes*. San Francisco: W. H. Freeman, 1982. This book is written for undergraduate college students in physics and the Earth sciences and for nonspecialists interested in a more detailed summary of knowledge of the Earth's interior. The text is relatively free of mathematics and is clearly and well illustrated. It offers a concise and readable treatment of the use of seismic waves to discover and interpret the Earth's interior.

Fowler, G. C. *The Inaccessible Earth: An Integrated View to Its Structure and Composition*. 2d ed. New York: Chapman and Hall, 1993. A commonly used introductory geophysics text. Readers should have knowledge of basic calculus. Includes a series of questions and problems.

Tarbuck, Edward J., and Frederick K. Lutgens. *Earth: An Introduction to Physical Geology*. Illustrated by Dennis Tasa. 9th ed. Upper Saddle River, N.J.: Pearson Prentice Hall, 2008. This college-level textbook for introductory geology courses is well written and illustrated. There is a very good chapter on the Earth's interior, as well as several chapters on volcanoes, earthquakes, and plate tectonics that describe their relationships with the interior structure and processes.

Van der Pluijm, Ben, and Stephen Marshak. *Earth's Structure*. 2d ed. New York: W. W. Norton, 2003. An introductory text on structural geology and tectonics. Designed for undergraduate students.

Vogel, Shawna. *Naked Earth: The New Geophysics*. New York: Plume, 1996. Covers the field of geophysics from 1960 with the theory of plate tectonics through the mid-1990's. Written with little technical jargon and many examples.

See also: Earth-Moon Relations; Earth-Sun Relations; Earth System Science; Earth's Age; Earth's Atmosphere; Earth's Composition; Earth's Core; Earth's Core-Mantle Boundary; Earth's Crust; Earth's Crust-Mantle Boundary: The Mohorovičić Discontinuity; Earth's Differentiation; Earth's Magnetic Field: Origins; Earth's Magnetic Field: Secular Variation; Earth's Magnetic Field at Present; Earth's Magnetosphere; Earth's Mantle; Earth's Oceans; Earth's Origin; Earth's Rotation; Earth's Shape.

Eclipses

Categories: Earth; The Solar System as a Whole

Eclipses, occultations, and transits occur when three celestial bodies line up, causing the middle body to block the path of light between those on the two ends. In particular, solar and lunar eclipses witnessed from Earth are spectacular phenomena that have been objects of awe, study, and speculation since ancient times. Once they were understood, they became powerful tools of science used to investigate topics as diverse as geodesy and general relativity.

OVERVIEW

Eclipses of the Sun and Moon are impressive events. They have captivated people since before recorded history and continue to excite us today. Once considered great omens or portents, they have become among the most powerful means with which science tests theories in a remarkable variety of areas.

A lunar eclipse (or eclipse of the Moon) occurs when the Moon passes through the shadow of the Earth. For this to happen, the Moon must be on the side of the Earth opposite the Sun, at the time when the Moon's phase is full. A solar eclipse (or eclipse of the Sun) occurs when the Moon passes between the Earth and the Sun, blocking all or part of the Sun as seen from Earth. For this to happen, the Moon's phase must be new.

A partial solar eclipse was visible from Earth as the Moon obscured the Sun in 1994. (©Sébastien Gauthier/NASA)

Not every full or new Moon results in an eclipse, since the orbit of the Moon lies in a plane tilted about 5° to the ecliptic plane, the plane containing Earth's orbit around the Sun and in which the Earth and the Sun always lie. Unless a full or new Moon occurs when the Moon is very close to crossing the ecliptic plane, the Earth's shadow will miss the Moon (at full Moon) or the Moon's shadow will miss the Earth (at new Moon), and no eclipse will take place. This condition has been known and used to predict eclipses since ancient times and is the source of the name of the ecliptic plane.

A lunar eclipse is visible from all points on Earth where the Moon is above the horizon. It may be either umbral or penumbral. During an umbral lunar eclipse, at least part of the Moon passes through the Earth's umbra, the dark inner shadow in which the Earth blocks light coming from all parts of the Sun. If the entire Moon passes through the umbra, it is called a total lunar eclipse; if only part of the Moon passes through the umbra, it is called a partial lunar eclipse. During a penumbral lunar eclipse, the entire Moon misses the umbra and passes only through the Earth's penumbra, a region of partial shadow surrounding the umbra in which

Earth cuts off light from some but not all parts of the Sun. An observer on the Moon during a penumbral lunar eclipse would see part of the Sun covered by the Earth and part of the Sun extending beyond the edge of the Earth.

The Moon is dimmed slightly while in the penumbra, but it does not darken appreciably unless it enters the umbra. When the Moon enters the umbra, the previously bright surface of the full Moon darkens to a much dimmer reddish glow, illuminated only by sunlight that has been refracted and scattered around the Earth by Earth's atmosphere. The brightness and color of this illumination can vary markedly from one umbral lunar eclipse to another (from orangish red to a dull reddish brown to a ghostly brownish gray), depending on the atmospheric conditions on Earth. Occasionally, some areas of the Moon will seem less illuminated than others.

A total lunar eclipse can last several hours. From the time the Moon begins to enter the umbra, it takes about an hour for the eclipse to become total. Totality can last for nearly two hours, with another hour required for the Moon to leave the umbra entirely. The limb (or edge) of the Moon nearest the observer's eastern horizon enters the Earth's shadow first, and at the end of the eclipse it is this limb that brightens first.

Solar eclipses are somewhat more complex. By coincidence, the Sun and the Moon both have nearly the same apparent size or angular diameter—about one-half degree of arc—as viewed from Earth. The Sun's actual diameter is about 400 times larger than the Moon's, but the Sun is also about 400 times farther away from the Earth than the Moon is. Because the orbit of the Moon around the Earth and the orbit of the Earth-Moon system around the Sun are both slightly elliptical, the apparent (angular) size of the Moon and Sun varies as seen from the Earth. On average, the angular diameter of the Moon, 0.518°, is slightly less than the angular diameter of the Sun, 0.533°, as seen from Earth.

However, when the Moon is at perigee (its closest approach to Earth), its angular diameter increases to 0.548°. When the Earth is at perihelion (its closest approach to the Sun), the Sun's angular diameter increases to 0.542°. Thus, when the Moon is near perigee, its angular size always exceeds that of the Sun, and if a solar eclipse occurs then, the Moon can completely cover the Sun.

The Moon's umbra is conical in shape, with its base at the Moon and narrowing to its apex (or tip) as it nears Earth. If the Moon is near perigee, the apex of its umbra will fall inside the Earth, and observers in the region of the Earth's surface within the umbra will see the silhouette of the Moon completely cover the Sun's visible surface (its photosphere); this is called a total

The shadow of the Moon darkens and moves across Earth during a solar eclipse; those on Earth's surface in the center of the shadow see a total solar eclipse, while those at the perimeters see a partial solar eclipse. (Centre National d'Études Spatiales)

solar eclipse. The region of the Earth's surface within the umbra at any moment is quite small, never more than a few hundred kilometers across. Due to the orbital motion of the Moon around the Earth and the Earth's rotation on its axis, its umbra sweeps a path (the eclipse track) thousands of miles long across the Earth's surface from west to east at speeds always exceeding 1,700 kilometers per hour relative to the Earth's surface. The maximum duration possible for totality is about seven and a half minutes, although the complete eclipse, including the partial phases at the beginning and end as the Moon slowly covers and then uncovers the Sun, may take more than four hours.

Because the Moon's average angular size is slightly smaller than the Sun's average angular size, the tip of the Moon's umbra will not always reach the Earth's surface during a solar eclipse. If this is the case, the Moon will not completely cover the Sun's photosphere. When the Moon is centered on the Sun, a narrow ring, or annulus, of the Sun's bright photosphere remains visible around the Moon's silhouette. This is called an "annular solar eclipse." Annular solar eclipses are about 20 percent more frequent than total solar eclipses.

Beyond the region of the Earth's surface in which a total or annular solar eclipse is seen, there is an area thousands of kilometers wide inside the Moon's penumbra. Within this area, the silhouette of the Moon covers part but not all of the Sun's photosphere. This is called a partial solar eclipse.

Occultations and transits are phenomena similar to eclipses in which the apparent angular size of the body in front is substantially larger or smaller than the apparent angular size of the body in back. An apparently large body moving in front of an apparently smaller one is called an occultation. The Moon frequently occults bright stars, which are seen to wink out instantly when they pass behind the limb of the Moon. The Moon also occasionally occults planets, and planets are seen to occult their satellites and stars. (Actually, other planets can both occult and eclipse their satellites as seen from Earth. The distinction between the two phenomena is that the satellite moving behind the planet is an occultation, while the sat-

ellite passing through the planet's shadow is an eclipse. The two events are not necessarily precisely coincident in time, because the planet's shadow cone will not be directly behind the planet as seen from Earth unless the planet is on the side of the Earth almost exactly opposite the Sun.) The Sun also occults objects, but because of the Sun's brightness, such solar occultations cannot be seen at visible wavelengths; however, they have been observed at radio wavelengths when the object being occulted is a source of radio emission. An apparently small body moving in front of an apparently larger one is called a transit. On rare occasions it is possible to witness from Earth the transit of the planets Venus or Mercury across the Sun. When this happens, the planet appears like a small, black dot moving across the face of the Sun. Transits of satellites and their shadows across their parent planets also can be observed.

Another, related phenomenon is that of eclipsing binary stars. Some of the stars seen in the sky are in fact pairs of stars orbiting about one another. If the Earth lies near the plane of their mutual orbit, the combined light of the system is seen to vary in brightness as the stars alternately block each other from Earth's view. By observing these variations in brightness, astronomers can determine some of the characteristics of the individual stars (such as relative sizes and surface temperatures) and study their interactions.

METHODS OF STUDY

Ancient peoples, who used astronomical observations to keep track of planting seasons and the like, usually imputed magical or spiritual significance to eclipses and consequently tried to predict them. They did this by watching the changing position of the Moon against the background of the stars and by recording patterns in the recurrence of eclipses.

A lunar, or synodic, month is the period from one new Moon to the next, about 29.53 days. The draconic month is the time required for the Moon to complete one cycle of crossing and recrossing the ecliptic plane (from south to north and from north to south), about 27.21 days. The coincidence of these two cycles produces eclipses, so the pattern of eclipses starts again when the two cycles return to the same relative matchup. This happens every 223 lunar months (equaling 6,585.32 days or 242 draconic months or a little more than 18 years), in a repeating pattern called the saros cycle. The cycle lasts 18 years and 11.32 or 10.32 days, depending on whether 4 or 5 leap years occur during that time. The extra third of a day (or about 8 hours) means that eclipses 223 lunar months apart, although similar in overall geometry, will occur about 120° farther west in longitude because of the Earth's rotation. After three such saros cycles, about 56 years and 1 month, eclipses repeat in nearly the same part of the Earth again.

These cycles, at least as they applied to lunar eclipses, were known to Babylonian astronomers by around the eighth century B.C.E. and may have been known to some peoples long before that (based on disputed interpretations of a circle of fifty-six pits around the neolithic monument at Stonehenge in England). Knowledge of these cycles enabled the Babylonians to predict the relative motions of the Sun and Moon in the sky. The saros cycle was of limited use in predicting solar eclipses, because the path of totality is so narrow. Very precise knowledge of the relative motions of the bodies involved was re-

quired. This was not possible before Sir Isaac Newton developed his laws of motion and gravity along with the calculus in the seventeenth century. One of the first tests he applied to his new methods was the calculation of the orbit of the Moon.

Centuries of refinement, both of mathematical methods and of measurements of the positions of the Moon, Earth, and Sun relative to one another, were necessary to achieve modern accuracy in eclipse predictions. Astronomers can now calculate eclipses, including exact times and paths of totality, many years into the future with almost total precision. However, even these calculations are limited by residual uncertainties in the motions of the bodies involved when extrapolations hundreds of years in the past or future are attempted. With three bodies gravitationally interacting, no exact solution for the orbits is possible, although modern approximation methods are very good. Furthermore, the rate of rotation of the Earth has varied over time and continues gradually to slow, complicating the calculation of eclipse times and locations far back into the past or far forward into the future.

A total solar eclipse is perhaps the most spectacular natural event that can be seen. During a total solar eclipse, the Moon appears as a dark

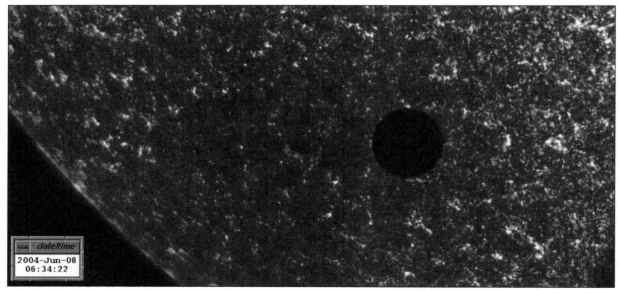

The planet Venus can be seen crossing the solar face during its 2004 transit. (NASA)

disk that slowly moves across and covers the bright disk of the Sun. Just before the Sun is completely covered, the remaining bright crescent narrows until it becomes a chain of bright spots along the edge of the Moon. These spots, called Bailys beads, represent a last glimpse of the Sun's photosphere between mountains at the edge of the Moon. Then for a few seconds, the Sun's chromosphere (a thin layer of transparent gas above the photosphere) can be seen as a red fringe along the leading edge of the Moon. At about this time, rapidly moving shadow bands, striations of light and dark a few centimeters across, can be seen rippling across the ground and along walls. These are believed to be due to atmospheric refraction. The sky turns dark during totality, but not completely dark; some light is scattered into the umbra from outside the region of totality.

The Hubble Space Telescope captures three of Jupiter's moons—Io, Ganymede, and Callisto—crossing the gas giant's face and casting shadows on its surface in a rare triple eclipse. (NASA/ESA/E.Karkoschka, University of Arizona)

The darkness at totality produces uneasiness among some animals, and birds are sometimes seen to go to roost, as at sunset. The solar corona, the Sun's outer atmosphere of hot ionized gases, is seen as a glowing white halo around the dark silhouette of the Moon. Smaller, fiery red solar prominences are often observed around the edge of the Moon. As totality ends, the shadow bands, chromosphere, and Baily's beads can be seen briefly again.

It is important never to look directly or through optical instruments such as telescopes, binoculars, or camera viewfinders at the uneclipsed or partially eclipsed Sun without using suitable filters manufactured for this purpose. Common sunglasses are insufficient to prevent severe, painful, and permanent eye injuries, which can occur nearly instantaneously. However, during totality, when the Moon completely covers the Sun's photosphere and only the co-

rona is visible, no filter is needed. Furthermore, no filters are needed to watch lunar eclipses. A telescope or at least binoculars generally are needed to enlarge the view of most occultations and transits.

CONTEXT

Eclipses, occultations, and transits have been powerful tools to derive useful information, to make new discoveries, and to test and confirm various theories and predictions. One of the oldest such uses can be traced back to the ancient Greeks. They understood how lunar and solar eclipses occurred, and the circular outline of the Earth's shadow seen on the Moon during lunar eclipses was cited by Aristotle (384-322 B.C.E.) and other Greek philosophers as evidence that the Earth must be spherical.

In 1675, the Danish astronomer Ole Rømer was studying the orbital motion of Jupiter's four largest moons (its Galilean satellites) by carefully timing their eclipses in Jupiter's shadow. He discovered that the eclipses occurred later than expected when Earth and Jupiter were farther apart and earlier than expected when Earth and Jupiter were closer. He realized that this phenomenon could be explained if light did not travel instantaneously but took time (about 16.6 minutes by his measurements) to cross the Earth's orbit (a distance of 2 astronomical units). Since the length of the astronomical unit was not known accurately then, he never actually calculated the speed of light in "everyday" type units, but this was the first demonstration that light traveled at a finite speed.

Eclipses of the satellites of Jupiter, as well as much less frequent lunar and solar eclipses, helped seafarers find their location at sea and map the Earth. Although latitudes can be determined easily by measurements of the maximum altitude above the horizon reached by the Sun in the daytime or specific stars at night, longitudes cannot be determined astronomically without a time reference. Pendulum clocks did not run accurately at sea because of the motion of the ship, but the calculated times of eclipses provided the necessary time reference and permitted reliable navigation and the construction of accurate maps.

Solar eclipses have helped resolve some of the most important questions in science. They provided an infrequent opportunity for observing some of the Sun's features—such as its corona, chromosphere, and prominences—which, before the development of modern instruments, were otherwise hidden most of the time by the brightness of the Sun. The element helium (from Helios, the Greek Sun god) was first discovered in the flash spectrum of the chromosphere during a total solar eclipse in 1868.

An observational test of general relativity was carried out first during the 1919 total solar eclipse. In 1915, Albert Einstein had published his then-controversial general theory of relativity. One of its predictions was that light passing by a massive object, such as the Sun, would be deflected by gravity. By photographing the star field around the totally eclipsed Sun and comparing the apparent positions of the stars near the edge of the Sun to a photograph of the same star field when the Sun was in a different part of the sky, astronomers verified this prediction of general relativity. It has been confirmed repeatedly at several total solar eclipses since then. Radio interferometry observations of a shift in the position of quasar 3C273 when it is occulted by the Sun have provided even more accurate confirmation.

In 1977, astronomers observed the occultation of a star by the planet Uranus to study its atmosphere by the way it absorbed light from the star. They were surprised when the star faded and brightened several times before and after being occulted by Uranus itself. This was attributed to a set of rings around Uranus, and was the first discovery of rings around a planet other than Saturn.

Firman D. King,
revised by Richard R. Erickson

FURTHER READING

Baker, Robert H. *Astronomy*. 7th ed. Princeton, N.J.: Van Nostrand, 1959. Baker's classic astronomy text provides a complete and lucid description of eclipse phenomena, along with related issues in spherical astronomy.

Brewer, Bryan. *Eclipse*. 2d ed. Seattle: Earth View, 1991. A good introduction to eclipse phenomena and their history. For the general reader.

Chaisson, Eric, and Steve McMillan. *Astronomy Today*. 6th ed. New York: Addison-Wesley, 2008. A well-written college-level textbook for introductory astronomy courses. Has several pages on eclipses.

Fraknoi, Andrew, David Morrison, and Sidney Wolff. *Voyages to the Stars and Galaxies*. Belmont, Calif.: Brooks/Cole-Thomson Learning, 2006. A well-written, thorough college textbook for introductory astronomy courses. Includes a section on eclipses.

Freedman, Roger A., and William J. Kaufmann III. *Universe*. 8th ed. New York: W. H. Freeman, 2008. College-level introductory astronomy textbook, thorough and well written. A major part of one chapter deals with eclipses.

Schneider, Stephen E., and Thomas T. Arny. *Pathways to Astronomy*. 2d ed. New York: McGraw-Hill, 2008. A thorough college textbook for introductory astronomy courses, divided into many short sections on specific topics. Eclipses are discussed in several sections.

Stevenson, F. Richard. "Historical Eclipses." *Scientific American* 247 (October, 1982): 170-183. This article discusses how historical records and astronomical calculations are compared to resolve questions in both history and astronomy.

Taff, Laurence G. *Celestial Mechanics*. New York: Wiley-Interscience, 1985. This book describes how orbits are calculated and makes it clear why the process becomes so complicated when long time intervals are involved.

Zirker, Jack B. *Total Eclipses of the Sun*. Expanded ed. Princeton, N.J.: Princeton University Press, 1995. Zirker explains how solar eclipses are observed and describes some of the scientific results that have been obtained from them.

See also: Auroras; Earth-Moon Relations; Earth-Sun Relations; Earth System Science; Earth's Age; Earth's Atmosphere; Earth's Composition; Earth's Core; Earth's Core-Mantle Boundary; Earth's Crust; Earth's Crust-Mantle Boundary: The Mohorovičić Discontinuity; Earth's Differentiation; Earth's Magnetic Field: Origins; Earth's Magnetic Field: Secular Variation; Earth's Magnetic Field at Present; Earth's Magnetosphere; Earth's Mantle; Earth's Oceans; Earth's Origin; Earth's Rotation; Earth's Shape; Earth's Structure; Greenhouse Effect; Solar Chromosphere; Van Allen Radiation Belts.

Electromagnetic Radiation: Nonthermal Emissions

Category: The Cosmological Context

Objects emit light in many ways. Different conditions cause different forms of light to be emitted. Understanding the different forms of light emission can help in understanding the conditions where the light is emitted by studying the light itself.

OVERVIEW

Electromagnetic radiation is the name given to the phenomenon of waves of coupled electric and magnetic fields. Visible light is one form of electromagnetic radiation. Other forms, such as radio waves and X rays, are different solely in the frequency at which they oscillate and their wavelengths. The energy of electromagnetic radiation is linearly proportional to the frequency of the radiation. Different physical processes produce different forms of radiation. Among the most commonly produced electromagnetic radiation is thermal radiation (sometimes called blackbody radiation). Thermal radiation is produced by the molecular motion of anything that is not at absolute zero temperature. A plot of intensity versus wavelength for thermal radiation produces a well-defined characteristic spectral profile. However, there are many other physical processes that produce electromagnetic radiation other than through thermal means.

One of the most common forms of nonthermal electromagnetic radiation is emission spectra. When an electron in an atom jumps from a higher-energy state to a lower one, it emits electromagnetic radiation. The particular wavelength of radiation emitted depends on just the energy states in which the electron started and finished its transition. The greater the energy difference, the higher the frequency and the shorter the wavelength of radiation. Different atoms have very specific permitted energy levels for their electrons, and therefore they emit only specific radiative wavelengths. This is unlike thermal radiation, which is continuous across many wavelengths. All atoms of a particular chemical element have the same possible energy levels, so they will all have the same possible spectral line emissions, or spectral "signature" or profile. Thus, analysis of the spectral lines emitted by an object can be used to determine its chemical composition.

Another form of electromagnetic radiation emission that is similar to the emission spectra from electron transitions occurs through electron spin-flip transitions. Electrons have a small angular momentum, called the spin of the electron. This gives the electron a small magnetic moment. If that magnetic moment is

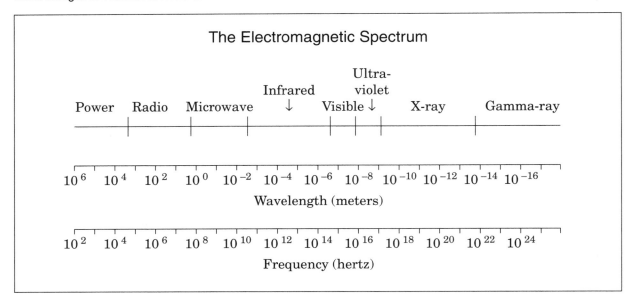

The Electromagnetic Spectrum

aligned with that of the atomic nucleus, then the atom has more energy than if it were aligned in the opposite direction. Therefore, if the electron flips its spin, the atom must emit a photon of light having energy equal to the difference in energy between the two states. One of the most important examples of this type of transition is the spin flip of hydrogen, which produces electromagnetic radiation with a characteristic wavelength of 21 centimeters.

Electromagnetic radiation comes from changing electric and magnetic fields, so any motion by a charged particle can produce electromagnetic radiation. In particular, certain types of electromagnetic radiation are produced from accelerated charges. One example of this type of radiation is synchrotron radiation. Acceleration is change in velocity. That change can be from speeding up or slowing down, or it can be from a change in direction of the velocity. When objects move around in circular or other curved paths, they change direction even if they do not change speed. If these objects are charged particles, then this curved motion results in the emission of electromagnetic radiation. This radiation is called curvature radiation or cyclotron radiation. However, if the charged particles are moving at or near the speed of light, the radiation is called synchrotron radiation and it takes on special properties. The synchrotron ra-

diation is beamed along the path of the electrons, which amplifies its intensity, and it is polarized (meaning that the electric fields of the various electromagnetic waves are aligned). A spectrum of synchrotron radiation shows a drop in intensity with frequency, making an easily recognized spectral profile. A common source of synchrotron radiation is electrons or other charged particles moving in a magnetic field. Jupiter's magnetosphere emits synchrotron radiation.

Another form of radiation produced by accelerated electrical charges is Bremsstrahlung radiation. The name comes from the German meaning "braking radiation" because it is produced when fast-moving charged particles abruptly slow down. Technically, the term Bremsstrahlung radiation can be applied to any radiation emitted by an accelerating charged particle, including synchrotron radiation; however, in practice the term is frequently limited to cases where the particles slow down (negative acceleration). Bremsstrahlung radiation can be generated, for example, during solar flares. Energy released in the flare event can accelerate electrons near the surface of the Sun to almost the speed of light. The electrons then interact with the outer layers of the Sun to slow down, producing a burst of X rays from the Bremsstrahlung radiation of their deceleration.

Like synchrotron radiation, Bremsstrahlung radiation has a spectral profile that decreases in intensity with increasing radiation frequency.

Another form of nonthermal emission is Cherenkov radiation, which is also produced by ultrafast charged particles. Nothing can travel faster than light in a vacuum. However, light travels more slowly through a medium than through a vacuum. When a charged particle moving near the speed of light enters a medium in which the speed of light is slower than the particle, the changing electromagnetic fields produced by the particle's motion augment each other to produce electromagnetic radiation moving in the same general direction as the particle's motion. The particle slows as a result. While this sounds much like Bremsstrahlung radiation, the physical process is different. High-energy cosmic rays produce Cherenkov radiation when they enter Earth's atmosphere.

Yet another nonthermal source of radiation is light amplification by stimulated emission of radiation (laser). An excited atom can be stimulated to emit radiation by the passage of a photon of light having the same wavelength as the light that would be emitted if the atom's electron were spontaneously to go from a high energy level to a lower one. The stimulated radiation is in the same direction and in phase with the stimulating radiation, amplifying its intensity. The atmosphere of Mars is capable of lasing infrared light. Laser light is characterized by the light's being coherent; that is, all of the light is the same wavelength, moving parallel and in phase.

Pair annihilation can provide yet another form of nonthermal electromagnetic radiation. All particles of matter have a corresponding particle of antimatter. When a matter particle comes together with its antiparticle, the particle and the antiparticle annihilate each other. The combined mass of the particle and the antiparticle is converted into energy through Einstein's famous relationship $E = mc^2$. The energy is realized by a pair of photons, or particles of electromagnetic radiation, moving in opposite directions from the annihilation site. The energy of the particles, and hence their wavelength and frequency, is determined by the mass of the particle and the antiparticle.

KNOWLEDGE GAINED

Electromagnetic radiation is produced in many different ways. The type of radiation produced is determined by the physical process producing the radiation. Measurement of that radiation and its characteristics can then be used to understand the source of the radiation.

For astronomers, one of the most important forms of electromagnetic radiation is the emission from electron transitions between energy levels in an atom or molecule. In fact, the electromagnetic spectrum produced by a collection of atoms or molecules doing this is simply called its emission spectrum. The emission spectrum of a body can be used to determine its chemical composition. Furthermore, if the electrons are in a low energy level, light shining on the body can excite electrons to higher energy levels. Only light having energy equal to the energy difference between energy levels can be absorbed. Thus, only certain wavelengths of light can be absorbed. These wavelengths are the same as those given off in emission spectra. Therefore, studies of the light absorbed and not reflected can also be used to determine chemical composition of a body.

The 21-centimeter radiation can be used to detect the presence of hydrogen atoms in space. Synchrotron radiation can be used to probe the magnetospheres of planets. Bremsstrahlung radiation is important in studies of planetary atmospheres and the Sun. Cherenkov telescopes have been constructed on Earth to study the nature of cosmic rays. Studies of the lasing properties of planetary atmospheres also lead to better understandings of those atmospheres.

Though some is visible light, much of the nonthermal electromagnetic radiation is in forms other than as visual light. Thus, studies of nonthermal radiation use techniques from all areas of astronomy: optical, radio, X-ray, infrared, and ultraviolet astronomy. In fact, many forms of this radiation could not be studied at all until astronomers had developed the tools to study nonoptical wavelengths of light. Now that it is possible to study all of these wavelengths and the character of many forms of nonthermal radiation is understood, our understanding of the universe has greatly improved.

CONTEXT

Astronomers study objects that are very far away. Unlike most other scientists, astronomers do not have the luxury of being able to have the object of their study in a laboratory setting, but rather have to look at it from afar. This means primarily studying the electromagnetic radiation of the objects. While robotic spacecraft are capable of conducting studies in situ on the surface of a world, only a very few planets have been probed in this manner. Other spacecraft have flown by planets or even been placed into orbit around those planets. However, even most of these spacecraft have conducted their studies using electromagnetic radiation. Understanding the nature of this radiation and how it is created can lead to a far greater understanding of the bodies that are producing that radiation.

Every object that is not at absolute zero temperature emits thermal radiation, and study of the thermal radiation emitted by a body is very important to understanding it. However, far more information can be gleaned from studies of the nonthermal radiation emitted. The tools and techniques of doing so have become among the most important tools in the astronomer's arsenal for understanding celestial bodies.

Raymond D. Benge, Jr.

FURTHER READING

Asimov, Isaac. *Understanding Physics*. London: Allen & Unwin, 1967. A quite old but still very good book that introduces the basic concepts of electromagnetism for the nonscientist. Even though it is dated, it does an excellent job of covering the basic principles without mathematics. Widely available in libraries.

Caroll, Bradley W., and Dale A. Ostlie. *An Introduction to Modern Astrophysics*. 2d ed. San Francisco: Pearson Addison-Wesley, 2007. A textbook designed for undergraduate astronomy majors. The book gives an excellent explanation of the uses of the information gained by studies of thermal and nonthermal radiation. A knowledge of calculus is assumed.

Freedman, Roger A., and William J. Kaufmann III. *Universe*. 8th ed. New York: W. H. Freeman, 2008. An excellent and thorough introductory college astronomy textbook. An entire chapter is spent on the nature of light (electromagnetic radiation), including emission and absorption spectra. Later chapters discuss synchrotron radiation.

Robinson, Keith. *Spectroscopy: The Key to the Stars—Reading the Lines in Stellar Spectra*. New York: Springer, 2007. Written for the serious layperson and amateur astronomer, this book does a good job of explaining spectroscopy. Prior knowledge of the physics involved is not necessary to follow the text.

Rybicki, George B., and Alan P. Lightman. *Radiative Processes in Astrophysics*. New York: Wiley, 1979. An advanced book aimed primarily at upper-division undergraduate students or first-year graduate students; however, anyone with a firm grasp of physics and mathematics can follow it. The book is a number of years old, but the basic processes that produce thermal and nonthermal radiation have been understood since before the book was written.

Verschuur, Gerrit L. *The Invisible Universe: The Story of Radio Astronomy*. 2d ed. New York: Springer, 2006. A well-illustrated work for the layperson, with many photographs. The book does not deal directly with a description of nonthermal electromagnetic radiation, but it does show how 21-centimeter and similar forms of nonthermal radiation are used in radio astronomy.

See also: Big Bang; Cosmic Rays; Cosmology; Electromagnetic Radiation: Thermal Emissions; General Relativity; Greenhouse Effect; Interstellar Clouds and the Interstellar Medium; Milky Way; Novae, Bursters, and X-Ray Sources; Optical Astronomy; Radio Astronomy; Space-Time: Distortion by Gravity; Space-Time: Mathematical Models; Telescopes: Space-Based; Universe: Evolution; Universe: Expansion; Universe: Structure.

Electromagnetic Radiation: Thermal Emissions

Category: The Cosmological Context

Electromagnetic radiation emitted from the surface of an object provides valuable information regarding the physical state of the object. Whenever the temperature of the object is above absolute zero, the intensity of the radiation emitted and its distribution with wavelength—its thermal spectrum—largely depend only on the absolute temperature of the object. Thus, the temperature of an object can be determined from its thermal emissions.

OVERVIEW

An isolated atom or ion possesses a unique set of discrete, quantized energy levels. Transitions from a higher energy level to a lower one result in the emission of a photon of electromagnetic radiation whose wavelength depends on the difference in energy between the two levels. Specifically, $E_{photon} = E_{higher} - E_{lower} = hc/\lambda$, where h is Planck's constant, c is the speed of light in a vacuum, and λ is the wavelength of the photon emitted. The inverse process, involving the absorption of a photon of the same wavelength, will cause the atom or ion to transition from the lower energy state to the higher one. Thus, if an atom or ion is excited to one of its high energy states by a collision with some other atom or by absorbing radiation, cascading down to lower energy states produces emissions of photons of various specific wavelengths corresponding to the energy gaps between pair combinations of energy levels. This emission (or absorption) spectrum identifies the atom or ion involved in the process.

A discrete spectrum, however, can be quite complex, because it depends not only on the temperature and chemical composition of the object but also on its density, the local gravity field, the object's speed, and whether all parts are in thermal equilibrium. When atoms crowd together at high density, however—as in a solid, a liquid, or even a gas in the interior of a star—the energy levels of each atom or ion shift from their isolated configuration. The resulting smearing of energy levels relative to their isolated configuration permits a continuum of transitions where formerly only discrete transitions existed. If the material is dense enough, no restrictions exist as to which wavelengths can be emitted or absorbed. The electromagnetic radiation emitted in this case is called thermal radiation. In an idealized case where the composition and surface texture of the object can be neglected, the electromagnetic radiation emitted is also referred to as blackbody radiation. The shape and characteristics of the resulting spectrum depend only on the temperature of the object emitting the radiation.

Experimental investigations in the last quarter of the nineteenth century showed that the thermal spectrum of a hot object is characterized by relatively low-intensity emissions at short wavelengths, a rapid rise in intensity peaking at some intermediate wavelength, followed by a gradual diminishing in intensity at longer wavelengths. Furthermore, Wilhelm Wien, by applying the laws of thermodynamics to electromagnetic radiation, showed in 1893 that the wavelength at which the peak in the thermal spectrum appears depends on the temperature of the object according to the relation $\lambda_{peak} = b/T$, where T is the absolute temperature of the object and b is a constant whose value is 2.898×10^{-3} meter kelvin. (Max Planck later showed that b is itself constituted from other fundamental constants of nature.)

In 1879, Józef Stefan used the experimental measurements of thermal spectra to deduce an expression giving the irradiance, the total energy radiated per unit time interval by a blackbody per unit area of the emitting surface, as a function of the object's temperature. In 1884, Ludwig Boltzmann provided the theoretical framework for the expression, which came to be known as the Stefan-Boltzmann law. This relation linking the irradiance with temperature, $R = \sigma T^4$, contains a new constant σ whose value is 5.67×10^{-8} joule/second/meter^2/kelvin^4 and indicates that the amount of energy radiated by a blackbody increases rapidly with temperature. Like the constant b, σ too was later shown to be composed of fundamental constants of nature.

By 1900 Max Planck had synthesized the

prior experimental and theoretical information with his quantum hypothesis to deduce the precise relation that describes the blackbody spectrum of an object. The intensity $I(\lambda,T)$ of light emitted at a wavelength λ by a blackbody at temperature T is given as

$$I(\lambda,T) = 2hc^2/[\lambda^5(e^{hc/\lambda kT} - 1)]$$

a relation known as Planck's law. The constant k is Boltzmann's constant. Though these relations strictly apply only to blackbodies, many objects, to varying degree, approximate blackbodies, especially if their temperatures are very high or very low.

Tracing an object's thermal spectrum involves measuring the intensity of light through different filters that transmit electromagnetic radiation in relatively small bands of wavelengths. Modern technologies permit the fabrication of filters and detectors for any region of the electromagnetic spectrum. Typically, however, measuring the intensity of thermal emissions in only three or four wavelength bands suffices to determine the shape of an object's thermal spectrum and therefore estimate its temperature. Because the Earth's atmosphere itself filters incoming electromagnetic radiation, intensity measurements in parts of the infrared and radio regions and nearly all measurements at wavelengths shorter than the violet are best made from satellites in space outside the atmosphere. The solar system contains objects that produce phenomena in a wide range of temperature.

The Sun, the hottest object in the solar system, generates enormous amounts of energy by means of nuclear reactions. Its hot, dense interior produces a spectrum which is nearly blackbody. Highly energetic phenomena in the Sun's photosphere, chromospheres, and corona superimpose nonthermal features on the blackbody spectrum, however. The planets, too, are sources of thermal emissions, not only because of their warmth in absorbing and then reemitting solar radiation, but also because of the residual radiating energy left over from their formation and internal heat generated by the decay of radioactive nuclei. Scattered throughout the solar system, dust fills much of the space between planets and is plentiful in its outer regions. Far from the Sun, the temperature of the dust falls to just a few tens of degrees above absolute zero, resulting in thermal emissions in the far-infrared region of the spectrum. Detecting thermal emissions from cold objects presents a challenge, however. In order to prevent the thermal emissions from the detector itself from overwhelming the signal from the object studied, the detector with its electronics must be cooled to temperatures very near absolute zero.

KNOWLEDGE GAINED

Thermal spectra provide a wealth of fundamental information regarding objects in the solar system. The Sun's spectrum is particularly informative. From satellite observations outside the complicating effects of Earth's atmosphere, the wavelength of the peak intensity is approximately 470 nanometers. Wien's law indicates that the corresponding temperature of the visible portion of the Sun, its photosphere, is 6,170 kelvins. Another estimate of the Sun's temperature may be obtained from the Stefan-Boltzmann law, which links the irradiance of the Sun to its temperature. From satellite measurements outside Earth's atmosphere, 1,368 watts of electromagnetic power are delivered to each square meter of detector surface. Because the distance to the Sun is known, the total power output of the Sun can be computed to be 3.85×10^{26} watts, and from its size, the irradiance is determined to be 6.34×10^7 watts/meter^2. The Stefan-Boltzmann law implies that the Sun's temperature is 5,780 kelvins. The temperatures determined by these two methods differ because the Sun is not a perfect blackbody. Nonthermal absorption features created by atoms and ions in its atmosphere remove energy from one part of the spectrum, and that energy reappears in other parts, thus producing a spectrum slightly distorted from that of a blackbody.

A combination of both Wien's law and the Stefan-Boltzmann law yields an equilibrium temperature profile within the solar system illuminated by the central Sun. Thus, objects farther from the Sun capture less of the Sun's radiant energy, are cooler than objects nearer the

Sun, and so possess thermal spectra whose peak wavelengths are longer than objects nearer the Sun. As an example of an application of this principle, the location of the majority of the dust in the solar system can be deduced. Visual observations from Earth have long recognized that the zodiacal light was likely due to dust lying in the plane of the ecliptic. Infrared satellite observations of the dust indicate that the peak in the thermal spectrum of the dust occurs at approximately 12 microns (1 micron = 10^{-6} meters). The 12-micron peak in the thermal spectrum of the dust in the solar system corresponds to a distance of 4.0×10^{11} meters, or about 2.7 AU (one astronomical unit, or AU, being the mean distance between Earth and the Sun), a location at the inner region of the asteroid belt. Most likely, collisions among objects within the asteroid belt generate the observed dust.

CONTEXT

An idealized blackbody spectrum quite closely mimics the spectrum of thermal electromagnetic radiation produced by a real object for the corresponding temperature. In this case, the wealth of theoretical results associated with blackbody radiation can be used to tease out information from the object under study. Within our solar system, images of rocky worlds taken through two or three filters in different parts of their thermal spectra produce a composite image that displays the distribution of minerals on their surfaces. In some instances, such a composite image identifies the minerals without having to retrieve a sample. This helps planetary scientists understand the relation between the mineralogy and the landforms, providing valuable clues as to the processes that produced the visible landscape. In other cases, thermal emission images of the surface of a planet, satellite, or asteroid serve as an important prelude to identifying the best target areas for a landing and site exploration, as in the case of many missions to the Moon and Mars.

The same information gleaned from the thermal spectra of objects within the solar system can be retrieved from a study of the spectra of objects and systems throughout the cosmos. Dust around young stars sometimes points to the formation of new solar systems beyond our own. Finally, the universe itself announces itself in subtle, omnipresent thermal emissions corresponding to a blackbody at a temperature of 2.725 kelvins, the radiation left over from the hot, dense creation of the cosmos 13.7 billion years ago.

Anthony J. Nicastro

FURTHER READING

Beiser, Arthur. *Concepts of Modern Physics*. 6th ed. New York: McGraw-Hill, 2002. An excellent text that presents a conceptual, accessible introduction to modern physics. Chapter 2 presents a unified overview of electromagnetic radiation and the development of theories to understand this phenomenon. Stresses the atomic nature of processes.

Bennett, Jeffrey, Megan Donahue, Nicholas Schneider, and Mark Voit. *The Cosmic Perspective*. 4th ed. San Francisco: Pearson/Addison-Wesley, 2007. A brilliantly illustrated, coherent, well-written introduction to astronomy. Chapter 5 contains a general overview of light and its interactions with matter. Thermal spectra, along with other types of spectra, are linked. Other parts of the book show applications to the solar system and the cosmos.

Cole, George H. A., and Michael M. Woolfson, eds. *Planetary Science: The Science of Planets Around Stars*. Bristol: Institute of Physics Publishing, 2002. Discusses an extensive array of applications of thermal and nonthermal spectra to objects in our solar system and in extrasolar systems. Primarily a book that covers descriptive planetary science, requiring very little prior knowledge of physics, astronomy, or mathematics.

Morrison, David, and Tobias Owen. *The Planetary System*. 3d ed. San Fransisco: Pearson/Addison-Wesley, 2003. With an early introduction to thermal radiation laws, this volume applies them to the study of all the objects in the solar system. Primarily descriptive and conceptual, the book makes occasional use of algebra.

Van der Meer, Freek D., and Steven M. De Jong, eds. *Imaging Spectrometry*. New York: Kluwer Academic, 2002. Provides a comprehensive description of the principles and applications of satellite and airborne spectrometry. Gives an extensive account of data collection and interpretation of images of geological and agricultural regions. Focuses primarily on terrestrial applications.

Zeilik, Michael, and Stephen A. Gregory. *Introductory Astronomy and Astrophysics.* 4th ed. Fort Worth, Tex.: Saunders College Publishing, 1998. A college-level text that provides broad coverage of astronomy and astrophysics. Results from an early cursory coverage of thermal radiation are used throughout an examination of the solar system. Later in-depth coverage of electromagnetic thermal and nonthermal radiation forms the basis for studying objects and phenomena across the cosmos. Makes heavy use of algebra and trigonometry.

See also: Big Bang; Cosmic Rays; Cosmology; Electromagnetic Radiation: Nonthermal Emissions; General Relativity; Greenhouse Effect; Interstellar Clouds and the Interstellar Medium; Milky Way; Novae, Bursters, and X-Ray Sources; Optical Astronomy; Radio Astronomy; Space-Time: Distortion by Gravity; Space-Time: Mathematical Models; Telescopes: Space-Based; Universe: Evolution; Universe: Expansion; Universe: Structure.

Enceladus

Categories: Natural Planetary Satellites; The Saturnian System

Enceladus is the brightest of the satellites located within the rings of Saturn. It has much in common with what scientists expected to find on comets, especially water ice. However, it appears to have been formed more than four billion years ago as a spinning mass of soft material.

OVERVIEW

Enceladus was discovered on August 28, 1789, by William Herschel. Its orbit around Saturn has a semimajor axis of 237,948 kilometers and an eccentricity of 0.0047, with a period of 118,386.82 seconds (nearly 33 hours). The orbit is inclined at 0.019° to Saturn's equator and located inside the E ring around Saturn. This places Enceladus at roughly 4 Saturn radii, located between the orbits of the moons Mimas and Tethys, which are about one Saturn radius on either side. Its rotation is synchronous and has no axis tilt, so that the same hemisphere always faces Saturn. It is nearly spherical, with a mean diameter of 504.2 kilometers, being a slightly flattened ellipsoid with dimensions of 513.2 kilometers along the orbit radius pointed at Saturn, 502.8 kilometers along the orbit path, and 496.6 kilometers between the north and south poles.

It is the sixth-largest moon of Saturn. It has a mass of 1.08022×10^{20} kilograms, and the mean density is approximately 1,609 kilograms per cubic meter. The value of acceleration due to gravity at the surface on the equator is 0.111 meter/second^2. The escape velocity at the surface is 0.238 kilometer/second, neglecting atmospheric drag. Enceladus has very high Bond albedo of 0.99 and geometric albedo of 1.375, the highest among the satellites embedded in the Saturnian rings, indicating strong reflection. Its apparent magnitude from Earth is 11.7.

Enceladus is tidally locked in synchronous orbit around Saturn, meaning that the hemisphere with the higher density always faces Saturn. Looking into the sky from the Saturn-facing side, the planet would occupy roughly 30° of the sky and appear to be spinning in roughly the same position at all times. From the side hidden from Saturn, the Sun would appear very small, rising and setting in roughly 17 hours as Enceladus completed half an orbit around Saturn.

The orbital eccentricity of Enceladus is attributed to a resonance with the satellite Dione, with Enceladus completing two orbits for each orbit by Dione. This resonance may also drive the tidal heating of Enceladus. The shape of Enceladus is very close to that of an equilibrium-flattened ellipsoid, hydrostatically balanced by gravity and spin, which is the shape that an object would have in space if it were composed of homogeneous and fluid material. However, simulations of orbital evolution and the tidal locking suggest some variation in its internal density. The higher average density than that of water indicates denser material, possibly silicates, inside Enceladus.

Most of the data about the surface and orbital environment of Enceladus come from close flyby observations by the Voyager and Cassini spacecraft. The Voyager mission revealed evidence of a complex thermal history of Enceladus and showed several provinces with distinct geographical features. Short periods of intense heating and geological activity appear to have been separated by long periods of inactivity. Surface features include long, narrow depressions (fossae), ridges with cliffs of several hundred meters (dorsa), plains (plantia), long, parallel canyons (sulci), and craters. The terrain of the northern latitudes appears to be more than 4.2 billion years old, with more than one crater every 5 square kilometers, most of them bowl-shaped. Some craters are as large as 35 kilometers in diameter, while the majority are very small, less than a meter in diameter. The equatorial plains, named the Sarandib Plantia, show striations and folding, with about one crater every 70 square kilometers. Craters in these younger regions show viscous relaxation, indicating mechanisms for melting or distortion of the surface.

The ridged and grooved plains of the Samarkand Sulcus, at 55° to 65° south latitude, are 100,000 to 500,000 years old, and the fractured regions south of that show few if any craters. The Cassini orbiter approached the south polar region of Enceladus to within 168 kilometers on July 14, 2005, taking images with a resolution of 4 meters per pixel. House-sized boulders believed to be made of ice littered the polar landscape, but craters were nearly absent. A set of parallel "tiger-stripe" fractures, roughly 500 meters deep, 2 kilometers wide, and 130 kilometers long, are flanked by 100-meter high ridges. Temperature is as high as 175 kelvins. South of

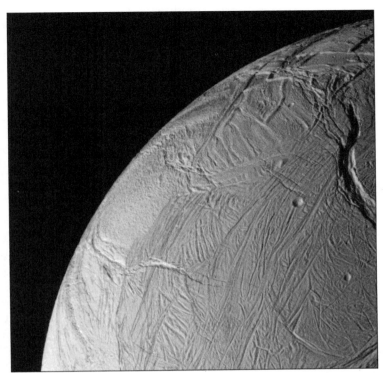

The smooth, young surface of Enceladus is seen in this October, 2008, image from the Cassini spacecraft. (NASA/JPL/Space Science Institute)

55° south latitude, a chain of fractures and ridges circumscribes the moon. Some fractures intersect and overlay others. Scientists associate these fractures and ridges with the flattening and extension due to gravitational interactions with Saturn and other moons.

The Cassini flyby of Enceladus in 2005 revealed a water-rich plume ejecting as narrow jets from vents in the tiger-stripe sulci of the south polar region. The fine sizes of the particles in the plume, which freeze to ice or sublimate soon after ejection, suggest that the plume originates in a subsurface body of liquid water. Some of the water jets reach exit speeds of 600 meters per second, well above the 238 meters per second needed to escape the gravity of Enceladus. The mass flow rate of the plume is on the order of 100 kilograms per second, comparable to that of air through a modern supersonic jet fighter engine. Much of this mass may initially escape Enceladus into the ring around Saturn and may be the origin of the mass in the E ring of Saturn.

Enceladus also captures mass as snow falling to the surface from the E ring. Some scientists think that less than 1 percent of the mass in the plume can escape into the E ring. Numbers cited in different papers vary greatly on this subject.

Cassini approached to within 52 kilometers of the southern middle latitudes on March 12, 2008, in its E3 close approach, then dropped behind Enceladus in its orbit. It grazed the edge of the plume, which trails the satellite like the tail of a comet. The density of these portions of the plume had been predetermined from Earth by observing the dimming of starlight as the plume crossed in front of a star with an ultraviolet imaging spectrograph (UVIS). Roughly 70 seconds after closest approach, the craft was 250 kilometers above the surface, when the onboard ion and neutral mass spectrometer (INMS) encountered a peak particle density of nearly 10 million particles per cubic centimeter. This was still in the outer edge of the plume. The sharply defined plume edge provides further strong evidence that the plume comes out as a supersonic jet, as opposed to a diffuse subsonic plume caused by friction heating of ice due to the tidal stresses at the fractures. Measured gas density was twenty times higher than that predicted based on thermal expansion.

KNOWLEDGE GAINED

Data from the INMS suggest that the plume contains, beyond nearly pure water (ice), significant amounts (ranging from 1 to 10 percent) of carbon dioxide, methane, and other organic molecules, both simple and complex. There was a strong signal from something of molecular weight near 28, but there is debate over whether this substance is nitrogen (per the INMS data) or carbon monoxide (per the Cassini plasma spectrometer data). The values were similar to those predicted for many comet tails. The organic molecules detected include acetylene (C_2H_2), ethane (C_2H_6), hydrogen cyanide HCN, formaldehyde (H_2CO), propyne (C_3H_4), propane (C_3H_8), and acetonitrile (C_2H_3N). The cosmic dust analyzer instrument on the craft did not succeed in capturing the particle sizes. The Cassini plasma spectrometer (CAPS),

An artist's conception of the surface of Enceladus, showing one of the moon's ice geysers and Saturn in the background. (NASA)

which measures ions, detected much larger particles as well, on the order of nanograms. If these were ice particles, they may have been as large as 0.01 to 0.1 millimeter in diameter, comparable to the particle sizes calculated from surface observation data of the south polar terrain. CAPS also detected positively and negatively charged ions, segregated in different regions of the plume.

Enceladus has no measurable internal magnetic field, but it has a significant influence on the magnetosphere of Saturn, strongly deflecting magnetic lines. This is now attributed to the water plume. Ions accelerated to energy levels of 20 kilo-electron volts by Saturn's strong magnetic field collide with the molecules in the plume, breaking them up into atoms and ionizing them. These fresh ions are again accelerated by the magnetic field, and, in turn, Enceladus substantially deflects the magnetic field lines of Saturn. Scientists also associate clouds of oxygen and hydrogen observed around Saturn with the atoms and ions generated when the water molecules in the plume from Enceladus collide with high-energy ions in the E ring.

Enceladus shows a trace atmosphere with surface pressure varying significantly in spatial location and perhaps in time, composed of about 90 percent water vapor, 4 percent nitrogen, 3.2 percent carbon dioxide, and 1.7 percent methane. Surface temperature varies from 32.9 to 145 kelvins, with a mean of 75 kelvins. The surface appears to be covered in clean water ice, accounting for the high reflectivity. The atmosphere is hypothesized to be an expanding, supersonic neutral gas cloud emanating from the surface after molecules are "sputtered" from the surface by collisions with high-energy ions.

In late 2008, presentations given at the American Geophysical Union meeting in San Francisco revealed that data from Cassini's observations of Encedalus strongly indicated that the satellite's surface displays action similar to the action of Earth's ocean floor, where new crustal material emerges from slits in the crust. Cassini imaging team leader Carolyn Porco proposed that liquid water was present on Enceladus's surface and that that surface splits and spreads apart. On Earth, molten rock rising up from deep in the planet causes the crustal

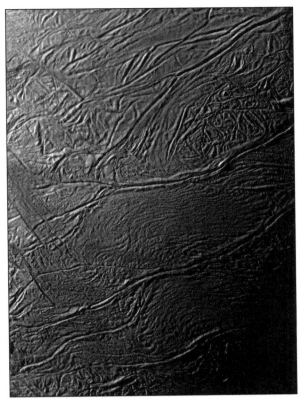

Enceladus's "tiger stripes" are visible in this Cassini image from 2008. (NASA/JPL)

spreading, whereas on Enceladus the surface spreading originates with upwelling of liquid, presumably water. Evidence suggested that the Tiger Stripes formations near the satellite's south pole are akin to the mid-ocean ridges found on the Earth's seafloor. Close flybys of that region resulted in more data on eruptions of water through vents in the Tiger Stripes formation. Combined with evidence of crustal spreading, these data have revealed Enceladus to be a surprisingly active world.

CONTEXT

The presence of water ice in a low-gravity body within seven years' travel time of Earth excited planners of deep space missions, since water is an excellent future propellant. The discovery of high-speed water jets from the south polar region provides strong evidence of liquid water below the surface, and continuing tectonic processes. Complex organic molecules discovered in the jet plume fuel speculation about precursors of life. Enceladus is one of three known planetary bodies in the solar system (besides Earth and the Jovian satellite Io) that has an internal heat made visible by remote sensing. The relations between Enceladus and the E ring, the magnetosphere, and clouds of oxygen and hydrogen around Saturn are subjects of intense study.

The similarity between Enceladus and comets has raised questions about the origin and evolution of the solar system. Comets were thought to have originated far outside the orbit of Pluto, independently of the planets, while the planetary satellites formed from the same cloud as the Sun and planets. However, Enceladus appears to have nearly pure water ice, as predicted for comets. How such a large, nearly spherical cometary body could have been captured in an orbit so close to Saturn is an unanswered question.

Narayanan M. Komerath and David G. Fisher

FURTHER READING

Benna, M., and W. Kasprzak. "Modeling of the Interaction of Enceladus with the Magnetosphere of Saturn." *Lunar and Planetary Science* 38 (2007). Discusses results of different numerical models of the magnetosphere interaction and the atmosphere of Enceladus, comparing against the results from the Cassini instruments.

Khurana, K. K., M. K. Dougherty, C. T. Russell, and J. S. Leisner. "Mass Loading of Saturn's Magnetosphere Near Enceladus." *Journal of Geophysical Research* 112, A08203, doi:10.1029/2006JA012110, 2007. Reports on modeling of magnetic field data on the interaction between Saturn's magnetosphere and Enceladus. Gives results on mass pickup and current generated.

Porco, C. C., et al. "Cassini Encounters Enceladus: Background and the Discovery of a South Polar Hot Spot." *Science* 311 (March 10, 2006): 1401-1405. Discusses the initial discovery of the relatively warm regions near the south pole of Enceladus.

Verbiscer, A., R. French, M. Showalter, and Paul Helfenstein. "Enceladus: Cosmic Graffiti Artist Caught in the Act." *Science* 315 (February 9, 2007). Discusses the albedo of Enceladus compared to the albedos of other moons of Saturn.

Wilson, D., et al. "Cassini Observes the Active South Pole of Enceladus." *Science* 311 (March 10, 2006): 1393-1401. Presents images and discusses the reasoning regarding the features around the south pole of Enceladus, based on the Cassini spacecraft's close approaches in 2005. Also discusses how shape and size considerations are used to form hypotheses on the evolution of Enceladus.

See also: Callisto; Eris and Dysnomia; Europa; Ganymede; Iapetus; Io; Jupiter's Satellites; Planetary Satellites; Saturn's Satellites; Titan; Triton.

Eris and Dysnomia

Category: Small Bodies

The scattered disk object Eris is the largest dwarf planet and the most distant solar-system object astronomers have identified. Observation of Eris so far has revealed only one satellite, Dysnomia. Eris was initially hailed as a possible tenth planet when discovered in 2005, but debate within the astronomical community regarding what constitutes a planet led the International Astronomical Union to categorize it as a dwarf planet.

OVERVIEW

Eris, formally designated (136199) Eris, is the largest of the known trans-Neptunian objects. As of 2009, only one satellite, Dysnomia, had been found in orbit around Eris. The most distant solar-system object observed to date, Eris is classified as a scattered disk object. Despite its remoteness, it can be viewed with powerful amateur equipment. At aphelion, Eris lies beyond the outermost region of the Kuiper Belt; at perihelion, it passes within the range of Neptune's influence. As of 2009, Eris was one of only three known plutoids (the others are Pluto and Makemake) and four known dwarf planets (the three plutoids plus Ceres). Eris is larger than

An artist's view of dwarf planet Eris and its satellite Dysnomia, based on data from the Hubble Space Telescope. (NASA/ESA/Adolph Schaller for STScI)

Pluto, its orbital period is more than twice Pluto's, and at aphelion it is roughly three times farther from the Sun than Pluto.

Researchers have calculated Eris's orbit using archival data collected before the dwarf planet's discovery. Once Eris was identified in 2005, the bright, slow-moving object was easy to spot in images going back to 1989. Eris takes 557 years to travel around the Sun, following a highly elliptical path with a semimajor axis, or mean orbital radius, of 67.9 astronomical units (AU), or more than 10 billion kilometers. When Eris was discovered, it was near its aphelion of 97.5 AU (more than 14.5 billion kilometers) from the Sun. It will reach perihelion at 38.2 AU (more than 5.7 billion kilometers) in the mid-twenty-third century. For the most part, the orbit of Eris is typical of a scattered disk object. However, Eris's orbital inclination is an atypi-

cally high 44° from the solar system's orbital plane. The inclination and period of Eris's rotation have yet to be ascertained, although it is currently thought that a day on Eris is about eight hours long.

Images from the Hubble Space Telescope indicate that Eris has a diameter of roughly 2,400 kilometers, making it slightly larger than Pluto. Spitzer Space Telescope observations suggest that Eris's diameter may be as great as 2,600 kilometers. Calculations based on Keck Observatory and Hubble Space Telescope observations of Dysnomia's orbit around the dwarf planet indicate that Eris has a mass of 1.66×10^{22} kilograms, or 3.66×10^{22} pounds (27 percent greater than Pluto's) and a bulk density of 2.3 grams per cubic centimeter (0.83 pound per cubic inch).

Using Eris's density, researchers have determined that the dwarf planet's interior is most likely similar to Pluto's—that is, about half rock and half ice. Where Pluto's surface appears reddish and partly rocky, however, Eris has a uniform, highly reflective, almost gray surface. Near-infrared images obtained at the Gemini North Observatory in Hawaii indicate that this surface is predominantly frozen methane. In Pluto's case, darker surface hues are attributed to tholins, reddish-brown breakdown products formed when methane and similar organic compounds are subjected to solar ultraviolet irradiation. Where tholin deposits make Pluto's surface darker, the albedo is lower and the temperature higher. Methane ice melts away from these comparatively warm patches. By contrast, methane ice appears to envelop Eris in a bright, near-uniform coating. This suggests not only that Eris remains cold enough that its methane stays in a frozen state but also that there may be a subsurface source of methane that replenishes Eris's surface coating of methane ice and covers up whatever tholins are deposited.

As Eris moves from aphelion to perihelion, its temperature increases from its current value of 30 kelvins (−243° Celsius) to 56 kelvins (−217° Celsius). It is possible that, when Eris approaches the Sun, some of its surface ices become warm enough to sublimate and form a thin atmosphere. Whatever gases do not escape the atmosphere freeze once again as the dwarf planet moves toward aphelion.

To date, only one satellite has been observed in orbit around Eris: Dysnomia, known technically as (136199) Eris I. Hubble Space Telescope and Keck Observatory data indicate that Dysnomia is roughly 150 kilometers in diameter. It takes about 16 days for Dysnomia to complete its near-circular orbit around Eris at a distance of approximately 37,000 kilometers. The small satellite is believed to be composed largely of frozen water.

Astronomers Mike E. Brown of the California Institute of Technology, Chad A. Trujillo of Gemini Observatory, and David Rabinowitz of Yale University discovered Eris through an ongoing survey conducted at Palomar Observatory in southern California using the Samuel Oschin telescope. Images taken on the night of October 21, 2003, showed the large, bright object traveling slowly across the sky. Its movement was slow enough, in fact, that Eris went undetected when the images were first analyzed. The researchers' discovery in November, 2003, of Sedna, another large and slow-moving trans-Neptunian object, led them to adjust their detection scheme and reanalyze their survey data. On January 5, 2005, they identified the planet-sized scattered disk object that would later be known as Eris, designating it 2003 UB313.

Brown and his team intended to follow standard scientific protocols by verifying their discovery, studying it, documenting it thoroughly, and making it known through a scientific paper published in a reputable journal. However, in July, 2005, they learned that detailed records of their telescope use had inadvertently been made accessible to anyone with Internet access, and that an abstract they had recently published unwittingly contained clues about where in the sky to look for their recent—and unannounced—trans-Neptunian discoveries. When, five days after the abstract was issued, researchers in Spain announced the discovery of 2003 E161—a trans-Neptunian object Brown and his colleagues had also found—it appeared the team had to lay claim to 2003 UB313 or risk having someone else take credit for finding it.

The California Institute of Technology, the Jet Propulsion Laboratory, and the National Aeronautics and Space Administration (NASA) announced the discovery of 2003 UB313 on July 29, 2005, in press releases that referred to the object as the tenth planet. The media was quick to adopt the team's nickname for the newly discovered member of the solar system: Xena, so called for the heroine of the television series *Xena: Warrior Princess*. (Some overenthusiastic journalists, seeing "planetlila" in Brown's Web address, pounced upon Lila as the new planet's name, only to learn that the URL was a whimsical tribute to Brown's newborn daughter.)

That same year, Brown, Trujillo, and Rabinowitz collaborated with the engineering team at the Keck Observatory on Mauna Kea, Hawaii, to search 2003 UB313 and three more of the brightest trans-Neptunian objects for satellites. Using the Keck's new Laser Guide Star Adaptive Optics system, which enabled the researchers to view details as precise as those seen from the Hubble Space Telescope, they found S/2005 (2003 UB313) 1—the faint satellite that would come to be called Dysnomia—on September 10, 2005. The team dubbed the satellite Gabrielle, after Xena's television-show sidekick.

In the summer of 2006, at a meeting of the General Assembly of the International Astronomical Union (IAU), criteria were developed regarding what constitutes a planet, and a new category of "dwarf planet" was established. Under the IAU's new definitions, 2003 UB313 was not a planet but rather a dwarf planet. On September 6, 2006, the discovery team proposed to the IAU that dwarf planet 2003 UB313 be named Eris and its moon be named Dysnomia. The IAU accepted and announced the names one week later on September 13, 2006.

KNOWLEDGE GAINED

The 2005 discovery of 2003 UB313 added fuel to a long-standing and heated debate in the astronomical community over Pluto's status as a planet. The newly discovered object had a greater diameter and was more massive than its distant neighbor Pluto. If Pluto's size and mass were sufficient to qualify it for "planethood," then the new object should likewise be classified as a planet.

The controversy came to a head in August, 2006, at IAU's General Assembly in Prague. At the unusually contentious meeting, members reached the non-unanimous conclusion that a celestial body should be considered a planet if it (1) orbits the Sun (satellites are not planets); (2) is massive enough for self-gravity to shape it into a sphere; and (3) has accreted or scattered other bodies in its neighborhood to clear its orbit. Objects meeting the first two criteria but not the third would be classified as dwarf planets.

Both Eris and Pluto orbit the Sun, and both are spherical, yet neither has cleared the vicinity around its orbit. With the IAU's official acceptance of the new definitions on August 24, 2006, 2003 UB313 ceased to be a possible tenth planet and instead became the largest known dwarf planet. Likewise, Pluto was "demoted" to dwarf planet status after more than seven decades of being regarded as a planet. Ceres was determined to be a dwarf planet that might also be an asteroid.

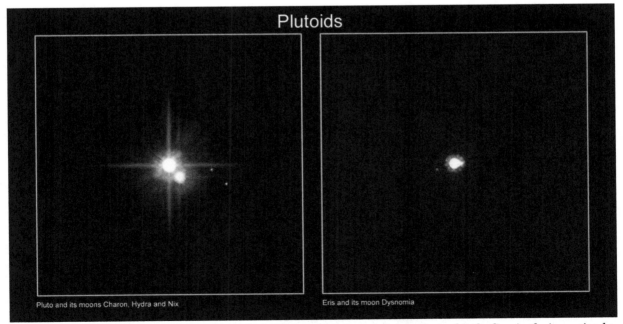

Plutoids

Pluto and its moons Charon, Hydra and Nix

Eris and its moon Dysnomia

Pluto and Eris, among the first bodies classified as "plutoids": dwarf planets that orbit the Sun in the icy region beyond Neptune. (International Astronomical Union)

2003 UB313 received its official name, Eris, on September 13, 2006. In Greek mythology, Eris was the goddess of discord and strife. Given the uproar the object's discovery caused among astronomers—and a public accustomed to a nine-planet solar system—it was aptly named. The dwarf planet's satellite was named after the mythological figure Dysnomia, the goddess Eris's daughter and the spirit of lawlessness. The name Dysnomia was also a tip of the hat to Eris's old nickname—the television actress who played warrior princess Xena was Lucy Lawless.

Almost two years later, Eris was assigned to a new subcategory of dwarf planet. On June 11, 2008, the IAU announced that dwarf planets orbiting the Sun at a semimajor axis greater than that of Neptune would be known as "plutoids." Eris and Pluto qualified; Ceres, located in the asteroid belt, did not. In July, 2008, the Kuiper Belt object known as 2005 FY9 received its official name, Makemake, and became the third largest dwarf planet and plutoid.

CONTEXT

Discovery in the early 1990's of small objects beyond Neptune inspired researchers such as Brown to look to the Kuiper Belt and beyond for larger celestial bodies. Brown and Trujillo made their first major trans-Neptunian find in 2002 with Quaoar. In 2003, they discovered a larger object, Sedna, then the remotest solar-system object that had ever been found. Sedna's slower motion across the heavens led the team to look for objects moving at even lower rates, which revealed Eris.

Theoretical models suggest that Eris and other comparatively large trans-Neptunian objects in high-inclination orbits were originally near the inner edge of the Kuiper Belt. When they were subsequently scattered into the outer belt and beyond, they achieved orbits with higher inclinations than objects originating in the outer belt. More massive objects that had their origins in the inner belt may now occupy remote, high-inclination orbits. Researchers are now looking for as-yet-undiscovered large objects orbiting at these high inclinations.

At present, there is no space mission planned to Eris. While the uncrewed New Horizons will conduct a flyby of Pluto, the spacecraft's limited maneuvering capability keeps it from passing close enough to observe Eris.

Karen N. Kähler

FURTHER READING

Brown, M. E., C. A. Trujillo, and D. L. Rabinowitz. "Discovery of a Planetary-Sized Object in the Scattered Kuiper Belt." *The Astrophysical Journal* 635, no. 1 (2005): L97-L100. The journal article in which Eris's discovery team first presented their findings. Explains how the team discovered 2003 UB313 and summarizes the initial physical and orbital characterizations of the object. Includes tables, figures, and references. Technical but straightforward.

_____. "Satellites of the Largest Kuiper Belt Objects." *The Astrophysical Journal* 639, no. 1 (2006): 143-146. The scientific paper in which the discovery team introduced Eris's moon. Describes the survey of 2003 UB313 and three other trans-Neptunian objects for possible satellites. Includes tables, figures, and references. Challenging for the lay reader.

Chang, Kenneth. "Dwarf Planet, Cause of Strife, Gains 'the Perfect Name.'" *The New York Times*, September 15, 2006, p. A20. A brief but informative article on how dwarf planet 2003 UB313 and its satellite came to be named Eris and Dysnomia—and nicknamed Xena and Gabrielle.

_____. "Ten Planets? Why Not Eleven?" *The New York Times*, August 23, 2005, p. F1. Reprinted in *The Best American Science Writing 2006*, edited by Atul Gawande. New York: HarperPerennial, 2006. An engaging profile of team leader Mike Brown, published shortly after the announcement that 2003 UB313 had been found. Includes a chart summarizing five of Brown's major finds, among them 2003 UB313.

Faure, Gunter, and Teresa M. Mensing. *Introduction to Planetary Science: The Geological Perspective*. New York: Springer, 2007. Designed for college students majoring in Earth sciences, this textbook provides an application of general principles and subject material to bodies throughout the solar system.

Excellent for learning comparative planetology.

Freedman, Roger A., and William J. Kaufmann III. *Universe*. 8th ed. New York: W. H. Freeman, 2008. College-level introductory text covering the field of astronomy. Contains descriptions of astrophysical questions and their relationships.

Fussman, Cal. "The Man Who Finds Planets." *Discover*, May, 2006, 38-45. A highly accessible account of the discovery of 2003 UB313 (then still popularly known as Xena) and other trans-Neptunian objects, as told by team leader Mike Brown. Includes photographs.

See also: Callisto; Ceres; Dwarf Planets; Enceladus; Europa; Ganymede; Iapetus; Io; Jupiter's Satellites; Lunar Craters; Lunar History; Lunar Interior; Lunar Maria; Lunar Regolith Samples; Lunar Rocks; Lunar Surface Experiments; Mars's Satellites; Miranda; Neptune's Satellites; Planetary Satellites; Pluto and Charon; Saturn's Satellites; Titan; Triton; Uranus's Satellites.

Europa

Categories: The Jovian System; Natural Planetary Satellites

Europa is one of the four "Galilean satellites" that orbit the giant planet Jupiter. Only slightly smaller than the Earth's moon, Europa is covered by a relatively smooth layer of highly reflective fractured ice. Tidal forces exerted by Jupiter cause internal heating on Europa that apparently results in the periodic resurfacing of watery flows, which have over time obliterated most impact craters and other blemishes. Heat flow may be sufficient to maintain a liquid water subsurface layer that could harbor simple life-forms.

OVERVIEW

Europa is one of the four large satellites of the planet Jupiter known as the Galilean satellites after their discoverer, Galileo Galilei. These (according to their distance from Jupiter) are Io, Europa, Ganymede, and Callisto. Jupiter has at least sixty-three satellites, but only the Galilean satellites are large enough to be observed from Earth by small telescopes. With a diameter of 3,138 kilometers, Europa, the smallest Galilean satellite, is slightly smaller than Earth's Moon (3,476 kilometers). By contrast, the largest Galilean satellite, Ganymede, measures 5,260 kilometers in diameter, larger than the planet Mercury (at 4,878 kilometers). Thus, if the Galilean satellites orbited the Sun instead of Jupiter, they would be considered full-fledged planets. Despite its relatively small size compared to its Galilean companions, Europa is nevertheless the sixth largest planetary satellite in the solar system. It is located about 780 million kilometers from the Sun, about 5.2 times the Earth-Sun distance.

Europa orbits Jupiter at an average distance of 670,900 kilometers; its orbital period (time to complete one orbit) is 3.55 Earth days. Its rotational period around its axis is also 3.55 days, which means that Europa always shows the same face toward Jupiter. The other Galilean satellites and Earth's own Moon follow this 1:1 ratio of orbital to rotational period, termed a "synchronous" relationship.

Galileo discovered Europa and two of the other three large Jovian satellites (Io and Callisto) on January 7, 1610, using a crude homemade telescope. At first he believed the tiny points of light in line with Jupiter were small stars, but later he realized that they in fact orbited Jupiter as if in a miniature solar system. Galileo originally called the moons the Medician planets (after the powerful Italian Medici family) and numbered each satellite with a Roman numeral beginning with the one closest to Jupiter. Europa in this scheme was designated II. Another observer, Simon Marius (Simon Mayr), who claimed to have discovered the Jovian satellites prior to Galileo in November, 1609, but was tardy in publishing his results, later named the bodies as we know them today.

The name Europa comes from a Phoenician princess, one of many mortal consorts of the supreme Greek god Zeus, whose Roman name graces the planet Jupiter. (The other Galilean satellites are similarly named for mythological characters associated with Zeus.) The most intriguing aspect of Europa is its unusual and unique surface. Images beamed to Earth in 1979 by the Voyagers 1 and 2 spacecraft as they flew through the Jupiter system showed a relatively smooth ice ball that some scientists compared in appearance to a fractured, antique ivory billiard ball. The satellite is covered by a globally encompassing shell of water ice, frozen at 128 kelvins, that gives Europa an extremely high albedo. While 64 percent of the light striking the surface is reflected back in all directions (giving Europa an albedo of 0.64), rocky surfaces like that of Earth's moon or Mercury reflect only about 10 percent.

Europa's density, 3.04 grams per cubic centimeter, suggests that most of the planet is composed of rocky silicate material like Earth. The icy surface layer, therefore, must be relatively thin; most estimates lie in the range of 75 to 100 kilometers thick. The surface shows relatively little topographic relief, nothing higher than 1 kilometer, and displays only a few small scattered impact craters, in dramatic contrast to its highly cratered neighbors, the two outer satellites Ganymede and Callisto. Large craters on the order of 50 to 100 kilometers in diameter are virtually absent on Europa but plentiful on Ganymede and Callisto. Most craters on Europa do not exceed about 20 kilometers in diameter. This suggests that Europa's icy surface is relatively young, indicating that resurfacing by liquid ice flows or other processes has covered over any large craters formed during early, heavy meteoroid impacting in the Jovian system. Estimates of the surface age of Europa range from a high of 3.0 to 3.5 billion years old to more recent estimates of only 100 million years. The younger age, if true, suggests significant resurfacing of the planet in the later stages of its history.

In December, 1995, the Galileo spacecraft entered orbit about Jupiter. Over the course of thirty-eight orbits, Galileo not only investigated the giant planet Jupiter but also flew by its many satellites, paying particular attention to Europa, Ganymede, and Callisto. For example, in 1997 Galileo produced images of a large, multiringed impact crater on Europa probably buried beneath the ice crust. Evidence for the crater consists of diffuse, dark, concentric,

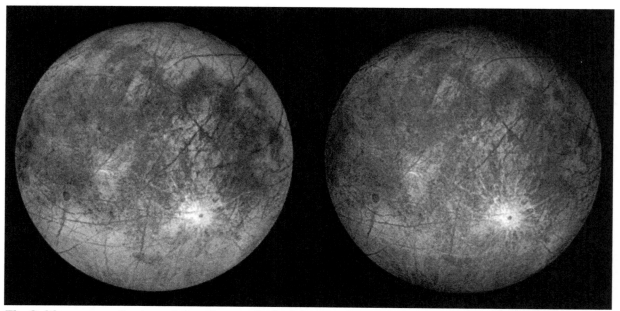

The Galileo spacecraft returned these images (both of the same hemisphere, the one on the right enhanced to emphasize details) of Jupiter's satellite Europa, whose surface is primarily water ice. Darker areas are rocky material, lines are crustal fractures, and the bright spot on the lower right is the crater Pwyll. (NASA)

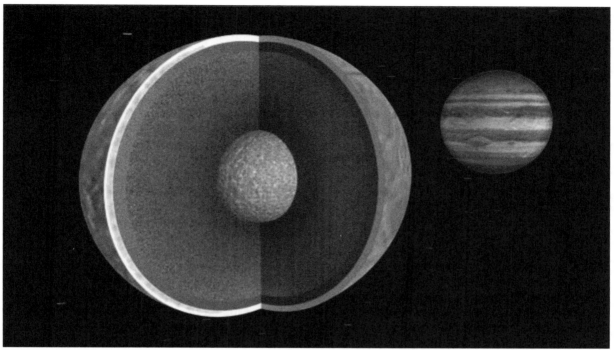

An artist's cutaway showing the probable interior structure of Europa, with an iron core, a rocky mantle, and an outer ocean of salty water capped with ice. (NASA/JPL)

arclike bands and associated fractures that define a structure more than 5,000 kilometers in diameter. The presence of this buried crater shows that the rocky surface below the ice layer was subjected to significant impacting early in Europa's history. It further suggests that the ice crust formed at some later time, probably after heavy meteoroid bombardment had greatly diminished.

The most striking aspects of Europa's surface are the mottled, colored terrains and linear fractures that crisscross most of its globe. Mottled terrains, based on color and subtle topographic expressions, are of two varieties: brown and gray. Brown terrains contain numerous pits and depressions from 1 to 10 kilometers in diameter. Several large "plateaus" occur that range from a few kilometers up to a few tens of kilometers wide and up to nearly 100 kilometers long. Some circular depressions, missing raised crater rims, may represent degraded impact craters. Gray terrains are similar to the brown but are generally smoother and less hummocky. The relationship between the two terrains is un-

known, but their differences may result from contrasting ages, degree of surface development, or both. The ultimate origin of these mottled terrains remains unknown, but a reasonable hypothesis is that they represent the effects of hydrothermal upwelling, causing heating and expansion of affected crustal areas. The "nonmottled" areas on Europa are very light in color and have very smooth topographies. These icy plains contain most of the observed linear surface fractures.

Linear features on Europa's surface may extend for thousands of kilometers. They are classified into three categories: (1) dark triple bands, some containing dark outer bands with a white strip down the center, thought to represent icy geyser deposits erupted along the axis of the fractures; (2) older and brighter lineaments that are crosscut by the triple bands and resemble them in some cases; and (3) very young cracks that crosscut the other two fracture types. Detailed analysis of the orientation of these three fracture types indicates that each type shows a distinct orientation that can be correlated with the relative age of the fractures. The data show that the direction of tidal stresses in Europa's crust has rotated in a clockwise direction over time. This observation has been used to suggest that Europa's rotation is not perfectly synchronous. Over time Europa may rotate faster than the synchronous rate, causing the surface to be progressively reoriented relative to tidal forces.

High-resolution images from the Galileo orbiter show places on Europa resembling ice flows in the Earth's polar regions. Large, angular pieces of ice have shifted away from one another, some rotating in the process, but reconstructions show that they fit together like puzzle pieces. This evidence for motion involving fluid flow, along with the possibility of geyser eruptions, shows that the ice crust has been, or is still, lubricated from below by warm ice, or even liquid water. The source of heating to produce this watery fluid is tidal forces by gravitational interaction with massive Jupiter, along with some escaping heat produced by radioactive minerals in the underlying silicate crust.

The Galileo spacecraft's mission was expanded to include the Europa Extended Mission, as it flew a number of close flybys to focus its instrumentation and cameras specifically on Europa. On one close encounter with Europa, Galileo came within 200 kilometers of the icy surface of the satellite. Ultimately, the Galileo spacecraft was purposely directed to plunge destructively into Jupiter's atmosphere on September 21, 2003. The reason for this was to safeguard any possible life-forms on Europa against the plutonium inside the Galileo spacecraft's radioisotope generators in the event that it might have crashed into the satellite.

Europa remains a high-priority location within the solar system for astrobiology studies. With the demise of the Galileo spacecraft, plans were proposed to send another spacecraft, this time to orbit Europa for prolonged and repeated studies. A more ambitious plan arose, called Jupiter Icy Moons Orbiter or JIMO. JIMO would have been the flagship mission of a more extensive program to develop nuclear propulsion as a means to cut down the time of travel between Earth and the rest of the solar system. That program was called Project Prometheus, but after

initial funding was granted the National Aeronautics and Space Administration (NASA) was forced to postpone, if not cancel, this futuristic propulsion system in favor of other expenditures arising from the Vision for Space Exploration program under the George W. Bush administration. Returning to Europa with a robotic spacecraft was therefore put on hold for the early portion of the twenty-first century. Data from the Cassini spacecraft in orbit about Saturn revealed aspects about the icy satellite Enceladus and Titan (with its thick atmosphere and organic compounds) that diverted the attention of many astrobiologists away from Europa.

METHODS OF STUDY

Jupiter and its four largest satellites have been studied using telescopes since Galileo first trained his on the system in 1610. Prior to the advent of interplanetary space probes, telescopic observations resulted in a remarkable treasure trove of data on the Galilean satellites.

For example, in the 1920's the astronomers Willem de Sitter and R. A. Sampson succeeded in obtaining reasonably accurate data on their masses. Calculations involved observing how each satellite disturbed the orbits of the others and by noting the nature of the resonant orbits of the inner three (first described by Pierre-Simon Laplace in the late eighteenth century). These resonant orbits dictate that for every one orbit of Io around Jupiter, Europa revolves two times and Ganymede four. This orbital resonance scheme implies a specific ratio for the masses of the bodies, which assisted de Sitter and Sampson in their calculations.

Diameters of the satellites were not accurately known until the advent of stellar occultation studies in the 1970's and later, when spacecraft imaging produced precise values. Prior to that, Europa was described by a popular 1950's-era science text as having a diameter of 1,800 miles (2,880 kilometers), only a bit less than the currently accepted value of 3,138 kilometers.

Although the first Earth-launched space probes encountered the Jovian system in 1973 (Pioneer 10) and 1974 (Pioneer 11), they paid scant attention to the Galilean satellites. The community of planetary scientists at the time viewed the satellites of the gas giant planets in the outer solar system to be nothing but rather boring ice balls. Even on the Voyagers, few planetary science studies were devoted to any of the icy satellites. Only imaging studies of Io and Titan, and to a lesser extent Europa, were planned as major portions of Voyager flyby operations in the Jupiter or Saturn system.

In 1979, however, knowledge of these bodies dramatically expanded as images of all four satellites were beamed back to Earth by Voyagers 1 and 2. The first pictures of Europa showed a previously unknown world, with a highly reflective, smooth surface mottled by brown and tan patches and crisscrossed by a complicated network of curved and straight lines. Four months later, higher-resolution images from Voyager 2 confirmed the presence of even more linear structures, which were interpreted as fractures but having virtually no relief associated with them. In addition to its imaging work, the Voyager probes also made precise measurements of the mass of the Galilean satellites by analyzing the gravitational effects of the planets on spacecraft trajectories, which, combined with improved size determinations, allowed for more accurate calculations of density. Density, in turn, is used to assess planetary composition.

In 1995 the Hubble Space Telescope discovered a thin oxygen atmosphere on Europa; this was later confirmed by the Galileo orbiter during its Europa Extended Mission. The official term for this rarefied atmosphere is a surface-bound exosphere. Hubble used its highly sensitive spectrometers to analyze the energy spectrum of light reflected from the moon's surface. Europa's atmosphere is so tenuous that its surface pressure is only one-hundred-billionth that of Earth. It is estimated that if all the oxygen on Europa were to be compressed to the surface pressure of the Earth's atmosphere, it would fill about a dozen Houston Astrodomes.

The Galileo spacecraft investigated the Jupiter system from late 1995 through 2003. After launching an atmospheric probe into Jupiter itself relatively early in its mission, the Galileo orbiter assumed an elliptical orbit that allowed it to make several close passes to all four Galilean satellites. The resolution of Voyager images of Europa made it possible to view surface features no smaller than about 4 kilometers across. In contrast, Galileo swooped down closer than either Voyager spacecraft, and with its

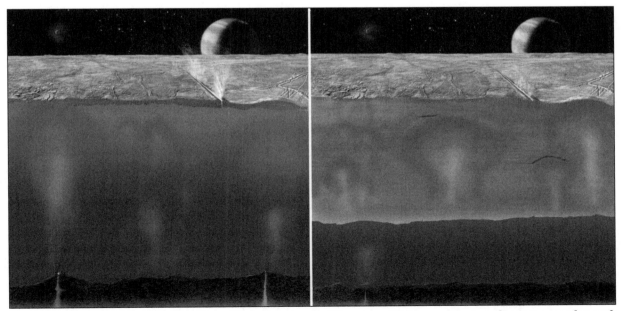

An artist's cross-section rendering of two likely structures of Europa, with ice overlying a saltwater ocean beneath and heat rising through the rocky mantle, possibly volcanically. (NASA/JPL)

more sophisticated cameras it achieved resolutions of around 10 meters per pixel, allowing objects the size of earthly buildings to be discerned. From these high-resolution images scientists have observed evidence of both tensional and compression ridges and have documented features like water-ice geysers, possible ice volcanoes, and jumbled ice flows that resemble puzzle pieces. These observations paint a picture of a dynamic planet in which tectonic faulting and flooding by liquid water occur periodically. The dark color of many surface fractures may result from the injection of water or warm ice mixed with darker silicates that well up into the fractures and freeze. Galileo images have generated renewed interest in the idea that a layer of liquid water exists below the ice or existed some time in the recent past.

Galileo also carried a magnetometer to detect the existence of a planet's magnetic field and to measure its strength. During a December, 1996, pass of Europa this magnetometer detected the first evidence of a magnetic field. Ganymede, the next moon out from Europa, also has a magnetic field. Although it is about four times weaker than that of Ganymede, Europa's field is still of substantial magnitude. Combined with gravity data suggesting a dense core, the Europa magnetic measurements indicate the probable existence of a sizable metallic core, as well as a layered internal structure similar to that of Earth. The magnetic field data also, however, provided constraints on the nature of the water on Europa. The magnetic field could not be explained by assuming pockets of salty water within a crust of ice, but rather required a spherical shell of liquid water.

CONTEXT

The Jovian system has long been of interest to scientists as a possible model analogous to the larger solar system of the Sun and planets. In this model Jupiter is a substitute for the Sun and the Galilean satellites represent the planets, particularly the rocky planets from Mercury to Mars. The considerable masses and stable orbits of the Galilean satellites suggest that they probably originated along with Jupiter during its formation from the gaseous solar nebula. If so, do these satellites show evidence for

having evolved in a manner that parallels that of the inner planets of the solar system, including Earth?

In the early 1970's planetary scientist John Lewis pointed out that the densities of the two outer satellites, Ganymede (1.93 grams per cubic centimeter) and Callisto (1.83 grams per cubic centimeter), were consistent with condensation of solar-composition gas (the solar nebula), where water ice is a stable compound. He predicted that these two bodies should be composed of about equal parts water ice and silicate rock, a view generally accepted today. The two inner bodies, Io and Europa, however, have higher densities (3.55 and 3.04 grams per cubic centimeters, respectively) and thus would be expected to contain less in the way of low-density materials like ice, with a density of 1.0 gram per cubic centimeter. In fact, these bodies show evidence of being largely composed of rocky material, with no ice on Io and only a thin crust of ice over an ocean of liquid water on Europa. What processes could have produced such a density distribution?

In the early 1950's astronomer Gerard P. Kuiper suggested that Jupiter had been very hot during its early history. Building on Kuiper's early work, current hypotheses confirm that Jupiter was probably hot enough in its infancy to have forced low-mass, volatile gaseous materials to the outer fringes of the Jovian region, leaving heavier compounds to accrete as planetoids closer to Jupiter. The lighter volatile gas would contain a high proportion of elements that would eventually freeze as ice, compared to denser silicate minerals. These materials would eventually accrete to produce Ganymede and Callisto, while the volatile-poor inner gas would eventually accrete as Io and Europa. Europa, being farther from Jupiter than Io, has more volatiles such as ice than Io, which has virtually no ice on its surface. Io probably also lost much of its volatile component as a result of long-term volcanic activity, the result of tidal heating produced by Jupiter's gravitational field.

In a similar fashion the solar system as a whole exhibits a composition distribution with high-density "rocky" (or terrestrial) planets near the Sun and more volatile-rich bodies in the outer regions. The inner planets show a similar density distribution. Thus, the Jovian satellite system shows that any evolving planetary system on a scale large enough to have a hot central "star" predictably develops a density distribution where low-density, high-volatile planets dominate the outer regions and high-density "rocky" bodies dominate the inner regions.

The study of Europa is important in terms of its possible role as a site of extraterrestrial life-forms. Images from the Galileo space probe have confirmed ideas spawned after the Voyager flybys that Europa may have a globally encircling layer of liquid water beneath its surface ice layer. Where water exists in liquid form on a planet, life as we know it can theoretically evolve. Europa has joined the ranks of Mars and Saturn's moon Titan as possible sites where primitive life could exist.

Comparative planetology could indeed be actively studied inside the Jovian system by robotic spacecraft. Voyager provided only tantalizing images and raised many questions. Galileo data strongly suggested that the four large Jovian satellites share essentially similar cores in terms of size. Io is volcanically active, essentially taking material deep inside and turning the satellite inside out by resurfacing itself with sulfur and sulfur compounds; therefore Io is devoid of an icy shell. Europa has internal heat that is insufficient to melt the ice cover in total but supplies warmth to a subsurface layer of liquid water. Ganymede and Callisto have thick, icy shells and retain more of their primordial character because they lack significant heating from their cores. Ganymede's magnetic field is something of a paradox in that its character is akin to the dipole field produced by convection within a molten iron core.

Europa once held the primary attention of astrobiologists as the favored place within the solar system for finding some type of life beyond Earth. As a result of extensive robotic investigation of other portions of the outer solar system, it must now share that hopeful spotlight with Titan and Enceladus.

John L. Berkley

FURTHER READING

Bagenal, Fran, Timothy E. Dowling, and William B. McKinnon, eds. *Jupiter: The Planet, Satellites, and Magnetosphere*. Cambridge, England: Cambridge University Press, 2007. A comprehensive work about the biggest planet in the solar system, offering a series of articles by recognized experts in their fields of study. Excellent photographs, diagrams, and figures about the Jupiter system and the various spacecraft missions that unveiled its secrets.

Beatty, J. Kelly, Carolyn Collins Petersen, and Andrew Chaikin, eds. *The New Solar System*. 4th ed. Cambridge, Mass.: Sky, 1999. A chapter on the Galilean satellites by Terrence Johnson gives a comprehensive overview of these satellites based on Voyager and Galileo data. The volume is amply illustrated with color images, diagrams, and informative tables. Aimed at a popular audience, this book can also be useful to specialists. Contains an appendix with planetary data tables, a bibliography for each chapter, planetary maps (including Europa), and an index.

Cole, Michael D. *Galileo Spacecraft: Mission to Jupiter*. New York: Enslow, 1999. Provides a full description of the Galileo spacecraft, its mission objectives, and science returns through the primary mission. Particularly good at describing mission objectives and goals. Suitable for a younger audience.

Fischer, Daniel. *Mission Jupiter: The Spectacular Journey of the Galileo Spacecraft*. New York: Copernicus Books, 2001. Suitable for a wide range of audiences, this volume thoroughly explains all aspects of the science and engineering of the Galileo spacecraft. Particularly good discussions about the nature of the Galilean satellites.

Geissler, Paul E. "Volcanic Activity on Io During the Galileo Era." *Annual Review of Earth and Planetary Sciences* 31 (May, 2003): 175-211. The definitive work describing the physics and planetary geology of volcanoes on Io. Provides a complete picture of Voyager and Galileo spacecraft results.

Greely, R. *Planetary Landscapes*. 2d ed. Boston: Allen and Unwin, 1994. This book concentrates on the nature and origin of planetary surface features. It is packed with excellent diagrams, tables, maps, and monochrome images of planets taken by robotic spacecraft.

A chapter on the Jovian system includes a detailed section on Europa, but it is dated prior to the arrival of the Galileo spacecraft into orbit about Jupiter. Contains an extensive reference section and index.

Harland, David H. *Jupiter Odyssey: The Story of NASA's Galileo Mission*. New York: Springer Praxis, 2000. Includes virtually all of NASA's press releases and science updates during the first five years of the Galileo mission, with an enormous number of diagrams, tables, lists, and photographs. Also provides a preview of the Cassini mission. Although the book was published before completion of the Galileo mission unfortunately, what is missing can easily be found on numerous NASA Web sites.

Hartmann, William K. *Moons and Planets*. 5th ed. Belmont, Calif.: Thomson Brooks/Cole, 2005. An updated version of a classic text on planetary science. The chapter on Jupiter covers all aspects of the Jovian system and spacecraft exploration of it.

Irwin, Patrick G. J. *Giant Planets of Our Solar System: An Introduction*. 2d ed. New York: Springer, 2006. Suitable as a textbook for upper-level college courses in planetary science. Focuses on Jupiter, Saturn, Uranus, and Neptune and their satellites, rings, and magnetic fields. Filled with figures and photographs.

Leutwyler, Kristin, and John R. Casani. *The Moons of Jupiter*. New York: W. W. Norton, 2003. Written by the original Galileo program manager, this heavily illustrated work provides discussions of the Galilean satellites and a number of the lesser known Jovian moons. The authors attempt an artful text to accompany the scientific findings, which may or may not be to the taste of all readers.

McBride, Neil, and Iain Gilmour, eds. *An Introduction to the Solar System*. Cambridge, England: Cambridge University Press, 2004. A complete description of solar-system astronomy suitable for an introductory college course. Filled with supplemental learning aids and solved student exercises. A Web site is available for educator support. Accessible to nonspecialists as well.

See also: Callisto; Ganymede; Io; Jovian Planets; Jupiter's Satellites; Neptune's Satellites; Planetary Satellites; Saturn's Satellites; Uranus's Satellites.

Extrasolar Planets

Category: Planets and Planetology

Indirect methods of observation have revealed the existence of an ever-increasing number of planets orbiting other stars. These extrasolar planets (or exoplanets) have surprised astronomers and led to new theories about planet formation, because they differ from the planets in our solar system. This new evidence suggests the uniqueness of our own solar system.

OVERVIEW

The discovery of extrasolar planets orbiting Sun-like stars has excited the imaginations of astronomers and laypersons alike. If it can be demonstrated that planetary systems are a common occurrence among the billions of stars within our galaxy, the possibility of extraterrestrial life in the universe takes on greater credibility. The idea that intelligent civilizations may exist on other planets could become more compelling.

Early in the twentieth century, spectroscopic evidence from Barnard's star, a nearby red dwarf one-seventh the mass of the Sun, indicated a slight wobble that seemed to imply gravitational interaction by one or two Jupiter-mass planets with decade-long orbits. However, by 1980 further work showed that the wobble of Barnard's star was more likely the result of a companion star too small to observe. The mass of an unseen companion can be estimated from the amount of wobble detected from a visible star.

Double-star systems like Barnard's tend to rotate around their common center of mass in larger orbits than the tiny wobble of a star with a planetary system. Masses between about ten and eighty Jupiter masses usually qualify as brown dwarfs, defined as objects that formed like other stars by gravitational collapse of a dust cloud rather than by accretion from a stellar disk. However, they are too small to sustain the nuclear fusion processes that energize the cores of most stars.

The first confirmed extrasolar planetary system was discovered in 1991, but it was a far cry from a Sun-like solar system that could support life as we know it. Pennsylvania State University radio astronomer Alex Wolszczan was observing a millisecond pulsar (PSR 1257+12) that he and Dale Frail had just discovered using the 305-meter Arecibo radio telescope in Puerto Rico. This pulsar resulted from the collapse of a massive star about a billion years ago. It is now

a neutron star that spins 161 times each second, generating a radio pulse about every 6.2 milliseconds. However, Wolszczan found that these pulses varied periodically from the usual high degree of regularity exhibited by other pulsars.

Analysis revealed two periods in the pulse variations from PSR 1257+12: one lasting sixty-six days and the other ninety-five. Wolszczan and Frail proposed that two Earth-size planets orbit the pulsar, gravitationally tugging on it and causing its radio pulses to arrive slightly earlier and then later than expected. Calculations showed that one planet had at least 3.4 Earth masses at an orbital distance of 0.36 astronomical unit (1 AU is the distance from

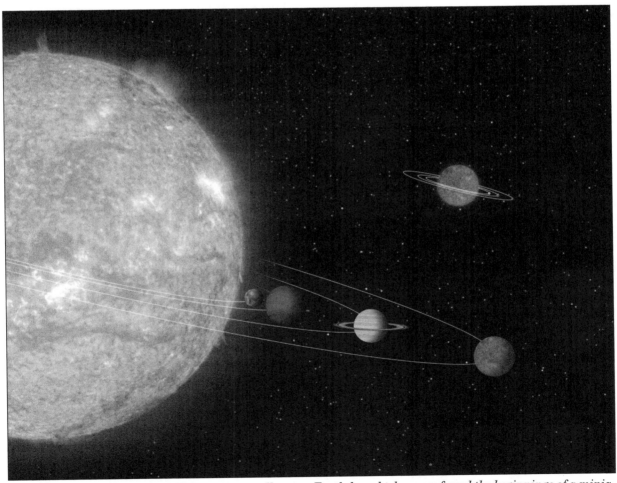

The Spitzer and Hubble Space Telescopes as well as two Earth-based telescopes found the beginnings of a miniature solar system centered on the star 55 Cancri, located in the Chamaeleon constellation, 500 light-years from our own solar system and depicted in this artist's rendering. (NASA/JPL-Caltech)

Earth to the Sun). The other was at least 2.8 Earth masses and 0.47 AU from the pulsar. By 1994, additional observations revealed a third planet with a period of twenty-three days that had about 0.015 Earth mass located at 0.19 AU.

Planet discoveries about Sun-like stars began in 1995, revealing two new and unexpected types of planetary objects: small-orbit, hot-Jupiter-type planets, and eccentric-orbit Jupiter-like planets. In October of 1995, Swiss astronomers Michel Mayor and Didier Queloz of the Geneva Observatory announced evidence of a companion object orbiting 51 Pegasi (in the constellation Pegasus), which is about 40 light-years away. A new generation of optical instruments and computers revealed a periodic Doppler shift of the light coming from the star. This suggested a tiny wobble caused by a planet of at least 0.46 Jupiter's mass and a period of only 4.2 days in a circular orbit at an orbital distance of 0.05 AU. At this small distance, the planet orbiting 51 Pegasi has a surface temperature of about 1,000 kelvins. While officially named 51 Pegasi b, the planet is informally referred to as Bellerophon.

In a 1996 survey of 120 nearby Sun-like stars, Geoffrey Marcy of San Francisco State University and Paul Butler of the University of California, Berkeley, used a refined form of Mayor and Queloz's method to discover six new Jupiter-size planets. The existence of the first two planets, announced in January of that year, were discovered from the tiny wobbles of stars in Virgo and Ursa Major, located 46 and 80 light-years away, respectively. The planet around the star 47 Ursae Majoris has a minimum mass of 2.3 Jupiter masses with an orbital period of 3.0 years and an orbital radius of 2.1 AU (less than half of Jupiter's distance of 5.2 AU). The planet orbiting 70 Virginis has a minimum mass of 6.6 Jupiter masses and a highly eccentric orbit (0.40 eccentricity) of 117 days at an average orbital radius of 0.43 AU. In 2002, Marcy and Butler, along with Debra Fischer, announced their finding of 47 Ursae Majoris c. The planet has an orbital period of 2,594 days and has roughly the mass of Jupiter.

The four other planets included three hot-Jupiter planets similar to 51 Pegasi with nearly circular orbits. At 46 light-years, 55 Cancri has a planet with mass at least 0.8 that of Jupiter and an orbital period of about 15 days, with an orbital radius of 0.11 AU. At 60 light-years, Tau Bootis has a planet with a minimum mass of 3.87 Jupiter masses, a period of 3.3 days, and an orbital radius of only 0.046 AU. Located 55 light-years away from Earth, Upsilon Andromedae has a planet with mass at least 0.68 that of Jupiter, an orbital period of 4.61 days, and an orbital radius of 0.06 AU. Eleven years later, in 2007, astronomers confirmed the existence of two other planets in the Upsilon Andromedae system. The middle planet has an approximate orbit of 242 days and is at least twice the mass of Jupiter. The outermost planet, at 2.5 AU, is at least four times as massive as Jupiter, with an orbital period between 3.5 and 4 years.

Marcy and Butler also announced a possible second planet orbiting 55 Cancri with a minimum mass of about 5 Jupiters, an orbital period of about 20 years, and an orbital radius of 5 to 10 AU. In 2002, 55 Cancri d was found orbiting the star at a distance of about 5 AU, with a mass 4.8 times that of Jupiter. At that time, a third planet—named 55 Cancri c and having a mass roughly that of Saturn's and a highly eccentric orbit—was also speculated to exist. In 2004, 55 Cancri e was discovered. This planet is about the size of Neptune. It is either a small gas giant or a large terrestrial planet. At this time, the existence of 55 Cancri c was confirmed.

The following year astronomer Jack Wisdom questioned the accuracy of naming 55 Cancri e as an exoplanet. After reexamining available data, Wisdom believed that planet e has an orbit of 261 days instead of the proposed 2.8-day orbit. In 2007, 55 Cancri f was announced to be about half the mass of Saturn and to have an orbital period of 260 days, placing it in the habitable zone of the star. 55 Cancri f is not believed to contain life, but hypothetically any satellites the planet has could contain at least microbial life-forms.

Evidence for the nearest planetary system was also announced in 1996 by George Gatewood of the University of Pittsburgh. He collected photometric data on many of the nearest stars with the 30-inch refractor telescope at Allegheny Observatory. The dim red dwarf star Lalande 21185, the sixth nearest star to the Sun

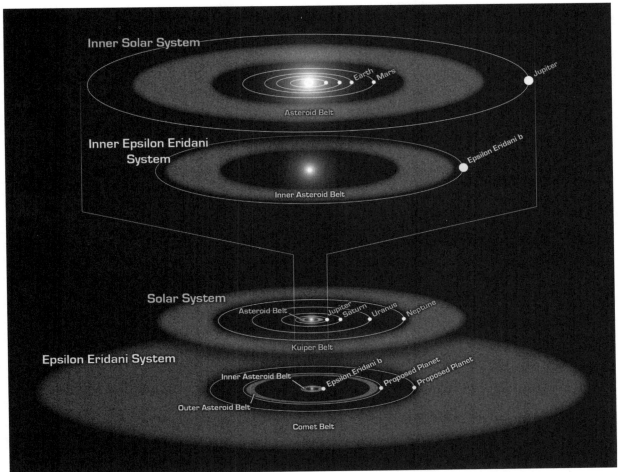

A comparison of the Epsilon Eridani system, the planetary system closest to Earth's, with Earth's solar system. (NASA/JPL-Caltech)

at 8.2 light-years away, appears to have two Jupiter-size planets in orbits similar to those of the gas giants in our solar system. Gatewood analyzed data from fifty years of photographic observations and eight years of photoelectric measurements, revealing tiny accelerations of the star that suggest one planet of about 0.9 Jupiter mass with a period of about 5.8 years in a circular orbit with an orbital radius of about 2.2 AU (similar to the asteroid belt) and a second planet of about 1.1 Jupiter mass with a period of about 30 years in a circular orbit with a radius of about 11 AU (similar to Saturn). A third, unconfirmed planet may orbit beyond these two Jupiter-like planets. The proximity of Lalande 21185 suggests the possibility of eventually cap-

turing an image of its planets with the Hubble Space Telescope.

Two more eccentric planets were announced in 1997. A group of Harvard astronomers led by David Latham discovered an object in 1988 with a mass of at least 9 Jupiters orbiting the star HD 114762 in an 84-day eccentric orbit that varies from 0.22 AU to 0.46 AU (0.35 eccentricity). For eight years, this object was classified as the smallest known brown dwarf, but after Marcy and Butler announced the 70 Virginis planet with a very similar eccentric orbit varying from 0.27 to 0.59 AU and a minimum of 6.5 Jupiter masses, the companion of HD 114762 appeared to qualify as a possible planet.

A third eccentric planet has by far the greatest eccentricity (0.67) of any known planet. Discovered by Marcy and Butler, it was also independently discovered by William Cochran and Artie Hartzes of the University of Texas at Austin. The planet orbits the star 16 Cygni B, a near solar twin that belongs to a triple star system 100 light-years away. It has a mass of at least 1.5 Jupiters and a 2.2-year orbit that varies between 0.6 and 2.8 AU, giving it wild seasonal variations.

Another hot-Jupiter planet orbiting the star Rho Coronae Borealis appears to fill a gap between the very close 51 Pegasi-like planets (less than 0.11 AU) and the 47 Ursae Majoris planet (2.2 AU). It was discovered in 1997 by a Harvard University team of astronomers led by Robert Noyes and has a nearly circular orbital radius of 0.23 AU, a period of 39.6 days, and a minimum mass of 1.1 Jupiter masses. Given the existence of giant planets with orbits from 0.046 AU (Tau Bootis) to 2.2 AU in a relatively continuous distribution, planet formation theories face dramatic challenges, especially since existing theories predict that Jupiter-size planets cannot form within 5 AU of their host stars.

In 2005, Osiris (HD 209458b), an exoplanet of 0.69 Jupiter mass orbiting at 0.046 AU around its parent star, was directly detected using the Spitzer Space Telescope. The telescope was able to observe infrared light coming from the planet itself. Astronomers noted differences in the light being produced when the planet was transiting in front of the star and also when the planet was blocked by the star. By factoring out the star's constant light, scientists were able to isolate the planet. From this, they were able to estimate the temperature of Osiris to be at least 1,023 kelvins.

HD 189733b was discovered in October, 2005, at a distance of about 63 light-years from Earth. The planet is considered a hot Jupiter, with a mass 15 percent greater than Jupiter's and an orbital period of just 2.2 days. In 2007, using the Spitzer Space Telescope, astronomers in Switzerland detected water vapor within the atmosphere of that planet.

The first "super Earths" were found in 1991 orbiting a pulsar. The two planets were only four times the mass of the Earth, too small to be considered gas giants. The general definition of a super Earth is a terrestrial exoplanet that has a mass between one and ten times that of the Earth. Other scientists say a super Earth must be five to ten times the mass of the Earth. In 2007, Stéphane Udry and his team announced the finding of two super Earths orbiting around Gliese 581. Both planets are within the habitable zone (the area where liquid water could potentially exist) of their star. Gliese 581 c orbits the star at 0.073 AU and has about five times the mass of the Earth.

In 2007, the Mount John University Observatory in New Zealand discovered the smallest extrasolar planet to date. The planet, MOA-2007-BLG-192LB, is only 3.3 times as massive as the Earth and orbits a brown dwarf. The first group of super Earths within the same planetary system was found orbiting HD 40307 in 2008. The three planets are all orbiting the star at a distance less than that of Mercury from the Sun, with nearly circular orbits.

METHODS OF STUDY

Detecting extrasolar planets from Earth is extremely difficult, requiring a new generation of computers and optical instruments. Planets are about a billion times fainter than their host star, making them virtually undetectable by direct methods. An indirect method involves searching for a tiny wobble in the motion of a star as it and any companions orbit about their common center of mass. Although the gravitational interaction between a star and a planet-sized object is too small to observe directly, the radial velocity (back and forth along the line of sight) alternately increases and decreases the wavelength of light from the star, causing an alternating Doppler shift toward the red and then blue end of its spectrum.

The velocity of a star can be determined from the magnitude of its Doppler shift. The shift in wavelength due to a Jupiter-size planet is only one part in ten million. An absorption cell (consisting of a bottle of iodine vapor placed near the focus of the telescope) absorbs certain known wavelengths of light, producing dark lines in the spectrum that act as a reference for measuring the Doppler shift accurate to within one part in a hundred million. These shifts are recorded by sending light from a star into complex spectrometers consisting of prisms, mirrors, and gratings costing several million dollars.

Periodic variations in the Doppler shift reveal the period of a planet's orbital motion. The velocity of the star and the period of its motion can be analyzed to determine the radius of the orbit (from Kepler's law) and the minimum mass of the planet (from Newton's laws). However, the unknown inclination of its orbit allows for a larger wobble than its apparent radial motion and thus a larger possible mass by a factor of about two. The periodic variation in Doppler shift also reveals the shape of the orbit, since a circular orbit produces a perfect sine wave, while an eccentric orbit produces an irregular variation that can be analyzed by computer to determine the orbital shape.

Using these methods, Marcy and Butler detected radial motions accurate to within ±3 meters per second, compared to at least 10 meters per second required to detect a planet. Since Jupiter, which contains most of the mass of the solar system at 318 times the mass of the Earth, causes the Sun to move at a speed of up to 12.5 meters per second, Jupiter-size planets can be readily detected. Most of the new planet discoveries have been based on stars wobbling at speeds between about 10 and 300 meters per second. Planets much smaller than Jupiter cannot be detected accurately with this method, and those with periods of several years require that data be collected over a long enough time span to determine their periodic variations.

Marcy and Butler began collecting Doppler-shift data in 1987 for their survey of 120 Sun-like stars, using Lick Observatory's three-meter telescope; but it was the computer methods used by the Swiss in their discovery of the 51 Pegasi planet that finally yielded results. Their first discoveries resulted from running six computers day and night at the University of California, Berkeley, to analyze data from sixty stars. These methods revealed a variety of planets that shocked astronomers because their orbits were so unexpected. Hot Jupiters and eccentric orbits have initiated a new generation of theories about planetary formation and the uniqueness of our solar system.

The Spitzer Space Telescope (SST) was launched in 2003. It consists of three main instruments: the Infrared Array Camera (IRAC), an infrared camera that operates simultaneously on four different wavelengths; the Infrared Spectrograph (IRS), a spectrometer able to observe at four wavelengths; and the Multiband Imaging Photometer for Spitzer (MIPS), which is made up of three different far-infrared detector arrays. In 2005, the SST was the first telescope to detect light from exoplanets HD 209458b and TrES-1. However, the light was not turned into actual images.

In 2004, a group of astrophysicists in France captured the first photograph of an extrasolar planet orbiting a brown dwarf. The planet appeared only as a small red dot. It is speculated to have a mass two to five times that of Jupiter, but it orbits the star at a distance greater than Pluto's average distance from the Sun. This "exoplanet" did not form from

An artist's concept of a young solar system in which gas giant planets are forming. (NASA/JPL-Caltech/T. Pyle, SSC)

an accretion event the way scientists currently believe planetary formation occurs. Also, a brown dwarf would not have enough material to form a Jupiter-sized planet, especially at such distances. Because of these objections, many astronomers do not consider the photograph to be of a real planet.

With the number of known extrasolar planets totaling more than three hundred, the next challenge is to capture a photograph of an actual exoplanet. Two of the programs dedicated to imaging a planet are located in Chile. In 2007, the Gemini South Observatory installed the first optics system specially designed to photograph exoplanets. In 2008, it started a two-year program to conduct a survey of young stars using its Near-Infrared Coronagraphic Imager (NICI). The NICI consists of a coronagraph and two cameras that can simultaneously photograph the star and its surroundings at two different infrared wavelengths. The two photographs would then be subtracted, leaving behind an image of the planet. This method will also help eliminate false planets that are actually background stars or stray starlight.

A possible first photograph of an extrasolar planet (1RXS J160929.1-210524 b) was announced in September, 2008. The image was taken using the Gemini North Telescope on Mauna Kea in Hawaii. The extrasolar planetary system is about 500 light-years away from the Earth. 1RXS J160929.1-210524 b has a mass eight times that of Jupiter and orbits its star at a distance of 330 AU. Some scientists are skeptical of its being an extrasolar planet because of its great distance from its star (Neptune, by comparison, is only 30 AU from the Sun).

CONTEXT

The discovery of extrasolar planets may seem at first to offer new hope for the existence of planetary systems like Earth's solar system that could support extraterrestrial life. However, the unexpected nature of these planets has raised new challenges for planet formation theories and new doubts about the possibility that any of them might harbor life. Pulsar planets were probably formed from the remnants of a companion star during a supernova explosion

that produced a spinning neutron star, and they are bathed with high-energy radiation that would make life impossible. The other new planets orbit more Sun-like stars but have either extremely small or highly eccentric orbits that also make them unlikely candidates for life. Evidence so far seems to indicate that our solar system is highly unusual, if not completely unique, in harboring water-based life-forms.

Joseph L. Spradley,
updated by Jennifer L. Campbell

FURTHER READING

Casoli, Fabienne, and Thérèse Encrenaz. *The New Worlds: Extrasolar Planets.* New York: Springer Praxis, 2007. The author discusses the history of the search for extrasolar planets, techniques used, and early discoveries. Also examines research and discoveries since 1996, during which the number of extrasolar planets increased to more than three hundred. The authors speculate about the possibility of life on these planets, as well as what they can teach us about planetary formation.

Dick, Steven J. *The Biological Universe: The Twentieth-Century Extraterrestrial Life Debate and the Limits of Science.* Cambridge, England: Cambridge University Press, 1999. Chapter 4, "Planetary Systems: The Limits of Theory," provides a good history of the search for extrasolar planets before 1995, with several illustrations.

Dvorak, Rudolf, ed. *Extrasolar Planets: Formation, Detection, and Dynamics.* Weinheim, Germany: Wiley-VCH, 2008. This work explains not only how extrasolar planets are detected but their formation, dynamics, and atmospheres as well. It also discusses habitable zones, along with plans to locate and study new extrasolar planets.

Goldsmith, Donald. *Worlds Unnumbered: The Search for Extrasolar Planets.* Illustrations by Jon Lomberg. Sausalito, Calif.: University Science Books, 1997. The first book to discuss the new extrasolar planet discoveries in detail, including theories of formation, methods of observation, and possibilities of life. Includes several color plates and an index.

Jet Propulsion Laboratory, National Aeronautics and Space Administration. *Planet Quest:*

New Worlds Atlas. http://planetquest.jpl.nasa .gov/atlas/atlas_index.cfm. A searchable database listing all currently known extrasolar planets and all known data about the planets and their stars, including scale diagrams comparing them to bodies in our solar system. Regularly updated by NASA's Jet Propulsion Laboratory in California.

Mammana, Dennis, and Donald McCarthy. *Other Suns, Other Worlds? The Search for Extrasolar Planetary Systems*. New York: St. Martin's Press, 1996. A comprehensive history of the search for extrasolar planets through 1995, with several plates, including one describing the January, 1996, discoveries of Marcy and Butler.

Ollivier, Marc, et al. *Planetary Systems: Detection, Formation, and Habitability of Extrasolar Planets*. New York: Springer, 2008. Explores the information known about the three hundred exoplanets currently known, including their sizes, atmospheres, locations, and habitability. The authors also explain the current and possible future methods for detecting exoplanets, as well as the importance of studying young star systems. For the enthusiast.

Scharf, Caleb. *Extrasolar Planets and Astrobiology*. Herndon, Va.: University Science Books, 2008. An advanced technical explanation of extrasolar planets designed for undergraduates or first-year graduate students. Includes helpful chapter summaries and problem sets.

See also: Extrasolar Planets: Detection Methods; Extraterrestrial Life in the Solar System; Infrared Astronomy; Jupiter's Interior; Planetary Formation; Protostars; Solar System: Origins.

Extrasolar Planets: Detection Methods

Category: Planets and Planetology

The search for extrasolar planets, or exoplanets, beyond our solar system and orbiting other stars has yielded several hundred such objects. Detection methods have been mostly limited to finding large Jupiter-like planets, but as such methods improve there is special interest in finding small Earth-like exoplanets in habitable zones that could support extraterrestrial life.

OVERVIEW

Interest in planets orbiting other stars has a long history, but highly sensitive detection methods are required and the first extrasolar planets were not confirmed until the 1990's. After Nicolaus Copernicus introduced his heliocentric theory of a Sun-centered planetary system in the sixteenth century, astronomers began to realize that space might be endless, with an infinite number of stars. At the end of the century, Giordano Bruno proposed that the stars were also suns with their own planets and suggested that there might be an infinite number of other populated worlds.

In 1855, an astronomer at the Madras Observatory of the East India Company claimed that a revolving double-star system in the constellation Ophiuchus (Ophiuchi 70) had orbital anomalies in its eighty-eight-year period that were probably caused by a planet around one of the stars. This claim was repeated in 1896 by American astronomer Thomas See, who calculated that the anomalies were caused by a planet with a thirty-six-year period. These claims were refuted in 1899 by Forest Moulton, who proved that such a three-body system would be highly unstable.

In the 1960's Peter van de Kamp of the Swarthmore College Observatory claimed to have discovered possible planets around Barnard's star, also in Ophiuchus and the second closest star to our Sun. This faint star, which is moving rapidly toward the Sun at about

140 kilometers per second, appeared to have a tiny wobble in its motion consistent with two Jupiter-size planets. However, this apparent wobble was not found by other observers, and it was later shown to be caused by lens adjustments. Since the reflected light of a planet is much dimmer than its parent star, most extrasolar planets have been discovered by indirect detection methods beginning in the 1990's. However, since 2004 astronomers using the European Southern Observatory's Very Large Telescope array in Chile have produced direct images of several brown dwarf stars with companions, and in 2005 one of these was confirmed as a planet with a mass several times larger than Jupiter's. Six indirect methods have been used to discover most of the known extrasolar planets, and other detection methods continue to be developed.

Three search methods try to detect the tiny elliptical wobbling of a parent star caused by the gravitational influence of an orbiting planet: astrometry, pulsar timing, and radial-velocity detection. The oldest method of searching for extrasolar planets is by astrometry, which requires precise measurements of tiny variations in the position of a star. Several astrometric discoveries of exoplanets were claimed in the 1950's and 1960's, but none was confirmed. Such movements are probably too small to observe with ground-based telescopes but were demonstrated with the Hubble Space Telescope in 2002. Future plans to search with the National Aeronautics and Space Administration's (NASA's) Space Interferometry Mission may reveal many new planets by astrometry. This method is most sensitive to planets with large orbits and long periods, complementing other methods that are more sensitive to small orbits with short periods.

The first confirmed discovery of an exoplanet used a pulsar timing detection method. Pulsars are rapidly rotating neutron stars that emit rapidly pulsed radio waves at highly regular rates matching the rotation rate. In 1992, radio astronomers Alexander Wolszczan and Dale Frail detected slight periodic changes in these millisecond pulse rates and recognized that they were caused by wobbling of the star due to three planets. This method is so sensitive that it

can detect planets smaller than those detectable by any other method—down to a tenth of the Earth's mass—but is limited by the limited number of known pulsars. Although the existence of such small pulsar planets is of interest, they do not offer the possibility of life as we know it since a neutron star emits radiation deadly to such life.

The radial-velocity or Doppler method detects back-and-forth variations in the wobbling of a star and has accounted for the majority of exoplanet discoveries. These radial motions relative to the Earth cause shifting of the star's spectral lines due to the Doppler effect, which decreases and increases the wavelength of the light as the star moves toward and away from the Earth respectively. Modern spectrometers can detect velocity variations down to about 1 meter per second, including the High Accuracy Radial Velocity Planet Searcher (HARPS) spectrometer at the European Space Agency's 3.6-meter telescope in Chile. This method requires high precision and is limited to nearby stars within about 160 light-years. It is most sensitive to large planets with short periods, known as "hot Jupiters" because of their proximity to the Sun, while longer periods require many years of observation. From the period of the planet, the orbital radius can be found. Velocity variations permit an estimate of a planet's minimum mass, which can be considerably larger if the orbit is highly inclined to the line of sight.

A few exoplanets have been detected by three more problematic but developing methods. The transit method measures the tiny dimming of a star when a planet passes in front of it. This method can reveal the size of the planet and can be combined with data from the Doppler method to find its true mass and density. The gravitational microlensing method is based on observations of a star whose gravitational field functions like a lens, focusing light from a distant star directly behind it in the same line of sight. Anomalies in the lensing light curve can reveal planets orbiting Sun-like foreground stars down to the size of the Earth. However, the lensing observations cannot be repeated when the chance alignment of two such stars changes. The circumstellar disk method analyzes the infrared radiation emitted by dust disks that sur-

An illustration of a giant star surrounded by a dust ring, similar to those surrounding the hypergiant stars R 66 and R 126, located in the Large Magellanic Cloud, next-door to the Milky Way, and discovered by the Spitzer Space Telescope. Such dust disks may represent either the seeds of a planetary system or the debris left over after their demise. (NASA/JPL-Caltech/R. Hurt, SSC)

round many stars. Images of dust disks have been obtained by the Hubble and Spitzer space telescopes, and some of these have features that imply the presence of planets.

Three detection methods that may be used in the future are of more specialized interest. The eclipsing binary method looks for anomalies in the light variation as one star of a binary system passes in front of its companion, giving evidence of planets in the system. In the orbital phase method, future space telescopes may be able to detect light variations due to the reflected light from planets that produce phases like that of the Moon. The polarimetry method studies the tiny fraction of polarized light from a star if it passes through the atmosphere of an orbiting planet.

KNOWLEDGE GAINED

Discoveries of extrasolar planets since the first confirmed pulsar planets have increased rapidly and have added much to our knowledge of the varied nature of such planets. The first confirmed planet orbiting a Sun-like star (51 Pegasi) was discovered by the radial-velocity method in 1995 by Michel Mayor and Didier Queloz of the University of Geneva in Switzerland. They were surprised to find that its period was only about 4.2 days and that its orbital radius was much less than that of Mercury. These measurements gave a minimum mass of about half that of Jupiter, or at least 150 times the Earth's mass.

The discovery of a planet around 51 Pegasi was confirmed by a California team led by Geoffrey Marcy, who used the radial-velocity method to discover nearly two-thirds of about three hundred possible extrasolar planets found over the next dozen years. These are mostly hot Jupiters whose masses are assumed to be less than the limit of about thirteen Jupiter masses that distinguishes a planet from a brown dwarf star. Among their achievements was the discovery of the first multiple-planet system, with three Jupiter-size planets around a Sun-like star, and the first transit detection of a planet previously discovered by the radial-velocity method, giving its actual mass and confirming that it was a planet. About twenty multiple-planet systems have been found, and four pulsar planets are known around two separate pulsars.

Among approximately three hundred extrasolar planet candidates, the vast majority have Jupiter-size masses. Since most of these were discovered by the radial-velocity method, only their minimum masses are known and the actual masses of some could eventually show that they are brown dwarfs. About sixty extrasolar planets have been confirmed by various methods that have determined their actual masses. Early discoveries were mostly hot Jupiters with large masses situated very close to their parent star and with very short periods. They seem to defy theories of planet formation based on studies of our solar system, which suggest that large gas planets form farther from their parent star than the smaller rocky planets. However, it is now believed that hot Jupiters probably formed farther out and then migrated in due to larger amounts of dust in their circumstellar disks.

Frequent observations of these hot Jupiters appear to be the result of a selection effect, since such planets are easier to detect over shorter time periods because of the larger and faster wobbling of their parent stars. As detection methods have improved, many such large planets have been found with larger orbits comparable to those of Jupiter and Saturn. Since 2004, a few Neptune-sized planets have been discovered with masses in the range of about seven to fourteen Earth masses. Most exoplanets have much more eccentric orbits than those in Earth's solar system, which is not due to an observational selection effect and is still a major puzzle for astronomers.

CONTEXT

The primary interest in extrasolar planets arises from the possibility of finding Earth-like planets in the habitable zone about their parent stars, where water and thus life might exist. The hot Jupiters and eccentric orbits of most of the exoplanets found so far appear to have little possibility of life. However, a few discoveries are beginning to reveal Earth-like qualities and similarities to our solar system. About seven exoplanets have now been discovered by gravitational microlensing, including a two-planet system found in February of 2008 similar to the Jupiter-Saturn system, which plays the important role of sweeping up errant comets and asteroids that might otherwise make life on Earth impossible.

In 2008 the Optical Gravitational Lensing Experiment (OGLE) at Princeton University found two rocky "super-Earth" exoplanets with masses of only five and three times the Earth's mass, the latter orbiting a brown dwarf, suggesting that it might be a planet covered with water. In May of 2008 several dozen possible new super Earths were announced by an MIT group in Chile using the radial-velocity method with the HARPS spectrometer.

Several planned space missions offer the possibility of much better exoplanet detection from outside the Earth's atmosphere. NASA's Kepler mission is expected to use the transit method with a special space photometer that can simultaneously scan a hundred thousand stars and should be able to detect Earth-sized planets and their statistical frequency around Sun-like stars.

Joseph L. Spradley

FURTHER READING
Casoli, Fabienne, and Thérèse Encrenaz. *The New Worlds: Extrasolar Planets*. New York:

Springer Praxis, 2007. Provides a comprehensive discussion of extrasolar planets and the methods of their detection, with more than two hundred illustrations, mostly in color.

Deeg, Hans, Juan Antonio Belmonte, and Antonio Aparicio, eds. *Extrasolar Planets*. New York: Cambridge University Press, 2008. Eleven contributors to this book provide detailed and comprehensive discussions of detection, formation, statistical properties, and habitability of extrasolar planets.

Dvorak, Rudolf, ed. *Extrasolar Planets: Formation, Detection, and Dynamics*. Weinheim, Germany: Wiley-VCH, 2008. The book comprises eleven papers reviewing research on the detection and analysis research of extrasolar planets and their properties.

Goldsmith, Donald. *Worlds Unnumbered: The Search for Extrasolar Planets*. Illustrations by Jon Lomberg. Sausalito, Calif.: University Science Books, 2002. This introduction to extrasolar planet discoveries and their significance is written by an experienced and authoritative science writer.

Miller, Ron. *Extrasolar Planets*. Minneapolis, Minn.: Twenty-First Century Books, 2002. This book, written for middle school students, includes historical background in astronomy leading up to a description of extrasolar planet discoveries and is colorfully illustrated, including some planetary landscapes by the author.

See also: Extrasolar Planets; Extraterrestrial Life in the Solar System; Infrared Astronomy; Jupiter's Interior; Planetary Formation; Protostars; Solar System: Origins.

Extraterrestrial Life in the Solar System

Category: Life in the Solar System

Exobiology is the search for and study of life on celestial bodies other than Earth. Within the solar system, Mars, Jupiter's satellite Europa, and Saturn's satellites Titan and Enceladus are all considered possible sites where life, or the precursor chemistry needed for the rise of primitive living organisms, might have developed.

OVERVIEW

Understanding where life might have developed in the solar system requires comprehension of how life arose on Earth. The earliest evidence of life on Earth is the presence of organic matter derived from biological processes recorded in rocks that are about 3.2 billion years old. Life may have developed very early in Earth's history. However, much of the fossil record of early life on the Earth has been erased by subsequent geophysical activity. Biologists have pieced together some of that early history by examining the remaining fossil record and by performing a series of laboratory experiments.

Life on Earth is based on complex organic molecules, consisting of chains of carbon, hydrogen, nitrogen, and oxygen. However, organic molecules can be produced by simple chemical reactions as well as by biological activity. Thus, to determine if a process is truly biological, rather than simply a chemical reaction, it is necessary to define the criteria for life. The ability of an organism to reproduce itself is considered to be an essential feature of life. Deoxyribonucleic acid (DNA) and ribonucleic acid (RNA) are the organic molecules that control heredity in terrestrial life-forms. Thus, DNA and RNA are considered essential for reproduction of life on Earth. These two nucleic acids are produced only with the help of certain proteins. A major focus of exobiology is to understand how DNA, RNA, and the proteins essential in their production originated.

A major breakthrough occurred in 1953, when Stanley Miller, a graduate student at the

University of Chicago, and his research supervisor, Professor Harold Urey, produced amino acids, the basic building blocks of proteins, in a sealed environment simulating conditions believed to be present on the early Earth. Miller and Urey continuously passed electrical sparks through a chamber filled with a gaseous mixture of methane, ammonia, and hydrogen (a composition believed to be similar to that of the early atmosphere of the Earth) and water vapor (representing the water contributed by the Earth's oceans). After several days they extracted a mixture of organic molecules, including amino acids, from the bottom of the chamber. The Miller-Urey experiment suggested that lightning discharges throughout Earth's early atmosphere could have deposited amino acids onto the planet's surface. Other experiments demonstrated that bombardment of the gas mixture by high-energy particles, simulating cosmic rays, produced similar results. These experiments suggest that three sufficient conditions must be met to produce amino acids: a supply of carbon-rich material must be present, liquid water must be available, and some energy source (electrical discharge, high-energy particles, or possibly heat and sunlight) is required. Although sufficient, it remains to be proven that these three conditions are also necessary.

Scientists have examined the planets and satellites of the solar system, searching for locations where all three conditions are met. Water may be the most critical restriction, since it remains a liquid only over a very narrow range of temperatures. The surfaces of Venus and Mercury are too hot for liquid water to be present. Jupiter, Saturn, Uranus, Neptune, and Pluto are too cold to support liquid water. Thus, of these nine planets only Earth and Mars seem to be suitable candidates for life, because they are in the range of distances from the Sun such that they could support liquid water. Venus is within the Sun's habitable zone as well, but a runaway greenhouse effect of unknown origin has left it inhospitable to life as we understand and recognize it.

Where life is abundant, it can produce changes in the atmosphere of a planet, allowing astronomers to search for unusual signatures of biological activity. The present composition of the Earth's atmosphere, dominated by nitrogen and oxygen, is regulated by the life-cycle processes of respiration and photosynthesis of Earthly organisms. The atmosphere of Mars, on the other hand, is dominated by carbon dioxide, and it contains only a trace amount of oxygen. Thus, by the 1960's astronomers had observed that, at least in the present era, living organisms were not present in sufficient abundance to perturb the atmospheric chemistry of Mars.

The beginning of the space age made it possible to employ robotic spacecraft to perform direct measurements on the surface of some planets in the expanding search

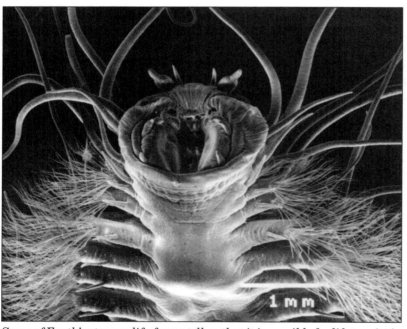

Some of Earth's strange life-forms tell us that it is possible for life to exist in extreme conditions. This 2-inch, centipede-like worm, for example, was found in 1997 living within mounds of methane on the floor of the Gulf of Mexico. (NOAA)

for evidence of life. The first search was performed on Mars by two Viking spacecraft, developed by the National Aeronautics and Space Administration (NASA), which landed safely in 1976. Each Viking spacecraft was equipped with instruments designed to examine the soils of Mars for evidence of Earth-like life.

During the 1980's and 1990's, developments in terrestrial biology changed how exobiologists looked at the essential conditions for the development of life. Single-celled organisms called archaebacteria, which may have developed very early in Earth's history, were discovered. These archaebacteria live in oxygen-deprived places, such as the hot springs at Yellowstone National Park. Archaebacteria take in carbon dioxide and give off methane, and they actually cannot thrive in the presence of oxygen. They have genetic material different from that of other terrestrial life-forms, suggesting that they possibly evolved independently from the more common life-forms very early in Earth's history at a time before the current oxygen-rich atmosphere arose. Other terrestrial microorganisms were discovered that live on sulfur from geothermal sources rather than by relying on the Sun to supply energy. The discovery of these unusual terrestrial life-forms suggests that conditions required for development of the common forms of life on Earth may not be required for the development of all life. Thus, some planets and/or their satellites previously believed to be unsuitable for the development of life may be habitable by organisms rather different from the common life-forms on Earth. This complicates the search for extraterrestrial life, because many experiments, such as those conducted by the Viking spacecraft on Mars in 1976, look only for signatures specific to common terrestrial life.

In the late 1990's and throughout the first decade of the twenty-first century, extrasolar planets were increasingly detected. Although the majority of the first hundred or so worlds were hot Jupiters or at least bizarre large planets in systems not conducive to life as we understand it, in time it became increasingly clear that technology would shortly be capable of picking up Earth-sized planets. A space-based observatory named Kepler was readied for

launch in late 2008. One of its planned objectives was to expand the list of extrasolar planets tremendously and perhaps detect the first Earth-like planets.

Three things must be noted, however. One, there are scientists who dispute the Miller-Urey experiment's validity in terms of the suggestion that production of organic materials in this fashion necessarily leads to the development of life. Two, there remain—even three decades after the Viking biological experiment produced data suggestive of superoxide reactions in the Martian soil rather than a biological metabolism—many scientists who believe that the Viking results were misinterpreted. Perhaps the dismissal of a biological result was premature. Third, many astrobiologists insist that if one restricts one's search to life as we know it based on DNA, one severely limits the possibility for a successful result. Some even question the need for water as a "universal" solvent for life. Organisms using a different solvent would be vastly different.

METHODS OF STUDY

One focus of the search for life is to identify the carbon-rich compounds available for life's development. Impacts of meteorites, asteroids, and comets are believed to have contributed a carbon-rich layer to the Earth's early surface and other planets and their satellites. One particularly carbon-rich meteorite, called Murchison, fell in Australia in 1969. Detailed studies of Murchison established that it contains numerous organic compounds, including amino acids.

In 1986 five spacecraft, two launched by the Soviet Union, two by Japan, and one by the European Space Agency, flew past Halley's comet. Dust analyzers on some of these spacecraft determined the chemical composition of individual dust particles emitted by the comet. These instruments detected a large number of carbon-rich particles, many of which also contained hydrogen, suggesting the presence of organic molecules in the dust. However, detailed analysis of organic molecules requires sophisticated scientific instruments too large and complicated to be flown on those spacecraft. NASA launched a spacecraft called Stardust to fly to Comet Wild 2 to collect dust emitted by that comet. It successfully returned samples to Earth in 2006. Laboratory study of the dust established the abundances and types of organic compounds present in Wild 2.

The second focus of the search for life is to perform direct tests for the presence of biological activity on other planets or satellites. Apollo astronauts collected the first samples from the Moon in 1969. When they returned to Earth, the astronauts, their spacecraft, and their prized lunar rocks were subjected to a twenty-one-day quarantine during which scientists searched for living microorganisms that might be hazardous to life on Earth. Fragments from lunar rocks were crushed and placed in a standard culture medium, a nutrient-rich soup that promotes the growth of microorganisms. Microscopic examination of these samples showed no evidence of living microorganisms. More detailed studies of the lunar rocks have shown no fossil evidence of life-forms that might once have developed on the Moon but are now extinct. Examination of lunar samples revealed them to be exceptionally dry, with none showing any evidence of liquid water. The absence of liquid water was taken to indicate that the Moon was always a lifeless body.

Initial experiments in the search for life on another planet were conducted in 1976 by the two Viking spacecraft that landed on Mars. Each Viking carried four instruments to examine the soil samples for evidence of such basic life-cycle processes as respiration or photosynthesis. The Gas Exchange Experiment deposited samples of Martian soil in a chamber containing a culture medium. This apparatus monitored the composition of gas within its chamber, looking for changes in the abundance of carbon dioxide, oxygen, or hydrogen that would signal metabolic activity by microorganisms in the soil.

In a second experiment, the Labeled Release Experiment, radioactive carbon atoms were incorporated into the culture medium. A detector looked for the appearance of radioactive carbon in released gas, signaling that the addition of Martian soil to the nutrient had resulted in a reaction of biological origin. Both experiments produced positive results, but the effects were much more dramatic than the scientists had ex-

pected. These positive results were eventually explained as chemical reactions initiated because of the highly reactive nature of the surface materials on Mars resulting from their exposure to ultraviolet light from the Sun, a superoxide chemical reaction.

The Pyrolitic Release Experiment provided an opportunity to test that explanation. It was also a labeled release experiment, but this apparatus had the additional capability of heating soil samples between experiments. Scientists heated soil to 548 kelvins, well above the temperature expected to kill any microorganisms present in the soil. Even then the Pyrolitic Release Experiment yielded positive results, suggesting that the release was produced by a chemical reaction involving superoxides rather than a biological process.

A fourth experiment, the Gas Chromatograph Mass Spectrometry Experiment, produced the most convincing evidence that the soils at the Viking landing sites contained no microorganisms. This instrument found no organic molecules within the soil down to a limit of a few parts per million. Even the organic molecules that would be expected in the soils from the accumulation of meteorites like Murchison were absent. Subsequent studies indicated that high chemical reactivity of the soils as well as intense ultraviolet radiation striking the surface would rapidly destroy most organic molecules. Thus, if there is life on Mars, the two Viking spacecraft, which were able only to sample the near-surface soils, were probably looking in the wrong places.

Although instruments on both Viking landers found no evidence of biological activity in their soil samples, the two Viking orbiters obtained high-resolution photographs of Mars's surface, producing results which excited exobiologists. Several regions on Mars revealed features similar to extensive water flow channels on Earth, leading many geologists to conclude that water had flowed freely on the surface of Mars at some earlier period in its history. Because of the assumed importance of liquid water in the development of life, some exobiologists suggested that life might have developed on Mars in that earlier era and that life might now exist in subsurface layers protected from ultraviolet radiation. Or perhaps such life had gone extinct, leaving only fossil evidence behind.

In 1996 scientists from NASA's Johnson Space Center reported that a meteorite called ALH 84001, one ejected from the surface of Mars and deposited in the Antarctic about thirteen thousand years ago, contained microscopic features that might indicate ancient Martian biological activity. This resulted in renewed interest in the search for life on Mars. These suspected fossils resembled wormlike creatures but their size was extraordinarily small. Many scientists pointed out that similar nanometer-sized structures could be produced geochemically and had nothing to do with life. This dispute has not yet been resolved.

After the 1997 Mars Pathfinder exploration of the Red Planet returned amazing images of rocks and terrain, NASA planned a series of robotic spacecraft to continue the exploration of Mars. Two of those spacecraft, the Mars Exploration Rovers named Spirit and Opportunity, launched in June and July, 2003, respectively. They successfully landed on Mars in early 2004 and spent at least the next four years moving about their landing sites searching for evidence of water. Later spacecraft were intended to be even more ambitious, leading to the ultimate desire of exobiologists and planetary scientists alike: a sample return mission from Mars sometime in the second decade of the twenty-first century.

The same techniques used to search for current or fossil life on Mars can be applied to other planets or satellites that are identified as suitable candidates for the development of life. The Galileo spacecraft, placed in orbit around Jupiter in late 1995, obtained close-up photographs of Jupiter's four largest satellites. One of these, Europa, emerged as another potential site for the development of life. One of Galileo's orbits around Jupiter took it within 363 miles of Europa's surface, allowing its cameras to photograph objects as small as 75 feet across. These images showed evidence of ice flows that had broken from a solid sheet and been displaced, suggesting that they had floated or slipped across a liquid ocean or on a layer of slush below. Calculations indicated that Jupiter's extreme gravitational pull could introduce tidal distor-

tions that produce sufficient heat to allow liquid water to exist beneath Europa's icy surface. Other photographs showed dark deposits, possibly carbon-rich material contributed by meteorites.

Titan, the largest satellite of Saturn, has a methane-rich atmosphere believed to be similar in composition to that of the early Earth. High-energy electrons and protons, trapped in the magnetic field of Saturn, continually bombard the upper region of Titan's atmosphere. This bombardment is believed to produce complex organic molecules that rain down onto Titan's surface. Titan is too cold to have liquid water. Titan remained the primary target of study for the Cassini spacecraft, which was launched in October, 1997, and arrived in the Saturn system in early July, 2004. Cassini dropped its Huygens probe, loaded with instruments to measure the types and abundances of the organic molecules, into Titan's atmosphere. The Huygens probe showed its surface may be covered with lakes of methane or ethane, which some scientists now speculate might be sufficient to allow primitive life to develop. Also, Titan's crust appears to move significantly as if floating on a subsurface ocean, adding another intriguing aspect to the possibility of organic chemistry and/or primitive life on Titan.

Even Enceladus displays unexpected geyser activity at its south polar regions. This suggested the possibility of liquid water underneath the surface and therefore the potential for primitive life. Neptune's Triton also exhibits cryovolcanism at an even lower temperature. More research is needed to determine the nature of this mechanism, and that investigation would likely have to await a Neptune orbiter.

Exobiologists were excited to see the possible existence of the three conditions believed necessary for the development of life: carbon-rich material, water, and energy from the Jovian tides. Several follow-on missions have been suggested. A spacecraft placed into orbit around Europa could use radar to see through several miles of ice, detecting any water below and providing a clear test of the ocean model. More ambitious proposals include a spacecraft that would fling a 9-kilogram projectile into the surface of Europa, catch some of the debris lofted by the collision, and return it to terrestrial laboratories for examination. Another common proposal would see a submersible vehicle melt its way through Europa's icy crust to reach a potential subsurface layer of liquid water and image the local environment directly.

CONTEXT

The possibility that life might have developed elsewhere in the solar system has been the subject of speculation for hundreds of years. In 1820, Carl Gauss, a German mathematician, suggested cutting geometrical patterns into the Siberian forest large enough to be seen by an observer using a telescope from the Moon or Mars. The idea was to motivate any inhabitants of the Moon or Mars to engineer similar geometrical patterns, initiating crude communication with the Earth. Other suggestions for communication with intelligent life included setting huge fires in the Sahara desert and constructing large mirrors to reflect sunlight back into space. These early ideas of how to communicate with intelligent life elsewhere in the solar system did not focus on particular sites where the conditions were expected to be appropriate for the development of life.

Although its origins go back as far as 1929, radio astronomy only gained respect within the astronomical community in the early 1950's in the aftermath of World War II, when radio equipment necessary to "listen" to the heavens became available as war surplus. Radio astronomers soon discovered that the natural universe was far from radio quiet. Some scientists, beginning with astronomer Frank Drake, wondered about and then tested the idea that intelligences beyond Earth might be transmitting recognizable radio signals. In due time a coordinated Search for Extraterrestrial Intelligence (SETI) program was developed. No verifiable signals of intelligence have yet been received from deep space.

Only in the second half of the twentieth century did biologists begin to develop an understanding of how life originated on Earth. This knowledge provided clues as to the conditions needed for similar forms of life to develop elsewhere in the solar system. The study of terrestrial life indicates that it originated as simple, single-celled microorganisms and that these simple microorganisms might develop quickly and easily on other planets and/or their satellites as well. Thus, the focus of solar system exobiology shifted from the search for intelligent life, which has not been seen on any planet other than Earth, to the search for simple microorganisms. However, SETI continued, although for a time Congress removed any support for the project through NASA's federal allocations. In time commercial funds supplemented federal funding for SETI projects. For a time in the period following release of the popular movie *Contact* (1997), based on a book by the late Carl Sagan, public interest in SETI increased dramatically.

The dawn of the space age inaugurated an era when spacecraft could be used to search for environments favorable to the development of life, perform experiments designed to detect living organisms on the surface of other planets and/or their satellites, and ultimately return samples to Earth so that scientists could examine them for evidence of biological activity or fossil evidence of past life. Although scientific interest in life elsewhere in the solar system reached a low point after the negative results of the Viking landers in 1976, there was a resurgence of interest by the end of the twentieth century. Discovery of river channels on Mars, possible fossil evidence for ancient microorganisms in a meteorite from Mars, hints of water ice on the Moon and Mercury, oceans on Europa and Enceladus, organic materials and an atmosphere on Titan, and cryovolcanism seen on Enceladus and Triton suggest that the solar system might not be as inhospitable to the development of life as was believed immediately following the results of the Viking landers.

George J. Flynn

FURTHER READING

Goldsmith, Donald, and Tobias Owen. *The Search for Life in the Universe*. 3d ed. New York: University Science Books, 2001. Speculates scientifically about the possibility of the existence of intelligent beings beyond Earth.

Greenberg, Richard. *Europa the Ocean Moon: Search for an Alien Biosphere*. Berlin: Springer, 2005. A complete description of current knowledge of Europa through the post-Galileo spacecraft era. Discussion of the astrobiological implications of an ocean underneath Europa's icy crust. Well illustrated and accessible to astronomy enthusiasts and college undergraduates.

Hansson, Anders. *Mars and the Development of Life*. New York: Ellis Horwood, 1991. A comprehensive, well-illustrated discussion of the conditions for the development of life and the search for life on Mars.

Harland, David M. *Cassini at Saturn: Huygens Results*. New York: Springer, 2007. Provides a thorough explanation of the entire Cassini program, including the Huygens landing on Saturn's largest satellite. Essentially a complete collection of NASA releases from the start of Cassini flight operations through the majority of Cassini's seventy orbits during its primary mission. Cassini's primary mission concluded a year after this book was published. Technical writing style, but accessible to a wide audience.

_____. *Mission to Saturn: Cassini and the Huygens Probe*. New York: Springer Praxis, 2002. A technical description of the Cassini program, its science goals, and the instruments used to accomplish those goals. Written before Cassini arrived at Saturn. Provides a historical review of pre-Cassini knowledge of the Saturn system.

Michaud, Michael A. G. *Contact with Alien Civilizations: Our Hopes and Fears About Encountering Extraterrestrials*. New York: Springer, 2006. Explores the possibility of extraterrestrial intelligence, speculates about human-extraterrestrial interactions, and discusses the impact on society that making contact could have.

Orgel, Leslie. "The Origin of Life on Earth." *Scientific American* 271 (October, 1994): 76-83. A comprehensive description of how the emergence of RNA is believed to have been critical to the development of life on Earth. Includes a good account of the Miller-Urey synthesis experiment.

Sagan, Carl. *Contact*. New York: Pocket, 1997. This novel by the Cornell astronomer and science popularizer provides an account of contact with a greatly advanced intelligent species beyond Earth. With all the real science that is included, the novel easily allows the reader to suspend belief in those areas where the science is highly speculative. The book prompted the production of a popular movie by the same name.

_____. "The Search for Extraterrestrial Life." *Scientific American* 271 (October, 1994): 92-99. A clearly written, well-illustrated account focusing on the scientific results of the Viking spacecraft and plans to investigate the atmosphere of Titan using the Cassini-Huygens spacecraft.

Squyres, Steve. *Roving Mars: Spirit, Opportunity, and the Exploration of the Red Planet*. New York: Hyperion, 2006. Written by the principal investigator for the Mars Exploration Rovers Spirit and Opportunity, this fascinating book provides a general audience with a behind-the-scenes look at how robotic missions to the planets are planned, funded, developed, and flown. A personal story of excitement, frustrations, the satisfaction of overcoming difficulties, and the ongoing thrills of discovery.

Ward, Peter. *Life as We Do Not Know It: The NASA Search for (and Synthesis of) Alien Life*. New York: Penguin, 2007. Written by a paleontologist, this book presents reasonable speculations about the nature of primitive life that may someday be found through space exploration or extrapolation from remote-sensing data.

Webb, Stephen. *If the Universe Is Teeming with Aliens . . . Where Is Everybody? Fifty Solutions to Fermi's Paradox and the Problem of Extraterrestrial Life*. New York: Springer, 2002. The great physicist Enrico Fermi declared that if intelligences greater than those of humans existed, then they would already be here visiting on Earth. This book looks at various solutions to surmounting the vast distances between star systems and the nature of extraterrestrial life in the context of answering Fermi's paradox.

See also: Habitable Zones; Life's Origins; Main Sequence Stars; Mars: Possible Life; Search for Extraterrestrial Intelligence.

Gamma-Ray Bursters

Category: The Stellar Context

Gamma-ray bursts for a long time were an unexplained phenomenon in high-energy astrophysics. A variety of spacecraft have detected and studied these random, brief, and intense bursts of gamma rays, which come from all parts of the sky. Most theories associate them with neutron stars in the Milky Way galaxy, but an extragalactic source cannot be excluded.

OVERVIEW

Gamma-ray bursts (GRBs) constitute a unique phenomenon in astronomy. During their brief appearance, they are brighter than all other objects in the sky, including the Sun. About a hundred strong GRBs occur every year. It is not known how many weaker bursts occur. For many years after their discovery, their source represented one of the greatest mysteries in astrophysics. They occur at random times and appear to be randomly distributed over the sky. There is no particular clustering of GRBs in any region, and until the Swift spacecraft was launched in 2004, they had not been associated with any known objects. GRBs are extremely difficult to study, since it is never known when or where a GRB will occur.

It has become recognized that there are distinct classes of GRBs, with different properties, that may be caused by entirely different objects or emission mechanisms. The situation is analogous to the recognition, long after the first telescopes came into use, that not all nonstellar objects should be classified simply as "nebulae," since they include objects as diverse as supernova remnants, galaxies, star clusters, and planetary nebulae. It is believed that, similarly, the phenomena that scientists call "gamma-ray bursts" will be found to be caused by more than one process or object.

Discovery of GRBs in 1972 by the Vela space-craft was a classic case of a serendipitous discovery, a discovery made while looking for something else. The Vela spacecraft were designed and operated to detect nuclear explosions from space. This series of small spacecraft, built by TRW and launched in the mid-1960's, contained a wide variety of sensors that "looked" in all directions. The spacecraft were launched into high, eccentric orbits so that they could even scan the area behind the Moon for clandestine nuclear explosions.

Gamma-ray detectors aboard the Vela space-craft were designed and built at the Los Alamos National Laboratory. They consisted of small scintillation detectors, the output of which was continuously monitored for an increased rate above background. After several years of operation, occasional triggers were detected, but they were dismissed since no other sensors on board the spacecraft recorded the events and because such "glitches" were common to detectors on other spacecraft. It was not until Los Alamos scientists began studying these triggers in greater detail that their nature became known. In many cases, two or more spacecraft would record a trigger at nearly the same time. It was first suspected that a source of gamma rays from Earth, the Sun, or another object or region within the solar system was causing the GRBs. When precise gamma-ray-burst timing analysis was performed, it became evident that triggers were caused by a plane wave of gamma rays striking the array of widely separated spacecraft. This type of wave could be caused only by a powerful point source of gamma rays far beyond the solar system.

Los Alamos scientists announced their discovery at a meeting of the American Astronomical Society in Columbus, Ohio, in 1973 and published their findings in an astrophysical journal. Almost immediately, there was a flurry of activity to try to explain GRBs and to obtain more experimental data. As experimenters began to look through old data and data from still-operating spacecraft, many confirmed GRBs were uncovered in addition to those detected by the Vela spacecraft. Among the earlier spacecraft that confirmed the existence of GRBs were the Orbiting Solar Observatories, the Orbiting Geophysical Observatories, the Small Astronomy Satellites, the Interplanetary Monitoring Platform, Kosmos 461, Apollo 16, and the German spacecraft TD 1. It should be noted that none of these spacecraft had detectors that were specifically designed to detect GRBs. It was only because of the intensity of the bursts and their coincidence with other observations that they were detectable by instruments designed for other purposes.

By the late 1970's, a network of small detectors on interplanetary spacecraft was established in an attempt to locate the source of the GRBs more precisely. Included in this network were instruments aboard the Pioneer Venus orbiter; International Sun-Earth Explorer 3 (ISEE 3); Veneras 11, 12, 13, and 14; Prognoz 7; and Helios 2. For the first time, these spacecraft provided long interplanetary baseline distances required to locate the GRBs within one arc minute. Unfortunately, with one important exception, no unusual objects were detected near the burst sources. The exception was the GRB of March 5, 1979. It occurred in or near a supernova remnant in the Large Magellanic Cloud. This burst was unusual in other respects, however, so it may have been part of a separate class of GRBs.

The Russians, in collaboration with Bulgaria, Denmark, and France, launched the Granat observatory late in 1989. Outfitted with two instruments to investigate high-energy astrophysical objects from X-ray to gamma-ray wavelengths, Granat lasted until 1998. However, five years into its orbital lifetime the observatory ran out of attitude control gas, making directional surveys no longer possible. Among Granat's greatest discoveries were the detection of electron-positron annihilation from a galactic microquasar, nineteen GRBs, and the identification of numerous objects that were candidates for black holes.

The Gamma Ray Observatory (GRO) was deployed into an independent orbit from the space shuttle Atlantis during the STS-37 mission in early April, 1991. The second in the National Aeronautics and Space Administration's (NASA's) Great Observatory series, once in orbit GRO was renamed the Compton Gamma Ray Observatory (CGRO or just Compton Observatory) after the Nobel Prize-winning physicist Arthur Compton, for whom an important effect involving an interaction of matter and electromagnetic radiation is also named. That interaction involves a shift in wavelength when a photon is "scattered" off a free electron. This effect is most pronounced in X-ray and gamma-ray photons.

CGRO was left in a low, nearly circular Earth orbit (at an altitude of 450 kilometers) to keep it out of the Van Allen radiation belts. Ideally it would have been deployed at an altitude well above those radiation zones, but it was far too heavy (17,000 kilograms) for the shuttle to put it up that high; use of an upper stage in concert with the shuttle carrying such a heavy payload was out of the realm of possibility as well.

Compton's most important improvement over previous gamma ray-detecting satellites was that its suite of four experiments covered energies ranging from as low as 20 kilo-electron volts (keV) to 30 giga-electron volts (GeV). No previous orbital detector had spanned six orders of magnitude in wavelength (or energy as a consequence) in the gamma-ray portion of the electromagnetic spectrum.

CGRO's four instruments were the Burst and Transient Source Experiment (BATSE), the Oriented Scintillation Spectrometer Experiment (OSSE), the Imaging Compton Telescope (COMPTEL), and the Energetic Gamma Ray Experiment Telescope (EGRET). Provided by NASA's Marshall Space Flight Center, BATSE was designed to search for short GRBs ranging in energy from 20 to 600 keV and also generate full-sky surveys for long-duration gamma-ray sources. Provided by the Naval Research Laboratory, OSSE was outfitted with four individually pointing detectors capable of picking up radiation from 0.05 to 10 mega-electron volts (MeV). A pair of these detectors would record emission from a source while the other two would record the background near those sources for contrast. COMPTEL was provided by a collaboration of the University of New Hampshire, the Netherlands Institute for

Space Research, the Max Planck Institute, and the European Space Agency's (ESA's) Astrophysics Division. COMPTEL was designed with a wide field of view and sensors capable of identifying sources (in an energy range of 0.75 to 30 MeV) to within one degree. EGRET was provided by a collaboration of NASA's Goddard Space Flight Center, Stanford University, and the Max Planck Institute. This was the highest energy detecting portion of CGRO, capable of recording emissions in the range of 20 MeV to 30 GeV and identifying the location of incoming radiation to within a fraction of a degree; that level of resolution was useful in having other satellites precisely locate sources picked up by EGRET.

Compton was not capable of being repaired in orbit like the Hubble Space Telescope. CGRO's systems began to degrade, especially threatening the loss of attitude control. It was therefore decided to drive the observatory into the atmosphere over a portion of the Pacific Ocean that was sparsely populated at best. Large debris that would survive reentry, such as a major portion of one of Compton's detectors, would drop harmlessly into the ocean. CGRO was deorbited on June 4, 2000, reluctantly ending nine years of unprecedented gamma-ray astrophysics research.

The Swift Gamma Ray Burst spacecraft was launched by a Delta II booster on November 20, 2004, and placed in an orbit 600 kilometers above the Earth's surface. Swift was designed to provide the best all-sky survey of gamma rays yet, to provide alerts to transient astrophysical events such as supernovae and GRBs of both short and long duration, to help identify the location of GRBs, and to assist in determining the distances to gamma-ray bursters at cosmological distances representing a time early in the universe's evolution. Swift was outfitted with just three instruments: the Burst Alert Telescope (BAT), X-ray Telescope (XRT), and Ultra-Violet/Optical Telescope (UVDT). The most important aspect of Swift's capability was the rapidity with which it could respond to a gamma-ray detector and precisely locate its source so that these three instruments and other assets available worldwide to the astronomy community could record and study the af-

terglow of a GRB. Within about fifteen seconds on average, Swift could identify the source of gamma rays to within approximately one arc minute of the sky. The Swift Mission Operation Center, located on the campus of Pennsylvania State University, serves as a clearinghouse alerting other astronomers around the world to GRB events.

NASA's Gamma-Ray Large Area Space Telescope (GLAST) was launched by a Delta II booster in June 11, 2008. GLAST involved an international team consisting of space agencies and research groups from the United States, France, Germany, Italy, Sweden, and Japan. GLAST was designed as a follow-on gamma-ray astronomical observatory to the lost Compton GRO; however, it was not considered a member of NASA's Great Observatory program. It was intended to investigate cosmological questions raised by Compton observations as well as energies far in excess of what can be produced in particle accelerators on Earth. Scientists expected to use GLAST to obtain better understanding of black holes, neutron stars, and high-speed gas and how they produce gamma radiation.

KNOWLEDGE GAINED

Since GRBs have not been identified with known objects, their distance is highly uncertain. This, in turn, makes it difficult to speculate on their origin. Since the distance to the burst sources is not known, the intrinsic luminosity of the source is even more uncertain. Many of the early theories of GRBs posited exotic phenomena or objects to explain them. In later years, most models have associated GRBs with explosive events near, or at the surfaces of, neutron stars within the Milky Way galaxy. These explosions could be caused by thermonuclear reactions resulting from the collision of interstellar material, comets, or asteroids with neutron stars, or from the annihilation of strong magnetic fields near such stars. Another theory attributes GRBs to a sudden shift of the solid crust that is thought to be present in neutron stars. There are also models of GRBs that attribute them to enormous explosions occurring at cosmological distances, or distances near the edge of the observable universe. At these distances, the luminosity of a GRB would be equivalent to that of a supernova, although all of its energy would be emitted at gamma-ray wavelengths and within the duration of a GRB.

The three observable properties of GRBs that are most often studied are their time histories, their energy spectra, and the statistical properties of their intensity and distribution over the sky. Attempts to locate a GRB and identify it with a known object have thus far been unsuccessful. Very sensitive optical, radio, and X-ray searches have been made of precisely located gamma-ray-burst "error boxes" (the region of uncertainty in the position of a celestial source). These searches have been either inconclusive or controversial. A search of old photographic plates from telescopes in the Southern Hemisphere, however, has shown two or three transient, starlike optical images at the locations of GRBs. The authenticity and the significance of these observations are still being debated.

The time history of a GRB refers to the intensity variations of the burst as a function of time. Some GRBs show extremely rapid fluctuations over their entire duration, which may encompass a minute or two. Other bursts last a few seconds, during which time only smooth variations are seen. Still others exhibit a single spike lasting only a fraction of a second. The rapid variations indicate that the source of a GRB is a very small region or compact object, such as a neutron star or a black hole. The GRB of March 5, 1979, was unique in that it had a single, intense spike that was followed by a lower-level emission with an eight-second period that lasted for more than two hundred seconds.

The spectra of GRBs indicate that the sources contain regions of extremely high temperatures—perhaps the highest in the universe. In many cases, the gamma-ray energies extend up to 100 MeV. Extremely rapid varia-

tions are observed in the spectra of most GRBs. In addition, gamma-ray line features are observed that may be explained by the gamma rays coming from regions with extremely high magnetic fields, such as those expected near neutron stars. There is also a class of GRBs that have softer spectra, or lower temperatures, and are observed to be repetitive. Soviet researchers also reported gamma-ray line features near 400 keV, which some interpreted as caused by the annihilation and redshift of electron-positron plasma near the surface of a neutron star.

Granat began working in a survey mode in late 1994. Before the demise of this Russian observatory, data from Granat detailed spectral and temporal variability of potential black holes, discovered the specific radiation emerging from electron-positron annihilation from the X-ray nova Muscae and a galactic microquasar, and improved imaging of the Milky Way's galactic center.

In February, 1997, telescopes sighted the source of a burst, a diffuse, elongated object with a bright core. Astronomers thought the ob-

This artist's rendering of GRB 050709 depicts a gamma-ray burst discovered July 9, 2005, by NASA's High-Energy Transient Explorer. The inset shows the X-ray afterglow visible three days later. (NASA/CXC/Caltech/ D. Fox et al./ Illustration: NASA/D. Berry)

ject might be a very distant galaxy, but they could not be certain without more data.

Compton provided a means whereby GRBs of short duration could be differentiated from longer-duration bursts. Its instruments picked up the first known gamma-ray repeaters. Over its lifetime, BATSE picked up at least one event per day, giving it a total of 2,700 gamma-ray detections. OSSE was used to survey the Milky Way's center and provided evidence suggesting that a cloud of antimatter existed exterior to the central region. COMPTEL observed gamma rays originating from the decay of the radioactive isotope aluminum 26 (^{26}Al) and generated a full-sky survey of that emission. EGRET's full-sky survey of high-energy gamma-ray emissions picked up 271 sources. However, only 100 of them were definitively identified. One surprise was the detection of terrestrial gamma rays originating from thunderstorms. CGRO instruments surveyed both pulsars and supernova remnants.

CGRO provided an enormous amount of data, and its instruments were still working at the time of its demise. Its data helped astrophysicists better characterize the nature of GRBs and provided the insight needed to design even better observations to follow, such as the Swift spacecraft and GLAST telescope.

The Swift observatory represented a major advance in the study of GRBs. Outfitted with detectors capable of observing from the visible through gamma-ray wavelengths, Swift was designed to scan the sky continuously for the signature of a GRBer. It had the capability to slew quickly and locate the source of the burst. At this point, the worldwide astronomical community would be alerted to use space-based and ground-based observatories to quickly measure and record the burster's afterglow. Although there were some anomalies with Swift's XRT instrument, the observatory was commissioned on February 1, 2005. It had detected an initial burster earlier on January 17, 2005, and was quickly able to identify the bright source in its field of view, thereby triggering an alert for further study by other observatories, such as the Chandra X-Ray Observatory. During its first year of operation alone, Swift found 90 GRBs. Swift was the first to detect the location of a

short-duration GRB. GRB 050509b, observed by Swift on May 9, 2005, had a burst duration of merely 50 milliseconds; this demonstrated the rapid response time of Swift.

On September 4, 2005, Swift discovered the most distant known GRB, GRB 050904, located 12.6 billion light-years distant. In 2006, Swift identified the location of GRB 060614 to be 1.6 billion light-years distant. This burster lasted 102 seconds and had a signature indicating it most likely resulted from the formation of a black hole. Another great discovery came on January 9, 2008. While Swift was studying a supernova in NGC 2770, it detected an X-ray burst in the very same galaxy. This triggered coordinated studies with Chandra, the Very Large Array, the Keck I telescope, the 200- and 60-inch telescopes at Palomar, and the Hubble Space Telescope, making this perhaps the most intensely studied supernova at a very early point in its development, a study that was performed with instruments across the electromagnetic spectrum. On March 19, 2008, Swift detected four GRBs, a record for one day's observations. Even more important, the second of the four discovered that day, GRB 080319B, turned out to be the brightest celestial object ever detected. Located 7.5 billion light-years from Earth, this GRB was 2.5 million times brighter than any other supernova.

GLAST only achieved first light by the time this article was composed; thus little in the way of results could be provided. However, among the scientific objectives of GLAST research were the examination of high-energy astrophysical objects at energies greater than could be duplicated on Earth in laboratories; the search for sources of dark matter to illuminate the identity and physics of such exotic matter; investigation of gamma-ray bursters; investigation of the mechanisms whereby black holes produce jets and accelerate the material in them to speeds very close to that of light; and investigation of solar flares, cosmic rays, and pulsars at high energy.

CONTEXT

The field of high-energy astrophysics is a product of the space age. It is necessary to carry instruments and telescopes above Earth's atmosphere in order to observe the universe at X-ray and gamma-ray wavelengths. This relatively new branch of astronomy not only has taught scientists more about objects that they already knew to exist but also has revealed new types of objects and phenomena, including X-ray stars, black holes, and GRBs. These objects are among the most energetic and violent in the universe. Most of them are associated with the final stages in the life cycles of massive stars.

GRBs once represented the greatest unsolved problem in high-energy astrophysics. Their distance and luminosity were unknown for a long time. They are difficult to study because of their random and transient nature. Although the initial discovery and studies of GRBs were made by groups in the United States, the Soviets, often with French collaborators, also obtained many of the gamma-ray-burst data. The establishment of an international gamma-ray-burst observation network, combining data from as many as nine spacecraft, became a model for international collaboration in space exploration. It is expected that the continued study of GRBs not only will help astrophysicists in their understanding of these objects but also will enable scientists to study conditions of extreme temperature, pressure, and density that are unavailable anywhere else.

Gerald J. Fishman

FURTHER READING

Arny, Thomas T. *Explorations: An Introduction to Astronomy.* 3d ed. New York: McGraw-Hill, 2003. A general astronomy text for the nonscientist. Includes an interactive CD-ROM and is accompanied by a Web site. Comets are covered.

Fabian, A. C., K. A. Pounds, and R. D. Blandford. *Frontiers of X-ray Astronomy.* Cambridge, England: Cambridge University Press, 2004. For the most serious astronomy reader or students of astrophysics. Covers contemporary research with space-based X-ray telescopes. Also relevant to the subject of gamma-ray astronomy.

Gregory, Stephen A. *Introductory Astronomy and Astrophysics.* 4th ed. San Francisco: Brooks/Cole, 1997. Suitable as a textbook for introductory college-level course or an advanced high school course in general astronomy. Covers all topics, from solar-system bodies to cosmology. Contains some errors and issues with mathematical presentations.

Hillier, Rodney. *Gamma-Ray Astronomy.* Oxford: Clarendon Press, 1984. This well-illustrated and well-referenced book provides a comprehensive, college-level overview of the objects and methods of gamma-ray astronomy.

Karttunen, H. P., et al., eds. *Fundamental Astronomy.* 5th ed. New York: Springer, 2007. A well-used university textbook in introductory astronomy. Contains some calculus-based treatments for those who find the standard treatise for typical introductory courses aimed at too low a level. Suitable for an audience with varied science and mathematical backgrounds. Covers all topics from solar-system objects to cosmology.

Katz, Johnathan I. *High Energy Astrophysics.* Reading, Mass.: Addison-Wesley, 1987. A well-organized book covering all aspects of high-energy astrophysics but concentrating on emission mechanisms. For advanced readers.

Maccarone, Thomas J. *From X-ray Binaries to Quasars: Black Holes on All Mass Scales.* New York: Kindle, 2006. Provides descriptions of high-energy processes that produce X-ray emissions. Describes the cosmological significance of quasars, black holes, and other high-energy objects.

Verschuur, Gerrit L. *The Invisible Universe: The Story of Radio Astronomy.* New York: Springer Praxis, 2006. Provides a history of developments in radio astronomy and along the way describes the discovery of pulsars, quasars, and radio galaxies. Suitable for a general college science course as well as for astronomy majors as background information.

Weeks, T. C. *Very High Energy Gamma Ray Astronomy.* New York: Taylor & Francis, 2003. Covers gamma-ray astronomy through results from the Compton Gamma Ray Observatory. For students of either theoretical or experimental high-energy astrophysics.

See also: Brown Dwarfs; Main Sequence Stars; Novae, Bursters, and X-ray Sources; Nuclear

Synthesis in Stars; Protostars; Pulsars; Red Dwarf Stars; Red Giant Stars; Stellar Evolution; Supernovae; Thermonuclear Reactions in Stars; White and Black Dwarfs.

Ganymede

Categories: The Jovian System; Natural Planetary Satellites

The Jovian moon Ganymede was the first natural satellite, other than Earth's moon, to be discovered. It is also the largest satellite in the solar system—large enough to generate its own magnetic field, an unusual characteristic for a satellite.

OVERVIEW

Ganymede, the largest satellite of Jupiter, was discovered by Galileo Galilei with a telescope in 1610. He published the information in *Siderius Nuncius* (starry messenger) and thereby initiated a dispute with the Church that would eventually lead him to be placed under house arrest for the remainder of his life; it was heresy to say that anything revolved around something other than the Earth. The satellite was named by Simon Marius for one of the lovers of the Roman god Jupiter.

Ganymede is 5,280 kilometers in diameter, and just over 1 million kilometers from Jupiter; it is the seventh of sixteen satellites. It is common for a satellite to present the same face to its planet at all times—a relationship called synchronous—and Ganymede does this, just as Earth's moon does. The rotation of Ganymede is prograde, that is, in the same direction as that of Jupiter. Ganymede's orbit is almost circular,

meaning that its eccentricity (the measure of how close to a circular orbit the satellite travels) is small. A circular orbit has an eccentricity of zero. Ganymede's angle of inclination is less than a degree, meaning that this moon rotates almost exactly in the plane of Jupiter's equator.

Ganymede's albedo, the amount of sunlight reflected, is large. This reflectivity is caused by ice mixed with carbon-rich soil on the surface of the satellite. When the ice underneath the surface is heated and melts, it erupts to the surface. The soil, which is denser than water, sinks below the water. The water freezes, causing a bright spot on the surface. The water is heated either by radioactive decay or by tidal flexing. Not only does the gravity of Jupiter and Callisto pull on Ganymede; the moon also has Laplace resonance, which occurs because of the forces from the satellites Io and Europa. Every time Ganymede rotates around Jupiter once, Eu-

This image of Jupiter's moon Ganymede shows the seismic activity in the crust and outer coating of water ice that has led to this body's nickname, the Moonquake World. It may be that Ganymede, like Earth, has plate tectonic activity. (NASA, Voyager, © Calvin J. Hamilton)

ropa, the satellite just inside Ganymede, goes around Jupiter twice, and Io, the moon inside Europa, goes around four times. Thus, during every orbit the three satellites are aligned, magnifying the gravitational effect. This increased gravitational pull and then relaxation not only causes the orbits to become elliptical but also causes stresses within the satellites themselves. This tidal flexing generates heat that melts ice and causes the surface of Ganymede to be smoother than expected. Many of Ganymede's impact craters have had their depths reduced by the changing surface of Ganymede.

The surface of Ganymede is a mixture of bright areas and dark regions. The dark regions are heavily cratered; the bright areas are less cratered but are

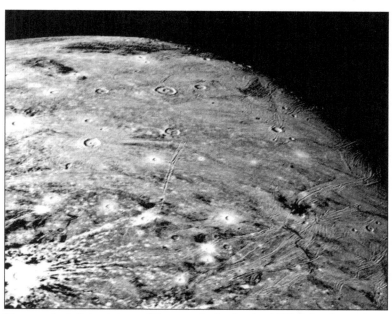

Voyager took this picture of Ganymede during its 1979 flyby, from about 250,000 kilometers away. Impact craters, icy materials radiating out from them, ridges, grooves, and other surface features are visible down to about a 5-kilometer resolution. (NASA/JPL)

often grooved. The bright areas consist of the soil left after the ice has sublimed away. The surface appears much like the surface would if there were lava flow. Water melted by the tidal flexing, then freezing as it reaches the surface, may still flow, much as glaciers flow on Earth. The heavily cratered regions are older and often show fractures and grooves through the craters, where water has broken through the surface at a crater. Ice on Ganymede causes the reflection of radar to be much greater than on most other satellites. Ice also allows radar to penetrate more deeply into Ganymede than if the surface were all silicates. The percentage of ice has been measured at 45-55 percent.

The bulk density of Ganymede is between that of ice and that of carbonaceous silicates, indicating a mixture of the two materials. Ganymede has differentiated; that is, the components have separated, producing a core of dense metals, probably iron or iron with sulfur. A molten iron core is usually the reason for a magnetic field, which Ganymede does have. One model, which agrees with the measured values, postulates a large core of iron with 10 percent sulfur.

The moon has a core with a radius of 695 kilometers, a silicate mantle, and a 900-kilometer-thick ice-water shell. Bombardment by meteors causes a change in the albedo of Ganymede in two ways. First, the meteor causes new, darker material to be thrown up onto the surface (the underlying silicates are darker than the ice or residue left by subliming ice). Second, meteor impacts cause the loss of volatile material, leaving an opaque, dark material.

Ganymede has a thin atmosphere, composed of electrically charged gases. One gas seen in the space around Ganymede is hydrogen. Water sublimed from the surface or escaping from a surface fracture condenses at the poles, producing a whitish polar cap down to latitudes of about 40°. Near-infrared spectra show the expected water and hydrated minerals. Unexpected are the indications of carbon dioxide, carbon bonded to hydrogen, carbon triple-bonded to nitrogen, sulfur bonded to hydrogen, and sulfur dioxide. The carbon dioxide appears to be trapped in the surface, perhaps in small bubbles. Jupiter's magnetic field causes ions to be swept along the orbit of the satellite, generating a current producing an auroral spotlight onto the poles of Jupiter.

Ganymede has an intrinsic magnetic field that is opposite to the field of Jupiter. It also displays an induced magnetic field caused by the strong rotating, angled field of Jupiter. The induced field is an indication of a conducting ocean deep under the icy surface. If the ocean has enough minerals dissolved in it to make it strongly conducting, it could generate the intrinsic magnetic field. Jupiter's strong magnetic field causes Ganymede to be bombarded by charged particles. This bombardment is thought to cause the molecular oxygen, O_2, and ozone, O_3, found in the surface of Ganymede.

Since Ganymede's orbit is in the same plane as Jupiter, it is thought that they were formed by the same process. Ganymede is out away from the very hot, dense region where the planet formed. Ganymede was formed in a cooler region, where water did not boil away but instead froze to form part of the satellite. Although Ganymede is locked into the same face toward Jupiter all the time, there are indications that this may not have always been true. One clue is that the number of meteor craters should be greater on the leading side of Ganymede, as is the case with Callisto, but this is not true of Ganymede. Another fact pointing to a change in the part of the ice shell facing Jupiter is the catenae that are found on the back side of Ganymede. Catenae are caused by a string of fragments from a comet that was broken up by the intense magnetic field of Jupiter but escaped capture to hit one of the satellites. They should occur only on the Jupiter-facing side of Ganymede.

KNOWLEDGE GAINED

The Jovian system, as Jupiter and its moons are called, has been visited by several space missions. Pioneer 10 (1973), Pioneer 11 (1974), Voyager 1 and Voyager 2 (both in 1979), and New Horizons (2007) flew through the system. They all used gravitational assists to gain speed to travel on toward the outer part of the solar system.

The Pioneer spacecraft provided the first visual images of the surface of Ganymede, as well as a much better estimate of the size and mass of Ganymede. The Voyager spacecraft improved these images to a resolution of a kilometer and provided color by means of six filters. Scientists were able to develop ideas of how the satellite formed and its structure. The Voyager data on craters indicated either that Ganymede's surface has changed, and thus erased the early craters from meteor hits, or that the surface was not firm enough to retain the early craters.

The Galileo mission arrived in orbit about Jupiter in 1996. The visual camera increased

The Galileo spacecraft flew by Ganymede more than two decades after Voyager, capturing this image of dark, heavily cratered terrain and scarps down to a resolution of about 20 meters. (NASA/JPL/Brown University)

the resolution of the surface to 20 meters. The model of the moon, showing its core, mantle, and shell of ice, was developed after data from Galileo provided Ganymede's mass, average density, and moment of inertia. The moment of inertia and average density required the data on gravitational fields produced by Galileo, and how the flight path was perturbed as the craft flew by the satellite. Galileo also provided information of the magnetic fields using a magnetometer. The dual purpose of the magnetometer was to determine if Ganymede had a magnetic field of its own and how the satellites interacted with the strong magnetic field of Jupiter. Galileo used its Near Infrared Mapping Spectrometer (NIMS) to make a compositional map of the surface.

While studying Jupiter was not the main focus of New Horizons, astronomers did not let the opportunity escape. Although as of 2008 only 70 percent of the data had been transmitted to Earth, and only part of those data had been analyzed, New Horizons added some interesting new information. The spacecraft's infrared Linear Etalon Imaging Spectral Array (LEISA) and its panchromatic Long-Range Reconnaissance Imager (LORRI) charge-coupled device camera mapped Ganymede's composition. These instruments' resolutions are better than any land-based instrument or Galileo's NIMS. Low-temperature crystalline ice was found as expected, but asymmetric bands of non-ice were found, especially in the darker regions. More ice is found in bright regions and in craters and ejecta from recent meteor hits. This correlates with darker material on the surface, except where meteor strikes have brought ice to the surface. New Horizons could also map parts of Ganymede that Galileo could not see.

Not all information is gathered by spacecraft. The Hubble Space Telescope (HST) has taken pictures of the auroras of Jupiter. Other types of data, such as those gathered by eclipse radiometry, can be used from Hubble or from Earth. Eclipse radiometry is the measurement of thermal radiation just as the satellite is eclipsed by the planet. For Ganymede, these studies suggest that heat is lost rapidly; therefore, the surface material must be porous, due to bombardment from meteors over millions of years.

CONTEXT

The more astronomers learn about large bodies like Ganymede, the more is revealed about how the solar system was formed and about Earth and its Moon. Ganymede may be showing the action of plate tectonics. Learning about the plate tectonics of Ganymede may explain what happened on Earth as the continents tectonically rearranged.

Each space mission has returned valuable information on how to survive in space. Not only are meteorites a danger but gravitational wells, and especially strong magnetic fields, can damage a spacecraft. When humans venture forth, all of those dangers will have to be considered. The number of craters on Ganymede gives scientists some indication of the chance of a meteor hitting the Earth.

C. Alton Hassell

FURTHER READING

Asimov, Isaac, and Richard Hantula. *Jupiter*. Milwaukee, Wis.: Gareth Stevens, 2002. The famous science-fiction author covers the planet and its satellites, the space missions that have studied them, and the comet collisions of 1994. Illustrations, bibliography, index.

Corfield, Richard. *Lives of the Planets*. New York: Basic Books, 2007. The author takes the reader through the different space missions. Divided by planets, the information gathered by each mission is discussed. Index.

Fischer, Daniel. *Mission Jupiter: The Spectacular Journey of the Galileo Spacecraft*. New York: Copernicus Books, 2001. The author takes the reader through each step of the journey of this amazing space probe. Illustrations.

Grundy, W. M., et al. "New Horizons Mapping of Europa and Ganymede." *Science* 318 (2007): 234. This article is one of the first published after the flyby of Jupiter by the Pluto-bound New Horizons spacecraft. Illustrations, bibliography.

Leutwyler, Kristin. *The Moons of Jupiter*. New York: W. W. Norton, 2003. Ganymede is covered in one of the main sections of this book. Illustrations, index.

McFadden, Lucy-Ann Adams, Paul Robest Weissman, and T. V. Johnson, eds. *Encyclopedia of the Solar System*. San Diego: Academic Press, 2007. The editors have collected articles written by many experts in one of the best scholarly surveys of material about the solar system. Illustrations, appendix, index.

Slade, Suzanne. *A Look at Jupiter*. New York: PowerKids Press, 2008. Written for the juvenile audience, this book covers the important information in an easy-to-read style. One big section is devoted to Ganymede and Callisto. Illustrations, bibliography, index.

See also: Callisto; Europa; Io; Jupiter's Magnetic Field and Radiation Belts; Jupiter's Ring System; Jupiter's Satellites; Neptune's Satellites; Planetary Satellites; Saturn's Satellites; Uranus's Satellites.

General Relativity

Category: The Cosmological Context

The general theory of relativity describes the effects of acceleration and gravity on bodies, as well as the structure of space and time. Developed by Albert Einstein in 1915, it supersedes Isaac Newton's laws of mechanics and gravitation. Although Newtonian physics still is sufficiently accurate in many situations in astronomy and physics, general relativity is needed for a deeper understanding of the universe.

OVERVIEW

The general theory of relativity is a model of how gravity works. Gravity is a major force in the universe. It holds stars, the solar system, and galaxies together. It affects the expansion of the universe as a whole. Any model of the universe, then, must have its basis in a theory of gravity. The general theory of relativity has been confirmed in repeated observational and experimental tests to become the basis of modern cosmological theory.

The motivation of Albert Einstein (1879-1955) in working out the principles of relativity was the belief that it is impossible to detect mo-

tion relative to any fixed point in space and that, therefore, there is no absolute motion. The general theory is the second of two stages of a larger picture of relativity theory. The first stage is special relativity, which deals with the laws of physics as seen by observers in uniform, or unaccelerated, motion. The general theory goes beyond the special theory to deal with accelerated motion and gravity.

General relativity describes the universe in terms of four dimensions: three dimensions of space, with time as a fourth dimension. Events occur at specific locations in space and at specific moments of time, thus requiring four coordinates (three of space and one of time) to identify each event uniquely. In general relativity, the four-dimensional fabric of the universe is referred to as space-time.

A fundamental concept in the general theory of relativity is the principle of equivalence, which asserts that the effects of acceleration and gravity are indistinguishable. This principle is often demonstrated with the use of a thought (or, from the German, "gedanken") experiment. Imagine that a person is in a chamber that has no windows. Now imagine that the chamber is in deep space far from any other mass, so there is no gravitational effect. As long as the chamber is motionless or moving in a straight line at constant speed, the person floats around freely inside, experiencing weightlessness. If a constant force is applied to the chamber, then it accelerates at a uniform rate and the contents of the chamber (the person) are pushed against the side of the chamber opposite to the direction it is accelerating. This accelerating force can be adjusted to match exactly the downward pull that the person would experience if the chamber were on the surface of the Earth. Without a window to determine the motion relative to the outside world, it would be impossible to tell if the force felt was attributable to gravity or to the acceleration of the chamber. From this type of thought experiment, Einstein concluded that gravity and acceleration are equivalent.

Following from the principle of equivalence, Einstein proposed that the Newtonian concept of gravity as a force operating at a distance between individual masses is entirely unnecessary. Instead, he described an alternative way of looking at gravity by asserting that the presence of mass changes the geometry of space-time by curving or warping it. Gravity then is the effect of curved space-time on mass. In the absence of matter, the shape or curvature of space-time is flat. Near massive objects, however, space-time is strongly warped. The larger the amount of matter at any location, the greater is the curvature of space-time around that location. The curvature is greatest near the massive object, and it becomes progressively less with increasing distance.

The curvature of space-time determines the path along which bodies move. In other words, the curvature of space-time causes moving objects to follow curved paths. The Moon, therefore, orbits the Earth not because of the gravitational force between the Earth and the Moon (as explained in classical, Newtonian physics) but because the mass of the Earth curves the space-time around it.

Another thought experiment is useful to picture the curvature of space. Imagine a pool table whose surface is not rigid but is made of a thin rubber sheet. When a large weight is placed on such a pool table, the normally flat sheet stretches, curving around the weight. The heavier the weight is, the more curved is the surface of the pool table. Attempting to play pool with this weight distorting the surface of the table, one finds that balls passing near the weight are deflected from their straight paths. This is a two-dimensional analogy demonstrating how moving objects are deflected by the curvature of space-time around massive objects. The size, shape, structure, and dynamics of the universe as a whole are determined by the net effect of space-time curvature produced by every mass it contains. Since matter and energy are equivalent (as summarized by Einstein's formula $E = mc^2$), energy produces and is affected by space-time curvature the same as mass. The interplay between matter-energy and space-time can be summarized by these two statements:

(1) Matter and energy tell space-time how to curve.

(2) The curvature of space-time tells matter and energy how to move.

While the basic principles of general relativity are straightforward, some of the implications of the theory defy common sense and ordinary experience. Einstein found that classical Newtonian physics, which applies in most circumstances, does not hold when very strong gravitational fields are involved or when velocities approach the speed of light. According to special relativity, for example, uniform motion at speeds approaching that of light affects measurements of length and time. In an accelerated frame of reference or a gravitational field, similar effects take place.

In 1917, two years after Einstein introduced his general theory of relativity, the German physicist Karl Schwarzschild used it to develop his metric equation describing space-time around a nonrotating spherical mass. One consequence of the Schwarzschild metric is that time slows as a result of the curvature of space-time near masses, and experiments have repeatedly shown this effect actually occurs. Any type of clock positioned near a massive object will run slower than an identical clock farther away. Researchers have found, for example, that clocks in Boulder, Colorado, a mile above sea level, gained about fifteen-billionths of a second per day as compared with clocks near sea level. The difference is attributed to Boulder's slightly greater distance from the Earth's center.

Another consequence that has been confirmed by observation and experiment is that the wavelength, frequency, and photon energy of electromagnetic radiation (light, radio waves, gamma rays, and so on) are affected by the space-time curvature around masses. Electromagnetic radiation has both wave-like and particle-like properties. As a wave, it can be characterized by wavelength and frequency. As particles, called photons, it can be characterized by the energy of the photons. Wavelength and frequency are inversely related, and photon energy is proportional to frequency and inversely proportional to wavelength. For example, radio waves have the longest wavelengths, the lowest frequencies, and the lowest photon energies. Blue light has a shorter wavelength than red light, and thus a higher frequency and higher photon energy. Photons falling or climbing through a gravitational field gain or lose energy, and thus are shifted to shorter or longer wavelengths. For example, if visible light is emitted from the surface of a very massive object, then its wavelength will be lengthened and the color of the light will move toward the red (longer wavelength) end of the visible spectrum as seen by a distant observer. Careful measurements of the wavelength or photon energy of light emitted by and climbing upward through the gravitational field of white dwarf stars and of gamma rays falling through the Earth's gravitational field have confirmed that this shift actually does take place.

APPLICATIONS

General relativity predicts that photons follow curved paths through curved space-time near masses. This was confirmed during a total solar eclipse in 1919. Stars were photographed near the edge of the darkened eclipsed Sun, and the same star field was photographed six months later in the night sky. The positions of the stars in the two photographs were compared, and it was found that their apparent positions near the eclipsed Sun were shifted by an amount consistent with the predictions of general relativity. When seen near the Sun during an eclipse, the positions of stars appear to shift because the path of their light is bent as it passes near the mass of the Sun. This bending of light in a gravitational field is also predicted classically, but the classical effect is only half as large as the relativistic effect. Repetitions of this test during many later solar eclipses all measured deflections that are consistent with the relativistic value, not the classical one.

The curvature of light in gravitational fields is also responsible for the phenomenon of gravitational lensing. If a relatively nearby galaxy lies almost on a line between Earth and a more distant object, light from the more distant object is deflected as it passes the nearer galaxy, and people on Earth will see one or more displaced images of the distant object. The first such gravitationally lensed image was discovered in 1979, and many more have been found since then.

Another successful prediction of general relativity is related to the orbit of the planet Mercury around the Sun. Mercury's orbit is slightly elliptical, and the whole elliptical orbit slowly precesses (or revolves) around the Sun, so that perihelion (the closest approach to the Sun) occurs slightly later on each orbit. Most of the measured advance of Mercury's perihelion is accounted for by the classical Newtonian gravitational influences of the other planets on Mercury, but a small residual is left over. This was the reason for the prediction that there was an unknown planet, Vulcan, inside the orbit of Mercury. According to general relativity, however, the warpage of space-time near the Sun causes Mercury's orbit to precess. The predicted amount of this relativistic perihelion advance for Mercury agrees within 1 percent with measurements of the actual residual perihelion advance.

A prediction arising from the extension of the wavelength-frequency-energy shift of electromagnetic radiation in gravitational fields surprised and disturbed even Einstein. If a mass were sufficiently compressed, then its gravitational field (in classical terms) or its space-time curvature (in general relativistic terms) would be so strong that photons trying to leave it would lose all their energy, implying that not even light could escape it. At first regarded as a mathematical curiosity of general relativity with no "real-world" application, almost fifty years later this prediction became the basis for describing the nature and properties of black holes. The name "black hole" was coined by the American physicist John Wheeler in 1969 for objects from which electromagnetic radiation cannot escape. Although black holes cannot be observed directly (because nothing, not even light, can escape), they can be detected by their effects on nearby matter.

Black holes usually are categorized according to their mass. Stellar-mass black holes have masses several times to several tens of times the mass of the Sun. This was the first type of black hole to be detected. They presumably form as an end state in the "lives" of massive stars. When massive stars (more than about eight times the Sun's mass) have exhausted all their available ways of generating energy by nuclear reactions, they collapse, rebound, and explode as supernovae. Much of their mass is expelled in the explosion. If what remains is more than about two or three times the Sun's mass, no known force can stop its final collapse becoming a stellar-mass black hole.

Super-massive black holes have masses hundreds of thousands to billions of times the mass of the Sun. Evidence for their existence has been found at the centers of many galaxies, including our own Milky Way galaxy. They probably formed in the early stages of galaxy formation, maybe by collapse of the central part of a protogalaxy, by the collision of lots of stars in the dense center of a galaxy, or by the merger of smaller galaxies.

"Mini" black holes have masses less than that of a mountain or small asteroid. They are still just hypothetical and have not yet been detected. However, they have been suggested as possible solutions to many problems in astronomy. If they exist, they may have been formed by some esoteric process in the aftermath of the big bang.

General relativity has provided the framework for modern cosmological models of the structure and evolution of the universe. It predicted the expansion of the universe before it was actually discovered observationally. Most modern cosmological models start with the big bang, which created space and time, matter and energy. As the universe expanded, galaxies, stars, and planets formed in it. According to general relativity, the average density of matter and energy determine the geometry and future of the universe. The universe appears to have flat geometry, but only a small fraction of the mass needed for this is observed. Most of the mass in the universe seems to be some exotic form of dark matter that has not yet been directly observed but that is beginning to be detected by its gravitational effects. Gravity alone predicts that the expansion of the universe should slow down, yet the expansion appears to be accelerating. The unknown cause of the acceleration has been dubbed dark energy, and there seems to be even more dark energy than dark mass.

CONTEXT

Relativity theory emerged as a logical step forward in the understanding of space and time. It arose out of the inability of Newton's laws of gravity and motion to describe the observed universe in certain circumstances, such as at very high velocities or in the presence of very strong gravitational fields. The general theory of relativity arose out of Einstein's earlier special theory of relativity. The concept of relativity is extremely important for the development of theories about the universe because it is based on two fundamental postulates: All motion is relative, and there is no preferred frame of reference in which space and time are defined absolutely; it is fundamentally impossible to detect motion relative to absolute space, and there is no basic universal time.

Einstein's goal was to develop a unified field theory. He spent the last years of his life searching for the link between gravitation, electromagnetism, and nuclear forces. This remains a major goal of physics, but it has been difficult to find a way to merge the general theory of relativity with quantum mechanics, the theory that describes how subatomic particles interact.

While it is impossible to prove that the general theory of relativity is correct, the observations and experiments that have been used to test it so far have failed to disprove it. It remains the most accurate theory by which to measure and predict the effects of gravity and has become the foundation of modern cosmology and physics. General relativity has altered the way we view space and time. As such it is an important tool in our continuing efforts to describe and predict events in the world and to understand more about the nature of the universe.

Divonna Ogier

FURTHER READING

Barbour, Julian B., and Herbert Pfister, eds. *Mach's Principle: From Newton's Bucket to Quantum Gravity*. Einstein Studies 6. Boston: Birkhäuser, 1995. Covers all aspects of relativity. Looks at historical, philosophical, and theoretical components in addition to experimental approaches. Includes a discussion of Mach's principle at a conference in Germany.

Calder, Nigel. *Einstein's Universe*. New York: Viking Press, 1979. A well-written book directed toward a general audience. Details the principles of special and general relativity in reference to Einstein's life and a general philosophy of the universe. Useful analogies and popular language make this difficult subject more easily understandable.

Chaisson, Eric. *Relatively Speaking*. New York: W. W. Norton, 1988. A highly readable book giving an overview of Einstein's work on special and general relativity and how it relates to modern astronomical and cosmological questions. Gives a nonmathematical analysis of cosmology and singularity theory.

Foster, J., and J. D. Nightingale. *A Short Course in General Relativity*. 2d ed. New York: Springer, 1995. A detailed text geared toward advanced undergraduates and first-year graduate students. Early chapters cover the needed math skills to understand the subject fully, including tensor calculus and differential geometry. Gives solutions to problems and exercises.

Glendenning, Norman K. *Compact Stars: Nuclear Physics, Particle Physics, and General Relativity*. New York: Springer, 1997. Discusses general relativity, nuclear and particle physics, stellar evolution, black holes, neutron stars, and related topics. Also includes several problem sets. For undergraduates and graduate students.

Hawking, Stephen. *A Brief History of Time: From the Big Bang to Black Holes*. New York: Bantam, 1988. The most celebrated physicist of his generation addresses many of the most important concepts in theoretical physics for a general audience. Discusses deep issues at the frontier of physics such as the nature of gravity, the nature of time, and the search for grand unification.

_____. *An Even Briefer History of Time*. New York: Bantam, 2008. An update of Hawking's celebrated first book, *A Brief History of Time*. Discusses string theory, dark matter, dark energy, and the ultimate fate of the universe.

Parker, Barry. *Einstein's Dream*. New York: Plenum Press, 1986. A perspective on relativity theory related to Einstein's ultimate goal of a unified theory of the universe. Gives a good historical analysis of relativity and unification theory. Devotes chapters to the origin of the universe, the ultimate fate of the universe, black holes, and quantum theory. Nonmathematical, but recommended for the reader with a background in physics as well as nonspecialists.

Sartori, Leo. *Understanding Relativity: A Simplified Approach to Einstein's Theories*. Berkeley: University of California Press, 1996. Addresses general relativity and cosmology with an emphasis on Einstein's theory of relativity. Designed for the general reader with basic algebra knowledge, this book provides historical and philosophical background for modern physics courses.

Schwartz, Joseph, and Michael McGuinness. *Einstein for Beginners*. New York: Pantheon Books, 1979. A delightful introductory book for those intimidated by the usual presentations of scientific theory. Written in simple language with comic-book-style drawings, it gives a basic overview of Einstein's life, the political and social environment from which he emerged, and the basic principles of his work.

See also: Big Bang; Cosmic Rays; Cosmology; Eclipses; Electromagnetic Radiation: Nonthermal Emissions; Electromagnetic Radiation: Thermal Emissions; Gravity Measurement; Interstellar Clouds and the Interstellar Medium; Milky Way; Optical Astronomy; Space-Time: Distortion by Gravity; Space-Time: Mathematical Models; Universe: Evolution; Universe: Expansion; Universe: Structure; White and Black Dwarfs.

Gravity Measurement

Category: Scientific Methods

Gravity, the most dominant universal force, attractive in nature, affects all forms of matter and even energy in spite of its extreme weakness. Traditional Newtonian gravitational theory, adequate for navigation and general astronomical purposes, requires modification when great precision in measurement is necessary.

OVERVIEW

Gravity measurement involves three separate aspects, some theoretical and others experimental or practical. First there is the basic investigation of the nature of gravity as either a force or a consequence of the curvature of space in extreme circumstances, specifically those requiring general relativity rather than classical physics. Second there is the determination of the gravitation field around planetary bodies in order to control satellites in orbit; this is the field of geodesy. Third there is the effort to detect gravitational radiation, with its implications for the relatively new field of gravitational astronomy and also high-energy astrophysics in general.

To the average person, gravity is the natural tendency of objects to move downward toward the Earth. Gravity has been traditionally described as a field with every particle of matter being a source of a gravitational field. The intensity of this field is affected by the distance from and position on the Earth's surface and the local mass distribution in relation to the total mass of the Earth.

The gravitational field produces an attractive force between bodies, which is directly proportional to the product of their masses and inversely proportional to the square of the distance between their centers. This statement summarizes the law of universal gravitation, as formulated by Sir Isaac Newton in 1687. This law assumes that the masses are distributed symmetrically about a sphere of constant radius and uniform density. Actual gravity surveys, however, demonstrate that no mathematical formula has been found that describes exactly the gravitational field of the Earth. The Earth's field is complicated by irregularities in the topography and mass distribution, combined with a pronounced flattening of the Earth at its poles, caused by rotation.

The force of gravity varies with position on the Earth's surface. The acceleration of free-falling bodies caused by the force of gravity is determined experimentally as greatest at the poles and smallest at the equator. The value for the acceleration of gravity, g, for example, is only 9.782 meters per second per second (m/s^2) in the Canal Zone of Panama but is 9.825 m/s^2 in Greenland, which is closer to the North Pole. This value of g near the equator is lessened by a centripetal factor, which is the square of the ve-

locity of a point on the Earth's surface divided by the radius of the Earth (V^2/R). Since points closer to the equator move with a greater velocity, the value of g will be smaller at the equator.

Gravitational acceleration diminishes with altitude; an object at the Earth's surface and near the equator that would have a value of g equal to 9.83 m/s^2 would have that value drop to 8.70 m/s^2 at an altitude of 400 kilometers. This decrease is observed because the effect of the gravitational force on acceleration follows an inverse square law; that is, the farther the object is from the center of Earth, the less the acceleration. At twice the distance, for example, the force and resulting acceleration would be only one-fourth of the original amount.

The equation for the law of universal gravitation has a constant G, the gravitational constant that was not known at the time Newton formulated the law. The constant G is assumed equal for all conditions and locations on the Earth and in the universe. The weight of an object W is equivalent to its mass m multiplied by the acceleration g ($W = mg$). If the weight is equated to the pull of gravity from the law of universal gravitation, then the gravitational constant G may be calculated directly. The value of G is found by squaring the radius of the Earth, R, then multiplying the result by g and dividing by the mass of the Earth, M ($G = R^2 g/M$). This result is a constant for a given location, which implies that g should be the same also for any location. One sees that the surface value of g depends on the mass of a planet or star or other body as well as how the object's mass is distributed within the object. For example, the surface value of g on the Moon is one-sixth that of Earth, and the surface value of g on Mars is 38 percent that of Earth.

The technical problems of the measurement of G were solved by Henry Cavendish in 1798. Cavendish devised a sensitive torsion balance composed of a light rod supported at its center by a thin wire approximately 1 meter long, with lead balls about 5 centimeters in diameter placed at the ends of the rod. If a force is applied to each lead ball in opposing directions and at right angles to both the wire and rod, the wire is subjected to a rotation that may be measured as an angular displacement. Cavendish initially applied small forces, measuring the amount of twisting that resulted. Carefully shielding the experimental equipment from air currents, Cavendish placed two large lead balls about 20 centimeters in diameter nearly in contact with the small lead balls but on opposing sides. Gravitational force between both sets of balls caused a twist in the wire, and from the angle displaced by the wire, Cavendish was able to measure the forces between the large and small balls. The force turned out, as expected, to be very small, only one two-millionth of a newton. The value of G could then be calculated directly, since Cavendish now knew the force involved as well as the masses of the lead balls and their distance of separation. The results of this experiment, as well as later determinations, have established that G has the same value whatever the composition of the masses or the location; the constant is truly universal in nature.

The gravity pendulum has been used to measure the differences in gravitational force on Earth. Modern gravity pendulums are governed by the principle relating to the period of oscillation discovered by the Dutch scientist Christiaan Huygens. The period of the pendulum, as he noted, varies directly with the square root of the length and inversely with the square root of the local value for the acceleration of gravity g. Gravity pendulums are built nearly friction-free, supported on knife-edge jewel bearings, and swing in chambers from which air has been evacuated.

Christiaan Huygens: Improving on Galileo

Christiaan Huygens's first plunge into scientific research took place in 1655, when he and his brother began to build improved telescopes, grinding their own lenses. With these instruments, Huygens found Saturn's moon Titan and discovered that Mars has a varied surface. He gradually discerned a ring around Saturn that nowhere touched the planet, thus improving on Galileo's more primitive observation. In order to protect the priority of this discovery while continuing his viewing, he announced by the publication of a coded message that he had found the ring. At about this time Huygens published his work on hyperbolas, ellipses, and circles, and in 1657 he published the world's first formal treatise on probability.

In addition to building the pendulum clock—his greatest original invention—Huygens enjoyed membership in France's Royal Academy of Sciences, residing in Paris from 1666 until 1681. Still, when Huygens left Paris in 1681 for a third trip to the Netherlands, he never returned. His patron, Louis XIV's chief minister Jean-Baptiste Colbert, died in 1683, and anti-Protestant sentiment was growing in France, making Huygens's position difficult, as he was nominally a Calvinist.

Huygens's philosophy of science was intermediate between those of the two giants of his day: René Descartes in France and Sir Isaac Newton in England. Descartes attempted to explain all

(Library of Congress)

phenomena by use of deductive logic alone. Newton, on the contrary, relied on observations and experiments as the bases for his laws. Huygens grew up a Cartesian but broke with his mentor over the latter's extreme devotion to the mathematical, or deductive, approach to science; his basic approach to the universe was mechanistic. He did prefer, however, Descartes's supposedly more tangible "vortices" of "subtle matter" to Newton's "gravity" in explaining the movements of heavenly bodies. In the matter of relativity, however, Huygens was in advance of Newton and anticipated Einstein. For Huygens, all motion in the universe was relative. Huygens also bested Newton in his understanding of light: Newton held to the corpuscular (particle) theory of light, whereas Huygens propounded a wave theory of light; modern quantum theory combines the two, but in Huygens's day the corpuscular theory was dominant.

Huygens remained in communication with Newton, although his relations with London's Royal Society dwindled after 1678. He visited England again in 1689, conversing with Newton and addressing the Royal Society on his non-Newtonian theory of gravity. Huygens's last years were spent in The Hague, where he died in 1695.

The period of oscillation is timed with precise chronometers enabling determinations to within a few parts per million. Unfortunately, gravity pendulums are unwieldy and difficult to transport; consequently, now most gravity measurements are made with portable instruments called gravimeters.

Gravimeters make use of the principle of a spring balance—that the distortion or strain is directly proportional to the applied stress or force, provided that the measurements are made within the elastic limits of the material. A small quartz fiber is distorted in the local gravitational field at an observed station with results compared to a measured pendulum station. The readings are generally so precise that distortions to one part in 10 million can be recorded. Through development of such instruments, variations in gravity over large areas can now be measured. Some instruments have been adapted to operate from aircraft in flight for aerial surveys of the Earth and for use on surface vessels at sea in regions not suitable for gravity pendulums because of wave disturbances and motions.

A torsion balance employs an arrangement similar to a Cavendish balance, but as opposed to measuring the deflection produced by large masses, the period of oscillation is measured when the large masses are placed perpendicular to the equilibrium positions of the small masses and next when the large masses are rotated another 90°. When the large masses are located at the first position, the period is less as a result of the additional restoring force. In the second position, the period is increased by the large masses, pulling the small ones away from the equilibrium position. The difference in the periods from both of these measurements gives the value of G. The advantage of this method is that the period may be measured more precisely than a corresponding deflection; the precision in error is estimated as 0.5 percent with this technique.

Although measurement of the gravitational constant G has improved, the value is not known with nearly the precision of other physical constants. Laboratory measurement of G is difficult because of the extremely small forces between the masses. Planet-sized objects are much larger, but the problem is not resolved, because the product of G and the mass of the attracting planet both appear in the equation. Planetary observations alone cannot determine the individual values of G or mass.

A torsion balance may also be used for ascertaining the equivalence between inertial mass and gravitational mass, known also as the principle of equivalence. For these experiments, one body, called the inertial mass, is defined with respect to a standard mass of 1 kilogram. The gravitational mass of the body is defined in terms of this acceleration and the distance between the objects. Experiments have tested a variety of materials, obtaining a ratio between the masses. Results indicate that various types of energy contribute to the inertial mass of a system to the same degree that they would contribute to the gravitational mass.

A gravity gradiometer is an instrument designed to measure local tidal fields. The instrument is portable and designed for use on an airplane or satellite and permits precise mapping of anomalies in the Earth's gravitational field. The instrument, in the shape of a Greek cross, has four masses at the ends of each arm, which are held together at the center by a torsional spring. When pressed together, the arms oscillate with a frequency of 32 hertz (Hz, or oscillations per second). If placed in a tidal field at right angles, the cross will be deformed. Rotated with an angular frequency of 16 Hz in the reference frame of the cross, a tidal driving force would appear at 32 Hz. Since the oscillation frequency matches the natural vibration frequency of the arms, a resonance condition is established, producing large amplitudes. Very small tidal fields have been detected with this instrument.

Detectors of gravitational radiation, or waves, were first built in 1966 at the University of Maryland by Joseph Weber. This type of detector consists of a large aluminum cylinder placed inside a vacuum tank and suspended on a wire. The cylinder is supported on rubber blocks as an insulation to external mechanical vibrations. Any oscillations of the cylinder in the fundamental or longitudinal mode are detected by piezoelectric strain transducers bonded to the outside middle section. The ability of the device to detect radiation at resonance is termed its cross section, which turned out to be quite small. Limiting the ability to detect this type of radiation is the thermal motion of the individual molecules and oscillations of the cylinder that interfere with observations. Experimental results obtained by Weber and his group have not been duplicated elsewhere, casting doubt on the technique's reliability.

More sensitive detectors were constructed and involved strains in solid bodies employing resonance; they were more sensitive to radiation of a given frequency and rejected all other frequencies. The biggest advance in gravitational wave detectors came with approval of large-scale facilities for detection, one on Earth (called LIGO) and one based in space (called LISA).

The Laser Interferometer Gravitational-Wave Observatory (LIGO) consists of two facilities working in coordination, one at Livingston, Louisiana, and another on the Hanford Nuclear Reservation in Washington State. The two sites are separated spatially by 3,002 kilometers and temporally by approximately 10 milliseconds, since gravitational waves would travel at the speed of light. Separating the facilities permits an interferometry approach for determination of a source of gravitational radiation. Gravity wave detection with LIGO involves looking for oscillations in lasers traveling down vacuum lines to mirrors. Essentially a higher-technology version of the classic Michelson interferometer, LIGO was designed to be able to detect a spatial oscillation of as little as 10^{-18} meter when a gravitation wave forces an oscillation as it passes through Earth. Note that this tiny spatial extent is approximately one-thousandth the size of a proton or neutron. If detected, a gravitational wave would produce a signal in LIGO that rises rapidly in oscillation amplitude, only to decay exponentially in time. An early test of LIGO was an observation of the gamma-ray burst GRB070201 in 2007. Although near the direction toward the neighboring Andromeda galaxy, the burst was determined not to have come from within that galaxy. Its closeness is such that LIGO should have picked up gravitational radiation if the object emitting the intense gamma-ray burst had actually been the

Andromeda galaxy, but it failed to pick up any signal.

Another gravitational radiation detection project under development is the Laser Interferometer Space Antenna (LISA). LISA will consist of three separate Michelson interferometers, each with two long arms arranged in the shape of the letter L positioned precisely in space about 50 million kilometers from Earth. The system, to be built by the National Aeronautics and Space Administration (NASA) and the European Space Agency (ESA), is planned for launch no earlier than 2018 for a minimum two-year mission. LISA's design includes a capability for determining spatial oscillations as small as 1 part in 10^{20}.

APPLICATIONS

Measurement of the gravitational field over Earth demonstrates that the field is not uniform. The Earth departs from being a perfect spheroid because of rotational effects and topographic variations. Regions that are topographically higher than a datum surface are located farther from the Earth's center and experience a smaller gravitational force. Other regions located below the surface, although closer to the Earth's center, may experience compensating effects from mass concentrations, thereby increasing the strength of the field over its expected value.

Gravity data indicate that the field is increased near mountain ranges because of the greater concentration of rock. Closer observations show that mountain masses do not deflect the field as much as expected if the mountain were a load resting on top of a uniform crust. If the mountain were merely a load on the rigid crust, the force of gravity (corrected for the effect of additional altitude) should be larger on top of the mountain than on the surrounding plains as a result of the increased gravitational pull of the mountain mass beneath the crust. Such observations led geophysicists to conclude that a rigid crust is not responsible for supporting the load of the mountains but is instead buoyed up by floating on a denser, deformable interior. The interior of the Earth must yield and be subject to lateral flow to compensate for loads on equal-size regions. Areas of depression in the crust, such as oceanic trenches, show lower values of the gravitational field, as there is less mass near the surface.

Newtonian mechanics, traditionally used to describe the behavior of bodies at the surface of the Earth, tend to break down outside the range of normal observable motion. The theory fails when gravitational fields become very intense near collapsed objects such as neutron stars or black holes. In 1915, Albert Einstein completely changed the understanding of gravitation with his general theory of relativity. According to theory, gravity is not a force in the usual sense but is the result of the curvature of space-time. Bodies then follow the easiest course through space-time, which is manifested in the shape of their orbits. Einstein theorized that gravity may be explained by geometry. Mercury's orbit could not be explained adequately by ordinary mechanics but only by the warping of space near the Sun. Time warps at the Earth's surface may be detected by using very precise clocks.

Gravitational time dilation—the slowing down of clocks in a gravitational field—may be used as a direct test of the curvature of space-time. For these experiments, cesium-beam atomic clocks are used. Small frequency shifts are measured in the clocks placed in a gravitational potential and are calibrated against clocks at rest in a stationary gravitational field. The pulses of the clocks are monitored to rule out the possibility of a frequency loss during the light-beam propagation. The clock tick rate is found to depend upon the strength of the gravitational field; therefore, space-time geometry is dependent upon the gravitational field. The possibilities that strong tidal forces would have an effect on the clock may be ruled out, because it is known that the atomic forces are stronger and resist tidal distortions.

Gravitational redshifts have been measured on light emitted from the atoms on stellar surfaces. It has been difficult to obtain reliable results for these measurements because of strong convection currents in the stellar photosphere masking the spectral lines, which are Doppler-shifted by gaseous motion. Measurements made above the photosphere of the Sun have given more definitive results. Gravitational redshifts are very prominent in light emitted by white dwarf stars because these stars have about 1 solar mass contained in a much smaller radius than the Sun and consequently have very intense gravitational fields.

LIGO was used to study the pulsar in the center of the Crab nebula. That pulsar's spin rate (30 revolutions per second) is slowing down considerably; indeed it is one of the most quickly slowing pulsars known. Pulsars slow down in their periodic spinning motion due to one of three mechanisms: (1) magnetic dipole radiation emission, (2) asymmetric emission of energetic particles, and (3) possible emission of gravitational radiation. General relativity indicates that if a rapidly rotating neutron star, a pulsar, were not perfectly smooth but distorted from a spherical shape, it should create gravitational waves. Since LIGO picked up no signals within detection capabilities during Crab pulsar observations, the conclusion was that to within well under a meter, the pulsar must be nearly perfectly spherical.

CONTEXT

Electromagnetic waves were a prediction of James Clerk Maxwell's field equations for electric and magnetic interactions. Heinrich Hertz verified that prediction's correctness with the first detection of radio waves; prior to that point, the only portions of the electromagnetic spectrum known, aside from visible light, were parts of the ultraviolet and infrared. In due course, astrophysics would be able to make use of observations across the electromagnetic spectrum, from high-energy gamma rays to the lowest-energy radio waves.

Gravitational waves are predicted by Einstein's field equations of general relativity. However, whereas production and detection of electromagnetic radiation came in rapid succession and opened up a vast expansion of astronomical investigations as a by-product, prediction and initial direct detection of gravitational radiation remain separated greatly in time. Despite null results, gravitation physicists remain steadfast in their belief that gravitational radiation should exist and that, in time, new technology such as LIGO and LISA will make that first definitive direct detection of such radiation. Thus far, the existence of gravity waves has been only indirectly inferred by examining the binary pulsar system PSR 1913+16, for example, and other results have been used to infer aspects of other high-energy exotic objects in the universe. When and if the technology exists to observe and examine gravitational radiation routinely, a brand-new branch of astronomy will provide a host of new insights into the nature of the universe and one of its most pervasive forces, gravity, which binds it together into complex structures and in part determines its evolution.

All of nature's events and activities can be explained in terms of four fundamental forces. In the historical context, gravity was the first of these four forces to be investigated scientifically. Although scientists have had an awareness of gravity and the direction in which it acts, the role of gravity as a force was not appreciated fully until Newton's law of universal gravitation was published. The importance of gravity is its universal nature. Everything in the cosmos is affected by it and every particle of matter is a source of gravity. The force of gravity as observed is always attractive, tending to pull matter together.

One of the surprising facts concerning gravity as the dominant universal force is its extreme weakness. Gravity is so weak that physicists generally ignore its effects completely when dealing with masses on the level of the subatomic particle. Gravity's strength on the atomic scale is vastly overwhelmed by the electromagnetic and nuclear forces at that distance.

The law of universal gravitation, which was adequate for more than two hundred years, was not effective in the twentieth century in explaining discrepancies in observations near very massive objects. In this respect, the law of universal gravitation conflicted with the relativity theory. In Newton's theory, gravitational force between two bodies should be transmitted instantaneously across space, but Einstein's theory rejects physical effects that travel faster than the speed of light. Gravitational fields around objects as massive as the Sun appear to distort space and time to a degree that is detectable. Observing stars near the Sun during solar eclipses indicates that they are not observed in their true positions; that is, the light from these

stars has been bent or deflected noticeably toward the Sun by its gravitational field. Black holes appear to distort severely the surrounding space and time. They are the final state of very massive stars that have collapsed into a singularity with a gravity so overpowering that not even light can escape.

Some physicists have postulated the existence of a fifth force in nature that may diminish the effectiveness of the gravitational force out to a limited range. Experiments performed in mines, for example, seem to show that the measured gravitational force does not agree with predicted values from theory. These observations as well as others, however, have not established conclusively the existence of a previously unknown repulsive force.

Despite all of its success, general relativity theory remains at odds with quantum theory (which describes subatomic particles on a statistical basis) and is at odds with superstring theory (which treats subatomic particles as very tiny vibrating loops). Physicists continue to search for a more comprehensive theory of quantum gravity that perhaps will be more useful in mapping out the very early history of the universe very near the moment of creation. The nature of just how gravitation is transmitted at a distance has not been resolved as to whether it has a wave or particle nature, or both. If gravity has a particle nature, then this particle must be extremely tiny because gravity as a force is very weak. Unification of gravity with the other three fundamental forces remains the biggest central problem in theoretical physics. After more than three decades, string theory has failed to achieve that unification. However, a large percentage of the physics community remains committed to string theory over other avenues of approach to the problem.

Michael L. Broyles

FURTHER READING

Beutler, G., M. R. Drinkwater, R. Rummel, and Rudolf von Steiger. *Earth Gravity Field from Space: From Sensors to Earth Sciences.* New York: Springer, 2003. Tells the story of NASA's GRACE satellite mission, in which gravity field measurements improved knowledge of Earth's interior and ocean dynamics. Technical.

Fowles, Grant R., and George L. Cassiday. *Analytic Mechanics*. 7th ed. New York: Brooks/Cole, 2004. A college textbook for a second course in Newtonian mechanics. Particularly strong on theory and applications of orbital motion. Requires knowledge of advanced calculus.

Gamow, George. *Gravity*. New York: Dover, 2003. A famed physicist in his own right, Gamow provides insight into the thinking of Galileo, Newton, and Einstein that led to contemporary understanding of the nature of the space-time continuum. Explains gravity as a consequence of the curvature of space. Accessible to the general reader.

Greene, Brian. *The Elegant Universe: Superstrings, Hidden Dimensions, and the Quest for the Ultimate Theory*. New York: W. W. Norton, 2003. A rather remarkable attempt to explain the most esoteric aspects of string theory and quantum field theory for the interested general reader. Some chapters require intense attention and multiple readings. Later chapters go into far more rigorous mathematical detail than the general reader may be able to follow.

Hawking, Stephen. *A Brief History of Time: From the Big Bang to Black Holes*. New York: Bantam, 1988. Hawking, perhaps the most celebrated physicist of his generation, provides a highly readable text that addresses many of the most important concepts in theoretical physics. Discusses deep issues at the frontier of physics, such as the nature of gravity, the nature of time, and the search for grand unification.

_____. *An Even Briefer History of Time*. New York: Bantam, 2008. An update to Hawking's celebrated *A Brief History of Time*, in which the physicist discusses string theory, dark matter, dark energy, and the ultimate fate of the universe.

Hawking, Stephen W., and William Israel. *Three Hundred Years of Gravitation*. New York: Cambridge University Press, 1987. A comprehensive treatise spanning the development of gravitation from Newton to concepts of quantum gravity and time asymmetry. Additional chapters discuss gravitational radiation, gravitational interaction of cosmic strings, inflationary cosmology, quantum cosmology, and superstring unification. An important reference in gravitational physics.

Hofmann-Wellenhof, Bernhard. *Physical Geodesy*. New York: Springer, 2006. An update to a text originally published four decades earlier. Offers thorough explanations of satellite methods for determining Earth's gravitational field. Highly mathematical.

Kaula, William M. *Theory of Satellite Geodesy: Applications of Satellites to Geodesy*. New York: Dover, 2000. A reprint of a classic book that was used by a generation of researchers involved in gravity field modeling. Technical.

Parker, Sybil P., ed. *McGraw-Hill Encyclopedia of Physics*. New York: McGraw-Hill, 1983. An excellent reference for the nontechnical as well as for the technical reader. Topics related to gravitation include Newton's law of universal gravitation, gravitational constant, mass and weight, gravity, gravitational potential energy, application and accuracy of Newtonian gravitation, relativistic theories, supergravity, and gravitational waves.

Thornton, Stephen T., and Andrew Rex. *Modern Physics for Students and Engineers*. 3d ed. New York: Brooks/Cole, 2005. A comprehensive presentation of the development of relativity, quantum mechanics, and nuclear and particle theory and experimentation. For undergraduate science majors and serious scientific researchers.

See also: Archaeoastronomy; Coordinate Systems; Earth System Science; General Relativity; Hertzsprung-Russell Diagram; Infrared Astronomy; Neutrino Astronomy; Optical Astronomy; Planetary Orbits; Planetary Orbits: Couplings and Resonances; Radio Astronomy; Solar Geodesy; Telescopes: Ground-Based; Telescopes: Space-Based; Ultraviolet Astronomy; X-Ray and Gamma-Ray Astronomy.

Greenhouse Effect

Category: Earth

"Greenhouse" gases absorb or trap infrared radiation emitted by a planet's surface. The absorbed or trapped energy is then released or reemitted by the greenhouse gases, resulting in additional heating of the planet's surface. Greenhouse gases include water vapor, carbon dioxide, methane, nitrous oxide, ozone, and a class of human-made molecules called chlorofluorocarbons. There is international concern that increasing atmospheric levels of these gases could lead to a global warming.

OVERVIEW

The Earth and the other planets in the solar system receive almost all their energy in the form of electromagnetic radiation from the Sun. While the Sun emits radiation over the entire electromagnetic spectrum, from shortwave X rays to longwave radio waves, the bulk of the solar radiation is in the ultraviolet, visible, and infrared parts of the electromagnetic spectrum, from about 0.15 to about 4 microns, or about 150 to about 4,000 nanometers. The peak of the Sun's radiated energy is in the visible part of the spectrum, at about 0.5 micron, or 500 nanometers, as that is the wavelength of maximum emission for an object at a temperature of about 6,000 kelvins, which is the temperature of the Sun's "visible surface," the photosphere. The intensity falls off toward shorter (bluer) and longer (redder) wavelengths, and the human eye perceives this distribution of intensity versus wavelength as a yellowish-white color. Hence, the Sun appears as a yellowish-white object in the sky.

The amount of solar radiation intercepted by a planet and potentially available for heating it depends on the Sun's luminosity, the planet's distance from the Sun, and the planet's cross-sectional area. For example, at Earth's distance from the Sun, about 150 million kilometers, the flux of solar radiation, also called the solar constant, is 1,368 joules per second per square meter multiplied by Earth's cross-sectional area of 1.28×10^{14} square meters; thus, the solar energy intercepted by Earth is about 1.75×10^{17} joules per second. Not all this intercepted solar radiation, however, is available for heating the planet's surface; some of the solar radiation is reflected back to space by the planet's surface and by clouds if the planet has an atmosphere.

The fraction of solar radiation reflected back to space is called the albedo of the planet. The solar radiation not reflected is absorbed by the planet and heats its surface. (To continue our example for Earth, its albedo is about 33 percent. Therefore, about 67 percent of the intercepted solar radiation is not reflected but instead is absorbed, heating Earth's surface.) The planet's surface in turn emits its own radiation at the same rate as it absorbs solar radiation. If the planet emitted radiation at a rate less than the rate it absorbed solar radiation, the planet would heat up and emit more radiation. However, if the planet emitted radiation at a rate greater than the rate at which it absorbed solar radiation, the planet would cool down and emit less radiation. The planet's rate of emitting radiation depends on its temperature. Equating the planet's rate of emitting radiation to the rate it absorbs solar radiation yields the planet's "effective temperature," which is the temperature of an ideal thermal radiator (or blackbody) that emits radiation at the same rate as the planet.

Earth's effective temperature is calculated to be 253 kelvins (−20° Celsius). At this temperature, Earth emits radiation in the infrared part of the electromagnetic spectrum, most of it between about 4 and 80 microns. Most of the incoming solar radiation with wavelengths between about 0.3 and 2 microns travels through the atmosphere without significant attenuation or absorption by atmospheric gases, and the infrared radiation flowing outward from Earth is absorbed by several atmospheric gases (mainly water vapor and carbon dioxide), called greenhouse gases (GHGs). After a fraction of a second, the GHGs reemit the infrared radiation randomly in all directions. About half of the reemitted infrared radiation is directed outward and about half is directed downward, back toward the surface. The downward component of the reemitted infrared radiation is absorbed by the surface, with an additional heating effect. Hence, Earth's surface is heated not only by incident solar radiation (wavelengths of 0.3 to 2 microns) but also by Earth-emitted infrared radiation (wavelengths of 4 to 80 microns) that was absorbed and then reemitted by atmospheric GHGs. The additional heating of the Earth's surface by the reemitted infrared radiation heats the surface about an additional 35 kelvins, raising the effective temperature of 253 kelvins to the actual average temperature of about 288 kelvins, or 15° Celsius. The temperature enhancement of 35 kelvins is termed the greenhouse effect. It is this temperature enhancement that makes the Earth habitable for life; without it, our average temperature would be well below the freezing point of water.

Venus is closer to the Sun than the Earth is, about 0.7 times our distance, but its albedo is about 0.65, meaning it reflects much more sunlight than Earth does; as a result, Venus appears very bright in the night sky. Consequently, the effective temperature of Venus is nearly the same as Earth's; it receives more solar radiation, but it also reflects more and thus absorbs less of it. However, while GHGs are only very minor constituents of the Earth's atmosphere, carbon dioxide is the major constituent of the atmosphere of Venus. About 96 percent of the molecules in Venus's atmosphere are carbon dioxide, with nitrogen accounting for the remaining 4 percent.

The surface pressure of the atmosphere of Venus is about 90 times the pressure of our atmosphere at sea level. As a result of the high percentage of carbon dioxide and the high surface pressure of the atmosphere, there is a very efficient and significant greenhouse effect on Venus. The greenhouse effect increases the temperature of Venus by about 480 to 500 kelvins, from its effective temperature of about 250 kelvins (close to the effective temperature of Earth) to the measured surface temperature of Venus of about 730 to 750 kelvins (about 855° to 890° Fahrenheit)—hot enough to melt lead.

There is international concern that the buildup of GHGs in Earth's atmosphere could lead to an enhanced greenhouse effect and global warming, resulting in record-high temperatures in many regions and severe droughts in others. Such global warming would lead to thermal ex-

pansion of the oceans, since water expands in volume when heated. More significantly, global warming could result in substantial melting of the ice and snow in Earth's polar regions, adding an even greater volume of water to the world's oceans. A rise in sea level would result in the flooding of many low-lying land areas.

The most important atmospheric GHGs are water vapor, carbon dioxide, methane, nitrous oxide, tropospheric ozone, and a human-made family of gases termed chlorofluorocarbons (CFCs), of which CFC-11 and CFC-12 are the most abundant species. All the GHGs are very minor constituents of Earth's atmosphere, which is composed primarily of nitrogen (78.08 percent by volume), oxygen (20.95 percent), and argon (0.93 percent). Water vapor is a variable constituent of the atmosphere and ranges from a small fraction of a percentage point to several percentage points. The concentration of carbon dioxide is about 0.038 percent, or 380 parts per million. The concentration of methane is about 1.7 parts per million. The nitrous oxide concentration is about 0.31 part per million. Ozone is a variable constituent of the troposphere and ranges from about 0.02 to 10 parts per million. The atmospheric concentrations of CFC-12 and CFC-11 are only 0.00038 and 0.00023 part per million, respectively. All these GHGs, with the exception of water vapor, are produced by human activities, such as the burning of fossil fuels and biomass (trees, vegetation, and agricultural stubble), the use of nitrogen fertilizers on agricultural areas, and various industrial activities. Each of the GHGs produced by human activities is increasing in concentration in the atmosphere with time. Carbon dioxide is increasing at a rate of about 0.4 percent per year; methane is increasing at a rate of about 1.1 percent per year; nitrous oxide is increasing at a rate of about 0.3 percent per year; and CFC-11 and CFC-12 are each increasing at a rate of about 5 percent per year.

Estimates suggest that tropospheric ozone is also increasing with time, with increases of between 1 and 2 percent per year over North America and Europe during the 1970's and 1980's. Collectively, methane, nitrous oxide, tropospheric ozone, and chlorofluorocarbons are now estimated to trap about as much infra-

The Greenhouse Effect

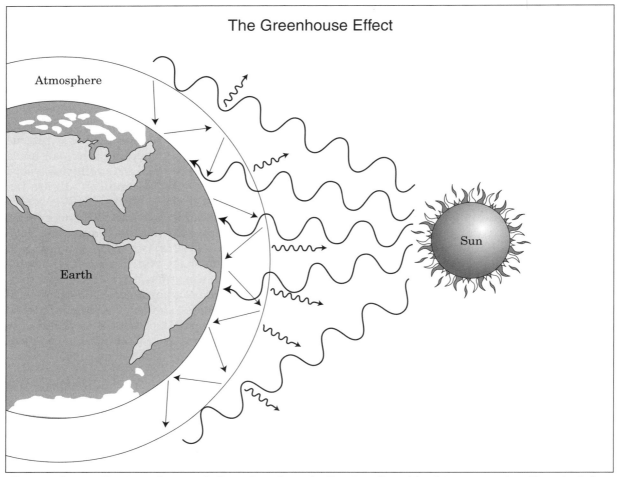

The greenhouse effect is aptly named: Some heat from the Sun is reflected back into space (small squiggled arrows), but some becomes trapped by Earth's atmosphere and re-radiates toward Earth (straight arrows), heating the planet just as heat is trapped inside a greenhouse.

red radiation as does carbon dioxide alone. These gases absorb infrared radiation in the spectral region from about 7 to 13 microns, known as the atmospheric infrared window, where water vapor and carbon dioxide do not absorb. Water vapor is a strong absorber below about 7 microns, and carbon dioxide is a strong absorber above about 13 microns. If the rates of increase of the GHGs persist with time, the greenhouse effect of the other gases, along with that of carbon dioxide, could amount to an effective doubling of carbon dioxide levels by the middle of the twenty-first century. This is many decades sooner than the level of carbon dioxide alone is likely to double.

A doubling of carbon dioxide in the atmosphere could lead to a global temperature increase of about 1.2 kelvins if there were no other changes in the climate system. However, a warming caused by an increase of carbon dioxide would lead to more evaporation of water vapor from the oceans and would permit the atmosphere to hold more water vapor, as the capacity of air to hold water vapor increases with increasing temperature. Because water vapor is itself a greenhouse gas, as its concentration increases in the atmosphere, the Earth would warm even further. The net result of these different feedback processes could be that for a doubling of atmospheric carbon dioxide, the surface temperature could increase by about 4 kelvins. It is estimated that the average surface temperature has increased globally by about 0.7 kelvin over the twentieth century, although this temperature increase cannot unambiguously be attributed to the buildup of GHGs over this period. Some temperature data sets also indicate that the 1990's was the warmest decade since the mid-1800's. On the other hand, satellite measures of global temperatures show that no warming has occurred in the first decade of the twenty-first century.

The sources of the GHGs, with the exception of water vapor, are mostly human-initiated. Carbon dioxide results from the burning of fossil fuels and of living and dead biomass, such as the burning of rain forests to clear land in the tropics and the burning of agricultural stubble after harvests. Methane is produced as a combustion product of biomass burning, natural gas leakage, and by the action of anaerobic bacteria in wetlands (such as rice paddies), in landfills, and in the stomachs of ruminants (such as cattle and sheep). Nitrous oxide is produced as a combustion product of biomass burning and by the action of nitrifying and denitrifying bacteria (which add and remove nitrogen) in natural, fertilized, and burned soils. Ozone results from atmospheric reactions involving methane, carbon monoxide, and oxides of nitrogen, which are produced by the burning of fossil fuel and by biomass burning. Chlorofluorocarbons (CFC-11 and CFC-12) are released into the atmosphere when they are used as propellants in aerosol spray cans, as blowing agents for foam insulation, and as refrigerants.

There are a number of other important players in the climate scenario whose roles must be more fully understood before scientists can completely assess the impact of increasing concentrations of GHGs on the future climate. Climate factors that must be studied in more detail include the formation and role of clouds, the heat capacity and circulation of the oceans, and the variability of the Sun's luminosity. The Earth's climate is a complex system affected by many processes and parameters. Experts need to learn more about the interplay of atmospheric GHGs, clouds, the ocean, and solar variability before they may accurately assess the future climate.

METHODS OF STUDY

Studies of the greenhouse effect and its impact on the Earth's climate are multidisciplinary and involve theoretical computer modeling, laboratory studies of the spectroscopic parameters and properties of GHGs, atmospheric measurements of GHGs, and aircraft and satellite measurements of parameters that control climate. Theoretical computer models of climate include zero-dimensional, one-dimensional, two-dimensional, and three-dimensional models. Zero-dimensional models give climate parameters that represent an average for the entire system, such as the mean temperature of the Earth's surface. One-dimensional models are used to study climate in either a horizontal (latitude) or vertical (altitude) direction. In these models, a latitude-dependent surface temperature is the climate parameter of major interest. The vertical one-dimensional model is known as a radiative-convective model and is used to study the effects of changes in concentrations of GHGs on the surface temperature.

Two-dimensional models involve characterizing the temperature variation as a function of both latitude and altitude. The most complex climate model is the three-dimensional model, or the general circulation model (GCM). This model gives a complete description of climate as a function of latitude, longitude, and altitude.

Major uncertainties exist in the understanding of several key parameters in the theoretical modeling of climate. A major uncertainty in these models is the role of cloud feedback, particularly as the Earth heats up. Calculations indicate that as the temperature of the Earth increases, clouds will appear in greater quantity. Clouds both reflect incoming solar radiation and trap Earth-emitted thermal infrared radiation. Hence, clouds can enhance or decrease global warming. The radiative properties of clouds and how they will vary with a warmer Earth are poorly understood and therefore are not included in most theoretical climate models. However, a number of studies have found that cloud cover has a negative feedback effect, suggesting that Earth's climate system is much less sensitive than most computer models indicate. If this negative feedback proves true, global warming would produce a temperature increase of less than 1 kelvin by 2100.

To understand the greenhouse effect of gases, scientists design laboratory experiments to measure the spectroscopic parameters and properties of GHGs. These studies provide information on the infrared wavelengths absorbed by these gases and the intensity of the absorption. These laboratory spectroscopic studies involve filling an absorption cell with a known concentration of a greenhouse gas and measuring the absorption of infrared radiation as a function of spectral wavelength using a scanning spectrometer.

To assess the impact of atmospheric GHGs on the Earth's climate, scientists must obtain accurate measurements of those gases that are found at trace levels in the atmosphere—for example, at atmospheric concentrations ranging from parts per million to parts per trillion by volume. Measuring atmospheric GHGs at these very low atmospheric levels involves a series of different analytical chemistry instruments. Carbon dioxide may be measured with a gas chromatograph equipped with a thermal conductivity detector or with nondispersive infrared instrumentation. Methane is measured with a gas chromatograph equipped with a flame ionization detector or with nondispersive infrared instrumentation. Nitrous oxide, CFC-11, and CFC-12 are measured using a gas chromatograph equipped with an electron capture detector. Ozone may be measured using various ultraviolet and infrared absorption techniques. Water vapor is measured using a hygrometer. Aircraft and satellite measurements have provided very important data on several of the parameters that control climate, including the flux of incoming solar radiation, the flux of outgoing Earth-emitted thermal infrared radiation, the flux of solar radiation reflected by clouds, and the geographical and temporal variability of clouds and surface temperature.

CONTEXT

The buildup in the atmosphere of GHGs such as carbon dioxide, methane, nitrous oxide, tropospheric ozone, CFC-11, and CFC-12, could result in a global warming of the Earth. Atmospheric GHGs are increasing with time. These GHGs result from a variety of human activities, including the burning of fossil fuels (carbon dioxide), the burning of living and dead biomass (carbon dioxide, methane, and nitrous oxide), and the application of nitrogen fertilizers and burning of agricultural and grasslands (nitrous oxide). GHGs are also produced from rice paddies, cattle, and sheep (methane), and from several industrial applications (CFC-11 and CFC-12).

A global warming of the Earth from the buildup of GHGs in the atmosphere would have a significant impact on people's daily lives. Most areas would experience more days per year when temperatures exceed 32° Celsius, and the growing seasons and patterns of rainfall would change. One of the most important effects would be a predicted increase in the height of the world's oceans. The increased height of the oceans would result from the thermal expansion of seawater because of the Earth's high temperature (water is a compressible fluid and expands in volume when heated) and because of the melting of polar ice and snow as the Earth becomes warmer. It has been estimated that a global temperature increase of about 4 kelvins might result in a 2-meter increase in the height of the world's oceans. This increase in water level would cause the flooding of low-lying land areas occupied by more than 40 million people worldwide. Higher temperatures resulting from the buildup of atmospheric GHGs would tax the world's air-conditioning facilities, which require the burning of fossil fuels for their operation. Ironically, the burning of fossil fuels is a major source of atmospheric GHGs.

It must be emphasized that there are fundamental uncertainties and deficiencies in scientists' understanding of climate and the processes and parameters that control it. More must be learned about the effect of clouds and oceans on climate and how these phenomena can affect the climate as the Earth begins to warm. Theoretical computer models of climate are not complete; more research is needed before the future climate of the Earth can be assessed with greater certainty.

Joel S. Levine

FURTHER READING

Ahrens, C. Donald. *Essentials of Meteorology.* 5th ed. Florence, Ky.: Brooks/Cole, 2007. A well-written, student-friendly introductory meteorology textbook. Includes a CD-ROM to aid self-instruction.

Andrews, David. *An Introduction to Atmospheric Physics.* Cambridge, England: Cambridge University Press, 2000. A well written college textbook. Includes information on the enhanced greenhouse effect, ozone depletion, and development of weather systems. Can be technical.

Cook, Alex. *The Greenhouse Effect: A Legacy.* Indianapolis: Dog Ear, 2007. A good introductory work for the general audience. Facts are presented through an easy-to-read narrative.

Environmental Protection Agency. "The Greenhouse Effect: How It Can Change Our Lives." *EPA Journal* 15 (January/February, 1989). Popular, nontechnical accounts of the impact of climate change on agriculture, forests, energy demand, and other areas in a special issue devoted to the greenhouse effect. The principles that control and regulate global climate, including the greenhouse effect, are presented simply for a general audience. Well illustrated with photographs and charts.

Frederick, John. *Principles of Atmospheric Science.* Sudbury, Mass.: Jones and Bartlett, 2008. A comprehensive work covering all areas of atmospheric physics, including the greenhouse effect. Geared toward undergraduate students.

Goody, R. M., and J. C. G. Walker. *Atmospheres.* Englewood Cliffs, N.J.: Prentice-Hall, 1972. A good nontechnical description of solar and infrared radiation in planetary atmospheres and how the radiation balance controls the temperature of a planet. Includes sections on the Sun and the planets, solar radiation and chemical change, atmospheric temperatures, winds of global scale, condensation and clouds, and the evolution of atmospheres.

Hansen, Joel E., and T. Takahashi, eds. *Climate Processes and Climate Sensitivity.* Geophysical Monograph 29. Washington, D.C.: American Geophysical Union, 1984. A collection of papers dealing with various aspects of the climate system, including atmosphere and ocean dynamics, the hydrologic cycle and clouds, albedo and radiation processes, polar ice, and ocean chemistry. Each paper was written by an expert in that particular area of climate research. The papers summarize what is known, along with the major uncertainties and deficiencies in understanding of the processes and parameters that control climate.

Henderson-Sellers, A., ed. *Satellite Sensing of a Cloudy Atmosphere: Observing the Third Planet.* London: Taylor and Francis, 1984. Each chapter is by an active researcher in the area covered, including radiation and satellite sensors, the Earth's radiation budget and clouds, water and the photochemistry of the troposphere, vertical temperature sounding of the atmosphere, cloud identification and characterization from satellites, and the remote sensing of land, ocean, and ice from space.

Levine, Joel S., ed. *The Photochemistry of Atmospheres: Earth, the Other Planets, and Comets.* Orlando, Fla.: Academic Press, 1985. A comprehensive textbook covering atmospheric composition, chemistry, and climate, the sources and sinks of GHGs, and the climate modeling of the Earth. The chapter on climate includes discussions of zero-dimensional, one-dimensional, and three-dimensional (global circulation) climate models and the underlying physical, radiative, and dynamic processes and parameters in each model.

National Research Council. *Changing Climate: Report of the Carbon Dioxide Assessment Committee.* Washington, D.C.: National Academy Press, 1983. Technical report that addresses the possible impacts of climate change on sea level, agriculture, plant growth, and society in general. Considers future carbon dioxide emissions from fossil fuels, the dissolution of carbon dioxide into the oceans, the biosphere storage of carbon dioxide, and the impact of increased carbon dioxide on climate, agriculture, and sea level.

See also: Auroras; Earth-Sun Relations; Earth System Science; Earth's Atmosphere; Earth's Oceans; Planetary Atmospheres; Titan; Triton; Van Allen Radiation Belts; Venus's Atmosphere; Venus's Volcanoes.

Habitable Zones

Category: Life in the Solar System

Habitable zones are the places beyond Earth where there is the best chance of finding life. As such, they are a major focus of scientific consideration and investigation that inspire exploratory efforts.

OVERVIEW

At root, the idea of a habitable zone around a star is simple. A planet must not be too close to a star to be too hot for life. On the other hand, it must not be too far from the star to be too cold for life. The spherical shell around a star where a planet will be "just right" for life is the habitable zone of that star. Another name is the circumstellar habitable zone; this name distinguishes it from the galactic habitable zone, the region of a galaxy most favorable to life.

Unfortunately, it is not a simple matter to calculate the limits of this shell for a given star. However, using Earth and the Sun as the basic measure and noting that energy concentration of light from a star decreases as the square of the distance, scientists can predict the rough average radius of the habitable zone around a star. The result is simply the square root of the ratio of the stellar brightness to that of the Sun. The radius will then be expressed in astronomic units (AU), where 1 AU is the average distance of Earth from the Sun. The brightness of the star must be rated at a standard distance, known as its bolometric luminosity.

To complicate matters, a planet might not stay in the habitable zone. To be habitable, a planet should have an orbit that remains in the habitable zone for billions of years. Only planets with relatively circular orbits can do that. Also, the habitable zone moves away from a star as that star ages and grows hotter. Stars are not all the same. Some burn their fuel quickly and do not last for billions of years. The habitable zone of a bright, hot star will not be the same as that

of a dim, cool one. Thus, the habitable zone is determined by the star, the planet, and the life-form under consideration.

The type of life involved is, of course, a critical consideration. We cannot discuss habitable zones without first establishing the expectations we have of life around a star. We can identify three very basic requirements for life. First, living things have bodies. Second, a life-form uses a flow of nutrients and energy to sustain its body and bodily processes. Third, life reproduces itself. Reproduction requires bodies (and, most likely, molecules) able to retain the complex and detailed information required for constructing more living forms. Life may be more than this, but it will surely never be less.

With these criteria we find, by an argument too extended to give here, that life in a habitable zone will be water- and carbon-chemistry-based. The habitable zone, then, can be calculated based on the requirement that a planet in the zone will be able to hold water in liquid form long-term. The actual calculation is complicated by many factors; chief among them is the fact that water vapor is a major greenhouse gas. Hence, one cannot simply find the incident energy from the star at various distances because the presence of water retains the heat supplied by the star and thereby expands the habitable zone. An early estimate by Michael Hart had the habitable zone of our Sun between 99 and 105 percent of the Earth's current distance. This was too conservative; a more likely estimate is 95 to 137 percent.

The habitable zone is unique to the star and is determined by the stellar mass and, to a lesser extent, the age of the star. In terms of spectral classes, stars such as Earth's Sun (in class G) and some K- and F-class stars can have habitable zones. The range also corresponds to the stellar surface temperature range of a bit less than 4,500 kelvins (K) to a bit above 7,000 kelvins. Our Sun, at 5,777 kelvins, is in the middle of this range. Stars with a mass 20 percent or more greater than that of the Sun (that is,

1.2 M_S) will not have habitable zones, because they emit deadly amounts of ultraviolet radiation (UV) along with their visible and infrared radiation (IR). UV destroys water molecules and, at high intensity, will eventually strip a planet of the water critical for life. About 1 percent of stars are so large that they consume their fuel and die long before life can form. Indeed, all stars larger than 1.5 M_S would turn into red giant stars and swallow up any life-bearing planets around them before intelligent life could appear. Stars with more than ten times the mass of our Sun are so intensely bright that planets cannot form around them, because light creates pressure on anything it strikes. This radiation pressure is usually too small to matter, but for these very large stars it is great enough that all the material around the star that might eventually form planets is pushed away from the star and is disbursed too quickly for planets to form.

On the other hand, stars with less than about 0.80 M_S do not produce enough high-energy UV light to support life on any planet; their UV output is insufficient for important atmospheric effects such as ozone creation. Any planet close enough to a star of less than about 0.65 M_S to receive sufficient heat will be so gripped by the stellar gravity that it will show the star one face, as the Moon does our Earth. If this is not the case, an effect called spin-orbit coupling will almost certainly force the rotation rate of the planet to be almost as slow as the planetary year, thus frying the planet on one slowly changing side while freezing it on the other. Mercury is such a case. It revolves around the Sun in 88 days but rotates once every 58.7 days, exactly two-thirds of the orbital period. Both these effects are due to the fact that no planet is perfectly spherical. In either case, the planetary face toward the star will be too hot for water, the side away from the star too cold.

Another factor in habitability is variability of stellar output. Our Sun has an eleven-year sunspot intensity cycle that causes a variation in solar luminosity of about 0.1 percent. However, 18 Scorpii, an almost identical star in the constellation Scorpius with a mass of 1.03 M_S and a temperature of 5,789 K, has a much greater variability over a 9 to perhaps 13 year cycle. If

great enough, this would make its habitable zone move in and out rapidly thereby negating its benefits for any planet in a basically fixed orbit.

KNOWLEDGE GAINED

The idea of a circumstellar habitable zone has stimulated wide-ranging research resulting in a significant extension of our knowledge of planetary systems generally, as well as of our own solar system in particular. A circumstellar habitable zone imposes quite severe limitations on where best to look for life in the universe. Responses to these limitations are likewise limited. One either accepts the limitations, at least tentatively, and looks for suitable planets around only suitable stars, or one must in some way challenge the limitations.

Tentative acceptance of the limitations takes us in the direction of what has become a successful search for exoplanets, planets orbiting stars other than our Sun. The list is large and growing. The primary technique used in this search detects stellar motion due to the stellar reaction to the orbital motion of the planet. Since large planets create more stellar reaction that is more easily detected, this technique is biased toward discovering large planets. It is no surprise then that most of the known exoplanets are large. It is a bit of a surprise that they tend to be relatively close to their stars and, hence, are sometimes called "hot Jupiters." If this trend continues, it may require revisions in the theories of how planetary systems form. Hot Jupiters are not expected to harbor life even if they are in a habitable zone.

Another puzzling result of these searches is that exoplanets seem to prefer highly elliptical orbits compared with those in our solar system. Such orbits are risky in that they may take the planet out of the habitable zone annually. On a more positive note, the work on exoplanets has confirmed that, as expected, planets tend strongly to be found around stars with high metal content. (In this context, "metal" means any element other than hydrogen or helium.)

Challenges to the idea of the circumstellar habitable zone have either been attempts to show there are niches of habitability outside the habitable zone or efforts to extend the habitable zone in size or to more types of stars, especially to red dwarfs. This later direction seems promising in light of the discovery of planets around the red dwarf Gliese 581. One of them, Gliese 581 c, is said to be the smallest planet yet discovered in the habitable zone of another star. That, of course, assumes that a red dwarf has a habitable zone.

Looking for niches of habitability in our solar system—and, hence, potentially elsewhere—offers the possibility of confirmation by direct examination in the not too distant future and is accordingly fairly popular. Thus, attention has become focused on Mars and some large satellites of Jupiter and Saturn.

Mars has received the most attention, as attested by missions such as the Mars Exploration Rover (which began operating on the Martian surface in early 2004). The Cassini-Huygens mission to Saturn (which entered into orbit around Saturn in mid-2004) included flybys of Titan, revealing its liquid methane oceans and dense atmosphere, while the Galileo mission to Jupiter in the late 1990's gathered a great deal of data on two of Jupiter's satellites, volcanic Io and Europa, whose deep ice sheet appears to have water in liquid form beneath it.

CONTEXT

The dream of "other races of men" on other worlds has been the currency of cosmological speculation at least since the ancient Greek atomists. Men on the Moon were described by the ancient Pythagoreans and in the seventeenth century by Johannes Kepler, and even the eighteenth century philosopher Immanuel Kant gave opinions on the inhabitants of Mars, Venus, and Jupiter. Modern science has tried to inform and thereby reduce this speculation. The concept of a habitable zone is a product of this effort, although it imposes limitations that would, no doubt, have disappointed earlier enthusiasts.

Enthusiasm for finding life elsewhere in the universe is by no means dead. The high profile of the Search for Extraterrestrial Intelligence (SETI) and the advent of the new academic discipline of astrobiology are proof of that. Both of these developments are inextricably connected with the concept of habitable zones and

are all but inconceivable without it. The prospect of habitable zones has also stimulated thinking and research in other areas. One such development is the idea of a galactic habitable zone.

The concept of habitable zones also connects to larger cosmological issues, such as questions of the "fine tuning" of the universe that makes life possible somewhere in the universe and the related issue of the anthropic principle, the notion that the universe must contain conditions that allow for the existence of an observing intelligent life-form.

John A. Cramer

FURTHER READING

Aczel, Amir D. *Probability 1*. New York: Harcourt Brace & Company, 1998. Aczel argues that the large number of stars outweighs the limitations of habitable zones to the point where intelligent life must occur throughout the universe.

Cohen, Jack, and Ian Stewart. *Evolving the Alien: The Science of Extraterrestrial Life*. London: Ebury Press, 2002. Cohen and Stewart dispute that alien life will be similar enough to terrestrial forms to frame a meaningful idea of a habitable zone. They also argue the case for various niches of habitability.

Cramer, John A. *How Alien Would Aliens Be?* Lincoln, Nebr.: Writers Club Press, 2001. The first half of the book shows how physical constraints limit where intelligent life might be found in the universe. Hence, it surveys many of the limitations that lead to the idea of a habitable zone and the possibility of habitable niches.

Dole, Stephen H. *Habitable Planets for Man*. 2d ed. New York: Elsevier, 1970. Something of a classic on habitable planets, this is one of the earliest discussions of habitable places for human colonization. It gives a good if somewhat dated account of what makes a place habitable for intelligent life, a more restrictive notion than a habitable zone for any life.

Gonzalez, Guillermo, and Jay W. Richards. *The Privileged Planet: How Our Place in the Cosmos Is Designed for Discovery*. Washington,

D.C.: Regnery, 2004. Gonzalez and Richards consider the idea that planetary habitability may be connected with the planet's suitability as a platform for observing the universe.

Grinspoon, David. *Lonely Planets: The Natural Philosophy of Alien Life*. New York: Harper-Collins, 2004. This is a wide-ranging and readable book covering habitable zones and many related topics.

Ward, Peter, and Donald Brownlee. *Rare Earth: Why Complex Life Is Uncommon in the Universe*. New York: Springer, 2000. Ward and Brownlee make the case that the limitations on habitable zones are severe to the point of making planets like Earth quite rare.

See also: Extrasolar Planets; Extrasolar Planets: Detection Methods; Extraterrestrial Life in the Solar System; Life's Origins; Mars: Possible Life; Red Dwarf Stars; Search for Extraterrestrial Intelligence; Venus's Surface Experiments.

Hertzsprung-Russell Diagram

Category: Scientific Methods

Between 1904 and 1915, Danish astronomer Ejnar Hertzsprung and American astronomer Henry Norris Russell independently discovered significant relationships between the luminosity and surface temperature of stars. The graph or diagram they developed that displays these relationships has become a powerful tool for summarizing many stellar properties and tracing the stages in the "lives" of stars.

OVERVIEW

The Hertzsprung-Russell diagram (or simply H-R diagram) is named for the two astronomers, Ejnar Hertzsprung (1873-1967) and Henry Norris Russell (1877-1957), who independently developed a powerful tool that quickly and easily summarizes many properties of a star.

In an H-R diagram, the ordinate (vertical axis) is used to plot some measures of a star's intrinsic brightness; these include absolute visual magnitude, absolute bolometric magnitude, and luminosity. The abscissa (horizontal axis) is used to plot some measures of a star's average photospheric or surface temperature; these include effective temperature, spectral type, and color index. Depending on the particular parameters used to plot stars, the diagram is also known as a color-magnitude diagram or a spectrum-luminosity diagram. However, all variations of the H-R diagram depict the same basic relationships; the slightly different forms of the diagram are due to the specific stellar parameters used in plotting it.

Stars plotted near the top of the H-R diagram are very luminous, while those at the bottom are very dim. Stars plotted on the left side of the diagram have the hottest surfaces and appear bluish, and those on the right side have the coolest surfaces and appear reddish. If spectral type is used, the spectral sequence O B A F G K M is plotted from left to right. The position of a star plotted on an H-R diagram not only indicates the relationship between its luminosity and surface temperature but also provides information about its radius, mass, and stage of development within the stellar life cycle.

The radius is found through application of the Stefan-Boltzmann law for thermal radiating bodies: the rate of emission of electromagnetic radiation per unit surface area is proportional to the temperature (on an absolute scale) raised to the fourth power. Thus the luminosity (L) of a star is proportional to its surface area (A) times its surface temperature (T) raised to the fourth power, or L is proportional to $A \times T^4$. Since the surface area of a sphere is proportional to radius squared, another way of stating this relationship is that luminosity (L) is proportional to radius (R) squared times surface temperature (T) raised to the fourth power, or L is proportional to $R^2 \times T^4$. A star's position in the H-R diagram gives its luminosity and surface temperature, and that information allows one to calculate its radius.

About 90 percent of all known stars fall along a broad band that extends diagonally across the H-R diagram in a "lazy S" shape, from bright, hot, blue stars at the upper left to faint, cool, red stars at the lower right. This band is known as the "main sequence" and is designated by luminosity class V. The distribution of stars along the main sequence band makes sense intuitively. The stars with the hottest surfaces emit the most light, while those with cooler surfaces emit much less light. Radius does not vary too much along the main sequence, ranging from about five to ten times the Sun's radius at the upper left (the luminous, hot, blue stars) and diminishing to about one-tenth to one-thirtieth the Sun's radius at the lower right (the faint, cool, red stars). It has been found that mass also varies along the main sequence, ranging from

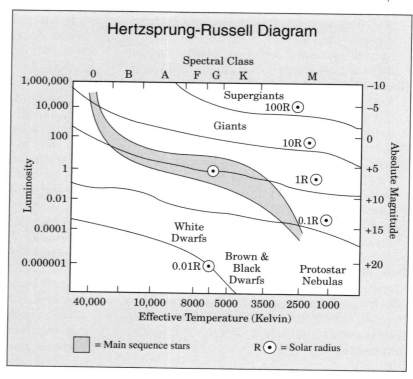

several tens to perhaps as much as one hundred times the Sun's mass at the upper left (the luminous, hot, blue stars) and diminishing to about one-tenth to one-fifteenth the Sun's mass at the lower right (the faint, cool, red stars). A star in this band (referred to as "on the main sequence") is in the most stable, longest-lasting chapter of its life. It generates energy by fusing hydrogen into helium in its core. Earth's Sun is a main sequence star with a spectral type and luminosity class of G2 V and an absolute magnitude of about +5, which means it is plotted about midway along the main sequence.

However, not all stars have luminosities and surface temperatures that place them on or near the main sequence. Stars plotted away from the main sequence represent other stages of the stellar life cycle. Some stars are located above and to the right of the main sequence. They are brighter than main sequence stars with the same surface temperature, so by the Stefan-Boltzmann law they must have much larger surface areas from which to radiate, and this means larger radii. Hence these stars are referred to as giants and supergiants. Giants

(luminosity class III) typically have about one hundred to ten thousand times the surface area and thus about ten to one hundred times the radius of the Sun. Supergiants (luminosity classes Ia, Iab, and Ib) are even larger, typically having surface areas ten thousand to one million times that of the Sun and radii one hundred to one thousand times that of the Sun. The very largest giants and supergiants are those with the coolest surfaces and hence are red in color. Giants and supergiants have masses ranging from several tens of times the Sun's mass down to about the Sun's mass. They represent stars that have left the main sequence because they exhausted the hydrogen in their cores. An easily seen orange giant (not quite as large or cool as a red giant) is Aldebaran (Alpha Tauri), with spectral type and luminosity class K5 III and absolute magnitude about minus 1, located in the constellation of Taurus the bull; if placed at the center of our solar system, it would fill more than half of Mercury's orbit around the Sun. An easily seen red supergiant is Betelgeuse (Alpha Orionis), spectral type and luminosity class M2 Iab, absolute magnitude about minus 5, located in the constellation of Orion the hunter; if placed at the center of our solar system, it would be about twice as big as the orbit of Mars and extend out into the asteroid belt.

Other stars are plotted below and to the left of the main sequence. They are quite hot but very faint, meaning that they must have small surface areas and radii. They are called white dwarfs and have about the mass of the Sun (about 300,000 Earth masses) packed into a volume about the size of the Earth. They are stars that are near the ends of their lives, having exhausted all their ways of generating energy and shining only because they are still hot.

APPLICATIONS

The Hertzsprung-Russell diagram illustrates the evolutionary stages of a star's life, and it has played a major role in stellar astrophysics. A star is "born" with a certain mass and chemical composition (typically about 75 percent hydrogen and 25 percent helium by mass, with traces of heavier elements). During most of its life, its mass remains nearly constant; only in the last stages of its life does it undergo significant mass loss. On the other hand, its chemical composition, at least in its central region, changes as the star generates energy through a series of nuclear fusion reactions, converting light atomic nuclei into heavier nuclei. A star is in a constant tug-of-war between its tendency to collapse under its own gravity and its tendency to expand due to its pressure. As the star evolves, tapping one energy source after another, its structure changes, and this affects its luminosity and surface temperature, and hence its position in the H-R diagram. By utilizing the H-R diagram, astronomers can easily depict these changes and can trace the evolutionary development of a star as a path followed on the H-R diagram.

Stars are born in interstellar clouds of gas and dust (nebulae) by gravitational contraction. The contraction may be triggered by shock waves from nearby supernova explosions or by an encounter with a galactic spiral arm density wave that compresses the density. As a part of the nebula contracts, it heats up and starts to shine as a protostar. A protostar first appears in the H-R diagram at the extreme right-hand

(cool) side. As the protostar continues to contract, it gets hotter and therefore brighter, and it moves diagonally up and to the left in the H-R diagram. When the star's central temperature reaches several million kelvins, hydrogen fusion ignites in the star's core, converting hydrogen to helium. The star stops shrinking and it stabilizes, landing on the main sequence. For a star with the Sun's mass, the protostar stage lasts about 30 million years. More massive stars have stronger self-gravity and contract more rapidly, while low-mass stars have weaker self-gravity and contract more slowly.

A star's location on the main sequence depends on its mass: higher masses toward the upper left and lower masses toward the lower right. The main sequence is the longest, most stable period in a star's energy-producing life. As long as it has hydrogen in its core to fuse into helium, it stays near the main sequence. The length of a star's main sequence stage depends on its mass. Massive stars have more fuel, but they consume it much more rapidly; that is why they are so luminous. Consequently they have very short main sequence lifetimes, no more than a few million years for stars with several tens of solar masses. Earth's Sun is about halfway through its 10-billion-year-long main sequence stage. Low-mass stars have less fuel, but they consume it very slowly; that is why they are so faint. Consequently they have long main sequence lifetimes, much longer than the Sun's and longer even than the present age of the universe; every low-mass main sequence star that ever formed is still a low-mass main sequence star. As a star consumes the hydrogen in its core, it becomes slightly hotter and brighter, and on the H-R diagram, it rises slightly above the zero-age main sequence.

When the hydrogen in the star's core is all transformed into helium, the changes accelerate. The helium core contracts and heats up, and hydrogen fusion is transferred to a shell surrounding the core where it proceeds at a faster rate. The star leaves the main sequence as it becomes more luminous and its outer layers expand and cool. A star like the Sun becomes a red giant. (When that happens to the Sun, in about 5 billion years, it will become about one thousand times brighter than it is now. Earth's oceans will boil away, its atmosphere will escape into space, the rocks of its surface will at least partly melt, and all life on the planet will be extinguished.) Eventually the core of a red giant gets hot enough to fuse helium into carbon. With a nuclear fusion reaction generating energy in the star's core, once again it enters a stable stage, but this phase is brief. A star like the Sun quickly consumes the helium in its core in no more than 1 billion years (compared to 10 billion years for hydrogen fusion in its core while on the main sequence), but it is not massive enough to shrink sufficiently to get hot enough to start any more nuclear fusion reactions.

Instead it puffs off its bloated atmosphere, losing as much as half of its mass. The ejected outer layers are called a planetary nebulae. (A "planetary nebula" has nothing to do with planets; the term developed in the nineteenth century, when these expanding bubbles of gas were seen as round, like planets, and fuzzy, like nebulae, through the telescopes then in use.) Ultraviolet radiation from the hot core often ionizes these gases, making them shine. Examples of planetary nebulae are the Helix nebula in the constellation Aquarius and the Ring nebula in the constellation Lyra. A star's own stellar wind, along with thermal pulses, will drive these outer shells away as expanding bubbles of gas, leaving an exposed hot core as the central star of the planetary nebula.

Such stars are located in a hook-shaped region at the extreme left side of the H-R diagram. The remaining core contracts as much as it can and enters the white dwarf stage. A white dwarf shines only because it is hot, but as it shines it radiates its heat away and becomes cooler and fainter, ending as a cooled-off black dwarf.

A very massive star (more than several times the mass of the Sun) becomes a red supergiant when it leaves the main sequence. It is massive enough for its core to shrink enough to get hot enough to initiate fusion reactions that form elements up to iron. Iron is the heaviest nucleus that can form in fusion reactions that release energy; heavier nuclei require the input of energy to form via fusion. The iron core collapses, sending shock waves through the star, which explodes as a supernova. The outer part of the star is blown away, leaving a collapsed core that becomes a neutron star or in some cases a black hole.

The position of the main sequence in the H-R diagram is affected slightly by the chemical composition of the stars that reach it. As supernovae explode and enrich the interstellar material with elements heavier than helium, new generations of stars will be born with slightly increased abundances of heavier elements, and this shifts the main sequence slightly farther to the right in the H-R diagram.

CONTEXT

Ejnar Hertzsprung's interest in astronomy was fostered by his father, who, although educated as an astronomer at the University of Copenhagen, worked for the Danish Department of Finance. He encouraged his son's interest in astronomy as an avocation, not a vocation, believing that it was not possible to make a living studying the stars. Hertzsprung graduated with a degree in chemical engineering in 1898 from the Polytechnical Institute in Copenhagen and began work in St. Petersburg, Russia. In 1901, Hertzsprung went to Leipzig and spent a year studying photochemistry in Friedrich Wilhelm Ostwald's laboratory. Photography was developing as a serious scientific tool, and Hertzsprung realized its inherent advantages in astronomy, particularly in studying the spectra of stars.

In 1902, Hertzsprung returned to Denmark, where he corresponded regularly with astronomer Karl Schwarzschild. In 1905 and 1907, he published two papers on stellar spectra and magnitudes. In these papers, he pointed out the distinction between red stars that were very luminous and those that were not. He realized that this indicated a significant difference in the size of the stars, and he named them giants and dwarfs respectively. Hertzsprung did not include a diagram with either his 1905 or 1907 paper, but in 1911 he published graphs of the relationship of color to magnitude based on the stars in several star clusters, including the Pleiades. Hertzsprung noted that stars in the diagrams could be divided into two groups: a more populous one, later known as the main sequence, and a smaller group currently recognized as giants and supergiants.

Henry Norris Russell, an American astronomer, began to measure stellar parallaxes using photographic techniques in 1903 in Cambridge, England. By 1910, he had accumulated hundreds of photographic plates. Russell's graphical analysis of the absolute magnitude and spectral type of different stars revealed an interesting correlation. The stars were not scattered randomly over the graph; for most stars, as the luminosity (as measured by absolute magnitude) decreased, so did the surface temperature (as determined from the spectral type). In December of 1913, Russell presented a graph of the relationship (later known the H-R diagram) to the American Astronomical Society. In his address, he also identified giant and white dwarf stars, for which he had laid the theoretical foundations in his papers of 1910 and 1912.

Through the filter of history, priority for the idea for the diagram goes to Hertzsprung. However, Russell was unaware of Hertzsprung's work when he presented his diagram in 1913, based on hundreds of stars he studied from 1903 to 1910, and began to relate the graph to theories of stellar evolution. The origin of identifying the graph as the "Hertzsprung-Russell diagram" is not clear. Hertzsprung himself often remarked that it should be called a color-magnitude diagram for clarification purposes. Calling it the Hertzsprung-Russell (or H-R) diagram seems to have evolved gradually, helped along in its use by the English astronomer Sir Arthur Stanley Eddington, who in 1924 discovered the mass-luminosity relation of main sequence stars. The H-R diagram was included in the articles and lectures of the Danish astronomer Bengt Strömgren during the 1930's, in which he provided an explanation of what the main sequence represents.

The early 1900's was a time of great advances in astronomy. The development of the H-R diagram was one of the more important. The recognition of a correlation between luminosity and surface temperature resulted in a domino effect for stellar research. The H-R diagram has proved to be a versatile astronomical tool. It provides a simple way to represent the structure and depict the evolution of stars.

Richard C. Jones and Richard R. Erickson

FURTHER READING

Chaisson, Eric, and Steve McMillan. *Astronomy Today*. 6th ed. New York: Addison-Wesley, 2008. Very well-written college-level textbook for introductory astronomy courses. Contains a thorough description of the H-R diagram and its use in the study of stellar evolution, complete with transparent overlays.

Fraknoi, Andrew, David Morrison, and Sidney Wolff. *Voyages to the Stars and Galaxies*. Belmont, Calif.: Brooks/Cole-Thomson Learning, 2006. A well-written, thorough college textbook for introductory astronomy courses. Has a good description of the H-R diagram and its use in the study of stellar evolution.

Freedman, Roger A., and William J. Kaufmann III. *Universe*. 8th ed. New York: W. H. Freeman, 2008. College-level introductory astronomy textbook. Thorough and well written, it includes a good description of the H-R diagram and its use in the study of stellar evolution.

Korn, Katherine G. "Henry Norris Russell (1877-1957)." *Vistas in Astronomy* 12 (1970): 3-6. A general biography of Russell, with insights into his life and work.

Leuschner, A. O. "The Award of the Bruce Gold Medal to Professor Ejnar Hertzsprung." *Publications of the Astronomical Society of the Pacific* 49 (1937): 65-81. A very readable summary of Hertzsprung's astronomical career with an emphasis on the significance of the H-R diagram, at least as it was understood at that time.

Nielsen, Axel V. "Ejnar Hertzsprung: Measurer of Stars." *Sky and Telescope* 35 (January, 1968): 4-6. This obituary notice explores the work and life of Hertzsprung. A good historical look at the man and how his contributions influenced astronomy.

Philip, D. A. G., and L. C. Green. "The H-R Diagram as an Astronomical Tool." *Sky and Telescope* (May, 1978): 395-400. Contains summaries of the International Astronomical Union's Symposium No. 80 of 1977, in which the H-R diagram was examined from various points of view. Describes how the data should be presented to illuminate particular problems, such as the distribution of stars and their chemical composition and age, as well as the variations from one star cluster or galaxy to another.

Russell, Henry Norris. "Relations Between the Spectra and Other Characteristics of the Stars." *Nature* 93 (1914): 227-230, 252-258, 281-286. The paper in which Russell illustrates the relationship between absolute magnitudes and spectral types.

Schneider, Stephen E., and Thomas T. Arny. *Pathways to Astronomy*. 2d ed. New York: McGraw-Hill, 2008. A college textbook for introductory astronomy courses, divided into many short sections on specific topics. Contains a thorough discussion of the H-R diagram, its interpretation, and its use in the study of stellar evolution.

Sitterly, Bancroft W. "Changing Interpretations of the Hertzsprung-Russell Diagram, 1910-1940: A Historical Note." *Vistas in Astronomy* 12 (1970): 357-366. A historical perspective addressing the evolving nature of this astronomical tool.

Struve, Otto. "The Two Fundamental Relations of Stellar Astronomy." *Sky and Telescope* (August, 1949): 250-252, 262. This is a fine historical article on the spectrum-luminosity relationship discovered by Hertzsprung and Russell.

See also: Archaeoastronomy; Coordinate Systems; Earth System Science; Gravity Measurement; Infrared Astronomy; Main Sequence Stars; Neutrino Astronomy; Optical Astronomy; Radio Astronomy; Telescopes: Ground-Based; Telescopes: Space-Based; Ultraviolet Astronomy; X-Ray and Gamma-Ray Astronomy.

Iapetus

Categories: Natural Planetary Satellites; The Saturnian System

The Saturnian satellite Iapetus is one of the most unusual satellites in the solar system. It has a dark side and a bright side, as well as a ridge along its equator that sits atop an equatorial bulge.

OVERVIEW

Iapetus was first noticed on one side of Saturn by Giovanni Domenico Cassini in October, 1671. Searching for it later on the other side of the planet was futile with the telescope that he was using. He tracked it many times over several years, but only when he had a better telescope in 1705 did Cassini finally see Iapetus on the other side of the planet. He concluded that Iapetus had a dark side and a bright side. It was named for the Titan god Iapetus, a brother of Cronus (the Greek name for the Roman god Saturn) in Greek mythology. Iapetus was originally called Saturn V, because the Saturnian satellites were originally numbered, and many scientists continued to use the numbers. With the discovery of other Saturnian satellites, Iapetus's scientific referent changed to Saturn VII and eventually Saturn VIII. It is one of the sixty satellites of Saturn that had been discovered by 2008. Iapetus's geological features, except for its dark region, are named after characters and places from the French epic poem *The Song of Roland*. The dark region is called Cassini Regio, in honor of Iapetus's discoverer, Cassini.

As Cassini deduced, Iapetus always presents the same face to Saturn, meaning it is synchronous with Saturn. Therefore the time of revolution and the time to circle the planet are identical, 79.32 days. Iapetus is prograde, meaning it turns in the same direction as Saturn. Iapetus's orbit is about 15.5° out of the plane of Saturn's equator; that is, its inclination is 15.5°. The inclination is large enough to cause questions about whether Iapetus was "captured" by Saturn's gravitational field or was generated by the same process that produced Saturn and the other satellites. Iapetus's elliptical orbit (its eccentricity is 0.029) and a synchronous satellite that is 3,561,000 kilometers from the planet also add to the doubt about whether Iapetus was formed by the same process that formed Saturn.

Iapetus is an oblate spheroid with a 1,494-kilometer diameter along the axis pointed at

The Cassini spacecraft captured this image of Iapetus in September, 2007. (NASA/JPL/Space Science Institute)

Uranus. The equatorial axis is 1,498 kilometers, and the pole-to-pole axis is 1,426 kilometers. With its equatorial bulge and squashed poles, Iapetus has been said to look like a walnut. A 20-kilometer high ridge that girds most of the satellite along its equator is another striking feature. No other known satellite has a ridge like the one on Iapetus. The ridge is triangular, with a base that is about 200 kilometers wide. The ridge is also cratered, proving it has been in existence for a long time. One theory about the formation of the ridge is that the crust formed while the interior was still flexible enough for the weight of the shell to crush the interior. The interior material forced its way to the surface, fracturing the shell at the equator and forming a ridge. The material cooled relatively quickly, locking it in the new shape that is Iapetus today. This theory requires that Iapetus had to spin much faster during its formation and cooling periods. The slowing of its spin, called despinning, had to take place near a large object such as a planet. This, along with the fact that Iapetus's composition is similar to that of the other Saturnian satellites, indicates that Iapetus was formed in the process that formed Saturn and the other large satellites.

The first feature of Iapetus that astronomers noted was its albedo, which is completely different on the two sides of this satellite. The pictures from the Cassini spacecraft (2007) show a tar-black leading hemisphere with a bright backside hemisphere. The albedo is about 0.05 for the leading side and about 0.6 for the trailing hemisphere. This variation in albedo is noted not only in the visible range but also in the ultraviolet and radio ranges. The Cassini images indicated that the dark material is not solid but is in streaks large enough to appear solid from a distance. Near-infrared spectra indicate that the bright material is ice. The moon's density of 1.1 grams per centimeter cubed indicates that the rock fraction can be no more than 22 percent. The composition of the dark side appears to be ice contaminated with materials such as ammonia, amorphous carbon, poly-hydrogen cyanide (poly-HCN), and hematite (Fe_2O_3).

Some scientists have theorized that the dark material originates from another satellite, Phoebe. Phoebe has a retrograde orbit, meaning that it is traveling in the direction opposite to that of Saturn's rotation and thus in the direction opposite to Iapetus's orbital motion. Material lost by Phoebe because of its retrograde motion can bombard the front face of Iapetus. However, data from the ultraviolet spectra collected by Cassini show that the composition of Phoebe is not the same as the composition of the dark material on Iapetus. There is the possibility that the dark material might be from Hyperion, since the composition of Hyperion does match that of Iapetus. A second option is that the material from Phoebe changes before it bombards Iapetus. The dark material is thought to be a thin layer containing only a small amount (5 percent) of ice. A third opinion is that the dark material is consistent with an external impact. This would cause the poles of Iapetus to be bright, which they

A landslide is seen in the Cassini Regio portion of Iapetus, imaged by the Cassini orbiter in 2004. (NASA/JPL/Space Science Institute)

are. The bright poles may be bright because of frost. The peak temperature of the dark side of Iapetus was measured at 130 kelvins, warm enough to allow ice to sublime and refreeze at the poles, producing bright areas. The dark material has a reddish color that might indicate organic compounds. Organics would darken with exposure to radiation. Indications are that the material is porous and in the form of fine particles.

The topography of Iapetus was not well known before the Cassini mission. Even pictures from the Voyager 1 mission did not show any feature on the dark side. The bright side was also largely unknown. Iapetus is cratered, but is not saturated with craters. Around the craters were tall, steep, wall-like features called scarps. It appears that the craters on Iapetus may not have retained the full height of the crater rim, which would indicate the lithosphere is not thick or totally solid. The largest crater is 800 kilometers across with a rim topography of 10 kilometers (that is, raised 10 kilometers from the crater floor). Some scientists believe that Iapetus was formed and its shell hardened at a very early time. The surface might be the oldest surface known.

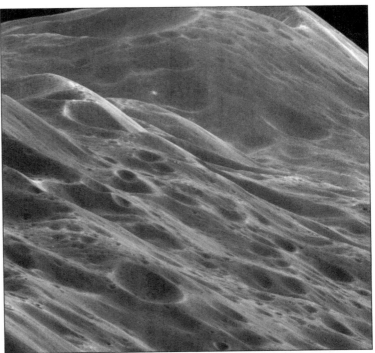

Iapetus's mountainous terrain rises about 10 kilometers above the surface in this Cassini spacecraft image of the moon's equatorial ridge, taken on September 10, 2007. (NASA/JPL/Space Science Institute)

KNOWLEDGE GAINED

Iapetus has been studied with land-based telescopes since it was discovered. Its dichotomous albedo made it an unusual object, and therefore an object of interest. Reflectance spectra taken using the McDonald Observatory produced data that showed reddening of the dark-side material, as dark-side reflectance was compared to that of the bright side. Other land-based studies determined the size of Iapetus, the albedo range, and the longitudinal symmetry of the dark material. One of the most in-depth studies of Iapetus's composition was done from the observatory at Hawaii's Mauna Kea, using both visible and near-infrared spectrometry.

The Voyager 1 mission revealed the diameter of Iapetus. Images of Iapetus were taken and revealed the great amount of cratering that exists on the bright side and at the poles. Perturbations in the path of Voyager allowed scientists to calculate the satellite's density.

In addition to the Image Science Substation (ISS), a version of which was on both Voyagers, the Cassini spacecraft (2004) had an ultraviolet imaging spectrograph (UVIS), which produces simultaneous spectral and spatial images. The visual and infrared mapping spectrometer (VIMS) was used to study the icy satellites by generating reflectance spectra and phase curves, as well as visual pictures. These data are especially important to compare with Earth-based data, providing an evaluation of the Earth-based data and possibly ideas of methods to correct the Earth-based data for two of the problems that occur using Earth-based instruments. Those two problems are the small phase angle seen from Earth and the extra light

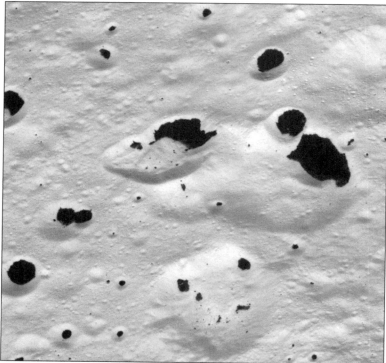

Iapetus's frozen surface, splotched with dark terrain, in an image captured by the Cassini spacecraft on September 10, 2007, at an altitude of about 6,030 kilometers. (NASA/JPL/Space Science Institute)

generated by reflection from the rings and from Saturn. The composite infrared spectrometer (CIRS) on Cassini studied the thermal infrared spectrum for emissivity features.

Certain compounds emit a thermal signature in the infrared range that is noticeable in the background thermal spectra. The thermal spectra from Iapetus did not show any strong features. The lack of emission features and the data from near-infrared spectra caused scientists to believe that the surface must be covered with small particles that have a high porosity. One bit of information gleaned from Cassini images is that scarps and crater walls are bright on their north sides and covered with dark material on their south sides.

CONTEXT

Iapetus is certainly an enigma. The black face on one side and the bright face on the other side remain unexplained. Scientists have ideas for explaining these phenomena, but none that seems to offer a complete answer. If the dichotomy is due to material from other satellites, exactly what is the transport mechanism and what changes to the material occur during transport? A second question about Iapetus concerns its formation. How did the equatorial ridge form and why? What does the existence of the ridge tell scientists about the formation of the solar system? Is the surface of Iapetus the oldest surface in the solar system? Can Iapetus lead scientists to determine how the solar system was formed?

Many scientists are surprised that Iapetus can have such a large angle of inclination and be so far away from Saturn, yet still be in synchronous relationship with the planet. Being synchronous requires a very slow revolution, and such slow revolutions are unusual. Iapetus's revolution can be explained only if one concludes that the gravitational force of Saturn has caused despinning. The gravitational force on Iapetus increases and decreases as Iapetus gets closer and then moves away from the planet, and this change in gravitational force causes tidal flexing in the satellite. The heat from this tidal flexing may explain the heat needed to generate some of the satellite's features, such as the ridge and the dichotomous albedo.

C. Alton Hassell and David G. Fisher

FURTHER READING

Bond, Peter. *Distant Worlds: Milestones in Planetary Exploration*. New York: Copernicus Books, 2007. The author discusses several systems, including the Saturnian system and its parts: planet, satellites, and rings. Also discusses how various interplanetary missions have developed our knowledge of the systems from which they have gathered data. Illustrations, bibliography, appendix, index.

Castillo-Rogez, J. C., et al. "Iapetus' Geophysics: Rotation Rate, Shape, and Equatorial Ridge." *Icarus* 190, no. 1 (September, 2007): 179. Using the better data from Cassini, the authors describe several models for the formation of Iapetus. Each model is then evaluated against the known data. Illustrations, bibliography, index.

Corfield, Richard. *Lives of the Planets.* New York: Basic Books, 2007. The author takes the reader through the different space missions. Organized by planet. Index.

Fendrix, Amanda R., and Candice J. Hansen. "The Albedo Dichotomy of Iapetus Measured at UV Wavelengths." *Icarus* 193, no. 2 (February, 2008): 344. The dual face of Iapetus is discussed in view of measurements taken in the ultraviolet range. The meaning of a UV dichotomy is discussed. Illustrations, bibliography, index.

Hartmann, William K., and Ron Miller. *The Grand Tour: A Traveler's Guide to the Solar System.* 3d ed. New York: Workman, 2005. Each major planet, then the major moons including Iapetus, is discussed. Outstanding illustrations; bibliography, index.

McFadden, Lucy-Ann Adams, Paul Robest Weissman, and T. V. Johnson, eds. *Encyclopedia of the Solar System.* San Diego: Academic Press, 2007. The editors have collected articles written by many experts in one of the best scholarly reference works about the solar system. Illustrations, appendix, index.

Thomas, P. C., et al. "Shapes of the Saturnian Icy Satellites and Their Significance." *Icarus* 190, no. 2 (October, 2007): 573. Discusses how measurements from the Cassini mission reveal the shapes of the Saturnian satellites. Illustrations, bibliography, index.

See also: Callisto; Enceladus; Eris and Dysnomia; Europa; Ganymede; Io; Jupiter's Satellites; Lunar Craters; Lunar History; Lunar Interior; Lunar Maria; Lunar Regolith Samples; Lunar Rocks; Lunar Surface Experiments; Mars's Satellites; Miranda; Neptune's Satellites; Planetary Orbits: Couplings and Resonances; Planetary Satellites; Pluto and Charon; Saturn's Satellites; Titan; Triton; Uranus's Satellites.

Impact Cratering

Category: Planets and Planetology

Space-age discoveries about the surface character of other terrestrial planets and satellites around planets throughout the solar system reveal that the early Earth must have been heavily scarred by impacts with planetesimals and minor bodies. Erosion processes and plate tectonics have obliterated most of these ancient craters, but new evidence that major impacts may have had a significant role in shaping the evolution of life has spurred a search for large impact craters.

OVERVIEW

Impact cratering is one of the most fundamental geologic processes in the solar system. Craters have been found on the surfaces of all the solid planets and natural satellites thus far investigated by spacecraft. Mercury and the Moon, bodies whose ancient surfaces have not been reworked by subsequent geologic processes, preserve a vivid record of the role that impact cratering has played in the past. It is inconceivable that the Earth somehow escaped the bombardment that caused such widespread scarring or that it does not continue to be a target for planetesimals still roaming the solar system.

As recently as the 1960's, only a handful of sites on Earth were accepted to be of impact origin. In the early 2000's, the number of confirmed astroblemes (circular surface features considered to have been large impact craters) was well in excess of one hundred and increasing at the rate of several per year. In addition, many "probable" and "possible" impact features are under study. Nevertheless, an enormous discrepancy exists between the number of identified or suspected impact sites on Earth and the number that might be expected.

It is assumed that the flux of incoming bodies is the same for Earth as it is for the Moon. Making allowances for the fact that the Earth is the largest "target" of any of the terrestrial planets and that more than two-thirds of its surface is covered by water, planetologists calcu-

A view of the Moon's Mare Imbrium, taken by the Apollo 17 crew in 1972, shows the cratered lunar surface. (NASA)

sources. In general, the number of large impact sites found in the well-explored areas of the Earth agrees with the accepted rate of crater formation on the other terrestrial planets in the past two billion years.

The obvious difference between the surface appearances of Earth and the Moon is explained not by any difference in the rate at which impact craters have formed but in the rate at which they are destroyed. Most of the tremendous numbers of craters on the Moon are more than 3.9 billion years old, while the Earth's oldest surviving astroblemes were formed less than 2 billion years ago. Studies have shown that erosion effectively removes all traces of a 100-meter (diameter) crater in only a few thousand years, and that a 1-kilometer-wide crater, such as the well-known Barringer meteor crater in Arizona, will disappear within a million years. Only craters with diameters greater than 100 kilometers can be expected to leave any trace after a billion years of erosion. This explains not only the absence of widespread cratering on Earth's landscape but also the fact that, among the astroblemes known to exist, medium and large scars are more common than small ones.

Significant craters can be produced only by objects having masses of hundreds of thousands to billions of tons. The Barringer crater, 1.2 kilometers wide and 200 meters deep, is believed to have been formed by a one-million-ton planetesimal that was perhaps 50 meters in diameter. A 27-kilometer-wide astrobleme known as the Nördlinger Ries crater in Germany required an impacting body greater than 1 kilometer in diameter with a mass in excess of 1 billion tons. Planetesimals as large as these two examples are not characteristic of the vagrant meteors that wander through the solar system and occasionally streak into the Earth's skies as shooting stars.

late that land areas of the Earth should have been scarred by at least fifteen hundred craters 10 kilometers or more in diameter. In actuality, only about half of the known astroblemes are in this size range. On a global scale, 99 percent of the predicted large impact craters seem to be missing. However, this statistic is not a valid indicator of the impact history of the Earth. Although the impact phenomenon is a geographic process, the probability for discovering impact sites is strongly modified by the geologic stability of various regions of the Earth and by the intensity of the search programs in those areas. Roughly one-half of all the confirmed astroblemes have been found in Canada, which constitutes only one percent of the Earth's surface. In part, this is owing to the stability of Precambrian rock of the Canadian Shield, which thus preserves more of the crater's features, but it also reflects a diligent research effort by Canada's Department of Energy, Mines, and Re-

Most of the past impacts on Earth and the Moon appear to be attributable to a family of asteroids known as the Apollo-Amor group (after two specific members of the family). Members of this group are in orbits that graze Earth's orbit and become subject to orbital perturbations that lead them across Earth's path periodically. It is estimated that the average Apollo-Amor object intersects Earth's orbit once every five thousand years, although usually the planet is at some other point on its orbit when this happens. The probability of a collision between Earth and any given Apollo-Amor object is small, but several studies have shown that this family contains between 750 and 1,000 asteroids larger than 1 kilometer in diameter. Statistical analysis suggests that such sizable bodies must collide with the Earth an average of once every 600,000 years.

Impact events involve tremendous transfers of energy from the incoming planetesimal to Earth's surface. A projectile's energy of motion increases only linearly with its mass but as the square of its velocity, so surprisingly large craters result from relatively small bodies traveling at hypervelocities. Depending on the directions of motion of Earth and of the planetesimal, impacts on the planet may involve relative velocities as high as 50 kilometers per second. At velocities surpassing 4 kilometers per second, the energy of the shock wave created by the impact is far greater than the strength of molecular adhesion for either the planetesimal or Earth. Therefore, on impact the planetesimal acquires the properties of a highly compressed gas and explodes with a force equivalent to a similar mass of blasting powder.

The shock wave from this explosion intensely compresses the target material and causes it to be severely deformed, melted, or even vaporized. In all but the smaller impacts, the entire projectile is also vaporized. The shock wave swiftly expands in a radial fashion, pulverizing the target material and intensely altering the nature of the target rock by extreme and almost instantaneous heat and pressure. This is im-

mediately followed by decompression and what is called a rarefaction wave that restores the ambient pressure. The rarefaction wave moves only over free surfaces, so it travels outward over the ground surface and into the atmosphere above the impact. It becomes the excavating force that lifts vast quantities of the pulverized target material upward and outward to create the crater cavity.

The rarefaction wave excavates a hole whose depth is one-third of its diameter and whose profile follows a parabolic curve, but this depression is short-lived and is therefore called the transient cavity. After passage of the rarefaction wave, a large amount of pulverized target material from the walls of the transient cavity slumps inward under gravity, and some of the ejecta lofted straight up into the atmosphere falls back into the excavation. Together, these sources contribute to a lens-shaped region of breccia that fills the true crater's floor and leaves a shallower, flat-floored apparent crater as the visible scar of the impact. Apparent craters generally exhibit a depth of only one-tenth to one-twentieth of their diameters. Meanwhile,

Earth's Largest Impact Craters

Diameter (km)	Location	Crater Name	Age (millions of years)
300	South Africa	Vredefort	2,023
250	Canada	Sudbury	1,850
170	Mexico	Chicxulub	65
100	Canada	Manicougan	214
100	Russia	Popigai	35
90	Australia	Acraman	590
90	United States	Chesapeake Bay	36
80	Russia	Puchezh-Katunki	175
70	South Africa	Morokweng	145
65	Russia	Kara	73
60	United States	Beaverhead	600
55	Australia	Tookoonooka	128
54	Canada	Charlevoix	357
52	Sweden	Siljan	368
52	Tajikstan	Kara-Kul	5

Source: Data are from the National Aeronautics and Space Administration/Goddard Space Flight Center, National Space Science Data Center.

the rarefaction wave carries ejecta particles outward over the surrounding landscape, where they fall to Earth as a blanket of regolith that is distinguishable from the local target rock by the effects of shock metamorphism.

METHODS OF STUDY

Impact phenomena are rare enough on the human timescale that no crater-forming events are known to have occurred in recorded history. Owing to this passage of time and to the fact that most existing astroblemes have been severely altered by erosion, impact cratering has been studied by the unique modifications that a powerful impact shock makes in the rocks and minerals at the site. Scientists study the deformation and structural damage to buried strata, and by looking for the presence of certain rare elements and minerals in the sediments surrounding suspected impact sites.

Much attention has been given to the effects of the shock wave on terrestrial rocks, since shock metamorphism is considered to be the most enduring and positive identifier of ancient astroblemes. Shock metamorphism differs from endogenic metamorphism by the scales of pressure and temperature involved and by the very short duration of the exposure to those pressures and temperatures. Endogenic metamorphism usually involves pressures of less than 1 gigapascal (100,000 atmospheres) and temperatures not greater than 1,250 kelvins. The pressures involved in shock metamorphism are exponentially greater, reaching several hundred gigapascals for an instant in the vicinity of the impact. Rock exposed to pressures in excess of 80 gigapascals and temperatures of several thousand kelvins is immediately vaporized. Lesser pressures and temperatures at increased distances from the point of impact produce signs of melting, thermal decomposition, phase transitions, and plastic deformation.

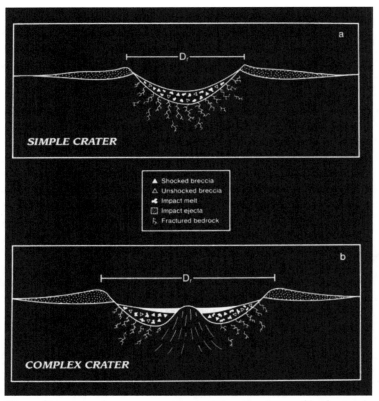

These diagrams show cross sections of the structures of both simple and complex craters. (NASA)

Pockets of melt glass up to several meters thick are commonly found in the breccia within the crater, indicating that pressures there reached 45-60 gigapascals. Coesite and its denser relative, stishovite, are forms of quartz that occur naturally only at impact sites. Shatter cones, conically shaped crystals created at pressures of from 2 to 25 gigapascals, are another prominent feature of shock metamorphism and are particularly well developed in fine-grained isotropic rock. Microscopic examination of impact-shocked porous rock reveals that quartz grains are deformed so as to fill the pores and interlock like the pieces of a jigsaw puzzle. Even at a considerable distance from the impact point, quartz grains tend to be elongated in the direction of the shock wave's passage.

Theories concerning cratering dynamics can also be tested by analogy to some of the craters produced by the detonation of nuclear devices.

This latter technique has adequately explained the morphology of the smaller astroblemes, those with diameters that do not exceed 2-4 kilometers. Larger impact events involve additional dynamics that are not mimicked by nuclear devices thus far tested. Astroblemes greater than 2 kilometers in diameter in sedimentary rock or 4 kilometers in diameter in crystalline rock display a pronounced central uplift owing to an intense vertical displacement of the strata under the center of the impact. An additional feature distinguishing complex craters is that their depths are always a much smaller fraction of their diameters than is the case with simple craters.

Photographic imaging of Earth from space has revealed some young and well-preserved astroblemes in remote and poorly explored areas of Earth, such as the Sahara Desert. More important has been the satellite's ability to reveal structures that still preserve a faint but distinct circularity when seen from orbit, although at ground level they are so eroded that their circularity has escaped detection. One of the largest astroblemes yet discovered was detected from Landsat satellite images in this way. New imaging technologies, including advanced radar and sonar mapping, promise to extend the capabilities of space surveillance and remote sensing in recognizing possible impact sites.

CONTEXT

The degree to which the Earth is in danger of being struck by a massive planetesimal began to be appreciated about the middle of the twentieth century. In 1980, a team led by Nobel Prize-winning physicist Luis Alvarez announced dramatic evidence suggesting that an asteroid impact that occurred 65 million years ago created such planetary stress that it explained a mysterious massive extinction of life-forms known to have occurred on the Earth at that time. At sites all around the world, researchers discovered that clay deposits at the boundary layer between the Cretaceous and Tertiary periods contained up to one hundred times the normal abundance of the metal iridium, which is rare in Earth's crustal rocks but 1,000 to 10,000 times more abundant in the makeup of many asteroids. This Cretaceous-Tertiary boundary layer is coincident with the point at which fully 70 percent of the life-forms then existing on the Earth, including the dinosaurs, became extinct. Further study has also revealed that this same sediment layer is rich in shock-metamorphosed quartz grains, known only to occur naturally from impact explosions.

Examples of impact craters on Venus, Earth's moon, Mars, and Ganymede. (Venus and Moon: Robert Herrick/Lunar and Planetary Institute; Mars: Calvin Hamilton/Los Alamos National Laboratory; Ganymede: Paul Schenk/Lunar and Planetary Institute)

Debate continues as to whether an asteroid impact was the primary cause of the mass extinctions at the close of the Cretaceous period or merely the final factor, but there is general agreement that a colossal impact occurred at that time. The volume of material represented in the boundary sediments suggests that the planetesimal was perhaps 10 kilometers in diameter and would have created a crater of as much as 200 kilometers in width. An astrobleme in the Gulf of Mexico near Belize, called the Chicxulub Crater, closely fulfills these criteria. Many scientists accept it as the impact site for the K-T (German for Cretaceous-Tertiary) event. Meanwhile, several other iridium spikes (abnormally high concentrations of the metal) have been found in the sedimentary beds coinciding with other recognized mass extinctions.

Early in the twenty-first century several researchers put forward candidate craters to mark an impact event dated to the time of what paleontologists often call the Great Dying. At the boundary between the Permian and the Triassic (the P-T boundary), which also marks the end of the Paleozoic and the start of the Mesozoic era, life on Earth was very nearly exterminated. A conservative estimate is that 95 percent of all species died out at that time. Life rebounded and the dinosaurs went on to rule the Earth, until they too were wiped out catastrophically. Of the various craters proposed to have resulted from a P-T boundary impact event, the one that appears most likely to turn out to be correct (if any of them are correct) is a crater located in Antarctica, buried unfortunately under 1.5 kilometers of ice. What provides extra confidence that this crater could be the result of a P-T boundary impact event is the fact that at its antipode is located the Siberian Traps. Energy from the impact would have undergone antipodal focusing through the Earth's core to ravage the area on the planet 180° away from the impact site. The Siberian Traps experienced tremendous amounts of volcanic activity around 248 million years ago, the very time of the P-T boundary and the Great Dying. This scenario remains controversial but, if true, would represent an even larger impact event than the accepted K-T boundary event that gave rise to the Chicxulub Crater.

Three related discoveries suggest the possibility that impact cratering may not be an entirely random process, so far as its distribution through time is concerned. Paleontologists David Raup and J. John Sepkoski, Jr. have shown evidence, based on a rigorous analysis of the marine fossil record, that mass extinctions appear to occur with regularity every 26 million years. Independently, the team of Walter Alvarez (also a member of the team that discovered the K-T iridium anomaly) and Richard Muller have discovered evidence that the ages of major known terrestrial astroblemes seem to be periodically distributed at intervals of roughly 28 million years. For some time, researchers have sought a mechanism that could account for the numerous polarity reversals in Earth's magnetic field over geologic history, and some have suggested that major impact events may be the cause. Several studies have reported an apparent fine-scale periodicity in Earth's magnetic field reversals with a cycle of 30 million years. Although the intervals are not in perfect agreement, they are very close, considering the difficulty of precisely dating extinctions and the exact ages of astroblemes.

These discoveries suggest that there may be an as yet undiscovered member of the solar system that moves in such a way as periodically to disrupt the Oort Cloud, the cloud of comets believed to exist on the fringes of the solar system, causing a barrage of planetesimals to descend upon the inner planets. Although the existence and location of such a body remain speculative and controversial, it has been characterized as a dwarf companion star of the Sun and is called Nemesis.

Richard S. Knapp

FURTHER READING

Consolmagno, Guy. *Worlds Apart: A Textbook in Planetary Sciences.* Englewood Cliffs, N.J.: Prentice Hall, 1994. A text accessible to college-level science and nonscience majors alike. Presents most topics using low-level mathematics; involves integral calculus where required. Demonstrates how the area of planetary science progresses by questioning previous understandings in the light of new observations.

De Pater, Imke, and Jack J. Lissauer. *Planetary Sciences*. New York: Cambridge University Press, 2001. A challenging and thorough text for students of planetary geology. Covers extrasolar planets and provides an in-depth, contemporary explanation of solar-system formation and evolution. An excellent reference for the most serious reader with a strong science background.

Dixon, Dougal. *The Practical Geologist: The Introductory Guide to the Basics of Geology and to Collecting and Identifying Rocks*. New York: Fireside, 1992. A beginner's guide to the physical processes that formed the Earth and modified its surface over geologic time. Heavily illustrated with guides for rock and mineral identification.

Encrenaz, Thérèse, et al. *The Solar System*. New York: Springer, 2004. A thorough exploration of the solar system from early telescopic observations through the space missions of 2003 that have investigated all planets. Takes an astrophysical approach to place the solar system in a wider context, as just one member of similar systems throughout the universe.

Faure, Gunter, and Teresa M. Mensing. *Introduction to Planetary Science: The Geological Perspective*. New York: Springer, 2007. Designed for college students majoring in Earth sciences, this textbook provides an application of general principles and subject material to bodies throughout the solar system. Excellent for learning comparative planetology.

Grieve, Richard A. F. "Terrestrial Impact Structures." *Annual Review of Earth and Planetary Sciences* 15 (May, 1987): 245-270. A thorough summary of what is known about the cratering process on the Earth, written by a leading authority on the subject. It is intended for the scientific reader, but its illustrations, extensive bibliography, and introductory and summary sections are of value even to those who are not familiar with the concepts and terminology in the body of the article.

Hartmann, William K. "Cratering in the Solar System." *Scientific American* 236 (January, 1977): 84-99. Dated, but a comprehensive explanation of the role attributed to impact cratering in shaping the surfaces of all of the terrestrial planets. The author explains the basis for estimating the frequency of impacts for various sizes of planetesimals and the logic behind using crater counts to estimate the ages of planetary surfaces. The article also explains the theory that the first half billion years of solar-system history involved an extremely heavy bombardment of all the inner planets.

_____. *Moons and Planets*. 5th ed. Belmont, Calif.: Thomson Brooks/Cole, 2005. An updated version of a classic text on planetary science. Includes chapters on all planets and their systems. Discusses the role of impact cratering in shaping planets and their satellites.

Kerr, Richard A. "When Disaster Rains Down from the Sky." *Science* 206 (November 16, 1979): 803-804. Taking a descriptive approach easily comprehended by laypersons, this article summarizes research by several investigators attempting to compute the frequency with which the Earth is struck by crater-forming bodies. The article places particular emphasis on the Apollo asteroid group and examines suggestions that the Apollo family is supplied with new planetesimals by the decay of former comets.

Morrison, David, and Tobias Owen. *The Planetary System*. 3d ed. San Francisco: Pearson/Addison-Wesley, 2003. Planetary atmospheres are treated as important physical features of the various members of the Sun's family. They are discussed individually in the context of what is known about each planet's characteristics and with regard to theories about their evolution and the evolution of the entire solar system. Geared for the undergraduate college student.

Muller, Richard. *Nemesis: The Death Star*. New York: Weidenfeld & Nicolson, 1988. Despite its tabloid title, this is an excellent discussion of the chain of discoveries leading to the Nemesis theory by the Berkeley physicist who developed it. The volume is organized in two parts: The first recaps the evidence for a major impact at the K-T boundary, and the second tells how further research led Muller to postulate the existence of Nemesis. Intended for lay readers, the book gives insight into how the scientific discovery process works, as well as explaining the theory.

Murray, Bruce, Michael C. Malin, and Ronald Greeley. *Earthlike Planets*. San Francisco: W. H. Freeman, 1981. Although terrestrial impact craters are not specifically discussed, the impact mechanics that produce craters are presented here in terms that are suitable for general readers. A somewhat dated but nevertheless excellent discussion of cratering as a ubiquitous aspect of the surfaces of all the inner planets.

Raup, David M. *The Nemesis Affair*. New York: W. W. Norton, 1986. The author is a significant figure in the field of paleontology and has done leading research on the apparent periodicity of extinctions and magnetic reversals. His narrative is a fascinating personal account of the ideas and the individuals who led the scientific community from extreme skepticism to general acceptance that impact "catastrophism" may have played a major role in the Earth's evolution and its life-forms.

See also: Lunar Craters; Lunar History; Lunar Maria; Mars's Craters; Meteorites: Achondrites; Meteorites: Carbonaceous Chondrites; Meteorites: Chondrites; Meteorites: Nickel-Irons; Meteorites: Stony Irons; Meteoroids from the Moon and Mars; Meteors and Meteor Showers; Venus's Craters.

Infrared Astronomy

Category: Scientific Methods

Infrared astronomy explores the universe by focusing on wavelengths of the electromagnetic spectrum that are longer than those of visible light. This region of the spectrum is useful for studying the process of star formation, for studying objects that are obscured by clouds of interstellar material, and for studying lower-temperature objects that do not radiate as prevalently in the visible portion of the spectrum.

OVERVIEW

Infrared astronomy focuses its study on wavelengths of electromagnetic radiation that are a little longer than those of visible light. The infrared region of the spectrum covers a wide range of wavelengths from waves slightly longer than those of visible light (0.7 micron, or 0.7 millionth of a meter) to those as long as 1,000 microns. The longest infrared wavelengths are about 1 millimeter in length and mark the boundary with the microwave radio spectrum.

Infrared radiation from distant sources is very difficult to detect. The Sun is so close that the infrared radiation it emits can be detected in the form of heat. The Moon also emits easily detected infrared radiation. However, to detect emissions from other stars, planets, nebulae, or galaxies, very sensitive detectors are needed. The shortest infrared waves are known to astronomers as the "photographic infrared" because they are very similar to visible light and can be detected with certain types of photographic emulsions and other types of optical detectors. At longer wavelengths of infrared radiation, objects can be detected that are not visible at optical wavelengths. Nevertheless, at these wavelengths, the detectors used for the photographic infrared are no longer useful.

Modern infrared detectors often use a substance called indium antimonide, which changes its electrical conductivity when exposed to infrared radiation. In order to be effective, however, it must be kept very cold. Solid nitrogen or liquid helium is used to surround the material to bring its temperature from 50 kelvins down to within a few kelvins of absolute zero. Another long-wavelength infrared detector uses a crystal of the semiconducting material germanium that contains traces of the rare metal gallium. This detector must be kept to a temperature only 2 kelvins above absolute zero.

Earth's atmosphere provides advantages as well as disadvantages to infrared astronomy. Some infrared observations can be done during the day as well as at night, allowing infrared detectors to be mounted on large optical telescopes for daytime use. The disadvantage posed by the atmosphere is that water vapor and carbon dioxide absorb certain wavelengths of infrared radiation, making them invisible to astronomers.

The Spitzer Space Telescope views objects in the infrared wavelength range; this artist's conception shows the spaceborne observatory against the sky as it would appear in the infrared, the bright band being the Milky Way. (NASA/JPL-Caltech)

Infrared astronomers therefore have to choose particular wave bands at which the atmosphere allows a clear window. To see through these windows is often a challenge, as common objects—such as telescopes and even the sky—can radiate at these same wavelengths if they are at the appropriate temperature.

Infrared astronomers have designed ways of partially overcoming the problems posed by the atmosphere. Infrared instruments are designed so that no stray radiation from the instrument itself can enter the detectors. To overcome sky brightness in the infrared, astronomers take measurements of the observational target. Measurements include the infrared brightness of both the object and the sky. The telescope is then moved slightly so that it is no longer point-ing at the source, where it takes an infrared measurement from the background sky only. When the second measurement is subtracted from the first, it is possible to determine the brightness of the object itself. This technique works well for stars but works less well for objects such as nebulae, which cover a wider field of view. The technique can be modified to scan a wider portion of the sky, making images of larger areas possible. Detectors have been developed that can record such images in a single exposure.

The ideal earthbound infrared observatories are located at very high altitudes and in arid atmospheric conditions. The best site is on Mauna Kea in Hawaii, 4,200 meters above sea level, where two of the world's largest infrared telescopes reside: the National Aeronautics and Space Administration's (NASA's) 3-meter-diameter telescope and the United Kingdom Infrared Telescope (UKIRT), with a 3.8-meter-diameter mirror.

At wavelengths longer than about 30 microns, the atmosphere begins to absorb so much infrared radiation that ground-based observation is impossible. To observe these longer (or far-infrared) wavelengths, astronomers conduct their observations remotely. In the 1970's, ten rocket flights carrying infrared detectors performed a survey of nine-tenths of the sky. During these early flights, it was discovered that the center of the Milky Way and other galaxies were strong sources of far-infrared radiation. High-altitude balloons have also been used to make observations; NASA converted a C-141 transporter plane into a flying infrared observatory, the Kuiper Airborne Observatory, complete with a 0.9-meter telescope. The observatory carries scientists to altitudes of about 12,500 meters, where they can make observations free of about 99 percent of the atmosphere's water vapor.

These types of observations are valuable, but the best way to solve the observational problems posed by the atmosphere is to observe outside the atmosphere completely. Although astronomers have flown many satellites to measure other types of radiation from space, the infrared band has presented difficulties because of the necessity of keeping the detectors at extremely low temperatures. In 1983, a fully dedicated infrared satellite was finally launched. The Infrared Astronomical Satellite (IRAS) was a joint project by the United States, the Netherlands, and England. IRAS investigated the sky from an orbital altitude of 900 kilometers. Throughout its development, this mission proved to be one of the most difficult ever attempted. The infrared detectors had to be designed so that even in orbit they were cooled to within a few degrees of absolute zero with nearly 90.7 kilograms of liquid helium. The lifetime of the satellites was limited because the liquid helium slowly boils away. IRAS was able to function efficiently for a total of ten months. The principal instrument aboard IRAS was an array of sixty-two semiconductors that were sensitive to the majority of the infrared spectrum. The satellite was roughly the size of a small automobile and weighed 1,076 kilograms.

In spite of the complexity of keeping the instruments cold, the mission was highly successful. IRAS scanned 95 percent of the sky a total of four times at the middle and far-infrared wavelengths. It was able to detect and catalog about 250,000 celestial sources of infrared radiation. IRAS produced such an enormous catalog of infrared sources, significantly expanding the field of infrared astronomy. Numerous space-based detectors and specially dedicated observatories followed. One example of an instrument on an observatory primarily taking data in the

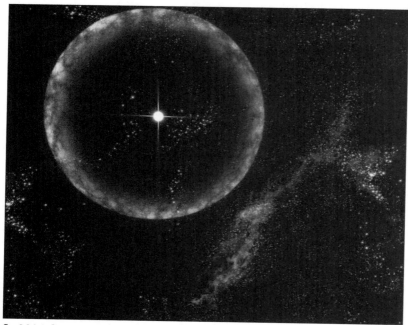

In 2004 this magnetar, a blast of energy from a flare caused by the twisting magnetic field of a neutron star, was so bright that it lit up Earth's upper atmosphere. (NASA)

visible was the Near Infrared Camera and Multi-Object Spectrometer (NICMOS) installed on the Hubble Space Telescope (HST) by shuttle astronauts in 1997 during the STS-82 servicing mission. Two examples of dedicated observatories that went far beyond the early discoveries of IRAS were the Space Infrared Telescope Facility (later renamed the Spitzer Space Telescope), a member of NASA's Great Observatories program, and ESA's Infrared Space Observatory (ISO).

The European Space Agency (ESA) launched the Infrared Space Observatory (ISO) on November 17, 1995, and placed it in an elliptical orbit ranging from as close to the surface of the Earth as 1,000 kilometers to as high above the surface as 70,500 kilometers; this gave ISO a twenty-four-hour orbital period. ISO was designed to detect infrared radiation ranging from 2.5 to 240 microns in wavelength. To achieve that, ISO was equipped with four separate scientific instruments cooled by liquid helium. The instruments were disturbed by energetic particles when ISO dropped through and rose out of Earth's Van Allen radiation belts, but fortunately the observatory spent 70 percent of each orbit well beyond those disruptive belts. The ISO mission concluded on May 16, 1998, after the cryostat's helium had boiled off, thereby raising the temperature of the instruments sufficiently high to render them useless for infrared measurements.

The Hubble Space Telescope was placed into orbit from the space shuttle Discovery on mission STS-31 on April 25, 1990. One of NASA's four Great Observatories, Hubble was designed to be adaptable and have rotating instruments in its science bay to conduct investigations from part of the infrared through the entire visible portion of the electromagnetic spectrum to portions of the ultraviolet. As first deployed, Hubble suffered from a precise but inaccurate optical prescription for its main mirror. Astronauts on STS-61 in December, 1993, installed a corrective device called Corrective Optics Space Telescope Axial Replacement (COSTAR) which brought incoming light into proper focus, saving Hubble from an otherwise dismal outcome. On a second servicing mission, STS-82, in February, 1997, astronauts removed an instrument

from Hubble and replaced it with the Near Infrared Camera and Multi-Object Spectrometer (NICMOS), designed to perform infrared studies involving wavelengths of 0.8 to 2.4 microns.

NICMOS was outfitted with a unique thermal management system making use of a block of solid nitrogen to keep the instruments cooled to about 40 kelvins. Unfortunately, after NICMOS was incorporated into HST in space and afterward put through early commissioning activities and subsequent science observations, it was clear that a heat leak had developed that would raise the solid nitrogen's temperature to a point, after two years or less, where it would no longer provide sufficient cooling for the sensors in the instruments to produce accurate results. Meanwhile, scientists and engineers developed a mechanical cryocooler that would in turn be installed on NICMOS during the next shuttle servicing mission to HST. It was capable of monitoring temperatures between 75 and 86 kelvins, low enough for the instruments to function. In March, 2002, astronauts on space shuttle mission STS-109 saved NICMOS. Further repairs to NICMOS became necessary, and plans were made for another mission to effect those repairs and allow HST to continue front-line infrared observations before its planned replacement by the James E. Webb Space Telescope.

The last member of NASA's Great Observatories, the Space Infrared Telescope Facility (SIRTF), was launched into a solar orbit on an expendable Delta II booster on August 25, 2003 (thirteen years to the day after Hubble was placed in Earth orbit), rather than being deployed from a space shuttle by astronauts. After launch the observatory was renamed the Spitzer Space Telescope (SST) in honor of astronomer and longtime proponent of placing large telescopes in orbit Lyman Spitzer. The Spitzer telescope was designed to observe infrared radiation from 3 to 180 microns for at least five years, being outfitted with an infrared camera (operating from 3 to 180 microns), an infrared spectrometer (operating between 5 to 40 microns), and far-infrared detector arrays. To achieve these goals, the observatory was cooled by liquid helium to 5.5 kelvins, aided greatly by passive cooling with a sunshield. Spitzer was going strong after five years, with much of its liquid helium remaining.

The James E. Webb Space Telescope (JWST) is designed to be set up in an operation position at the L2 Lagrangian point, a spot 1.5 million kilometers from Earth where the gravitational influence on the observatory from Earth will be balanced by that from the Sun. This observatory, fully devoted to infrared astronomy, is designed to detect emissions from 0.6 to 28 microns. JWST's design is such that it will be half as massive as Hubble, but with a folded optical system of 18 individual hexagonal segments with a light collecting area six times greater than HST once fully deployed. To achieve its intended mission and remain cooler than 40 kelvins, JWST is desiged to rely primarily on sunshields. Launch of JWST is targeted for 2013 by the European Space Agency atop an Ariane V booster; JWST will be a NASA observatory with ESA instruments, launched by ESA in exchange for time using the telescope. If successfully set in place at the L2 spot, JWST will begin to fulfill four primary objectives: (1) search for light from the earliest stars and collective structures formed in the first hundreds of millions of years after the big bang; (2) produce images and data that will assist in expanding our understanding of galactic formation and subsequent evolution; (3) produce images and data that will assist in expanding our understanding of planetary formation and subsequent evolution; and (4) search for the existence of organic material essential for the development of life in the universe.

APPLICATIONS

Infrared radiation can give astronomers valuable information about the formation of stars. Stars are believed to be formed from large clouds of rotating dust and gas that condense under their own gravity. Energy released in the collapse causes the forming star, or "protostar," to increase in temperature until nuclear reactions begin. It is not until the star "turns on" in this way that it begins to emit radiation in visible wavelengths. As a result, the process of star formation is difficult to study optically. As the star begins to shine, newly created energy warms the surrounding dust, which radiates

the energy away as infrared waves. The process is still not understood completely, and astronomers have learned much by the study of infrared and radio wavelengths.

Many infrared sources are clouds of dust heated by a nearby star. Infrared stars generally are either very young or very old stars, those that are associated with dust clouds. One of the early infrared discoveries was of a giant cloud of gas and dust in the constellation of Orion: the Kleinmann-Low nebula, named for its discoverers. It was found to have a mass greater than two hundred times that of the Sun, yet it is invisible at optical wavelengths. In the infrared, it outshines the Sun more than 100,000 times. It was determined to be a relatively close area of active star formation (within 1,600 light-years). Detailed studies of the Kleinmann-Low nebula in the infrared and radio bands indicate that it contains a number of young stars and clouds of dust and gas that may be in the process of collapsing to form new stars. By studying this nebula, astronomers are learning more about the process of star formation. ISO detected such "stellar nurseries" in the Milky Way and other galaxies, where, its data suggest, star formation occurs at a higher rate than astronomers expected.

One of the most exciting discoveries made by IRAS was a disk of dust grains around the star Vega. Scientists believe this disk of material may be remnants of the dust cloud from which the star formed. The theory in 1990 of planet formation suggests that a similar but smaller disk of material around the Sun provided the raw material from which Earth and other planets were formed. If the disk of material around Vega follows the same pattern, it could eventually form asteroid or planet-sized bodies. The findings from IRAS suggest that such material is common around other stars as well.

A year after the disk of material was found around Vega, a small companion object was found orbiting a faint star. The object was between thirty and eighty times the mass of Jupiter. It was too small to sustain nuclear reactions, as a star would, and some astronomers suggested initially that this object heralded the discovery of the first planet outside our solar system. It was theorized that the object was a

This infrared image of the Carina nebula, taken by the Spitzer Space Telescope, when compared with the same view in the optical range (inset) shows a stark difference in the phenomena that can be detected. The infrared image shows dust pillars where stars are forming at the "tips." These features are obscured in the optical image because of interstellar dust that absorbs visible light. (NASA/JPL-Caltech/N. Smith, University of Colorado at Boulder)

brown dwarf, an object between a star and a planet. It was an important discovery, as astronomers were finding there may be many more brown dwarf-type objects than expected. It was thought that, if these objects outnumbered visible stars, astronomers' theories regarding the amount of matter in the universe (and its eventual fate) would be in need of revision. In due course, however, it was realized that brown dwarfs could not account for all the missing mass in the universe, even as more brown dwarfs were being found by the Spitzer Space Telescope.

IRAS examined many peculiar galaxies, one of which is a galaxy known as Arp 220. IRAS found that the galaxy was emitting eighty times more energy in the infrared than in all other wavelengths. Although the object is not excessively bright at optical wavelengths, its infrared brightness would make it about as energetic as some quasars. (Quasars are extremely powerful, bright sources of energy located in a very small area at the center of a galaxy that outshine the entire galaxy around them.) Arp 220 is actually two galaxies that are colliding. While the individual stars of the galaxies are not likely to collide, huge clouds of dust and gas would collide, generating shock waves and heat by compression. This energy would be radiated in the infrared.

Researchers using IRAS employed a very rigorous observational screening process to weed out any stray infrared detections caused by charged particles. They screened out all but the sources that remained stationary over time and were repeatable. This method of observation lent itself to the discovery of some fast-moving objects that were eliminated because they moved too quickly from one observation to the next. In studying the rejected observations, scientists discovered a comet in 1983—the IRAS-Araki-Alcock—named for the satellite and for G. Iraki and E. Alcock, the independent discoverers. The comet passed closer to the Sun than any other comet in the last two hundred years, and IRAS was able to study it in detail, along with other ground-based observations. In total, six comets were discovered by the satellite, and five other known comets were studied.

Among the most important discoveries made by scientists using ISO was the signature of water around planets within the solar system and in regions of stellar formation. With regard to the latter, such a signature had been hidden from detection by the presence of dust within such forming star systems. With regard to the former, ISO determined that as much as 10 kilograms per second of water "rain" down in the upper atmosphere of Jupiter, Saturn, and Uranus. ISO's data did not answer the question as to the origin of that water. Water was also found in the thick atmosphere of Titan, one of Saturn's satellites—a tantalizing result that Cassini-Huygens scientists eagerly hoped to verify and investigate further when that probe entered the Saturn system beginning in July, 2004. Cassini found evidence of water clouds in Saturn's lowest cloud deck at a distance about 130 kilometers under the tropopause. At that atmospheric level, the local temperature is near the freezing point of water. As for the water in the gas giant planets' upper atmosphere, one leading hypothesis for the source was influx of small cometary nuclei.

ISO was also used for extragalactic studies. For the first time evidence of dust was con-firmed for the otherwise rather empty space between galaxies. One particularly outstanding extragalactic finding from ISO was that intergalactic dust in a large group of more than five hundred galaxies clustered together within the constellation Coma Berenices was heavily concentrated toward the cluster's center. Determining intergalactic space is laden with very low-density dust concentrations held implications for cosmological models.

NICMOS provided a means for HST to observe celestial objects in infrared, which could be then contrasted with images of the same sources taken in visible light. One major comparison involved deep-sky observations of distant dim galaxies. The visible light images and infrared images of the same tiny areas of the sky were combined to indicate the differences between different classes of galaxies. The false-color scale used in making these survey images indicated galaxies with strong infrared emissions as reddish, while galaxies glowing more strongly in visible light appeared bluish. These sorts of survey images included blue dwarf galaxies, red elliptical galaxies, and spiral disk galaxies. Examination of those images provided insights into the populations of dust-obscured galaxies at the earliest times in galaxy formation, shortly after the big bang.

Closer to home, NICMOS was used to examine Uranus's cloud features. In visible light, Uranus reveals little of its atmospheric structure. Using infrared, however, HST's NICMOS instrument found as many as twenty clouds near a bright band in the planet's atmosphere. Wind speeds in the region were determined to be between 300 and 500 kilometers per hour.

Naturally, infrared astronomers expected great performance from the Spitzer Space Telescope, but they were pleased beyond those preliminary expectations when initial images taken by Spitzer of the dust disk surrounding a forming star, the glow of a stellar nursery, and the swirling dust in a large galaxy revealed tremendous detail. The spectroscopic capability of Spitzer revealed the signature of organic material. SST was the first observatory to detect light directly from extrasolar hot Jupiters (specifically HD 209458b and TrES-1). Views of our neighboring galaxy, the Andromeda galaxy (or

M31), taken by Spitzer clearly show the spiral arms by noting the dust lanes in them. SST studies were the first to examine the core of the Milky Way in such a way as to determine that the galaxy core has a barred structure.

Spitzer was used to determine the atmospheric temperature of the extrasolar planet HD 189733b, the first time such a measurement was made. Astronomers used Spitzer to perform surveys with long time exposures. In early summer 2008, at the American Astronomical Society meeting held in St. Louis, infrared astronomers presented an infrared image of the Milky Way that consisted of a collection of 800,000 individual images. This composite image showed the distribution of dust within the galaxy in greater detail than had previously been possible.

In early 2008 the Spitzer Space Telescope's infrared spectrometer recorded the first evidence proving the existence of water in protoplanetary disks. Observations were made of DR Tau and AS 205A, which are 457 and 391 light-years distant from Earth, respectively. Water is an essential ingredient in an evolving solar system for the possibility of an Earth-like planet forming, one that might permit the development of life. Water is also, however, important in a protoplanetary system for the formation of icy satellites around large planets in the outer fringes of the system. Water closer to the young star at the center of the protoplanetary disk could be in gaseous or perhaps even liquid form. Both of the aforementioned protoplanetary systems produced large numbers of water emission lines.

Spitzer provided insight into numerous astrophysical phenomena, one of the most bizarre being magnetars. These are stars with magnetic fields approximately fourteen orders of magnitude more intense than that of typical stars on the main sequence. Magnetars are the highly compressed remains of massive stars that went supernova, but they are hardly dead stars. In addition to their intense magnetic fields, magnetars pulsate in the X-ray portion of the electromagnetic spectrum. In late May, 2008, the Spitzer Science Center, run by the California Institute of Technology, reported on observations of the magnetar SGR 1900+14. In addition to having the usual attributes of a magnetar, this object was surrounded by a ring of material 7 light-years across that was energized by the energetic X-ray pulsations. Heated dust in the ring resulted in the ring glowing strongly in the infrared.

Although results from IRAS, HST, ISO, and Spitzer have been spectacular, ground-based telescopes are still useful for observing many infrared phenomena. Infrared observations from NASA's Infrared Telescope Facility on Mauna Kea have revealed volcanic eruptions on Jupiter's moon Io. A volcano that had been erupting at the time of the Voyager flybys in 1979 was found to be erupting still, and a new volcano was detected. Observations such as these help to gather valuable information over time that can elaborate on the findings of other missions.

CONTEXT

Infrared astronomy is part of the revolution that has been called the "new astronomy." Instruments of modern astronomers give them access to information from the entire range of the electromagnetic spectrum. This revolution has occurred mostly since the early 1960's, when it became possible to place remote detectors above Earth's atmosphere. Before then, astronomers relied for the most part on the optical range of wavelengths for their information about the universe.

It was not until 1800 that the first sign of another way to look at the universe was discovered. While analyzing sunlight by separating the white light into a spectrum, English astronomer Sir William Herschel noticed that a thermometer placed in the dark area just outside the red limit of the spectrum registered an increase in temperature. In 1881, American astronomer Samuel Pierpont Langley developed the bolometer, an electrical detector that measures heat over a broad range of wavelengths. In measuring the Sun's energy from a high altitude, Langley found that the radiant energy of the Sun extended far past the visible portion of the spectrum and far past the region that Herschel had discovered previously. Herschel had discovered the near-infrared, whereas Langley was detecting the longer-wavelength middle-infrared band.

Infrared radiation from the Sun was fairly simple to detect, but more sensitive instruments had to be developed before it was possible to detect the infrared from far distant sources. In 1856, near-infrared radiation was detected from the Moon, but it was not until the 1920's that it began to be detected from the other planets and bright stars. Available instruments were still unable to see into the far infrared. While working on superconductivity experiments in the late 1950's, physicist Frank Low began the development of more sensitive instruments. By the early 1970's, Low was among the first to attempt observations of the far infrared by leading observations aboard high-flying jets.

An infrared satellite was first proposed in the mid-1970's. NASA was facing troubled times with budget cuts, inflation, and cost overruns in other projects. It might have scrapped the project entirely except for the interest of the Dutch space agency. The Dutch had completed several successful satellite programs and were interested in collaborating on an infrared satellite. England then joined the project, which came to be known as the Infrared Astronomical Satellite program. The project was a difficult one, but the diplomatic aspects of an international collaboration helped to give the program stability, and the satellite was launched successfully in 1983. A succession of increasingly sophisticated space-based observatories expanded upon the groundwork laid by IRAS. Hubble, ISO, Spitzer, and eventually the James E. Webb Space Telescope and turned infrared astronomy into an integral component of astrophysical investigations, leading to insights into the interstellar medium, thermal processes on planets in the solar system, protoplanetary disks, nebulae, galaxy formation, brown dwarfs, and many other phenomena.

Divonna Ogier

FURTHER READING

Chaisson, Eric, and Steve McMillan. *Astronomy Today.* 6th ed. New York: Addison-Wesley, 2008. This traditional treatise on introductory astronomy includes up-to-date spacecraft information such as that from NASA's Great Observatories program and rovers on Mars. Includes images that span the electromagnetic spectrum.

Dinwiddie, Robert, et al. *Universe.* New York: DK Adult, 2005. A remarkable collection of articles written by science writers and professional astronomers on a wide range of topics that span the discipline of astronomy. Heavily illustrated and filled with high-quality photographs. For the general reader.

Gregory, Stephen A. *Introductory Astronomy and Astrophysics.* 4th ed. San Francisco: Brooks/Cole, 1997. Suitable as a textbook for an introductory college-level or advanced high school course in general astronomy. Covers all topics from solar system bodies to cosmology. Some errors and issues with mathematical presentations.

Henbest, Nigel, and Michael Marten. *The New Astronomy.* New York: Cambridge University Press, 1983. Compares optical, infrared, ultraviolet, radio, and X-ray observations of well-known astronomical objects. For general readers.

Karttunen, H. P., et al., eds. *Fundamental Astronomy.* 5th ed. New York: Springer, 2007. A well-used university textbook in introductory astronomy. Contains some calculus-based treatments for those who find the standard treatise for introductory astronomy classes too low-level. Suitable for an audience with varied science and mathematical backgrounds. Covers all topics from solar-system objects to cosmology.

Kwok, Sun. *Physics and Chemistry of the Interstellar Medium.* New York: University Science Books, 2006. Although this text emphasizes physical processes, it is suitable for undergraduate courses in advanced astronomy covering infrared astronomy. Also addresses astrochemistry and mathematical theory on the interstellar medium.

Seeds, Michael A. *Horizons: Exploring the Universe.* New York: Brooks/Cole, 2007. A general astronomy text that also asks big-picture questions. Examines humanity's place in the universe, including physical and biological evolution.

Spitzer, Lyman, Jr. *Physical Processes in the Interstellar Medium.* New York: Wiley, 1998. Written by the astronomer after whom the

Spitzer Space Telescope is named, this work covers the physics and chemistry of the interstellar medium by frequently referencing infrared observational data. Accessible to the general reader as well as the astronomy enthusiast and student.

Verschuur, Gerrit L. *The Invisible Universe: The Story of Radio Astronomy.* New York: Springer Praxis, 2006. Provides a history of developments in radio astronomy and along the way describes the discovery of pulsars, quasars, and radio galaxies. Suitable for a general college science course as well as for astronomy majors as background information.

See also: Archaeoastronomy; Coordinate Systems; Earth System Science; Gravity Measurement; Hertzsprung-Russell Diagram; Neutrino Astronomy; Optical Astronomy; Protostars; Radio Astronomy; Telescopes: Ground-Based; Telescopes: Space-Based; Ultraviolet Astronomy; X-Ray and Gamma-Ray Astronomy.

Interplanetary Environment

Category: The Solar System as a Whole

Far from empty, the vast spaces between the Sun and its planets and out to the edge of interplanetary space constitute a dynamic environment suffused with fields of forces, crossed by swiftly moving particles, littered with debris, and penetrated by cosmic rays from outside the solar system. These phenomena endanger human technology, both in space and on Earth, but they also tell scientists much about the Sun and the origin of the solar system.

OVERVIEW

The interplanetary environment principally contains materials ejected from the Sun, and debris. Space debris comprises particles of great variety in size and composition. Anything smaller than 0.01 millimeter, astronomers call micrometeoroids or interplanetary dust. Anything larger is a meteoroid, of which many are tens of meters in size. By far the largest proportion of the debris originates when asteroids collide in the asteroid belt between the orbits of Mars and Jupiter, or from comets that swing into the inner solar system and leave behind a trail of particles eroded from them by the solar wind. Accordingly, most dust lies in the plane of the ecliptic (the disk-shaped region of the Sun and planets) in two bands: in the inner solar system, out to about 3 astronomical units (AU, the average distance from the Sun to Earth), and in the Kuiper Belt, about 10 to 40 AU. Additionally, a small amount of matter infiltrates from interstellar space as the solar system drifts through galactic gas clouds or from volcanoes on satellites, such as Jupiter's Io. Found both as chondrites (clusters of particles) and as solid chunks, interplanetary dust, like meteoroids, is rich in carbon, iron, sulfur, nickel, and silicates, but many other elements in mineral combinations have been found, including tiny spheroids of glass embedded with metal sulfides (known as GEMS). In a clear, dark sky of late evening or early morning, interplanetary dust is visible near the horizon in the direction of the Sun as a faint glow of reflected sunlight, called zodiacal light.

Although the interplanetary debris is largely gathered into two clouds, it is not static. Depending on size and location, particles move at different speeds and in different directions. The pressure exerted by solar photons pushes micrometeoroids between 0.5 and 1 micron slowly outward and eventually out of the solar system altogether. Particles larger than 1 micron are affected by radiation pressure that creates drag, which causes them to decelerate so that they fall inward toward the Sun until they are vaporized. Asteroid collisions and comets replenish the clouds. It is also the case that debris left behind by comets retains the parent body's velocity and orbit as a meteoroid stream. When the Earth passes through one of these streams, the vastly increased numbers of particles entering the atmosphere can create spectacular displays of streaking light in meteor showers, notably the Leonid showers each November and the Perseid showers each August (named after the constellations Leo and Perseus, from which they appear to arrive).

In addition to electromagnetic radiation, such as the photons of visible light, the Sun generates the gravitational field that holds together the solar system and magnetic fields. Magnetic fields accelerate a steady stream of plasma (electrons, protons, and ionized atoms) in all directions. That stream of particles is the solar wind. Near Earth the solar wind has an average density of about ten particles per cubic centimeter, a velocity of about 400 kilometers per second, and a temperature of about 100,000 kelvins, although these properties are highly variable. Particle velocity remains fairly constant beyond Earth's orbit, but density decreases as the solar wind spreads outward.

The solar wind defines the extent of the Sun's influence within its galactic neighborhood. As the plasma attenuates in the outer solar system, it eventually is slowed by the pressure of gas between stars and is deflected. The boundary where this occurs is called the termination shock. It fluctuates with the intensity of the solar wind and density of interstellar gas at between 90 and 100 AU. In December, 2004, the Voyager 1 spacecraft confirmed its existence, passing through it at about 94 AU. Voyager 1 then passed into a region of turbulence where the solar wind mixes with interstellar particles, called the heliosheath. Its outermost edge, the heliopause, marks the limit of the heliosphere,

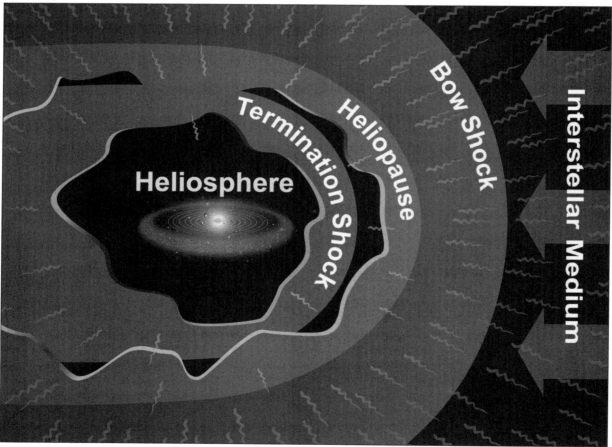

An artist's diagram of the heliosphere, created by the solar wind, which suffuses the interplanetary environment and is slowed at the point called the termination shock, where it meets the incoming interstellar medium. This region, along with the heliopause and external bow shock, define the outer boundary of our solar system and may fluctuate somewhat with the competing pressures of solar wind and incoming radiation of the interstellar medium. (NASA)

the total area the solar system. The heliosphere is something of a misnomer, however, because it assumes a teardrop shape as the solar system moves through interstellar gas clouds; its forward edge is thought to extend to between 115 and 150 AU.

Occasionally the Sun hurls outward immense bubbles of matter called coronal mass ejections (CMEs). Also known as solar storms, they typically involve between one hundred trillion and one quadrillion grams from the Sun's outer layer. By the time the typical CME reaches Earth's orbit, it has a speed, on average, of about 280 kilometers per second. It also carries with it its own magnetic field. The frequency of CMEs varies during an eleven-year cycle of solar activity. During solar maximum, Earth happens to be in the way of about seventy-two CMEs per year; at solar minimum, eight CMEs. This constitutes a minority of storms that the Sun ejects in all directions; during solar maximum the average is 3.5 CMEs per day. They can vary enormously in size and speed, the slowest moving a few dozen meters per second and the fastest 2.5 kilometers per second.

Planets with magnetic fields, such as Earth, are protected from the solar wind and CMEs. Earth's magnetic field, for instance, deflects the ionized particles so that it resides at the center of a tear-shaped bubble of relative calm. The interaction between the solar wind and Earth's southern and northern magnetic poles produces auroras, eerily rippling sheets of color in the sky. Satellites, asteroids, and planets with little or no magnetic field, such as Mars, are subject to a bombardment of their surfaces, which is deadly to life.

Sudden eruptions on the Sun's surface called solar flares emit X rays that can affect the properties of a planet's upper atmosphere. The Sun also broadcasts a constant blizzard of neutrinos. Some 70 billion of them strike every square centimeter of Earth's surface every second. However, these particles, nearly without mass, seldom interact with other matter. Although the solar wind deflects low-energy cosmic rays, the high-energy variety penetrate the interplanetary environment from unknown, distant sources; additionally, like everything in space, the solar system is bathed in the cosmic background radiation, the fading glow from the universe's origin.

KNOWLEDGE GAINED

Forces and particles of the interplanetary medium provide information about the origin of the solar system, its composition, the structure and behavior of the Sun, the evolution of planets, and the nature of other planetary systems.

The pervasive streams of neutrinos, for example, confirm theoretical calculations about the conversion of hydrogen into helium during fusion, the nuclear reaction that produces the Sun's radiant power. The gusty solar wind enables scientists not only to sample the constituent elements of the radiation; it also provides clues to the behavior of magnetic fields in the solar corona. The production of high-energy X rays by solar flares reveals the extent of those fields' power, as do CMEs. The frequency of CMEs and flares characterizes the Sun's eleven-year cycle of activity.

Most of the matter left over from the formation of the Sun and planets was long ago ejected from the solar system. However, enough remains mixed in with interplanetary debris, particularly in particles shed from comets, to provide clues about the relative abundance of elements, and specific isotopes, in the presolar gas cloud. With this information, astronomers can distinguish the unique chemical makeup of the solar system. Moreover, the Earth receives a steady rain of particles from the interplanetary dust clouds. Geologists estimate about 40,000 metric tons falls to the surface yearly.

Because the solar system's present dust clouds result from collisions between asteroids and material sloughed from comets, it is reasonable to infer that dust clouds around other mature stars evolved from the same processes. Astronomers have detected such dust clouds around about one-third of stars. Planetary systems appear to be common.

CONTEXT

Knowledge acquired about the interplanetary medium is of more than scientific concern. Understanding the interplanetary medium helps protect humanity from the dangers posed by its various contents. Meteoroids and micrometeor-oids streak past Earth at speeds of 11 to 70 kilometers per second, so even the smallest carries enough kinetic energy to damage space vehicles. Many satellites and the International Space Station suffered minor punctures even though the number of micrometeoroids in near-Earth orbit is low.

The solar wind is also dangerous. Astronauts and sensitive electronic equipment must be shielded from it. Still more dangerous is the sudden hurricane of particles and magnetic fields unleashed in a CME. Astronauts then must retreat to heavily protected areas in their spacecraft, and unprotected satellites may be rendered useless. Especially large CMEs can penetrate Earth's protective magnetosphere to disrupt communications and cause power outages. Because of the Earth's tilt toward the Sun and the amount of landmass in the Northern Hemisphere, power grids there are particularly at risk. A surge of X rays from a solar flare can likewise affect Earth's ionosphere and drown out radio communications. Additionally, exposure to cosmic rays increases the risk of cancer for astronauts, a fundamental problem to overcome if there are to be long voyages, as from Earth to Mars.

The combination of relative particle densities, solar flares, the solar wind, and CMEs is called space weather. As humanity grows more dependent on technology, it also becomes more vulnerable to the vagaries of space weather. Should a fierce solar storm strike unexpectedly, it could cause chaos in civilian and military telecommunications, weather prediction, and global positioning systems. It could also cause massive power outages that would bring modern life to a virtual standstill.

As a result, space agencies in the United States, the European Union, Russia, and Japan launched a variety of space probes and orbiting observatories to monitor the interplanetary medium, particularly as it is influenced by solar activity. Some of these, such as the Solar and Heliospheric Observatory (SOHO), can detect a solar storm before it reaches Earth, giving technicians time to protect or power-down sensitive equipment and giving astronauts time to seek shelter. Ground-based observatories offer simi-lar vigilance, some of them watching especially for large meteoroids or meteoroid steams that cross Earth's orbit.

The perils awaiting in the tenuous space between planets, as well as the wealth of information and potentially exploitable resources there, offer a lesson. Humanity exists not only in the framework of civilization and the physical environment of Earth's biosphere; its survival also requires understanding of interplanetary space.

Roger Smith

FURTHER READING

Bradley, J. P. "Interplanetary Dust Particles." In *Meteorites, Comets, and Planets*, edited by Andrew M. Davis. San Diego, Calif.: Elsevier, 2005. A thorough survey of knowledge about interplanetary particles by a physicist, with helpful graphics.

Jones, Barrie W. *Discovering the Solar System*. New York: John Wiley & Sons, 1999. This moderately technical introduction to the solar system includes a chapter on its formation and small bodies, such a micrometeoroids. Graphics and other illustrations, glossary, and bibliography.

Lang, Kenneth R. *The Cambridge Guide to the Solar System*. New York: Cambridge University Press, 2003. Many color and black-and-white photographs accompany the information-rich but generally nontechnical text in this introductory book, which includes information on the solar wind and interplanetary dust. With bibliography and sidebars devoted to special topics.

_____. *Sun, Earth, and Sky*. 2d ed. New York: Springer, 2006. Provides a thorough introduction to the Sun-Earth relationship, without recourse to mathematics, including sections on solar particles and magnetic fields. Includes a wealth of color illustrations and graphics, sidebars on special topics, a generous glossary, and a bibliography.

McBride, Neil, and Iain Gilmour, eds. *An Introduction to the Solar System*. New York: Cambridge University Press, 2004. Written on the level of an introductory text suitable for high school science students, this collection has a section on minor bodies of the solar system that discusses interplanetary particles and other sections on solar phenomena. Color illustrations and graphics, glossary, sidebars on special topics, and bibliography.

Peuker-Ehrenbrink, Bernhard, and Birger Schmitz, eds. *Accretion of Extraterrestrial Matter Throughout Earth's History*. New York: Kluwer Academic/Plenum, 2001. A collection of professional articles by a variety of specialists. Especially valuable, and accessible to lay readers is "The Origin and Properties of Dust Impacting Earth," by Donald E. Brownlee.

See also: Earth-Moon Relations; Earth-Sun Relations; Earth's Origin; Eclipses; Kuiper Belt; Oort Cloud; Solar System: Element Distribution; Solar System: Origins.

Interstellar Clouds and the Interstellar Medium

Category: The Cosmological Context

Although the space between stars appears quite empty by terrestrial standards, careful observation yields evidence for the existence of interstellar dust grains and more than one hundred different kinds of molecules present under a variety of conditions. Study of the interstellar medium provides important information on the life cycles of stars and evolution of the cosmos.

OVERVIEW

From the terrestrial standpoint, "outer space" begins where Earth's atmosphere ends. One thinks of the solar system as a group of eight planets together with a collection of dwarf planets, smaller bodies, moons, asteroids, and comets, which orbit the Sun. Similarly, the Sun and several hundred billion other stars orbit the center of our galaxy, the Milky Way. The fact that starlight can be seen from stars in this galaxy hundreds or thousands of light-years away would suggest that the space between the stars—the interstellar medium—is essentially empty. Careful observation has shown, however, that a variety of atoms and molecules can be found almost anywhere in the galaxy. Furthermore, the density and temperature of the interstellar medium vary greatly from place to place; in many regions, this medium forms "interstellar clouds," that may emit or scatter light or obscure the view of the stars behind them.

By terrestrial standards, interstellar regions are very nearly empty, seemingly a vacuum. Even the best vacuums attainable in the laboratory, however—about one one-hundred-trillionth of an atmosphere—contain about 100,000 molecules per cubic centimeter. This is perhaps the upper limit of density for interstellar matter. The space between stars in a galaxy is so immense, compared to the space actually occupied by the stars, that 10 percent or more of a galaxy's mass can be contained in the interstellar medium. The important characteristics of the interstellar medium are density, or number of particles per cubic centimeter; chemical composition; and temperature, which provides a measure of the speed of the molecules in the medium. The interstellar medium is said to be cloudlike in character, in that different regions have different densities. The least dense material, called the intercloud gas, has from one to ten particles per 100 cubic centimeters and an absolute temperature of about 10,000 kelvins.

The known components of the interstellar medium have been identified either through the electromagnetic radiation they emit or by their absorbing effects on the electromagnetic radiation from stars that pass through them. While Earth-based studies of the interstellar medium have had to rely on the microwave region of the electromagnetic spectrum—observed by radio telescopes—and on the visible region, satellite and rocket or balloon-based observations in the infrared, ultraviolet, and X-ray regions have contributed much to the understanding of the interstellar medium.

A few basic physical principles provide the basis for the interpretation of the electromagnetic radiation from the interstellar medium. The atoms or molecules of a gas at low density will emit or absorb electromagnetic radiation only at those energies (proportional to the frequency and inversely proportional to the wavelength) that correspond to the difference in energy between allowed quantum mechanical states. If a body of gas lies in front of a star, its chemical composition can be determined from the absorption spectrum, or pattern of dark lines seen against the continuous background

spectrum of starlight when viewed through a spectroscope. The temperature of a gas cloud can be determined from the frequency at which the greatest amount of emission occurs. If the gas is in motion toward Earth, the emitted radiation is shifted to slightly higher frequencies and shorter wavelengths. If the gas is moving away from Earth, the emission is shifted to slightly lower frequencies and longer wavelengths. This Doppler shift makes it possible to determine the speed of an interstellar cloud relative to Earth.

By far, the most common chemical elements found in interstellar space are hydrogen and helium. Astronomers distinguish between H I regions, in which the hydrogen exists primarily in the form of isolated atoms, and H II regions, in which the hydrogen exists primarily in ionized form, as separated protons and electrons. H II regions can be quite warm, with temperatures of 10,000 kelvins or more, and are characterized by a red emission produced by excited hydrogen atoms. Such regions include emission nebulae such as the Trifid nebula, which is one of the most beautiful objects in the sky. H I regions are somewhat cooler than H II regions and are characterized by the emission of radiation at 21-centimeter wavelengths, which can be detected by radio telescopes.

At somewhat lower temperatures, one finds a variety of cloud types in which much of the material exists as molecules rather than as separated atoms. Diffuse clouds have a temperature of about 100 kelvins and a density of about one hundred particles per cubic centimeter; they contain a mixture of hydrogen atoms and hydrogen molecules, along with some partially ionized carbon atoms, a few types of neutral atoms, and some small molecules, including carbon monoxide (CO) and formaldehyde (H_2CO_3). Dark clouds have a temperature of about 10 kelvins and a density of ten thousand particles per cubic centimeter, with most of the hydrogen appearing in molecular form. A wide variety of molecules is found in dark clouds. So-called giant molecular clouds exist at comparable temperatures and can be up to 400 light-years in diameter.

The number of molecular species that have been identified in molecular clouds is quite

A now famous 1995 image from the Hubble Space Telescope that redefined the popular image of the interstellar medium: Large clouds of cool interstellar hydrogen gas and dust, part of the Eagle Nebula, act as stellar incubators. (NASA/ESA/STScI/J. Hester and P. Scowen, Arizona State University)

large. Compounds of hydrogen, carbon, nitrogen, and oxygen are most common, with sulfur and silicon appearing in a few compounds. Molecules identified include those such as ethanol (C_2H_5OH), which are stable and even common under terrestrial conditions, as well as those such as the ethynyl radical, C_2H, which are too reactive to be isolated in the laboratory but can exist for a significant length of time at the very low densities present in molecular clouds. The largest molecule identified before 1990 is the thirteen-atom linear molecule cyanopentaacetylene ($HC_{11}N$). Since then there has been an explosion of the list. Every year it seems that more new molecules are identified than during the past year.

One region of the galaxy that is particularly rich in interstellar molecules is Sagittarius B2, near the galactic center. Nearly half of the molecules identified in the interstellar medium also are found in Sagittarius B2. Some of the most complex molecules ever detected include trans-ethyl methyl ether ($CH_3OC_2H_5$), propylene (CH_3CHCH_2), and cyanodecapentayne ($HC_{10}CN$). Some molecules are even observed in deuterated form, where ordinary hydrogen (in which the nucleus consists of a single proton) is replaced by its isotope deuterium (in which the nucleus consists of one proton and one neutron); these include heavy water (D_2O), three forms of deuterated ammonia (NH_2D, NHD_2, ND_3), and two forms of deuterated formaldehyde (HDCO, D_2CO).

About 1 percent of the total mass of the interstellar medium exists in the form of microscopic solid particles, generally called "dust" by astronomers. Dust grains are believed to be about one-millionth of a meter or less in size, and are irregularly shaped. The presence of interstellar dust is indicated by the absorption and scattering of starlight. When a dust cloud is illuminated from the front by a nearby star, one can see the bluish scattered light. Such an object is called a reflection nebula, the best-known example of which is associated with the star group known as the Pleiades. Scattering properties of the interstellar dust grains allow astronomers to make an estimate of their size and shape. Dust particles are more efficient in scattering blue light than red light; thus, distant stars appear somewhat redder than their actual color. Although dust grains have been recovered from the interplanetary space in the solar system, it is not certain that they are representative of interstellar grains. The relatively low concentrations of certain elements in the interstellar gas, as compared to the concentrations in stellar atmospheres, indicate that these elements—magnesium, iron, and silicon—may be prevalent in the interstellar grains. One remarkable effect of the interstellar dust is the polarization of starlight that passes through it; the interstellar medium has somewhat the same effect on starlight as a polarized sunglass lens has on light. In the case of the lens, the polarizing effect is the result of long, thin light-absorbing molecules, which are held parallel to one another in the polarizing film. The dust grains must therefore be elongated in shape, and it is probable that they are held in a parallel orientation by an interstellar magnetic field, which is relatively constant in direction over large areas of space.

The amount of matter to be found in the interstellar medium varies from one galaxy to another. Elliptical galaxies appear to contain substantially less interstellar matter than spiral galaxies, such as the Milky Way. Irregular galaxies can contain very large amounts. The Magellanic Clouds, satellites of the Milky Way visible in the Southern Hemisphere, are nearly 40 percent interstellar matter. Observations of clusters of galaxies indicate the presence of an intergalactic medium with a density of less than one particle per thousand cubic centimeters and at temperatures of more than 10 million kelvins. Some heavy elements, including iron, have been identified in this medium.

APPLICATIONS

The principal reason for astronomers' interest in the interstellar medium is that it provides insights into the formation and subsequent history of stars and hence the solar system as well. According to the big bang theory, the explosive event that marked the origin of the universe produced hydrogen and some helium, but only traces of any of the heavier elements. Almost all the carbon, nitrogen, oxygen, and other elements found in interstellar molecules, the planets, and Earth's atmosphere were produced by

The spiral galaxy Messier 81, imaged here by the Spitzer Space Telescope in the infrared, displays old stars, superheated interstellar clouds and dust, and the regions of star formation. (NASA/JPL-Caltech/S. Willner, Harvard-Smithsonian Center for Astrophysics)

nuclear reactions that took place in the interiors of stars; these molecules formed from the interstellar medium and were later returned to it. Theories of the structure of the galaxy and the life cycles of the stars must account for the present composition of interstellar matter as a result of nucleosynthesis occurring in the stars and the exchange of matter between stars and the interstellar medium.

Much of the interstellar medium exists in the form of very low-density gas at a very high temperature. A density fluctuation, perhaps induced by the "shock wave" from an interstellar explosion, can trigger a collapse of a portion of this medium into a smaller region. At first, a cooling of the material occurs as the energy released by collisions of particles is radiated into space. Collisions of atoms with one another and with the dust grains result in the formation of molecules. In the Milky Way, this process appears to occur primarily in the spiral arms, which is also the location of the most recently formed stars. When cloud density becomes suf-

ficiently great, the cloud becomes opaque to the passage of electromagnetic radiation, causing the released energy to become "trapped"; the cloud begins to warm and eventually forms a protostar, usually with an associated H II region and emission nebula. Gravitational collapse and the associated warming continue until the temperature becomes sufficiently great to allow nuclear fusion to ignite.

Nearly all nuclei in the universe, other than hydrogen and helium, were formed in the interiors of stars. The mechanism of formation is somewhat different for elements containing up to about sixty protons and neutrons (that is, up to the iron-nickel-cobalt group) and for larger nuclei, which are generally much scarcer. For the greater part of a star's lifetime, the principal nuclear reaction is the fusion of hydrogen to form helium. Once hydrogen has been substantially depleted, further gravitational collapse leads to additional warming, igniting the helium to form carbon, oxygen, and other small nuclei. If the star has sufficient mass, it may go through several additional stages of collapse and ignition, with the formation of still heavier elements. While the core of the star becomes warmer and more compact, the outer regions expand and cool so that eventually the star enters a red giant phase. The temperature in the outer extremes of a red giant are cool enough to allow the formation of molecules and possibly dust grains, which, since they are so far from the stellar core, may be able to escape the relatively weak gravitational field at the stellar surface. Planetary nebulae are H II regions that may represent a late stage in this process.

A number of other processes result in the release of matter from stars back to the interstellar medium. Stars of the Sun's mass or smaller typically have coronas, outer gaseous layers with temperatures of about a million kelvins, in which particles have sufficient speed to escape into space. For very large stars, the radiation pressure of light leaving the star is responsible for the stellar wind, a release of matter from the outer layers of a star. In binary star systems, the capture of matter from one star by the other can result in a nova, an explosion that ejects much stellar material. The sudden gravitational collapse of a large stellar core results in a supernova, which produces elements heavier than iron and returns them to the interstellar medium.

One other significant source of nucleosynthesis is the collision of existing nuclei with cosmic rays. Cosmic rays are particles, almost always protons or other nuclei, traveling at immense speeds. The interaction of cosmic rays with the interstellar medium provides a means of studying nuclear reactions occurring at very high energies, including the fragmentation of heavy nuclei, which appears to be the only source of some of the less abundant isotopes.

CONTEXT

The existence of interstellar dust and gas has played an important role in astronomers' study of the Milky Way. The belief that the broad band of light in the night sky, called the via galacta by the ancient Romans, is actually a collection of an immense number of stars of which the Sun was a member has been generally accepted since the beginning of the twentieth century. The problem was to determine the shape and size of this collection and the Sun's position in it. In the early 1900's, the Dutch astronomer Jacobus Cornelis Kapteyn conducted a survey of the distribution of stars of different magnitudes in different parts of the sky and concluded that the Sun appeared to be at the center of this distribution and that the number of stars in a given volume of space diminished with increasing distance from the Sun in any direction. This view was challenged, however, in 1917 when the American astronomer Harlow Shapley published a study of the distribution of globular clusters, large groupings of up to hundreds of thousands of stars, and showed that these clusters appeared to be centered on a point several thousand light-years from the Sun. The discrepancy between these observations was resolved

in 1930 when the Swiss American astronomer Robert Julius Trumpler showed that the interstellar dust obscured the view of more distant stars and that Shapley's method, based on large collections of stars, was more reliable.

Absorption lines caused by interstellar molecules were first identified in 1904 by the German astronomer Johannes Franz Hartmann, but the first identification of an interstellar molecule, the methylidyne radical, CH, did not occur until 1939. In 1951, the American William Wilson Morgan made the first observations of H II regions in the Milky Way, identifying them with the spiral arms. At the same time, Edward Mills Purcell and Harold Ewen at Harvard University were able to detect the 21-centimeter radiation of atomic hydrogen using the techniques of radio astronomy. By measuring the Doppler shift of the 21-centimeter radiation, astronomers have been able to construct a map of the Milky Way.

Interest in the chemistry of the interstellar medium increased substantially with the discovery in 1963 of the first oxygen-containing species, the hydroxyl radical, OH, and the subsequent discoveries of interstellar water, H_2O, and ammonia, NH_3, in 1968. Later years saw the discovery of increasingly complex molecules, giving credibility to the notion that interstellar chemicals may have played some part in the origin of life on Earth or possibly elsewhere in the universe. Ever more complex molecules are being discovered primarily through radio astronomy and infrared observations.

Donald R. Franceschetti

FURTHER READING

Arny, Thomas T., and Stephen E. Schneider. *Explorations: An Introduction to Astronomy.* 5th ed. New York: McGraw-Hill, 2007. A general astronomy text for the nonspecialist. Includes an interactive CD-ROM and a companion Web site.

Dudly, W. W., and D. A. Williams. *Interstellar Chemistry.* New York: Academic Press, 1984. Provides a fascinating record of the extensive information that has been accumulated about the chemical and physical characteristics of the interstellar clouds. For more advanced readers.

Freedman, Roger A., and William J. Kaufmann III. *Universe*. 8th ed. New York: W. H. Freeman, 2008. College-level introductory text covering the field of astronomy. Contains descriptions of astrophysical questions and their relationships.

Karttunen, H. P., et al., eds. *Fundamental Astronomy*. 5th ed. New York: Springer, 2007. A well-used university textbook in introductory astronomy. Contains some calculus-based treatments for those who find the standard treatise for typical astronomy 101 classes too low-level. Suitable for an audience with varied science and mathematical backgrounds. Covers all topics from solar-system objects to cosmology.

Kwok, Sun. *Physics and Chemistry of the Interstellar Medium*. New York: University Science Books, 2006. Although this text emphasizes physical processes, it is suitable for undergraduate courses in advanced astronomy. Also provides astrochemistry and mathematical theory on the interstellar medium.

Sagan, Carl. *Cosmos*. New York: Random House, 1980. This classic volume by the prominent astronomer and popular writer includes a chapter on the lives of the stars. Presents the formation of the chemical elements in a highly entertaining and memorable fashion. A companion video series of the same name is available.

Spitzer, Lyman, Jr. *Physical Processes in the Interstellar Medium*. New York: Wiley, 1998. Written by the astronomer after whom the Spitzer Space Telescope is named, this work covers the physics and chemistry of the interstellar medium by frequently referencing infrared observational data. Accessible to the general reader as well as the astronomy enthusiast and student.

Tielens, A. G. G. M. *The Physics and Chemistry of the Interstellar Medium*. New York: Cambridge University Press, 2005. A comprehensive presentation of the physics and chemistry of the interstellar medium for undergraduates through professional astronomers. Includes a detailed bibliography.

Verschuur, Gerrit L. *The Invisible Universe: The Story of Radio Astronomy*. New York: Springer Praxis, 2006. In addition to providing a history of developments in radio astronomy, this volume covers a great deal of infrared astronomy. Discusses the nature of interstellar nebulae and the detection of molecules in the interstellar medium. Suitable for a general science course in college as well as for astronomy majors as background information.

Wynn-Williams, Gareth. *The Fullness of Space: Nebulae, Stardust, and the Interstellar Medium*. Cambridge, England: Cambridge University Press, 1992. A comprehensive survey of interstellar matter and the interstellar medium, including astronomical observations in regions of the electromagnetic spectrum besides the visible. Written for laypersons, astronomy buffs, and students.

See also: Big Bang; Cosmic Rays; Cosmology; Electromagnetic Radiation: Nonthermal Emissions; Electromagnetic Radiation: Thermal Emissions; General Relativity; Milky Way; Space-Time: Distortion by Gravity; Space-Time: Mathematical Models; Universe: Evolution; Universe: Expansion; Universe: Structure.

Io

Categories: Natural Planetary Satellites; The Jovian System

Io is the innermost of four large satellites orbiting Jupiter, the largest planet of the solar system. Io is the most volcanically active body in the solar system. Tidal friction occurs constantly on Io, heating its core. Internal thermal energy is vented through immense volcanoes that spew sulfur and sulfur components into space which fall back, resurfacing the satellite. Io has one of the youngest surfaces in the solar system.

OVERVIEW

Until 1979, Io was known as little more than a pinpoint of light, even when seen through large telescopes. Io is one of the four satellites of Jupiter that were first observed telescopically by Galileo in 1609. With roughly the same size, density, and surface gravity as Earth's Moon, Io was expected to have similar features. Earth-based spectroscopic observations during the 1970's, however, raised speculation that Io was quite different. Suspicions were dramatically confirmed by the flybys of Voyagers 1 and 2 in 1979 and by multiple encounters of the Galileo spacecraft between December, 1995, and September, 2003. The closest encounter with Io, early in the Galileo mission, was at a distance of 22,000 kilometers; the probe approached no closer because of concerns about the intense radiation environment the spacecraft would encounter. During the late stages of the extended Galileo mission, however, the spacecraft passed within a mere 180 kilometers of Io's surface; at this late point in the mission, the science return was worth the risk to the spacecraft's health.

Voyager 1 made the remarkable, and completely unsuspected, discovery that active volcanoes dot the surface of this planet-like Jovian moon. From Voyager and Galileo evidence, augmented by Earth-based observations, scientists identified Io as a world dominated by volcanic eruptions. Most of the moon's surface features are transformed daily by heated liquid, with gaseous emissions onto the surface and into the otherwise nonexistent atmosphere. On Io, impact craters are not formed as they are on the other, extremely cold, satellites of the solar system. Io's impact features are apparently absorbed by molten lava flows that extend across Io's surface. Volcanic activity is not sporadic but virtually continuous. Of the eight active volcanoes observed by Voyager 1, seven were still spewing gaseous plumes when viewed by Voyager 2 four months later. Pictures taken by the two spacecraft portrayed these huge eruptions against the backdrop of the black sky over Io's limb (its visible horizon) and from above the red-orange surface with its active calderas. However, Io's volcanic activity changed in the time between the Voyager flybys and the arrival in orbit of the Galileo spacecraft on December 7, 1995.

With its image enhanced by spacecraft instruments, Io looked like a giant pizza, with wide plains of different hues punctuated by darker and lighter active regions. The latter are

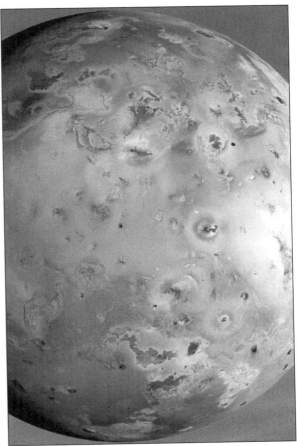

The Galileo spacecraft captured this high-resolution image of Jupiter's moon Io from about 294,000 kilometers in March, 1998. (NASA/JPL/University of Arizona)

calderas, which dot the surface, at least two hundred of them, each having a diameter of more than 20 kilometers. The largest include eleven observed plume-emitting volcanoes named by the International Astronomical Union for mythological gods of early and primitive religions. Two of these, separated along the Loki fissure, are associated with a lava lake 200 kilometers wide, known as Loki Patera, which apparently is the major outlet for the planet's internal heat. Its temperature, like those of other Io "hot spots," averages about 300 kelvins. The hot spots contrast with the remaining 98 percent of the surface, which at 130 kelvins is considerably colder, as would be expected for an atmosphere-poor body so far from the Sun.

Mountains tend to cluster near the polar regions on Io. Some have peaks as high as 10 kilometers, but they do not appear to have been formed by plate tectonics (the shifting of continental geologic structures). They lack cone-like tops and could not have been formed by recent volcanism. It is speculated that some of Io's upper crust may detach and float about the molten plains in a manner analogous to icebergs. Also, erosion scarps form near emission calderas. The fluid surface suggests that Io might receive a new surface 10 micrometers thick every year, making it unique in a solar system of much older, inactive, heavily cratered planets and satellites.

The chemistry of Io helps explain this satellite's dynamic volcanism. Previously thought to have a solid interior, like other satellites of the solar system, Io appears to have a molten silicate core. Planetary geologists hypothesize that about four billion years ago, heated sulfur dioxide lying just below the surface became the driving force for the volcanoes, ejecting Io's internal heat in gaseous eruptions similar to geysers. Long-lived eruptions, like Loki, eject materials ballistically at 0.5 to 0.6 kilometer per second. Short-duration powerhouses, typified by Pele

(and possibly Aten and Surt, seen by Voyager 2) do so at twice that velocity. These eruption rates are significantly higher than that of Earth's volcanoes (0.1 kilometer per second). Sulfur compounds are ejected as majestic mushroom-shaped plumes to a maximum height of 300 kilometers, enabling the lighter compounds, such as water and carbon dioxide, to escape into space. The heavier compounds, such as sulfur and sulfur dioxide, fall back to the surface as frozen, whitish, snowlike matter. Flows of molten matter therefore are low-viscosity sulfur and sulfur compounds rather than silicate rock lavas typical of Earth's volcanoes. Io's surface color results from the various sulfur compounds.

Io's geologically active behavior is caused by its proximity to massive Jupiter and to sister Galilean satellites Europa and Ganymede. Jupiter's gravitational pull causes Io to "flex" along its axis 10 kilometers toward Jupiter, while the combined attractions of Europa and Ganymede cause torques that give Io a slightly eccentric, noncircular orbit. The result is two opposing tidal forces stretching Io from within as it orbits Jupiter every 1.77 days. The ensuing friction raises Io's internal heated power to 60 to 80 trillion watts, partially melting the silicate compounds of the crust and generating volcanic eruptions. Furthermore, because Io's orbit lies entirely within Jupiter's radiation belts, the satellite is bombarded by charged particles and affected by the powerful electrical currents produced by Jupiter's magnetic field. These phenomena also influence Io's internal heating, as does the spontaneous radioactive decay of isotopes, which is typical of all planetlike bodies.

Io's volcanic emissions account for its irregular atmospheric pressure, first detected by Pioneer 10 in 1973. Atmospheric pressure variations result from the heat differential associated with the anomalous

This amazing image of Io, taken by Voyager 1 in March, 1979, shows the eruption of a huge volcano at the upper left horizon—marking the first sighting of an active volcano beyond Earth. (NASA)

hot spots and the typically cooler surface. Earth-based observations in 1974 and 1975 detected a "cloud" of neutral sodium and potassium extending along Io's orbit for more than 100,000 kilometers. This observation was explained as a "sputtering" process whereby potassium atoms are ejected into space from Io because of the impact of charged particles from Jupiter's magnetosphere striking Io's surface. Their rate of ejection is greater than 10 kilometers per second, well above the necessary escape velocity of 2.5 kilometers per second. By comparison, Io's most active volcano, Pele, has an ejection speed of only 1 kilometer per second. Volcanoes, then, are an indirect rather than a direct cause of this sodium-potassium cloud; they bring the elements to the surface in molten and gaseous states for emission into space.

In 1976, additional Earth-based observations revealed a plasma "torus," or faint ring of excited glowing gas, belonging to Io's orbit. This torus occupies space within Jupiter's magnetosphere and results from the sputtering process. New electronic cameras and filters carried aboard the National Aeronautics and Space Administration's Kuiper Airborne Observatory recognized sulfur in 1981 and oxygen in 1982, as well as sodium and potassium, escaping from Io. The entire cloud assemblage supplies the torus with raw materials for further breakdown into discrete atoms and ionization. Thus energized, these ionized elements join the Jovian radiation belt that helped create them. The discovery of the torus was as unexpected as the volcanoes. Pioneer 10 and Voyager 1 flew directly through the torus in 1973 and 1979, respectively, but provided only knowledge supplementary to the major data obtained through continuous Earth-based monitoring. On its way to orbit insertion in December, 1995, the Galileo spacecraft flew a relatively safe distance from Io, one that had originally been considered to be the closest that

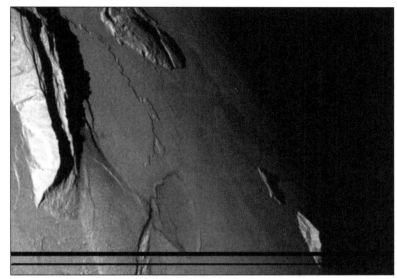

In February, 2000, the Galileo spacecraft took this image of Io at a resolution of about 335 meters. To the left, Mongibello Mons rising about 7 kilometers from the surface. (NASA/JPL/University of Arizona/Arizona State University)

Galileo would ever get to this innermost Jovian satellite, and then passed quickly through the plasma torus. No significant radiation damage was incurred by the spacecraft.

The volcanoes of Io offer the greatest promise for resolving the details of the complex relationship between Jupiter and Io. Of the eleven volcanoes discovered by Voyagers 1 and 2, the two-part vent of Loki is the most important. With a height of about 225 kilometers and a width of more than 430 kilometers, Loki and its lava lake, Loki Patera, appear to be the major outlet for Io's internal heat, as suggested by thermal emission polarization measurements in 1984. The thermal output from Io's greatest volcanoes, Pele (305 kilometers high and 1,200 kilometers wide), Surt, and Aten, is also significant. One of the three apparently ceased eruptions in 1986, according to Earth-based observations. The rest are 100 kilometers or less in height. Known calderas make up about 5 percent of Io's surface. Some, like Loki and Pele, have asymmetrical plumes and surface flows, which probably are consequences of irregular vent shapes. Others, like Prometheus, have symmetrical, fountainlike plumes and circular flows.

The Galileo spacecraft confirmed the massive scale of Io's volcanism, detecting a fresh volcanic deposit the size of Arizona and establishing that the satellite is rich in silicates. Increasingly sophisticated Earth-based instruments and orbiting telescopes have provided additional analysis. During eclipses in 1985, ice-covered Europa passed through Io's shadow, reflecting sunlight that revealed the uniform distribution of sodium about Io. The Hubble Space Telescope spotted an active plume in June, 1997, which Galileo also detected.

KNOWLEDGE GAINED

Principally from Voyager 1 and Earth-based observations, Io was revealed to be the only volcanically active satellite of Jupiter. Volcanic gases and molten lava flows were seen being emitted from eleven major fissures. The most notable is the large dual vent of Loki Patera, which appears to be the focal point of the satel-

lite's heat emissions from its interior. Because these eruptions are continuous, the ejected heavier compounds of material steadily move across the surface, eroding low-lying scarps and erasing the craters formed by objects striking Io. At least one lava lake was discovered, along with mountains near the poles. Emissions appear to be generated by Io's molten interior, from which sulfur compounds are emitted onto the surface and into the atmosphere. The heating mechanism for this activity is apparently internal friction caused by the gravitational pull of massive Jupiter on one side, and the satellites Europa and Ganymede on the other.

Because Io lies within Jupiter's magnetosphere, charged particles strike the satellite's volcanic surface, causing jets of potassium, sodium, sulfur, and oxygen to be ejected into its atmosphere—a "sputtering" process that feeds a cloud of those neutral elements. This cloud in

An artist's conception of the role of Io's sulfur, which emerges from volcanoes and lands on the surface (see the arching arrow), where it is cooled into S_3, S_4 (both pinkish in color), and sulfur, S_8, which gives the moon its characteristic yellow hue. (NASA/JPL/Lowell Observatory)

turn supplies raw materials for a torus of excited gas along Io's orbital path.

Determination of Io's bulk density and moment of inertia has revealed the satellite to be a differential body composed of a silicate mantle with a metallic core that makes up as much as half of Io's radius. Comparative studies of the four Galilean satellites indicate that they share a number of similarities in their cores as well as differences due to their environment. Io, being active, has lost any significant icy shell it might have had over its core. Ganymede and Callisto, being colder and far less active, have thick, icy shells over their cores with craters on the icy surfaces of each. Europa has activity that gives it an icy crust that breaks up and flows over what is believed to be a liquid ocean beneath that crust.

As a result of Galileo's repeated observations of Io, hundreds of volcanic sites were identified, as were about two hundred dark surfaces believed to be fresh silicate lava. The orbiter's near-infrared mapping spectrometer and solid-state imager identified a hundred active hot spots through thermal emission.

CONTEXT

Io emerged from the Voyager 1 and 2 missions as unique, not only among the satellites of Jupiter but also among all planets and satellites within the solar system. Io is recognized as the most volcanically active planetary body in a solar system where only a few other worlds display volcanism, and many of those display a type of cryovolcanic activity quite different in nature from Io's volcanoes. Io's volcanism is of particular interest because, like Earth, it is a dry body with a molten interior, and its sulfur-enriched chemistry may mimic volcanic conditions that existed during Earth's early history. Earth's active volcanoes convert water to steam for geothermal output. Io's sulfur-based volcanism provides an active laboratory for the study of planetary evolution, because volcanic eruptions constantly resurface its crust.

Io exists well within the Jovian magnetosphere. Electromagnetic fields affect Io's surface, allowing lighter elements emitted through its volcanic vents to escape into the atmosphere and feed the torus that encircles Jupiter as part of the radiation belt. The Jupiter-Io connection serves as a laboratory for the study of large-scale magnetic forces in the solar system. One of Io's effects on the Jovian system is a very gradual slowing of the rotation of Jupiter and erosion of the orbits of Europa and Ganymede.

Because active satellites like Io had not been anticipated before the 1979 Voyager flybys, their study has enhanced the evolving field of comparative planetology. Io was joined by Enceladus and Triton as satellites in the outer solar system that display unexpected volcanic activity.

Naturally, Io remains a high-priority target in planetary science for further study by robotic spacecraft. However, the radiation environment makes it difficult to dispatch probes into close proximity to the satellite. Human exploration is highly unlikely. There is a scene in Peter Hyams's film *2010: The Year We Make Contact* (1984) in which the wayward *Discovery* and its deactivated HAL 9000 computer are found adrift near Io. Spacewalking astronauts then transfer from a Russian spacecraft called the *Leonov* to enter and activate *Discovery*. In reality, the intense radiation environment would have provided such spacewalking astronauts a lethal dose well before they could return to the safety of another ship.

Clark G. Reynolds

FURTHER READING

Bagenal, Fran, Timothy E. Dowling, and William B. McKinnon, eds. *Jupiter: The Planet, Satellites, and Magnetosphere*. Cambridge, England: Cambridge University Press, 2007. A comprehensive work about the biggest planet in the solar system. A series of articles provided by recognized experts in their fields of study. Excellent repository of photography, diagrams, and figures about the Jupiter system and the various interplanetary missions that have unveiled its secrets.

Beatty, J. Kelly, Carolyn Collins Petersen, and Andrew Chaikin, eds. *The New Solar System*. 4th ed. Cambridge, Mass.: Sky, 1999. A richly illustrated summary of early space-age discoveries that radically revised knowledge of the solar system, particularly useful in tracking initial reactions of scientists to those discoveries. Major features of Io and its volcanoes are covered.

Cole, Michael D. *Galileo Spacecraft: Mission to Jupiter*. New York: Enslow, 1999. Provides a full description of the Galileo spacecraft and science returns through the primary missions. Particularly good at describing mission objectives and goals. Suitable for a younger audience.

Fischer, Daniel. *Mission Jupiter: The Spectacular Journey of the Galileo Spacecraft*. New York: Copernicus Books, 2001. Thoroughly explains all aspects of the science and engineering of the Galileo spacecraft. Particularly good are the discussions about the nature of the Galilean satellites. Suitable for a wide range of audiences.

Geissler, Paul E. "Volcanic Activity on Io During the Galileo Era." *Annual Review of Earth and Planetary Sciences* 31 (May, 2003): 175-211. The definitive work describing the physics and planetary geology of volcanoes on Io. Provides a complete picture of Voyager and Galileo spacecraft results.

Greeley, Ronald. *Planetary Landscapes*. 2d ed. London: Allen and Unwin, 1994. A brief but instructive photographic examination of imaged planetary surfaces of the solar system, including a treatment of volcanism that sheds light on Io's unique volcanic processes.

Harland, David H. *Jupiter Odyssey: The Story of NASA's Galileo Mission*. New York: Springer Praxis, 2000. This book provides virtually all of NASA's press releases and science updates during the first five years of the Galileo mission. Contains enormous numbers of diagrams, tables, lists, and photographs. Provides a preview of the Cassini mission. Published before the completion of the Galileo mission.

Hartmann, William K. *Moons and Planets*. 5th ed. Belmont, Calif.: Thomson Brooks/Cole, 2005. An updated version of a classic text that covers all aspects of planetary science. Takes a comparative planetology approach rather than including separate chapters on individual planets of the solar system.

Lopes, Rosaly M. C., and John R. Spencer. *Io After Galileo: A New View of Jupiter's Volcanic Moon*. Heidelberg: Springer, 2007. A volume in the Springer Praxis Space series, this book summarizes the knowledge gained by Galileo. Suggests new investigations needed to explain those questions that remain about the volcanism of this Jovian moon. Technical.

Morrison, David. "The Enigma Called Io." *Sky and Telescope* 69 (March, 1985): 198-205. An updated summary of original Voyager 1 and 2 data collected between 1979 and 1984, including contemporary information from Earth-based instruments, by a leading authority on the subject. Special attention is given to Loki Patera and the Jovian nebula, or thin gaseous torus, generated by Io.

Morrison, David, and Jane Samz. *Voyage to Jupiter*. NASA SP-439. Washington, D.C.: Government Printing Office, 1980. The official account of the Pioneer and Voyager flybys of the Jovian system, covering the day-to-day revelations from each mission. The most notable is the dramatic discovery of Io's volcanoes by Voyager 1. Lavishly illustrated.

See also: Callisto; Europa; Ganymede; Jovian Planets; Jupiter's Magnetic Field and Radiation Belts; Jupiter's Satellites; Neptune's Satellites; Planetary Satellites; Planetary Tectonics; Saturn's Satellites; Uranus's Satellites; Venus's Volcanoes.

J

Jovian Planets

Categories: Planets and Planetology; The Jovian System

Jupiter, Saturn, Uranus, and Neptune are called the Jovian planets. These "gas giants" have a mass 15-320 times greater than Earth, are of very low relative density, are mainly fluid (gas and liquid), and are composed of relatively light elements such as hydrogen and helium. All of them are surrounded by ring systems, a host of diverse satellites, and complex magnetospheres.

OVERVIEW

The four Jovian, or "gas giant," planets are totally different geologically and physically from the terrestrial planets, Mercury, Venus, Earth, and Mars. Massive gaseous and liquid bodies composed primarily of hydrogen and helium, the Jovian planets are relatively rapid rotators. Each rotates about its own axis in less than twenty-four hours. The atmospheres of the Jovian planets—the feature that dominates observational work done on these planets—are very similar to one another in composition. Hydrogen represents about 90 percent of the atoms present, with helium making up the bulk of the remaining atmospheric gases. Methane and ammonia are also present, although ammonia on the two colder planets Uranus and Neptune has most likely precipitated out of the atmosphere. Weather systems that dominate these atmospheres, particularly in the case of Jupiter and Saturn, consist of rapidly rotating belts and zones that are visible from Earth. In the case of Jupiter, wind speeds on the order of 300 kilometers per hour are common, while on Saturn winds of two to three times that speed have been measured. Note that on Earth, hurricane-force winds rarely exceed 150 kilometers per hour. Ironically, wind speeds in the colder Uranus and Neptune are even higher than those seen on Jupiter and Saturn.

Both Jupiter and Saturn are much hotter than might be expected in view of their distances from the Sun. They have their own heat sources deep within their planetary interiors and thus are able to produce extensive thermal cells to drive high-speed winds. The nature of the heat within these planets is not entirely evident, and there is some evidence that even in the case of Uranus there may also be a heat-driven weather system resulting from a much more modest heat source on the planet. Uranus displays little atmospheric structure in visible light, but in ultraviolet there are some features. Images from Voyager 2 have shown that Neptune has an actively driven weather system. Dark storms and white streakers are seen to evolve in short time frames.

Although there are no measurements to indicate what lies below these turbulent atmospheres, there is indirect evidence that toward the center of a typical Jovian planet, pressures become higher. A portion of the interior is liquid. Toward the center of both Saturn and Jupiter, a very unusual state exists, that of liquid metallic hydrogen. This liquid metallic state would enhance both the thermal and electrical conductivity of the planetary interiors and no doubt is largely responsible for the strong magnetic fields associated with Jupiter and Saturn. Pressures necessary to create liquid metallic hydrogen are on the order of millions of times the atmospheric pressure at the surface of Earth. Although they can be re-created in tiny cells in the laboratory, no such pressures have been sustainable in large-scale systems for prolonged periods of time on Earth. Thus, the liquid metallic hydrogen layer in the interiors of Jupiter and Saturn have effects that are not yet fully described. Uranus and Neptune probably have no such layers, for their masses are not great enough to produce such enormous interior pressures.

If one were proceeding inward toward a Jovian planet's center, one would next approach the core of that planet. Theorists disagree as to what might exist there, but the dominant opinion is that the cores would be largely solid and would contain relatively heavy metals such as iron, or they may be silicon in some high-pressure phase. In the case of Jupiter, such a solid core might have a mass 20 times greater than that of Earth, but this is a small fraction of the total mass of this planet, which is 320 times that of Earth. Uranus and Neptune might have solid cores on the order of several Earth masses, while that of Saturn would be about 5 to 10 Earth masses.

Little or nothing has been learned experimentally about planetary interiors; this is true even for Earth and its fellow terrestrial planets. Modeling planetary interiors, particularly on the scale necessary in the cases of Jupiter and Saturn, requires knowledge of pressure effects on bulk matter at pressures of millions of atmospheres and at alleged temperatures of 50,000 kelvins or hotter. It is known, however, that all the Jovian planets have a very low density, that is, a low specific gravity. Specific gravity is a measure of relative density, using water as a unit of 1 gram per cubic centimeter or 1,000 kilograms per cubic meter. Saturn has a specific gravity or relative density of 0.7; this means that it would float if one could find an ocean of water big enough in which to place it. It is by far

the least dense of all the planets. Jupiter has an average density of 1.3 grams per cubic centimeter, while Uranus and Neptune have average densities of 1.2 and 1.7 grams per cubic centimeter, respectively. A typical terrestrial planet has a specific gravity of about 5. Earth's density is on average 5.5 grams per cubic centimeter. Overall, such relatively low density measurements indicate the predominance in Jovian planetary structures of light elements such as hydrogen and helium.

Jupiter has a sizable magnetosphere. Its strong magnetic field is about ten times as intense as that of Earth. Jupiter's magnetosphere, which consists of trapped charged particles in amounts that would be lethal to humans, is so large that Saturn, which is 9.5 astronomical units (AU) from the Sun, passes through it. Saturn is almost twice as far from the Sun as is Jupiter (at about 5.2 AU), yet its magnetosphere is very strongly influenced by that of Jupiter. Saturn itself has a magnetic field slightly larger than that of Earth.

All the Jovian planets rotate about their own axes rapidly in relation to the terrestrial planets, which take at least twenty-four hours to make one rotation. (Earth is the fastest rotating terrestrial planet.) All Jovian planets thus exhibit some degree of oblateness. Saturn has an

A composite image of the four gas giants known as the Jovian planets (left to right): Jupiter, Saturn, Uranus, and Neptune. (Lunar and Planetary Institute)

Facts About the Jovian Planets

	Jupiter	Saturn	Uranus	Neptune
Mass (10^{24} kg)	1,898.6	568.46	86.832	102.43
Volume (10^{10} km^3)	143,128	82,713	6,833	6,254
Equatorial radius (km)	71,492	60,268	25,559	24,764
Ellipticity (oblateness)	0.06487	0.09796	0.02293	0.01708
Mean density (kg/m^3)	1,326	687	1,270	1,638
Surface gravity (m/s^2)	23.12	8.96	8.69	11.00
Surface temperature (Celsius)	−140	−160	−180	−200
Satellites[a]	63	60	27	13
Mean distance from Sun millions of km (miles)	779 (483)	1,434 (891)	2,872 (1,785)	4,495 (2,793)
Rotational period (hrs)[b]	9.9250	10.656	−17.24	16.11
Orbital period	11.86 yrs	29.66 yrs	84.01 yrs	164.79 yrs

Notes:
a. Numbers are for *known* satellites as of the year 2009.
b. Retrograde rotational periods are preceded by a minus sign.
Source: Data are from the National Aeronautics and Space Administration/Goddard Space Flight Center, National Space Science Data Center.

oblateness of about 0.1, which means that its equatorial diameter is about 10 percent bigger than its polar diameter, and thus it appears noticeably flattened at the poles. Saturn takes 10 hours and 13 minutes to make one complete rotation; Jupiter spins even faster, taking only 9 hours and 55 minutes to complete a rotation. Uranus takes 17 hours for one complete rotation, while Neptune takes about 16 hours (a figure that remains subject to conjecture in the aftermath of Voyager 2 studies in 1989).

These rapid rotations are surprising for such gaseous and liquid planets, because an angular momentum principle of elementary physics would have bigger bodies rotate more slowly than smaller ones. Such rapid rotation rates, then, are a mystery of the first magnitude in solar physics and geophysics. The correlation between magnetic fields and rotation rates is not very strong. Why some planets have powerful magnetic fields and others negligible ones is unknown. In general, however, if magnetic fields result from dynamo currents deep within planets, then rapid rotators should have strong magnetic fields. This is largely true in the case of both Jupiter and Saturn. Uranus has a magnetic field weaker than that of Saturn. Neptune has a magnetic field roughly comparable to that of Uranus.

Neptune and Uranus seem to be smaller and colder versions of Saturn and Jupiter. Uranus has a pale blue, almost greenish-blue appearance, undoubtedly because of the presence of methane. Imaging from Voyager photographs has shown that Uranus has belts and zones, although they are not as spectacular as those of Jupiter and Saturn. Ammonia within the Uranian atmosphere and that of Neptune is thought to have precipitated to the surface. In August, 1989, Voyager 2 flew by Neptune and returned images of that planet and its atmosphere. No good photographs of Neptunian surface features can be made from Earth's surface because of its distance (30 AU from the Sun). However, the Hubble Space Telescope has been used for intermittent studies of both Uranus and Neptune, some of the most productive research being performed by astronomer Heidi Hammel. Voyager 2 showed that Neptune is also a pale blue planet with irregular marked bands in its atmosphere. It has a gigantic Dark Spot, somewhat analogous to Jupiter's Great Red Spot, which seems to cause a tremendous sinking and upswelling of its atmospheric winds. This spot has a diameter about the same as the diameter of Earth. In addition, extremely high clouds, about 50 kilometers above the (normal) Neptunian atmosphere, make its atmosphere different from those of the other Jovian planets. Winds on the order of 200 meters per second have been measured in the Neptunian atmosphere, with unique streams and band systems.

All Jovian planets possess numerous moons, or satellites. By 2008, Neptune was discovered to have at least thirteen satellites. Uranus has at least twenty-seven detected moons, while Saturn has sixty and Jupiter has at least sixty-three. It is a bit surprising that these planets rotate so rapidly while at the same time each has such a large family of satellites. The presence of such satellites should have slowed the rotation rates, if indeed the satellites and the primary planets had common origins. Many of the satellites present the same face toward their primary planet, and thus are tidally locked.

All Jovian moons have rings as well, although they vary considerably in texture and content. The Voyager 1 spacecraft discovered a very thin ring around Jupiter in 1979. Since that time research has revealed that there is more structure to Jupiter's ring system. Presently it is referred to as the Halo Ring, the Main Ring, and the Gossamer Ring; the Gossamer Ring itself has two parts, the inner Almathea Gossamer Ring and the outer Thebe Gossamer Ring. Jupiter's rings are dark and composed of dust, and therefore they cannot be seen from Earth. Saturn's ring system of mainly icy particles is incredibly dynamic, with a number of gaps due to gravitational resonances and shepherding by small, embedded moonlets. Uranus has at least nine complete rings, some considerably brighter than others; it must be noted that five of Uranus's rings were discovered in the late 1970's by ground-based observations, not by the Voyager 2 mission, although the Voyager probe provided the first intense study of the entire ring system. It was Hubble-based research that found four additional dark rings long after the Voyager flyby. Neptune has only partial arcs. Astronomers are still debating whether partial arcs are ring systems being formed or in late stages.

METHODS OF STUDY

Galileo's discovery in 1610 of the four large natural satellites of Jupiter (Io, Europa, Ganymede, and Callisto, known as the Medician and then later as the Galilean moons) launched the age of modern science. By the 1650's, Christiaan Huygens in Holland and other astronomers in Italy had established conclusively that Saturn had rings and at least one large satellite, Titan. The Great Red Spot on Jupiter has been observed ever since 1660, and the zones and belts on both Jupiter and Saturn had been clearly detected by enterprising visual astronomers. For three centuries, astronomers around the globe have tracked the Great Red Spot and noted changes in the belts and zones of these two gigantic planets. Uranus, too, was viewed by many from the 1700's onward, but it was not clearly designated as a planet until the late eighteenth century.

With the advent of spectroscopy in the nineteenth century, helium was discovered first on the Sun and shortly thereafter on Jupiter and Saturn. In the early twentieth century, it was learned that methane and ammonia were present in the Jovian atmospheres as well. In 1955, radio astronomers detected radio signals coming from Jupiter's magnetosphere.

It was not until the 1970's, however, that the greatest discoveries about the Jovian planets were made. Data from Voyager 1 revealed the unexpected existence of rings around Jupiter. Earth-based observations showed rings around Uranus and Neptune, and images returned from Voyager 2 in the 1980's revealed ten satellites circling Uranus and orbiting Neptune.

Radio astronomy probes the decimetric and decameter radio signals emitted from Jupiter, which signal the extent of its magnetosphere and the relationship of its halo and its volcanic innermost satellite Io, respectively. Radio astronomy conducted by the Cassini spacecraft in orbit about Saturn provided insight into the nature of the magnetic field on Enceladus. Infrared astronomy also has been very helpful in determining some of the features of the cold Jovian planets Uranus and Neptune. Many experimental techniques have been used to determine the size and extent of the ring systems surrounding these planets, and still there are many unanswered questions about these systems.

The atmospheres of Jupiter and Saturn have been probed with all sorts of sensitive spectrometers, yet experimental information is valid only for a penetration depth of a few tens of kilometers. What lies below the turbulent, fast-moving atmosphere has not been experimentally detected; all that scientists can do is rely on the best theories and modeling techniques presently available.

The Pioneer 10 and 11 probes found that Jupiter is a tremendous source of electrons and that it generates several times as much heat as it receives from the Sun. The origin of these electrons and heat is far from clear to the most discriminating theorists in both physics and geophysics. There are no comparable conditions on Earth or the nearby terrestrial planets to produce such effects. Voyager 2 passed Neptune in August, 1989, and its use as a planetary probe effectively ceased. Its next primary objective was to characterize the approach to interstellar space. The Hubble Space Telescope (HST) has produced much better images of Uranus and Neptune than had previously been available on a regular basis.

The Galileo spacecraft arrived at Jupiter in 1995. A special probe released from Galileo entered Jupiter's atmosphere on December 7, 1995, and its instruments detected a new radiation belt, fierce winds, lightning, and upper-atmosphere densities and temperatures much higher than expected. The Cassini spacecraft, launched in October, 1997, began exploring the Saturn system from an orbital vantage point beginning in 2004. The Huygens probe that Cassini carried along on its journey from Earth to Saturn was released and sent down through the atmosphere of Saturn's largest satellite, Titan, a moon with a thick atmosphere that obscures its surface. Cassini carried an imaging radar to map the satellite's surface during repeated close flybys. Huygens survived its plunge through the atmosphere and landed in a mushy, cryogenic surface. Huygens sent its data on two redundant channels, but, because of a software error, only one transmitted properly; fortunately, an alternative path recovered most of the data that otherwise could have been lost. Huygens and Cassini found evidence of complex hydrocarbons under cryogenic conditions on the surface of Titan. Cassini's primary mission was completed in 2008, and the program received a fully funded two-year extension.

CONTEXT

It was expected that early exploration of the Jovian planets and their extensive satellite systems would provide scientific clues as to how the

Jovian Planets' Atmospheres: Comparative Data

	Jupiter	Saturn	Uranus	Neptune
Surface pressure (bars)	>100	>100	>100	>100
Surface density (kg/m^3)	~0.16	~0.19	~0.42	~0.45
Avg. temperature (kelvins)	~129	~97	~58	~58
Scale height (km)	27	59.5	27.7	~20
Composition				
Ammonia	260 ppm	125 ppm	tr	tr
Ethane	5.8 ppm	7 ppm	—	1.5 ppm
Helium (%)	10.2	3.25	15.2	19
Hydrogen (%)	89.8	96.3	82.5	80
Hydrogen deuteride (ppm)	28	110	~148	192
Methane	3000 ppm	4500 ppm	~2.3%	1.5%
Water	—	~4 ppm	tr	tr

Notes: Composition: % = percentages; ppm = parts per million; tr = trace amounts.
Source: Data are from the National Space Science Data Center, NASA/Goddard Space Flight Center.

solar system formed. Instead, a whole series of new mysteries has appeared, spurring additional robotic exploration of the outer solar system.

For atmospheric physicists, the weather systems evident in the atmospheres of both Jupiter and Saturn have provided much material for study. The Great Red Spot and several of the lesser white spots on both Saturn and Jupiter have proved to be cyclonic or anticyclonic storms that are able to maintain themselves for decades. The Great Red Spot has been observed for at least for 350 years. Could such a massive storm system be maintained on Earth? What conditions on Jupiter contribute to the tremendous longevity of the Great Red Spot? In attempting to answer questions of this sort, scientists have modeled all manner of weather systems, which has proved to be useful in deciphering meteorological patterns on Earth. Thus Jupiter and Saturn have served as gigantic, high-pressure, turbulent laboratories for atmospheric modelers. Indeed, the greatest potential outcome of comparative planetology is a better understanding of complex geophysical and atmospheric physics processes right here on Earth. Such an understanding is fundamental to determining whether or not Earth is presently undergoing global warming of natural or human-made origin.

Even in the esoteric discipline of fluid mechanics, particularly in the study of turbulent flow, data from Jupiter and Saturn have been unexpectedly helpful. These studies are critical in air-frame design and, when coupled with modern computer modeling techniques, have proved to be very valuable in the design of supersonic air frames and high-speed hydrofoils. Neptune's Dark Spot should provide fodder for studies of both fluid-mechanics and meteorology well into the twenty-first century. Many scientists believe that solar-system locations most likely to host life or organic chemistry necessary for life are either Jupiter's ice-covered satellite Europa, or Saturn's satellites Titan and Enceladus. Some sort of life systems could be operating in either of these locations, for the energy and chemical conditions seem suitable. Should some sort of complex organic molecules or anaerobic bacteria be found on either Jupiter or Europa,

the perennial mystery as to how life formed on Earth and why it exists at all could be addressed intelligently, perhaps for the first time. Saturn's satellite Enceladus has cryogenic geysers in its south polar region. Neptune's satellite Triton has also appeared to contain some sort of cryogenic geyser activity. The unexpected detection of warm liquids in the outer solar system could drive biological networks. Titan has organic materials that are believed to be indicative of the primordial Earth, although the satellite is far colder than Earth was when life developed here. Thus Titan might be a frozen example of what the early Earth might have been when life first arose.

In 1955 radio astronomers Bernard Burke and Kenneth Franklin, while studying the Crab nebula, inadvertently discovered radio emissions coming from Jupiter. Decades later, Voyager 2 recorded the largest electrical current ever measured as it passed near Jupiter. In the first half of the twentieth century, most scientists did not realize that Earth, with its reasonably strong magnetic field, produced a magnetosphere just as Jupiter did. It was not until early American spacecraft discovered the Van Allen radiation belts that radio engineers, astronomers, and plasma physicists realized that Earth's magnetosphere was a smaller version of Jupiter's. The magnetospheres of Jupiter, Saturn, and even Earth are still not completely understood; what influence they might have had on planetary origins and developments is unknown. Eventually, studies of Jupiter and Saturn might provide clues regarding the forces and mechanisms behind electrical storms, violent atmospheric electricity, and radio blackouts that can have pronounced effects on life on Earth.

John P. Kenny

FURTHER READING

Bagenal, Fran, Timothy E. Dowling, and William B. McKinnon, eds. *Jupiter: The Planet, Satellites, and Magnetosphere.* Cambridge, England: Cambridge University Press, 2007. A comprehensive work about the biggest planet in the solar system, including a series of articles provided by recognized experts in their fields of study. Excellent repository of photography, diagrams, and figures about the Jupiter system and the various spacecraft missions that unveiled its secrets.

Bortolotti, Dan. *Exploring Saturn*. New York: Firefly Books, 2003. A look at the Cassini-Huygens mission for a younger audience. Full of charts, photographs, a section on observing Saturn, and an overview of the history of our understanding of the Saturnian system, from antiquity to the launch of Cassini.

Greenberg, Richard. *Europa the Ocean Moon: Search for an Alien Biosphere*. New York: Springer, 2005. A complete description of Europa through the post-Galileo spacecraft era. Discusses the astrobiological implications of an ocean underneath Europa's icy crust. Well-illustrated and readable by both astronomy enthusiasts and college students.

Harland, David H. *Jupiter Odyssey: The Story of NASA's Galileo Mission*. New York: Springer Praxis, 2000. Collects virtually all of NASA's press releases and science updates during the first five years of the Galileo mission, along with a preview of the Cassini mission. Includes an enormous number of diagrams, tables, lists, and photographs.

Harland, David M. *Cassini at Saturn: Huygens Results*. New York: Springer, 2007. Provides a thorough explanation of the entire Cassini program, including the Huygens landing on Saturn's largest satellite. Essentially a complete collection of NASA releases from the start of Cassini flight operations through the majority of Cassini's seventy orbits during its primary mission (Cassini's primary mission concluded a year after this book was published). Technical writing style but accessible to a wide audience.

_____. *Mission to Saturn: Cassini and the Huygens Probe*. New York: Springer Praxis, 2002. A volume in Springer's Space Exploration series, this is a technical description of the Cassini program, its science goals, and the instruments used to accomplish those goals. Written before Cassini arrived at Saturn. Provides a historical review of pre-Cassini knowledge of the Saturn system.

Hartmann, William K., ed. *Astronomy*. 5th ed. Belmont, Calif.: Wadsworth, 2004. Hartmann's section of this astronomy textbook, which should be accessible to high school and college students, examines the Jovian planets and other parts of the solar system. Besides discussing many late twentieth century findings, Hartmann lists various theories of planetary origins and natures, and he examines the strengths and weaknesses of each. One chapter focuses on Jupiter, and another compares Jupiter to the other Jovian planets. Well illustrated.

Irwin, Patrick G. J. *Giant Planets of Our Solar System: An Introduction*. 2d ed. New York: Springer, 2006. Suitable as a textbook for upper-level college courses in planetary science. Focuses on Jupiter, Saturn, Uranus, and Neptune and their satellites, rings, and magnetic fields. Filled with figures and photographs.

Karttunen, H. P., et al., eds. *Fundamental Astronomy*. 5th ed. New York: Springer, 2007. A well-used university textbook in introductory astronomy. Contains some calculus-based treatments for those who find the standard texts for introductory astronomy too low-level. Covers all topics from solar-system objects to cosmology.

Lovett, Laura, Joan Harvath, and Jeff Cuzzi. *Saturn: A New View*. New York: Harry N. Abrams, 2006. A coffee-table book with about 150 of the best images returned by the Cassini mission to Saturn. Covers the planet, its many satellites, and the complex ring system.

Russell, Christopher T. *The Cassini-Huygens Mission: Orbiter Remote Sensing Investigations*. New York: Springer, 2006. Provides a thorough explanation of the remote-sensing investigations of both the Cassini orbiter and the Huygens lander. Outlines the scientific objectives of all instruments on the spacecraft and describes the planned forty-four encounters with Titan. Given the publication date, only early science returns are discussed.

See also: Jupiter's Atmosphere; Jupiter's Great Red Spot; Jupiter's Interior; Jupiter's Magnetic Field and Radiation Belts; Jupiter's Ring System; Jupiter's Satellites; Neptune's Interior; Planetary Atmospheres; Planetary Formation; Solar System: Element Distribution.

Jupiter's Atmosphere

Categories: Planets and Planetology; The Jovian System

Jupiter's atmosphere differs greatly from that on Earth. It is composed mainly of hydrogen and helium and is far enough from the Sun that the temperature of the visible cloud deck is only 153 kelvins. Voyager spacecraft data revealed details concerning chemical composition, heat transport, and wind patterns within the atmosphere. The Galileo spacecraft's atmospheric entry probe sampled that atmosphere directly and forced a rethinking of the physical model of Jupiter.

OVERVIEW

Observed through an Earth-based telescope, Jupiter's most striking aspects are a pearly glow reflected from the planet; a series of east-west bands of pastel yellows, whites, browns, and blues; and the oblate aspect of its disk. The equatorial diameter is 6 percent larger than the polar diameter, a fact that is readily apparent to the observer. Closer inspection will reveal that distinctly bright and darker individual cloud features are visible within the banded structure of the atmosphere. Throughout an evening, an observer will notice that cloud features are rotating from west to east at a rate of about 36° per hour, indicating that Jupiter rotates on its axis in slightly less than ten hours. A careful student of Jupiter will discover that the planet's visible cloud deck does not rotate as a solid body; instead, eastward winds at the equator sweep the clouds past those at midlatitudes at a rate that displaces them 7° eastward each day. This type of motion indicates that the cloud deck is opaque. In order to understand the atmospheric winds, one must know how fast the core of the planet is rotating.

Radio astronomers collected data from Jupiter and realized that variations in the radio signals from Jupiter could be attributed to the interaction of the charged particles ejected from the Sun with the magnetic field of the planet. Although the signal that the radio astronomers were measuring was generated above the atmosphere of the planet, the signal varied as Jupiter's magnetic field rotated. Astronomers' understanding of magnetic fields led them to believe that the radio astronomers were measuring the rotation of Jupiter's core. They determined that one rotation took 9 hours, 55 minutes, and 29.771 seconds.

A ground-based observer equipped with an eyepiece outfitted with a crosshair can center the planet and record the time that a selected cloud feature takes to rotate past the crosshair. Because the planet rotates about its axis in approximately ten hours, while the observer is constrained to observe within a twenty-four-hour time frame, the feature will be visible on alternate nights. The rate of rotation of the feature across the visible disk of the planet, however, is such that the observer will find that the cloud rotates five times in slightly more than two days.

Accurate periods of rotation were determined by the British Astronomical Association and the American Lunar and Planetary Institute during the first half of the twentieth century. Careful measurements of photographic data by Elmer Reese from 1960 to 1974 refined these data. When the data were related to the planetary core using the radio period of rotation, an alternating pattern emerged. Strong eastward winds near the equator, as swift as 150 meters per second (more than 450 kilometers per hour), decreased poleward. Near 15° latitude, the prevailing wind in both hemispheres was westward. Between 10° and 35° latitude, the displacement of clouds revealed two westward and two eastward peaks in the horizontal wind speeds. Highly reflective regions called zones were bracketed on the equatorward side by westward winds and on the poleward side by eastward winds. The less reflective, browner regions, or belts, were nested between the zones. Horizontal wind flow of this type in Earth's atmosphere would generate conditions that would cause rising air in the zones, leading to the formation of ice clouds at high altitudes. Air would descend in the belts and cause ices to melt, allowing a longer line-of-sight through the atmosphere and more absorption of light—hence, less reflection. Recognition of this general circulation pattern in the 1960's led to questions concerning the nature of Jupiter's ices.

As light travels outward from the Sun, it spreads out equally; thus, its ability to heat a surface decreases rapidly via an inverse square relationship in all radial directions. By the time sunlight reaches Jupiter, at a distance five times greater than Earth's distance from the Sun, the intensity is diluted by a factor of twenty-five. This dilution leads to temperatures too low to allow melting of ice formed from water; therefore, the visible cloud deck must contain another kind of ice.

In an effort to refine their understanding of the temperature regime, astronomers began to make infrared measurements of Jupiter's atmosphere. They determined that the planet radiates one and a half times more heat than it absorbs from the Sun. These results imply that the interior of Jupiter is hotter than the cloud deck and that convection from the interior transports heat outward. The picture of a deep atmosphere dominated by east-west winds emerged. If Jupiter were composed of the same chemical mixture as the Sun, its atmospheric gas would be so strongly compressed that deep below the visible cloud deck it would form a sea of liquid hydrogen and helium. There would be an indistinct change between the surface of the cryogenic fluids and the atmosphere. It also became apparent that it was essential to understand the chemistry of the atmosphere.

Calculations carried out by John Lewis and Ronald Prinn led to a model of the atmosphere that posited the existence of an upper cloud layer of ammonia ice, underlain by an ammonium hydrosulfide layer and a cloud layer composed of water ice (where the pressure is greater than ten times Earth's atmospheric pressure at sea level). All these ices are white; therefore the calculations yielded no information about coloring agents in the Jovian atmosphere. Above the topmost cloud deck, the atmosphere is mainly hydrogen and helium gas, with traces of ammonia and methane.

Molecules of methane and ammonia can absorb enough energy from incident ultraviolet light to break bonds, which allows the hydrogen atoms to escape. Darrel Strobel performed calculations indicating that ionized molecules combine to form more complex molecules and possibly aerosols or hazes. Laboratory work by other investigators has shown that many of the compounds that are formed are yellow and brown. Much of the color variation in the Jovian atmosphere may arise from variations in the heights of the underlying clouds.

Removal of the smog occurs in two ways. Either the particles grow large and fall to lower levels, or convective clouds of ammonia are carried upward like thunderheads, and the ammonia ice encases the smog particles and again causes them to fall to lower levels in the atmosphere. Computer modeling of scattering and transmission of light through the layers of hazes, gas, and clouds, carried out by Martin

The Cassini spacecraft delivered four images in December, 2000, that were compiled to produce this picture of Jupiter with a resolution of 144 kilometers. (NASA/JPL/University of Arizona)

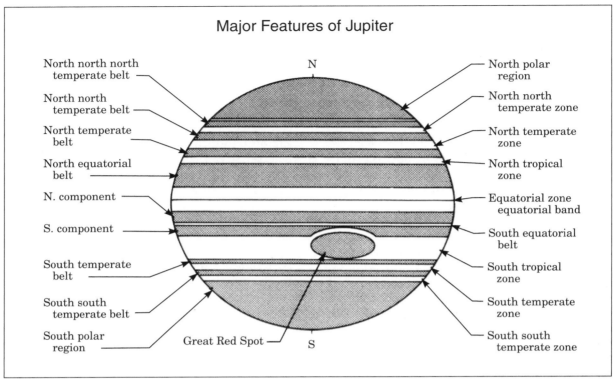

Major Features of Jupiter

North north north temperate belt

North north temperate belt

North temperate belt

North equatorial belt

N. component

S. component

South temperate belt

South south temperate belt

South polar region

Great Red Spot

N

S

North polar region

North north temperate zone

North temperate zone

North tropical zone

Equatorial zone equatorial band

South equatorial belt

South tropical zone

South temperate zone

South south temperate zone

Source: Morrison, David, and Jane Samz. *Voyage to Jupiter.* NASA SP-439. Washington, D.C.: National Aeronautics and Space Administration, 1980, p. 4.

Tomasko, Robert West, and others, supports the theory that colorization is dependent on the varying heights of underlying clouds.

These calculations cannot explain Jupiter's Great Red Spot—neither its colors nor its long life. This feature is the largest Jovian cloud system. North to south, it spans a distance slightly larger than the diameter of Earth, and it extends farther than two Earth diameters in the east-west direction. This large, unique cloud system is trapped between a westward wind on the equatorward side and an eastward wind on the poleward side. The winds are diverted around its perimeter, rotating it in a counterclockwise direction. In Earth's atmosphere, a weather system with similar motion would rise in the center, spiral outward at the top of the cloud deck, and descend around its perimeter. The degree of redness of the Spot varies with time. A unique property of the Red Spot is its ability to absorb ultraviolet, violet, and blue light. Apparently some constituent that has

been carried up from lower, warmer depths absorbs the blue light, causing the Spot to appear redder than the surrounding clouds.

Observations of Jupiter in the infrared indicate that brown and blue-gray regions are warmer than white areas. Thus, the white zones are colder than the brown belts. The clouds above the Red Spot are cold, but there is a warmer region around its perimeter. The general heat loss from the planet indicates that the interior is warmer than the upper cloud deck that radiates to space; hence, infrared maps allow astronomers to determine relative heights of clouds. A desire to obtain high-resolution maps of the infrared data and the structure of the cloud deck led to the development of the instruments that were on board the Pioneer and Voyager missions. Perhaps the best way to determine a planet's atmospheric character is to directly encounter it. That was the reasoning behind having the Galileo spacecraft carry an atmospheric probe that then was dispatched to hit the upper atmosphere and relay data until tremendous pressures destroyed it.

KNOWLEDGE GAINED

The Pioneer 10 and 11 spacecraft, which arrived at Jupiter in November, 1973, and November, 1974, respectively, each carried three instruments that sampled the Jovian atmosphere. The fact that the Pioneer spacecraft were spin-stabilized limited the types of instruments that could be placed on board. Voyagers 1 and 2 passed through the Jupiter system four months apart in 1979. Ultraviolet and infrared instruments and a spin-scan camera were trained on Jupiter's atmosphere. The Galileo probe also sampled the atmosphere beginning in 1995. The Cassini spacecraft in December, 2000, and the New Horizons spacecraft in February, 2007, conducted in-depth studies of Jupiter as they flew past on their way to Saturn and Pluto, respectively.

Infrared data revealed that there was little temperature variation between the equator and the pole at the cloud-top level. Data indicated there are limits to the role that equatorial solar heating plays in driving the zonal winds. Andrew Ingersoll proposed that solar heating at the equator could bring about a cloud structure that would act as an insulating blanket, causing the heat from the interior to emerge near the poles and resulting in little temperature variation at the level of the ammonia cloud deck. This hypothesis implies that the outward transport of the heat from the interior may dominate the atmospheric wind patterns.

Pioneer spin-scan cameras were equipped with blue and red filters and polarizers. The nature of the camera did not allow a large number of images to be obtained. Nevertheless, data provided valuable material for study of the scattering properties of the atmospheric smog and haze layers. Although a series of images with sufficiently high resolution to map cloud motions could not be obtained, images of the Red Spot and north polar regions confirmed information previously gained from ground-based observation. They also provided data on the scale of the cloud structures.

Voyagers 1 and 2 carried five instruments that were used to observe Jupiter's atmosphere:

two television cameras (one with a wide-angle view and the other with higher resolution and a narrow field of view), infrared and ultraviolet spectrographs, and a photopolarimeter. Multicolor high-resolution mapping of the visible cloud deck could be obtained at three-month intervals with each spacecraft. Near-encounter infrared measurements resolved temperature variations as a function of latitude and longitude on the planet. The ultraviolet spectrometer obtained data concerning high altitudes in the Jovian atmosphere. This extensive data set has been combined with the Pioneer and historical ground-based data sets in an effort to shed light on both short-term and long-term atmospheric variations. Cloud displacements were measured by Reta Beebe, Ingersoll, and others. Eastward winds near the equator were as powerful as 160 meters per second. Westward wind speeds at 15° north and 17.5° south latitude were both retrograde at 40 and 70 meters per second, respectively. Eastward wind maxima at 20° north and 24.5° south latitude were 170 and 60 meters per second, respectively. Voyager scientists found considerable differences between the magnitudes of wind jets in the northern hemisphere and those in the southern hemisphere; no change in the average zonal wind was detected at any latitude, however, during the five-month interval between the two encounters.

Infrared measurements indicated that the winds decrease with height above the deck, and that temperatures and abundance of ammonia above the cloud deck are consistent with an atmosphere that is driven by cloud motions at the level of the visible cloud deck.

Galileo's entry probe hit Jupiter's upper atmosphere on December 7, 1995, at a speed of as much as 170,000 kilometers per hour. The atmosphere decelerated the probe at an increased g-load of approximately 230 (that means the force was 230 times that of normal Earth gravity at sea level). During the probe's 57-minute-long plunge, it successfully relayed its findings to the Galileo orbiter for storage and eventual playback to Earth. Galileo was a little over 200,000 kilometers above the probe at the time.

The probe provided some surprising data. It had been hoped that the probe would find considerable amounts of water vapor in the atmosphere and detect extensive amounts of electrical activity, or lighting. In reality it found very little of either. It did find a new radiation belt just 50,000 kilometers above the cloud tops. As it descended through Jupiter's atmosphere, the probe registered very strong winds and experienced significant turbulence. Spectrometers found lower abundances of helium, neon, carbon, oxygen, and sulfur than had been expected. Helium was nearly half as abundant as expected in contemporary atmospheric models for Jupiter; Galileo researchers were expecting the probe to fall through a three-layered cloud structure. The probe did not experience anything like what had been predicted. The net flux radiometer on the probe did find some high-level ammonia ice clouds, and the nephelometer instrument provided some evidence of ammonium hydrosulfide clouds. Water ice was absent, suggesting the probe had entered one of the driest spots on Jupiter.

Wind strengths and atmospheric temperatures varied during the probe's descent. Winds reached 350 kilometers per hour with gusts up to 525 kilometers per hour. After the probe had plunged 156 kilometers through the atmosphere under its main parachute, the temperature and pressure environment destroyed it; most likely it was crushed, vaporized, or both nearly simultaneously. Essentially the probe's encounter forced planetary scientists to rethink the current model of Jupiter's atmosphere.

CONTEXT

By the mid-nineteenth century, astronomers had become aware that Jupiter was unlike Earth. Using the basic laws of motion, the apparent size of Jupiter, and the known distances within the solar system, they determined that even though the volume of Jupiter was more than eleven hundred times larger than Earth's volume, its mass was only 318 times larger than that of Earth. Thus, although this planet is much more massive than Earth, gravity has not compressed Jupiter's interior to the high densities that are present in the interior of Earth. Nineteenth century astronomers concluded that Jupiter could not have the same chemical composition as Earth.

By 1960, spectroscopists had determined that the atmosphere of Jupiter was cold and that temperatures at the level of the visible clouds were near 153 kelvins. Spectra revealed absorption by molecules of methane and ammonia. These observations are consistent with an atmosphere that is composed mainly of hydrogen and helium with small amounts of carbon and nitrogen. At the observed temperatures, oxygen would combine with hydrogen to form water, which would be trapped below the visible cloud deck. It became apparent that Jupiter was composed of a chemical mixture similar to that of the Sun, and that the small silicon- and iron-rich planets of the inner solar system were very different from the outer gas-rich planets.

Current models concerning the formation of a solar system

Voyager 1 captured this image of Jupiter's roiling atmosphere in 1979. (NASA/JPL)

propose that planets the size of Jupiter form first at distances far enough from the parent star that hydrogen and helium have not been expelled by the radiation from the star. The turbulence that this generates in the preplanetary gas and dust cloud leads to the formation of the other planets, with the inner ones forming from hydrogen-poor material. The importance of Jupiter-sized bodies in the formation of other planets that could possibly support other life-forms has stimulated interest in learning more about the nature of this gas giant. However, it must be noted that extrasolar planets have been found to have masses in excess of Jupiter and to be located extremely close to their stars.

Jupiter's atmosphere is chemically unlike that of Earth. The planet's huge depth and lack of irregular landmasses at its lower boundary contrast with conditions in Earth's atmosphere. There are, however, some similarities: The main constituents of Jupiter's atmosphere are hydrogen and helium. Like the nitrogen and oxygen molecules of Earth's atmosphere, these particles do not absorb sunlight readily. A large portion of solar energy passes through the upper atmospheres of these planets and, in the case of Earth, the surface absorbs the energy and is warmed. The atmosphere is then heated from the bottom, with trace constituents, carbon dioxide, and water absorbing and reradiating the energy. This leads to decreasing temperatures at increasing altitudes in the lower atmosphere.

Voyager infrared data confirmed that Jupiter has an internal heat source and that it emits 1.67 times more energy than it absorbs from the Sun. Infrared data indicate that the winds of Jupiter are driven, like those on Earth, by energy input in the lower atmosphere. Knowledge of planetary atmospheres has become more general and efforts to define the factors that lead to climate variations can be applied to more than one planet. Jupiter's dissimilarity to Earth provides checks and challenges in the search to understand Earth and the solar system.

Reta Beebe

FURTHER READING

Bagenal, Fran, Timothy E. Dowling, and William B. McKinnon, eds. *Jupiter: The Planet, Satellites, and Magnetosphere.* Cambridge, England: Cambridge University Press, 2007. A comprehensive work about the biggest planet in the solar system, covered in a series of articles provided by recognized experts in their fields of study. Excellent repository of photography, diagrams, and figures about the Jupiter system and the various spacecraft missions that unveiled its secrets.

Beatty, J. Kelly, Carolyn Collins Petersen, and Andrew Chaikin, eds. *The New Solar System.* 4th ed. Cambridge, Mass.: Sky, 1999. Filled with color diagrams and photographs, this popular work covers solar-system astronomy and planetary exploration through the Galileo missions. Accessible to the astronomy enthusiast.

Cole, Michael D. *Galileo Spacecraft: Mission to Jupiter.* New York: Enslow, 1999. Provides a full description of the Galileo spacecraft, its mission objectives, and science returns through the primary mission. Particularly good at describing mission objectives and goals. Suitable for a younger audience.

Fimmel, Richard O., James Van Allen, and Eric Burgess. *Pioneer: First to Jupiter, Saturn, and Beyond.* NASA SP-446. Washington, D.C.: Government Printing Office, 1980. A detailed review of the original Pioneer mission. Reproduces most of the images obtained by the spin-scan camera. Suitable for the general reader.

Gehrels, Tom, ed. *Jupiter.* Tucson: University of Arizona Press, 1976. A historic collection of scientific essays covering all aspects of Jupiter. The volume reflects the state of knowledge of the planet before the Voyager mission; a subsequent survey has not been published. Its ample documentation will, however, direct the serious student to journals and other sources that update the information available here. For the advanced reader.

Harland, David H. *Jupiter Odyssey: The Story of NASA's Galileo Mission.* New York: Springer Praxis, 2000. This book provides virtually all of NASA's press releases and science updates during the first five years of the Galileo mission. Provides a preview of the Cassini mission. Includes an enormous number of diagrams, tables, lists, and photographs. The book's description ends before completion of the Galileo mission unfortunately, but what is missing can easily be found on numerous NASA Web sites.

Hartmann, William K. *Moons and Planets*. 5th ed. Belmont, Calif.: Thomson Brooks/Cole, 2005. An updated version of a classic text that covers all aspects of planetary science. The chapter on Jupiter thoroughly addresses the Jovian system and spacecraft exploration of it.

Hunt, Garry E., and Patrick Moore. *Jupiter*. New York: Rand McNally, 1981. Reviews the original Voyager mission and describes the Jovian system. Photographs and illustrations are plentiful. Requires some background knowledge.

Irwin, Patrick G. J. *Giant Planets of Our Solar System: An Introduction*. 2d ed. New York: Springer, 2006. Focuses on Jupiter, Saturn, Uranus, and Neptune and their satellites, rings, and magnetic fields. Filled with figures and photographs. Suitable as a textbook for upper-level college courses in planetary science.

McAnally, John W. *Jupiter, and How to Observe It*. New York: Springer, 2008. An observing guide for the amateur astronomer that also provides detailed descriptions of the Jovian system. Discusses observational techniques, including a wide range of popular telescopes and ancillary equipment.

McBride, Neil, and Iain Gilmour, eds. *An Introduction to the Solar System*. Cambridge, England: Cambridge University Press, 2004. A complete description of solar-system astronomy suitable for an introductory college course. Accessible to nonspecialists as well. Filled with supplemental learning aids and solved student exercises. A companion Web site is available for educator support.

Morrison, David, and Jane Samz. *Voyager to Jupiter*. NASA SP-439. Washington, D.C.: Government Printing Office, 1980. A description of the events surrounding the Voyager missions to Jupiter. Illustrated with many color photographs. Accessible to the general reader.

Peek, Bertrand M. *The Planet Jupiter*. London: Macmillan, 1958. A detailed summary of ground-based observations recorded by the British Astronomical Association. This classic book provides an overview of the time-dependent aspects of the Jovian cloud deck and the history of the Red Spot. Although dated, this accessible presentation of the basics can be compared with more contemporary understandings.

See also: Auroras; Brown Dwarfs; Comet Shoemaker-Levy 9; Earth's Atmosphere; Earth's Composition; Earth's Magnetic Field: Origins; Eclipses; Europa; Extrasolar Planets; Io; Jovian Planets; Jupiter's Great Red Spot; Jupiter's Interior; Jupiter's Magnetic Field and Radiation Belts; Jupiter's Ring System; Jupiter's Satellites; Planetary Atmospheres; Planetary Magnetospheres; Planetary Ring Systems; Planetary Rotation; Planetary Satellites; Saturn's Atmosphere.

Jupiter's Great Red Spot

Categories: Planets and Planetology; The Jovian System

High-resolution data have been obtained concerning the nature of Jupiter's Red Spot, a weather system with horizontal dimensions comparable to the diameter of Earth and monitored behavior spanning centuries.

OVERVIEW

Excluding Jupiter's general east-west belt and zone pattern, the Great Red Spot is the most obvious, persistent, and continuously observed feature of Jupiter's visible cloud deck. Centered at about 20° south latitude, it spans about 14,000 kilometers in the north-south direction and about 26,000 kilometers in the east-west direction. Compared with the diameter of Earth, the Red Spot is a huge cloud structure large enough to span two Earths.

This well-defined oval feature has raised considerable curiosity ever since its discovery, which many credit to original observations made by Giovanni Cassini or Robert Hooke in the seventeenth century. Coordinated reports by the British Astronomical Society and original drawings maintained in the Royal Astronomical Society library collections in London definitively established that the Red Spot has been present since at least 1830. Earlier scattered reports of pink spots in the atmosphere of Jupiter extend back several centuries to the era of earliest efforts to improve the resolution of simple telescopes.

Realizing that they were seeing an opaque cloud deck and that the surface of Jupiter was never visible, some observers suggested that the Red Spot was caused by interference between an elevated surface feature and the prevailing winds. In the early 1960's Raymond Hide proposed a model for the driving mechanism. If this model had withstood scrutiny, it would have permitted scientists to calculate the planet's surface rotation rate and to interpret other cloud motions by extrapolating from this rate. Careful examination of measured periods of rotation indicated, however, that Hide's model was inconsistent with available data. The Red Spot circles Jupiter at an almost unvarying speed for a period of twenty to fifty years. At the end of each period, it suffers an acceleration or deceleration which occurs over a period of weeks. After this adjustment period, the Red Spot continues to circle at its new speed. This behavior indicates that it cannot be the result of an upwardly propagating disturbance above an elevated region on a hidden surface. Measured positions indicate that there is no rate at which the interior of the planet could rotate that would not force the Red Spot to drift freely either east or west within the atmosphere, yet historical observations indicate that the spot is trapped in the prevailing east-west wind pattern and is not free to move north or

The Great Red Spot is seen in this Voyager 2 image, appearing as the large, eye-shaped oval to the right of center. (NASA/JPL)

south like weather systems in Earth's midlatitudinal regions.

Along the southern edge of Jupiter's equatorial zone, winds blow eastward at 150 meters per second. West-to-east zonal winds decrease poleward, until at 17.5° south latitude they are moving westward at 70 meters per second. From 17.5° to 24.5° latitude, winds increase eastward to a maximum of nearly 60 meters per second. From 24.5° to 50°, winds alternate eastward and westward. This alternating east-west wind pattern, with four cycles between the equator and 50° south latitude, generates significant latitudinal wind shear. If local heating occurs below the cloud deck, causing the atmosphere to rise and clouds to form, maintaining a long-lived cloud system in the presence of strong horizontal shear would require the cloud to rotate about its center. If the cloud rotates in the same sense as the local horizontal shear, it can deflect the prevailing winds about its perimeter. The Red Spot displays this behavior. Not only is it trapped between westward winds at 17.5° and eastward winds at 24.5°, it also deflects westward wind flow around its equatorward perimeter, creating a large indentation, or hollow, in the poleward side of the dark adjacent belt. Other, smaller oval cloud systems are associated with the more poleward wind-shear regions. Three white oval cloud systems, noted in 1938, are located near 29° south latitude. The east-west dimension of each of these systems is about 12,000 kilometers. A series of smaller ovals circle the planet near 37° south latitude. Morphologically, the Red Spot is not unique. However, it is the largest example of a type of cloud system common to the southern hemisphere of the planet.

The Red Spot is notable not only in size but also in coloration. Jupiter's other oval clouds are white, indicating that their cloud decks are composed of highly reflective ammonia ices. When visible red and infrared reflection from the Red Spot is analyzed, data indicate ammonia ice is present there as well. The Red Spot has additional trace constituents in its cloud deck that are strong absorbers of ultraviolet, violet, and blue wavelengths. This gives the Red Spot its unique color. Small, short-lived ovals that form at similar latitudes in the northern hemisphere also absorb ultraviolet and blue light. This suggests that these absorbers are carried upward from below. Also the rate of vertical motion or the depth to which the convective motion reaches permits transport not present at the top of the cloud deck in storms located at more poleward latitudes.

A trip to a mountaintop on Earth's surface makes it clear that lower elevations of Earth's atmosphere are compressed. Jupiter's atmosphere must behave similarly. In order for the Red Spot to behave as an isolated system, its vertical dimension must be small in relation to its horizontal extent. Comparisons of the Red Spot with a hurricane are inappropriate. The Red Spot is a giant rotating cloud system, trapped in the prevailing winds. Reflectivity and degree of redness vary with time; still, deflection of the westward jet around the equatorward side of the Red Spot is always visible.

In 1878 the Red Spot suffered a deceleration. The surrounding cloud deck became highly reflective and white; however, the Red Spot remained dark and red. This sharp contrast made many casual observers aware of the phenome-

non. In 1901, a disturbance occurred in the South Tropical Zone, the white band south of the Red Spot. This event appeared to be a major weather disturbance that moved eastward and caught up with and then passed the Red Spot, thereby accelerating the Spot. This continued until 1938. Then the belt just south of the South Tropical Zone underwent a major disruption resulting in greatly increased reflectivity of the belt and the formation of three white ovals. In the early 1930's, the Red Spot drifted at a rate similar to that seen prior to 1878. After formation of these ovals, the Red Spot decelerated to its slowest drift rate ever observed. Since 1962, the Red Spot has been drifting at a rate similar to that of the 1878-1901 period.

This constitutes evidence that the Red Spot interacts with its surroundings and that variations in local temperature, pressure, and wind patterns occur. Even so, the entire range of variation in average Red Spot motion, with the average velocity derived from the annual longitudinal displacement of the Spot relative to the rotation rate of radio noise, lies between −4.4 and −0.6 meters per second. Although this variation is small when compared to daily wind speeds at midlatitudes on Earth, an annual increase of 2 meters per second in wind speed results in an eastward displacement of about 63,000 kilometers, or about two and a half times the Spot's length.

That the Spot's recovery from a given acceleration or deceleration takes years is expected. A body's heat loss rate depends heavily on the relative temperatures of the body and its surroundings. The Jovian cloud deck temperature is approximately 153 kelvins. Thus, the rate of heat loss to black sky is relatively slow. It is logical that once an excess amount of heat has been inserted into the atmosphere, it will be several years before the atmosphere returns to its previous state. One basic question that atmospheric scientists wanted to answer concerned the nature of acceleration mechanisms. Ground-based observations indicated that these events occurred over short time intervals.

During the 1960's and early 1970's, Elmer Reese made many detailed measurements using photographs of Jupiter. One result of this work was a measurement of Red Spot rotation.

The Red Spot rotates counterclockwise, completing one rotation every twelve days. A feature in Earth's atmosphere with this behavior would have air rising in the center, flowing outward at the cloud top, and descending around the perimeter. Measurement of divergent flow was one goal of the two Voyager spacecraft. Superimposed on the drift is an oscillatory motion of the whole feature, speeding up and slowing down so that its velocity oscillates every ninety days, causing the Spot to shift back and forth about 900 kilometers relative to its average path. This behavior apparently results from some natural period of response of the system to its surroundings.

Because the Red Spot had already been subjected to much scrutiny, planetary scientists eagerly looked forward in turn to obtaining high-resolution data from Pioneer and Voyager flybys in the 1970's and subsequently from the Galileo orbiter between December, 1995, and September, 2003. The Galileo spacecraft's science mission centered on investigations of the planet's particles and fields, various satellites, and general atmospheric dynamics; the latter

included an atmospheric probe, but that heavily instrumented payload was not flown into the Red Spot. Additional high-resolution images of the Red Spot were obtained from the Earth-orbiting Hubble Space Telescope and during flybys of probes passing through the Jovian system to gain a gravity asset from Jupiter. Those flybys provided opportunities for controllers to test scientific packages on those spacecraft. The Cassini orbiter in December, 2000, on its way to the Saturn system, obtained images of the Red Spot superior to those from the Voyager spacecraft, even though its closest approach to Jupiter was considerably farther out. New Horizons in February, 2007, on its way to the Pluto-Charon system, obtained high-resolution images of Jupiter. It also verified that observations from ground-based telescopes and the Hubble Space Telescope had indeed recorded a lightening phase of the Red Spot beginning in 2006. In addition, New Horizons studied a relatively new storm farther south of the Great Red Spot, one referred to as the Little Red Spot, a much smaller version of its long-lived sibling.

The Little Red Spot started as one of three

A near-infrared mosaic of Jupiter's Great Red Spot from images taken by the Galileo spacecraft. (NASA/JPL)

white storms that formed in the 1940's. Two of those merged in 1998, and the resulting white spot merged with the third in 2000. This feature then demonstrated a trend toward reddening in late 2005, and after 2006 it was referred to as the Little Red Spot. This storm continued to grow in wind speed and in reddish hue, providing a marvelous opportunity for in-depth study by the contemporary technology of the passing New Horizons spacecraft. Comparisons of the Little and Great red spots were expected to shed light on the complex atmospheric physics raging within these large-scale and unique storms. That research continues.

KNOWLEDGE GAINED

Both Pioneer spacecraft were spin-stabilized, so that the planet swept past their instruments' fields of view. This design feature placed strong constraints on the type of instrumentation that could be implemented. The imaging experiment utilized a scanning camera that sampled one point on the planet at a time. Images were constructed by scanning the planet row by row as it passed the field of view. This method limited the ultimate resolution of the data and severely curtailed the number of images that could be recorded. Nevertheless, Pioneer data were highly useful to astronomers. Tom Gehrels and his team observed Jupiter at a time when the belt adjacent to the Red Spot was highly reflective and white. The Red Spot was quite dark. Comparison with descriptions of the Red Spot in 1878 indicated that reflectivity of the Spot and its surroundings at the time of both Pioneer encounters was highly similar to its condition approximately a century earlier.

That Red Spot behavior continued until July, 1975, when a bright white cloud appeared west of the Spot that expanded rapidly and sheared out in the zonal winds. Considerable turbulence accompanied this event, and within a few weeks the Red Spot and belt had changed significantly. Material from the disturbance encountered the Red Spot from the west and formed a large white mass of clouds to the west of the Spot. Turbulent cloud masses also spilled into the westward wind jet along the south side of the belt. This material was carried around the planet and approached the Red Spot from the

east side. The contrast of the Red Spot decreased as it became whiter. Historically, Red Spot lightening has been fairly common. It appears that increased turbulence and vertical mixing in the belt that lies to the equatorward side of the Red Spot carry ammonia ices into the Spot. The Spot retained this appearance from 1975 to 1987 and was observed at high resolution in this condition by the Voyager spacecraft.

Voyagers 1 and 2 arrived at Jupiter in March and July of 1979. The two spacecraft were equipped with two television cameras each; one had a wide field of view and the other focused on a smaller field with higher resolution. The two were boresighted; thus, simultaneous views allowed detailed sampling, while defining the direction that the cameras were pointing. Ultraviolet and infrared spectrometers and a photopolarimeter allowed observations as a function of wavelength. Red Spot images with higher spatial resolution than could be obtained from Earth were taken over a period of three months with each spacecraft. The Galileo probe began collecting visible light and near-infrared images in 1996. The Hubble Space Telescope, Cassini orbiter, and New Horizons spacecraft also supplied data.

Ultimately, high-resolution sequences requiring as many as twenty-seven narrow-angle camera frames to map the Red Spot were executed. Resulting data revealed details of the flow pattern around the Red Spot. Winds were deflected around the Spot; small ammonia ice clouds, however, were observed to pass around the equatorward edge and to continue around the western cusp and along the feature's southern edge. When these clouds reached the Spot's southeast corner, they moved into the Red Spot and were sheared apart to form a high-velocity collar inside the Red Spot. Jim Mitchell, Reta Beebe, Andrew Ingersoll, and others analyzed the flow within the Red Spot and the white ovals. Velocities of rotation about the Spot's center as high as 150 meters per second were measured in the outer third of the feature. In the inner half of the Spot, reflectivity was lower, and motion of the cloud deck was small and random. No outward flow from the center toward the perimeter of the Spot was detected. When infrared data from the Red Spot and one of the white ovals were compared, no difference in absorption as a function of color could be detected; thus, the infrared data offered no clues to the identity of the ultraviolet absorber. This finding was not unexpected, because it was known that the ammonia ice would tend to dominate in the infrared.

CONTEXT

High-resolution spacecraft imagery has been combined with long-term, lower-resolution ground-based photography in an effort to understand the Red Spot's nature. Apparent motion of planetary atmospheric features can be attributed to mass motion when material is physically translated in the zonal wind or to wave motion. In the case of wave motion, variations in local pressure and temperature introduced by the wave cause local condensation or evaporation. Many of the small-scale patterns that add beauty to Earth's water clouds are of the second type. Thus, in an effort to elucidate the Red Spot's nature, models that consider different types of wave structures have been constructed. Not all wave structures are a series of oscillations with equal amplitudes traveling through space. By varying modeled environmental conditions within which the wave is formed, various researchers, including Tony Maxworthy, Andrew Ingersoll, and Gareth Williams, have investigated the characteristics of waves and related them to the morphology of the Red Spot.

In order to construct a realistic model of the Red Spot, information concerning the manner with which the zonal winds change with depth as well as with latitude is necessary. Because all required parameters are not available, a series of models must be constructed. Peter Read and others have attempted to shed light on atmospheric flow around the Red Spot by constructing cylindrical tanks of rotating fluids, within which they generate closed eddies that have characteristics similar to those of the Red Spot. High-resolution spacecraft data have provided astronomers with a wealth of information. Data have stimulated computer analysis and the gathering of additional understanding through a great deal of observation and experimentation.

It is not clear why the well-formed oval clouds in Jupiter's atmosphere preferentially form in the southern hemisphere. The fact that they are very long-lived is, however, consistent with their being large, closed eddies rotating in the local wind shear. Little is known concerning the rate of vertical motion associated with these features or the depth to which they extend below the cloud deck. Interplanetary spacecraft will continue to provide high-resolution data for researchers struggling to define the nature of Jupiter's Great Red Spot. In the planning phase for perhaps the second decade of the twenty-first century, a proposed Jupiter Icy Moons Orbiter would also provide prolonged observation of Jupiter's atmosphere, including the dynamics and evolution of the Great Red Spot.

The variability of Jupiter's atmosphere, as well as its ability to sustain prolonged features, was illustrated when a number of smaller red spots broke out beginning in early 2006. The first small spot continued into 2008, when in May a third spot appeared, this one located in the southern hemisphere farther south in longitude than the Great Red Spot. Both little red spots moved in a way that led scientists to expect them to merge. However, the influence of the Great Red Spot held the potential to push both little red spots to the side. Coordinated observations of these small storms from the Hubble Space Telescope and ground-based telescopes in visible and near-infrared light suggested that these little red spots started as white storms and then assumed a reddish color. Jupiter appeared to be undergoing a global climate change in which the equator was warming and the south pole was cooling. As a result, jet streams in the southern hemisphere were destabilized such that new storms could be generated.

Near the end of May, 2008, the first of two little Red Spots in the planet's southern hemisphere began to grow both in size and in wind speed. This new rival to the Great Red Spot developed winds of 172 meters per second, very nearly the same speed as winds in the Great Red Spot. It remained to be seen if this Little Red Spot would continue to grow independently or be swallowed up by the extremely long-lived Great Red Spot. Its general motion after its increase in size and wind speed was toward the Great Red Spot.

Reta Beebe and David G. Fisher

FURTHER READING

Beatty, J. Kelly, Carolyn Collins Petersen, and Andrew Chaikin, eds. *The New Solar System*. 4th ed. Cambridge, Mass.: Sky, 1999. The editors have assembled a collection of articles written by top researchers in the field as a survey of then-current knowledge of the solar system bodies. One chapter places the Red Spot in its atmospheric context. Includes many illustrations.

Consolmagno, Guy. *Worlds Apart: A Textbook in Planetary Sciences*. Englewood Cliffs, N.J.: Prentice Hall, 1994. A text accessible to college-level science students, using low-level mathematics as well as integral calculus where required. Demonstrates how the area of planetary science progresses by questioning previous understandings in the light of new observations.

Fimmel, Richard O., James Van Allen, and Eric Burgess. *Pioneer: First to Jupiter, Saturn, and Beyond*. NASA SP-446. Washington, D.C.: Government Printing Office, 1980. This detailed overview of the Pioneer mission includes reproductions of the Red Spot images obtained by the spin-scan camera.

Fischer, Daniel. *Mission Jupiter: The Spectacular Journey of the Galileo Spacecraft*. New York: Copernicus Books, 2001. A detailed overview of Galileo's investigations of the Jupiter system. Provides numerous full-color images taken from Galileo and other spacecraft.

Gehrels, Tom, ed. *Jupiter*. Tucson: University of Arizona Press, 1976. This collection of scientific essays on all aspects of Jupiter reflects the pre-Voyager state of knowledge of the planet. Still, its copious documentation will point the interested reader to the journals and other sources that present later findings. A challenging text; requires some background in astronomy.

Harland, David M. *Jupiter Odyssey: The Story of NASA's Galileo Mission*. New York: Springer, 2000. Provides a detailed account of all major Galileo spacecraft operations and the mission's scientific investigations of the Jupiter system. Filled with technical diagrams and images from Galileo and other spacecraft.

Hartmann, William K. *Moons and Planets*. 5th ed. Belmont, Calif.: Thomson Brooks/Cole, 2005. An excellent text for a course on planetary science, accessible to advanced high school students and undergraduates alike. Covers the entire solar system and includes much about the Red Spot. Takes a comparative planetology approach rather than including separate chapters on individual planets in the solar system.

Hunt, Garry E., and Patrick Moore. *Jupiter*. New York: Rand McNally, 1981. Reviews the Voyager mission and describes the Jovian system. Features numerous color photographs and illustrations of the Red Spot. Accessible to the general reader.

Irwin, Patrick G. J. *Giant Planets of Our Solar System: An Introduction*. 2d ed. New York: Springer, 2006. Suitable as a textbook for upper-level college courses in planetary science. Focuses on Jupiter, Saturn, Uranus, Neptune, and their satellites, rings, and magnetic fields. Filled with figures and photographs.

Morrison, David, and Jane Samz. *Voyager to Jupiter*. NASA SP-439. Washington, D.C.: Government Printing Office, 1980. A description of the events leading up to the Voyager missions. Communicates the excitement experienced by scientists as they received close-up views of the Red Spot. Contains many color reproductions.

Peek, Bertrand M. *The Planet Jupiter*. London: Macmillan, 1958. A classic detailed summary of ground-based observations recorded by members of the British Astronomical Association. This book provides an overview of time-dependent aspects of the behavior of the Red Spot.

See also: Comet Shoemaker-Levy 9; Jovian Planets; Jupiter's Atmosphere; Jupiter's Interior; Jupiter's Magnetic Field and Radiation Belts; Jupiter's Ring System; Jupiter's Satellites; Neptune's Great Dark Spots; Planetary Atmospheres; Planetary Rotation; Planetology: Comparative.

Jupiter's Interior

Categories: Planets and Planetology; The Jovian System

Jupiter, the largest planet in the solar system, is often described as a "gas giant" planet. However, the interior structure of Jupiter is far more complex than a big ball of gas. These interior characteristics are also responsible for many of the planet's observed properties.

OVERVIEW

Jupiter is the largest planet in the solar system, with a mean diameter of about 138,000 kilometers. Jupiter's mass, about 320 times more than Earth's mass. This is greater than that of the rest of the planets combined. However, most of Jupiter's mass is made up of the two lightest elements in the universe: hydrogen and helium. Jupiter is believed to be formed directly from the disk of material swirling together to make the Sun. Thus it is not a surprise to find that Jupiter's composition is very similar to that of the Sun (a star), which is also mostly hydrogen and helium. However, there are significant differences between the structure of Jupiter and that of a star, or even a failed star such as a brown dwarf. The interior structure of Jupiter has not been directly observed. It can only be inferred through mathematical modeling based on observations of the planet. Jupiter does not have a solid surface on which a spacecraft can land, and cloud layers in Jupiter's upper atmosphere shield the interior of the planet from view.

Jupiter is observed to have an oblateness of 0.065 (its equatorial diameter is 6.5 percent greater than the pole-to-pole diameter). This observation constrains the size of any solid core that the planet may have. Mathematical models suggest that Jupiter has a rocky core, with perhaps a nickel-iron inner core, of mass somewhere between 8 and 15 times the mass of the Earth. Due to Jupiter's large mass, though, that is only between 2.5 to 4.7 percent of Jupiter's mass. Despite the core being several times the mass of Earth, the extreme pressure inside Jupiter, nearly 70 million atmospheres, compresses the core to a size on the order of that of

Earth. The temperature at the core of Jupiter is believed to be perhaps 22,000 kelvins. It is not known if the core is solid or liquid.

Because Jupiter is in a region of the solar system in which a large number of icy bodies exist, a great number of these bodies must have impacted Jupiter over its history, starting at its formation. These ices, being heavier than the hydrogen and helium that make up the bulk of Jupiter, would have settled toward the deep interior of the planet. Planetary scientists believe that there may exist a layer of this material perhaps 3,000 kilometers thick on top of the core. The ices include frozen ammonia and methane rather than just water ice. The temperature and pressure deep inside Jupiter would ensure that this material is in a liquid state, though, rather than frozen. Therefore, the term "liquid ices" is often used to describe this layer. Great pressure at this depth would make this material behave in ways quite different from the way the same material would behave on Earth.

The bulk of Jupiter is composed of hydrogen and helium. The outermost parts of Jupiter are gaseous, with clouds in the upper 100 kilometers. At great depths inside the planet, the pressure becomes great enough to compress these gases into a liquid state. That pressure is reached at a depth of about 1,000 kilometers below the cloud tops. However, there is no vast ocean of liquid hydrogen under Jovian skies the way that Earth's water collects in oceans. On Jupiter, in fact, there is no clear boundary between the gaseous atmosphere and the liquid interior, because the temperature and pressure inside Jupiter are well in excess of hydrogen's critical point. Beyond the critical point of a substance, there ceases to be a definite phase transition between liquid and gas. Rather, the material takes on a state known as a supercritical fluid. At greater altitudes, the hydrogen in Jupiter is clearly gaseous. At much lower levels, it definitely has more liquid properties, but there is no obvious depth at which the hydrogen becomes liquid. Instead, with increasing depth, the hydrogen becomes more and more like liquid. Though Jupiter is called a "gas giant" planet, the majority of the planet's composition is actually liquid.

At sufficient pressure and temperature,

hydrogen takes on metallic properties. That means that it conducts heat and electricity like any other element on the left-hand column of the periodic table of elements. These conditions are met in Jupiter below a depth of about 7,000 kilometers below the planet's cloud tops. Liquid metallic hydrogen exists from that depth all the way down to the liquid ices at the core. That means that the bulk of Jupiter's mass is in a mantle composed of helium and liquid metallic hydrogen, possibly comprising about two-thirds of the planet.

Jupiter has the strongest magnetic field of any planet in the solar system. At its equator, Jupiter's magnetic field is nearly fourteen times stronger than Earth's magnetic field. Planetary magnetic fields are believed to be created by magnetohydrodynamics in a planet's interior. A dynamo model of planetary magnetic fields shows that a suitable conductor moving in a magnetic field can regenerate that magnetic field, producing a long-lived magnetic field. However, this dynamo effect requires a highly conducting fluid in order to operate. It may be possible for part of Jupiter's core to have a liquid iron region, but that would be not be large enough to account for Jupiter's magnetic field. Rather, Jupiter's magnetic field originates primarily in its liquid metallic hydrogen mantle.

Unlike Earth, Jupiter radiates nearly twice as much energy as it gets from the Sun. This surplus energy is produced by kelvin-Helmholtz contraction. When planets form, a large amount of gravitation energy is released as the materials that form the planet come together. For fluid bodies such as Jupiter, as they radiate thermal energy into space, they contract somewhat. This contraction then compresses the material making up the planet, heating it further. Other gas giant planets besides Jupiter probably also had kelvin-Helmholtz contraction after they formed, but they have long since stabilized at a point where such contraction no longer is a major source of thermal energy. Jupiter is at nearly the perfect mass to extend kelvin-Helmholtz contraction to the longest time possible. If Jupiter had more mass, then it would have compressed faster, until it reached a point of maximum compression. If Jupiter had less mass, like Saturn, then it would have not had sufficient gravity to keep contracting for as long as it has.

KNOWLEDGE GAINED

Much has been learned about Jupiter through observations from Earth. The interior of the planet, of course, cannot be directly measured. Understanding the nature of matter has allowed astrophysicists to make theoretical models of Jupiter's interior; however, it took measurements by spacecraft sent to Jupiter to actually begin to learn more about that planet's interior structure.

Since Jupiter's magnetic field is produced in the planet's mantle, it rotates with the planet's interior. Studies of the magnetic field show that Jupiter's interior rotates once every 9 hours, 55 minutes, and 30 seconds, somewhat more slowly than the rate of rotation of the cloud tops near the planet's equator. Until spacecraft were able to approach Jupiter to measure its magnetic field, astronomers could only guess at its interior rotational period. Jupiter's huge liquid me-

Jupiter Compared with Earth

	Jupiter	Earth
Mass (10^{24} kg)	1898.6	5.9742
Volume (10^{10} km^3)	143,128	108.321
Equatorial radius (km)	71,492	6378.1
Ellipticity (oblateness)	0.06487	0.00335
Mean density (kg/m^3)	1,326	5,515
Surface gravity (m/s^2)	23.12	9.78
Surface temperature (Celsius)	−140	−88 to +48
Satellites	63	1
Mean distance from Sun millions of km (mi)	779 (483)	150 (93)
Rotational period (hrs)	9.9250	23.93
Orbital period	11.86 yrs	365.25 days

Source: Data are from the National Aeronautics and Space Administration/Goddard Space Flight Center, National Space Science Data Center.

tallic hydrogen layer produces a magnetic field that is so powerful that Jupiter has a magnetic field stronger than any other planet in the solar system, and Jupiter's magnetosphere is the largest of any of the planets. The existence of Jupiter's powerful magnetic field provided evidence of the metallic nature of hydrogen well before it was produced in the laboratory.

To date, most of the extrasolar planets discovered have been gas giant planets. As the largest gas giant planet in the solar system, studies of Jupiter help to reveal the nature of these planets. Gas giant planets can have masses greater or less than that of Jupiter. However, in size, Jupiter is about as large as a gas giant planet can be. If it had much more mass, gravity would compress it to a smaller volume. Less mass would, of course, make a smaller planet, but its lower gravity would compress the planet less. For example, Saturn has almost 30 percent of Jupiter's mass but more than 80 percent of Jupiter's diameter and more than 55 percent of Jupiter's volume.

Studies of Jupiter's composition have led astronomers to believe that Jupiter may have formed somewhat farther from the Sun than the distance at which it currently orbits. Interactions with other planets, notably Saturn, could have caused Jupiter to migrate inward. This planetary migration also explains observations of extrasolar gas giant planets that appear much closer to their stars than can be explained through current understanding of planetary formation.

In late 2008 a research team reported the results of computer simulations based on the properties of hydrogen-helium mixtures exposed to the extreme conditions of temperature and pressure deep inside Jupiter. Their computer simulation also incorporated a

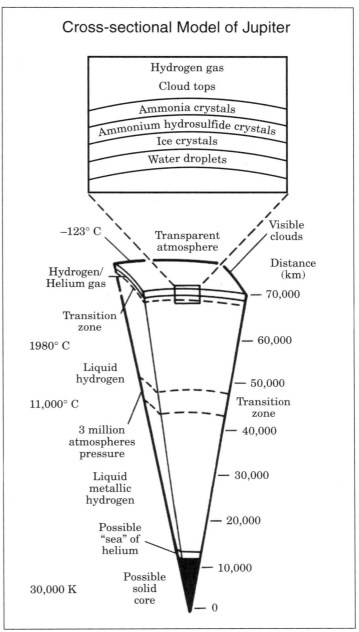

Source: David Morrison and Jane Samz. *Voyage to Jupiter.* NASA SP-439. Washington, D.C.: National Aeronautics and Space Administration, 1980, p. 4.

core accretion model. In a paper published in the November 20, 2008, issue of *Astrophysical Journal Letters*, these researchers presented an argument that Jupiter's core could be twice as big as previously thought. Their simulation predicted a rocky core perhaps amounting to 5 percent of Jupiter's total mass, making the rocky core equivalent to 14 to 18 Earth masses. That core would have layers of metals, rocky material, methane ice, water ice, and ammonia ice. Like Earth's core, the very center of Jupiter's core would be composed of iron and nickel. This computer simulation could be applied to attempts to understand the cores of the other gas giant planets as well. However, being a computer model, additional observational data and further analysis would be needed before this intriguing claim could achieve complete acceptance from the planetary science community.

CONTEXT

Jupiter and Saturn, the two largest planets in the solar system, probably share a common origin and similar structure. Both formed largely from material that was coming together to form the Sun. Thus, studies of these two worlds allow astronomers to learn more about conditions in the early solar system. Understanding these planets helps astronomers to understand how other planets, including the "rocky" planets such as Earth, form. Jupiter also seems to be similar to exoplanets that have formed around other stars and thus a sort of laboratory for understanding those extrasolar planets.

However, Jupiter is still nearly 600 million kilometers away from Earth, even at its closest approach. Thus, detailed studies have required visits by spacecraft. In total, seven spacecraft have studied Jupiter. Launched in the early 1970's, Pioneer 10 and Pioneer 11, followed in the late 1970's by Voyager 1 and Voyager 2, eventually flew past Jupiter on their way to the outer solar system. Early in 2007, the New Horizons spacecraft flew past Jupiter on its way to Pluto and the Kuiper Belt. The Cassini spacecraft passed by Jupiter December 30, 2000, on its way to Saturn. The Ulysses spacecraft flew by Jupiter in February, 1992, using that planet's gravity to send it into an orbit that permitted it to study the Sun's polar regions. All of these spacecraft studied Jupiter as they went past. The Galileo spacecraft, however, was sent specifically to study Jupiter, orbiting that planet from 1995 to 2003. Galileo also sent an

atmospheric probe into Jupiter, the only probe to enter the atmosphere of any of the gas giant planets.

Jupiter is the best studied of the gas giant planets, therefore, but it still holds many mysteries. Its interior must still be investigated through inferences from observations of Jupiter's exterior and of the planet's magnetic field. Debate continues over the exact nature of the planet's interior structure as astronomers pursue additional studies to develop a detailed understanding of this world.

Raymond D. Benge, Jr.

FURTHER READING

Bagenal, Fran, Timothy E. Dowling, and William B. McKinnon, eds. *Jupiter: The Planet, Satellites, and Magnetosphere.* New York: Cambridge, 2004. A collection of papers on the current scientific understanding of the planet Jupiter, including a chapter on Jupiter's interior.

Corfield, Richard. *Lives of the Planets: A Natural History of the Solar System.* New York: Basic Books, 2007. This book focuses less on the planets themselves than on the history of planetary exploration and the spacecraft that have studied the planets.

Fischer, Daniel. *Mission Jupiter: The Spectacular Journey of the Galileo Spacecraft.* New York: Copernicus Books, 2001. A good overview of the Galileo mission to Jupiter. Some of the initial findings are given, along with many color photographs of the planet and its moons.

Freedman, Roger A., and William J. Kaufmann III. *Universe.* 8th ed. New York: W. H. Freeman, 2008. An excellent college-level introductory astronomy textbook. An entire chapter is devoted to Jupiter and Saturn.

Irwin, Patrick. *Giant Planets of the Solar System: An Introduction.* New York: Springer, 2006. An overview of all four gas giants in the solar system, this text is written at the level of advanced students. It covers all aspects of the planets, including theories of formation, and has a very good bibliography.

McAnally, John W. *Jupiter and How to Observe It.* New York: Springer, 2008. Intended for amateur astronomers, this book focuses on observations of Jupiter, but it also includes information on the planet itself.

See also: Comet Shoemaker-Levy 9; Europa; Extrasolar Planets; Extraterrestrial Life in the Solar System; Io; Jovian Planets; Jupiter's Atmosphere; Jupiter's Great Red Spot; Jupiter's Magnetic Field and Radiation Belts; Jupiter's Ring System; Jupiter's Satellites; Planetary Interiors; Planetary Magnetospheres; Planetary Ring Systems; Planetary Rotation; Planetary Satellites; Planetology: Comparative.

Jupiter's Magnetic Field and Radiation Belts

Categories: Planets and Planetology; The Jovian System

An understanding of Jupiter's magnetic field has proved vital to furthering comprehension of this enormous planet's singular structure, and such knowledge also enriches Earth, planetary, and solar-system science.

OVERVIEW

Since 1610, when Galileo focused his telescope on Jupiter and four of its satellites, this immense planet, orbiting 628 million kilometers from Earth at its closest approach, has received much attention from astronomers. Jupiter's mean distance from the Sun is 5.2 times the mean distance between the Earth and the Sun, the latter being known as an astronomical unit (AU). After the Sun, Jupiter is the largest object in the solar system, possessing a mass 318 times greater than Earth's and a diameter 11 times longer. Its volume is thirteen hundred times that of Earth. Because of its prominence, many of Jupiter's basic characteristics were ascertained centuries ago. Sir Isaac Newton accurately calculated its mass and density, for example. Its radius, diameter, rate of rotation, chemical composition, and singular surface features have similarly been under study for centuries. However, this body, which is neither a star like the Sun nor a terrestrial planet like Earth,

has retained certain mysteries, many of which have to do with the composition of Jupiter's interior and the origins and behavior of its magnetic field and radiation environment.

During the 1950's and 1960's, prior to investigations by uncrewed spacecraft, radio astronomers gathered approximate data on Jupiter's magnetosphere, that zone of powerful magnetic influence that surrounds the planet. In 1955, Bernard Burke and Kenneth Franklin, both radio astronomers, found evidence that Jupiter's magnetosphere was a source of nonthermal radio activity (in contrast to the thermal radiation emitted by all objects with temperatures above absolute zero) at a frequency of 22.2 megahertz (MHz). Other astronomers later noted that this radio activity occurred, if not continuously, at least in a patterned way, at the same point in the planet's rotation. Such emissions distinguished Jupiter from the other planets and raised provocative questions about the validity of previous theories of radiation.

Several years later, additional radio bursts were picked up by Earth-based radio telescopes in a different portion of the radio spectrum (300 to 3,000 MHz). Unlike the emissions detected earlier, which originated at the planet's surface, these decametric radiations emanated from within Jupiter's toroidal region—a region encircling the planet, tilted about 10° from its equatorial belt and extending about 286,400 kilometers from it into space. Within this region and on both sides of the planet at about 140,000 kilometers from its surface there are two "hot spots," areas of intense radiation activity, which evoked considerable scientific curiosity.

In 1959, two astronomers, Frank Drake and S. Hvatum, identified the source of this toroidal radio activity as synchrotron radiation. Atomic particles within the region were being accelerated to very high speeds by a powerful magnetic field and by changes in the frequency of the electric field. There has been a growing consensus among astronomers that these high-energy particles, electrons that came from the Sun, have been trapped by Jupiter's external magnetic field, which is 19,000 times stronger than Earth's magnetic field. Forming radiation belts around the planet, these high-energy particles, moving at high velocities, may produce radio emissions when they strike the top of Jupiter's atmosphere.

Growing scientific curiosity about these prodigious emissions led to Jupiter's magnetosphere becoming a subject for investigation during the flybys of Pioneers 10 and 11. Pioneer 10, launched in March, 1972, flew to within 130,000 kilometers of Jupiter on December 3, 1973, securing remarkable photographs of Jovian cloud tops and measuring Jupiter's radiation belts before sailing farther into space. Pioneer 11, also referred to as Pioneer-Saturn, was launched in April, 1973. On December 2, 1974, it flew to within 42,000 kilometers of Jupiter, photographing the vast but previously unstudied Jovian polar regions. Among other measurements, Pioneer 11 obtained fresh data on Jupiter's complex magnetic field. The massive flow of data from the Pioneer flights greatly contributed to more precise assessments of the configurations of Jovian radiation belts and the magnetic field, as well as of the extent of the magnetosphere and the distribution of energetic electrons and protons within the planet's interior. This information, in turn, encouraged fresh ideas about

Jupiter's structure and rotation. In the mid-1970's, for example, John D. Anderson of the Jet Propulsion Laboratory in Pasadena, California, and William B. Hubbard of the University of Arizona made use of Pioneer data to devise a new model of Jupiter's internal structure that was consistent with knowledge about both its gravitational field and its magnetic field.

Anderson and Hubbard proposed that beneath the dense and apparently chaotic Jovian atmosphere lies a thick layer of liquid molecular hydrogen, beneath which an even thicker layer of liquid metallic hydrogen exists. The heart of the planet, they proposed, consists of a small, rocky core of iron and silicates heated to temperatures of nearly 30,000 kelvins. Although the presence of such a core could never be proved by gravitational studies, its existence was plausibly deduced from the assumption that if its composition were similar to the Sun's, it should contain some measure of the same elements. The liquid metallic hydrogen layer presumably extends 46,000 kilometers out from the core, is heated to 11,000 kelvins, and is compressed under a pressure of 3 million Earth atmospheres.

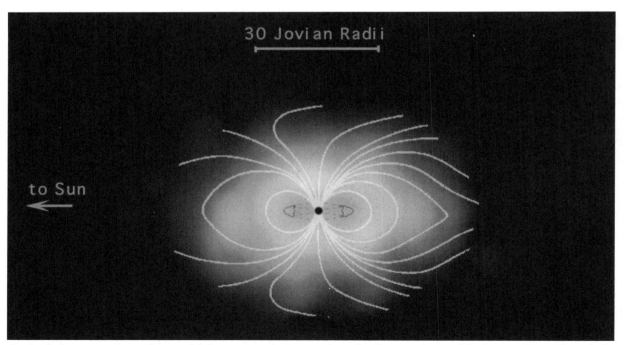

Jupiter's huge magnetosphere, which covers more space than the Sun, as imaged by instruments aboard the Cassini spacecraft on December 30, 2000. (NASA/JPL/Johns Hopkins University Applied Physics Laboratory)

This layer cannot as yet be experimentally modeled in bulk in a laboratory, yet the construct is plausible: Metallic hydrogen has been created in small amounts in the laboratory. The first to do it were researchers at Lawrence Livermore Laboratory in 1996. In a liquid metallic state, hydrogen molecules break down into atoms and become electrical conductors.

Information from the Pioneer flybys of Jupiter also led to revisions of the hypothesis that the planet's excessive radiation of heat results either from radioactivity or from heat generated by gravitational contraction of the largely gaseous mass of which Jupiter is composed. Since it now appeared that Jupiter was a liquid body (and liquids are all but incompressible), it seemed likely that its excess radiated heat was merely a residue of the heat generated when the planet coalesced from the solar nebula. The implication would be that the planet's original thermal energy is continuously finding its way to the surface and, in the process, creating convection currents, the rising of hot gases or liquids and the downward movement of cold liquids or gases in the planet's interior. Such grand-scale convection currents, as described by John H. Wolfe, who served as chief scientist for the Pioneer missions to Jupiter, could constitute a mechanism for generating the Jovian magnetic field. As primordial heat rose through Jupiter's liquid metallic hydrogen, stirring it, the Coriolis force affected the resulting convection currents. The Coriolis force arises from planetary rotation and deflects other forces in motion—depending on whether they are north or south of the body's equator—either to the right or to the left, much as a person walking across a moving carousel appears thrown off course as seen in an inertial reference frame. Deflected convection currents in such circumstances would set up loops of electric current, which could and may create a magnetic field.

The magnetosphere of Jupiter was determined to expand and contract under pressure from the solar wind. Where an equilibrium existed between magnetic forces and the solar wind's incident stream of charged particles, a planetary magnetopause developed. Pioneer data confirmed the magnetopause to be as far out from the Jovian atmosphere as 7,135,000 kilometers

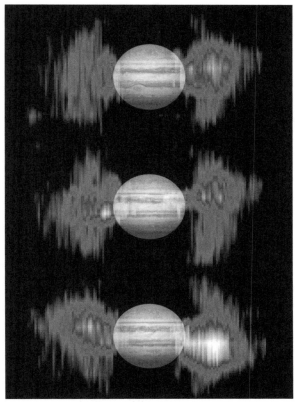

Mapped by the Cassini spacecraft, Jupiter's inner radiation belts, showing three views during a ten-hour rotation of Jupiter. The image of the planet has been superimposed. Because Jupiter's magnetic field is tilted in relation to the planet's poles, the radiation belts wobble during the planet's rotation. (NASA/JPL)

lometers and as close to its atmospheric layer as 3,565,000 kilometers. Just as there is a bow shock wave, there is also a magnetotail, which, much like a ship's wake, extends several hundred kilometers behind Jupiter in a direction away from the Sun.

Like Earth's magnetic field, the Jovian field is dipolar, but its magnetic axis is tilted between 9.5° and 10.8° from the planet's rotational axis, a displacement of about 7,000 kilometers from Jupiter's center. The strength of the magnetic field, as measured at Jupiter's cloud tops, varies from 3 to 14 gauss (a unit of magnetic field strength), extremely powerful by comparison with the 0.3- to 0.8-gauss strength of Earth's magnetic field at the surface. Probably because of the still unknown circulation patterns of the liquid metallic hydrogen in the Jovian interior, Jupiter's magnetic field is far more complex than Earth's. In addition to its dipolar field, the Jovian field, according to Pioneer 11 data, also has quadrupole and octupole movements. That is, components of the main field have four and eight poles, respectively, although these are much weaker than the main dipolar field and have been detected only close to the planet.

Further complications arise from the motion of at least five Jovian satellites that orbit within its magnetosphere. In the course of their orbits, Io, Ganymede, Europa, Callisto, and Amalthea absorb highly charged particles that otherwise might be trapped by Jupiter's magnetosphere. Thus these satellites clear channels through Jupiter's radiation belts. Io sputters material into space that forms a torus of charged particles along its orbit about Jupiter. This torus consists of sulfur, sodium, oxygen, and a few lesser constituents ejected from volcanic activity. Ganymede complicates Jupiter's magnetosphere since that satellite itself possesses a magnetic field and therefore has a magnetosphere within Jupiter's own magnetosphere.

Perhaps because Io in particular possesses an ionosphere that provides it with a conductive fluid, not only has it trapped charged particles otherwise destined for Jupiter's magnetosphere, but it also produces and accelerates charged particles. When Io is in a fixed position along an Earth-Jupiter line, radio emissions increase; thus, it is believed that its charged particles are an additional source of these emissions. Amalthea has also shown peculiarities as it orbits through the Jovian magnetosphere. Charged particles in its magnetic field unexpectedly do not increase in density toward Jupiter's magnetic equator or to its "surface." Instead, particle density varies widely at many different points.

One of the major reasons that the Galileo spacecraft was outfitted with a sophisticated magnetometer was to attempt to determine how plasma is transported through Jupiter's magnetosphere. Naturally the magnetometer was also intended to provide precise determinations of spatial and temporal variations of Jupiter's magnetic field, and the extent and shape of the planet's complex magnetosphere as well as its interactions with the solar wind.

KNOWLEDGE GAINED

Between the recording of Jovian radio emissions in the 1950's and the Pioneer and Voyager robotic space flights more than twenty years later, knowledge of the Jovian magnetic field increased by quantum leaps. New data not only have confirmed or disproved older theories about the field's origins, extent, and inner and outer complexities but also have led to more consistent and plausible theories about the structure of Jupiter itself—indeed, of other planets as well.

Sources and at least some causes of Jupiter's copious radio emissions have been identified. Based on hard, if still incomplete, evidence gathered by American spacecraft, the configurations of the magnetosphere, magnetopause, magnetosheath, and magnetotail have been delineated. The magnetic field's axis and its location have been defined. Quadrupole and octupole fields within the overall dipolar field have been discovered. The strength of the field and fluctuations within it have also been measured. Investigations of the magnetosphere have led to new theories concerning the makeup of the Jovian interior and some of the convective and conductive functions of the liquid metallic hydrogen that composes much of it.

Further, many of the special characteristics of the outer magnetosphere have been explained as the result of two principal mechanisms. First, as observed by James Van Allen (discoverer of Earth's Van Allen radiation belts), a mass of low-energy plasma trapped in the Jovian magnetic field has created pressures that have inflated the field as if it were a balloon. Second, because of interactions with the planet's rotating magnetic field, the plasma corotates (over a period of 9 hours, 55.5 minutes), creating a centrifugal force that contributes to the outward pressure. Plasma analyzers on the two Voyager spacecraft indicated that the plasma originated from gases—principally sulfur dioxide and hydrogen sulfide with sodium and oxygen in lesser amounts—vented by Io's vigorous volcanic activities. This plasma is responsible for the Io torus, a phenomenon unique to Jupiter's magnetosphere. That is, Io is surrounded by a doughnut-shaped band of excited charged particles, some captured from the solar wind as it orbits Jupiter and some produced and agitated by Io's own environment.

It is also understood now that Io and the other inner satellites, in the course of their orbits, attract many particles that enter Jupiter's inner magnetosphere and thereby limit that region's population of trapped particles. Particles trapped in the Jovian magnetosphere also affect the chemistry of the planet's environment. Andrew Ingersoll has shown that trapped particles rain down from the magnetosphere into the Jovian atmosphere. There lightning and other charged particles break down dominant chemical species, thus keeping a balance between production and breakdown of hydrocarbons in the Jovian atmosphere.

The Galileo spacecraft provided long-term measurements of Jupiter's magnetic field and the interaction of the larger satellites with that complex structure. Magnetic measurements combined with gravitational information refined the understanding of Jupiter's interior structure.

CONTEXT

Jupiter attracts scientific attention because it is big, stunningly dynamic in visual wavelengths, and noisy in the radio range. By far the largest body in the solar system after the Sun, it is a natural target of interest for the astronomer's gaze and telescopic observation. Consequently, the planet's general outward appearance—its cloud cover, its bands, and its fascinating Great Red Spot—have been closely observed over several centuries.

An emitter of abundant radio emissions, Jupiter was sufficiently noisy to arouse the curiosity of radio astronomers, who were then led to explore the spectrum of these emissions and theorize about the planet's structure. By the 1950's, thanks to analysis in the visible, infrared, and radio frequency portions of the electromagnetic spectrum, much was known of Jupiter's chemical composition. The light gases hydrogen and helium were dominant, and the sheer size of Jupiter meant that it would require billions of years to dissipate them. Some believed that beneath the planet's dense cloud cover there were mountains, while others believed the planet to be entirely gaseous. In the light of the vast quantities of information gained by the early 1980's, much of the pre-1950 understanding of the planet seems rudimentary, even laughable.

Between the 1950's and the early 1980's, several broad intellectual currents inspired scientific interest in planetary science, in Jupiter, and thereby ultimately in the planet's magnetosphere. One such current was general scientific acceptance of a theory of the solar system's origin that postulates condensation from a disk-shaped nebula of gas and dust. This theory proposes that about 4.6 billion years ago, a cloud of interstellar gas or dust, overwhelmed—as Carl Sagan and others have suggested—by an exploding star, collapsed and condensed to form the solar system. The central mass in the interstellar formation, contracting under its own gravity, produced such prodigious heat that it generated a thermonuclear reaction, from which the Sun evolved. Lesser masses—Earth, for example—experienced less heating and so, as planets, moons, asteroids, and comets, bathed in and reflected the Sun's light. Substantiation of this theory's accuracy, validity, and reliability required a fresh empirical examination of each major object in the solar system.

With the dawn of the space age, it became possible to make in situ observations or at least the nearest equivalent. Chronologically speaking, data collected by the flybys of Pioneers 10 and 11 and Voyagers 1 and 2, by Galileo in orbit about Jupiter, and by Cassini and New Horizons passing through the Jupiter system, en route to other destinations, vastly increased the storehouse of knowledge concerning Jupiter's magnetosphere. Contributions of these exploratory missions to astronomy and comparative planetology helped advance a greater understanding of Earth's relationship to the rest of the solar family.

Clifton K. Yearley

FURTHER READING
Bagenal, Fran, Timothy E. Dowling, and William B. McKinnon, eds. *Jupiter: The Planet, Satellites, and Magnetosphere.* Cambridge,

England: Cambridge University Press, 2007. A comprehensive work about the biggest planet in the solar system, comprising a series of articles by experts. An excellent repository of photography, diagrams, and figures about the Jupiter system and the various spacecraft missions that unveiled its secrets.

Beatty, J. Kelly, Carolyn Collins Petersen, and Andrew Chaikin, eds. *The New Solar System*. 4th ed. Cambridge, Mass.: Sky, 1999. Filled with color diagrams and photographs, this is a popular work on solar-system astronomy and planetary exploration through the Mars Pathfinder and Galileo missions. Accessible to the astronomy enthusiast, it provokes excitement in the general reader and communicates an appreciation for the need to understand the universe around us.

Cole, Michael D. *Galileo Spacecraft: Mission to Jupiter*. New York: Enslow, 1999. Provides a full description of the Galileo spacecraft, its mission objectives, and science returns through the primary mission. Particularly good at describing mission objectives and goals. Suitable for a younger audience.

Fimmel, Richard O., James Van Allen, and Eric Burgess. *Pioneer: First to Jupiter, Saturn, and Beyond*. NASA SP-446. Washington, D.C.: Government Printing Office, 1980. A classic, informative, and readable account of the first spaceflight into the Jovian environment, offering an excellent synthesis of its amazing findings. Includes illustrations, tables, a modest bibliography, and an index.

Harland, David H. *Jupiter Odyssey: The Story of NASA's Galileo Mission*. New York: Springer Praxis, 2000. This book provides, in a single work, virtually all of NASA's press releases and science updates during the first five years of the Galileo mission. Contains an enormous number of diagrams, tables, lists, and photographs. Provides a preview of the Cassini mission (which did not end until after publication).

Hartmann, William K. *Moons and Planets*. 5th ed. Belmont, Calif.: Thomson Brooks/Cole, 2005. An updated version of a classic text that covers all aspects of planetary science. Jupiter is discussed in comparison to the other planets, as the text takes a comparative planetology approach rather than providing individual chapters on each planet in the solar system.

Ingersoll, Andrew P. "Jupiter and Saturn." In *The Planets*, edited by Bruce Murray. San Francisco: W. H. Freeman, 1983. An authoritative discussion that is profitable for both specialists and educated laypersons. Includes superb photographs and graphics. At the end of the book, there are biographies of Ingersoll and other authors, as well as a select bibliography for this and other chapters. The volume covers only what was being learned from the Pioneer and Voyager spacecraft on their flybys of Jupiter, but it is an excellent source for early understandings of the two gas giants.

Irwin, Patrick G. J. *Giant Planets of Our Solar System: An Introduction*. 2d ed. New York: Springer, 2006. Suitable as a textbook for upper-level college courses in planetary science. Focuses on Jupiter, Saturn, Uranus, and Neptune and their satellites, rings, and magnetic fields. Filled with figures and photographs. Accessible to the serious general audience.

McBride, Neil, and Iain Gilmour, eds. *An Introduction to the Solar System*. Cambridge, England: Cambridge University Press, 2004. A complete description of solar-system astronomy suitable for an introductory college course but accessible to nonspecialists as well. Filled with supplemental learning aids and solved student exercises. A companion Web site is available for educator support.

Stone, Edward C., and A. L. Lane. "Voyager 2 Encounters with the Jovian System." *Science* 206 (November 23, 1979): 925-927. Technical, but readily understandable by readers with a basic background in mathematics and physics. Discusses data generated about the Jovian magnetosphere and other features of the planet (and its moons). Stone was chief scientist for the Voyager mission. Contains illustrations, tables, and a brief bibliography.

See also: Auroras; Brown Dwarfs; Comet Shoemaker-Levy 9; Europa; Io; Jovian Planets; Jupiter's Atmosphere; Jupiter's Great Red Spot; Jupiter's Interior; Jupiter's Ring System; Jupiter's

Satellites; Kuiper Belt; Planetary Atmospheres; Planetary Interiors; Planetary Magnetospheres; Planetary Ring Systems; Planetary Rotation; Planetary Satellites; Planetology: Comparative; Saturn's Atmosphere.

Jupiter's Ring System

Categories: Planets and Planetology; The Jovian System

Jupiter's ring system consists of four relatively dull, ethereal rings composed of submicron- to micron-sized dust grains. This system provides important clues and insights into the processes that are involved in the generation of circumstellar disks around planets. In the case of Jupiter, the primary mechanism that produces and replenishes its rings is dust generated when interplanetary meteoroids collide with four of Jupiter's small, inner satellites.

OVERVIEW

Trailblazing missions to explore Jupiter and Saturn were conducted by Pioneer 10 and 11 and Voyager 1 and 2. When Pioneer 11 flew by Jupiter in 1973-1974, observations of rather rapid variations in the number of charged particles orbiting Jupiter at specified distances from the planet suggested the possibility of a ring system that might be absorbing the particles. Although the Pioneer spacecraft were not sufficiently stabilized to facilitate taking images, Voyager 1 and 2 were. On March 4, 1979, an overexposed image from Voyager 1 finally confirmed the existence of a ring system around Jupiter, a result long anticipated by astronomers. Voyager 2 cameras captured numerous pictures of Jupiter's ring system at geometries and resolutions that were previously unobtainable. Three separate rings were discovered, the central Main Ring, the inner Halo Ring, and the outer Gossamer Ring. These rings exist within the Roche limit, the distance from the planet to where tidal forces prevent ring particles from forming into aggregates due to gravitational attraction.

On October 18, 1989, the 2.7-ton Galileo spacecraft, consisting of the main body orbiter and a probe, was launched. On December 7, 1995, the orbiter reached Jupiter and made thirty-four orbits around the planet before plunging into the planet's atmosphere in 2003. Using a solid-state imaging camera, high-quality images were taken of Jupiter's satellite-ring system. After careful analysis of the pictures, it was concluded that Jupiter's ring system is formed from dust generated as high-speed interplanetary micrometeoroids collide with the planet's four small inner satellites (sometimes called moonlets)—Metis, Adrastea, Amalthea, and Thebe—which orbit within the rings. A totally unexpected result was that the outermost Gossamer Ring consisted of two rings, one embedded inside the other. The outermost Gossamer Ring was bounded by the orbit of Thebe and named the Thebe Gossamer Ring, while the innermost one was bounded by the orbit of Amalthea and named the Amalthea Gossamer Ring.

The Cassini spacecraft, designed to obtain high-resolution images of planetary ring systems, made its closest approach to Jupiter on December 30, 2000. It imaged Jupiter's rings using different wavelengths that provided further constraints on the size, distribution, shapes, and composition of the particles within the rings. The reddish colors of the Jovian ring particles indicated a silicate or carbonaceous composition, just like that of the small embedded satellites. Images showed that the particles in Jupiter's rings were nonspherical. Cassini images also captured the motion of the two Gossamer Ring satellites, Thebe and Amalthea.

Images of Jupiter's rings taken by the New Horizons spacecraft in early 2007 confirmed that the dusty Jovian ring system was being replenished continually from embedded source bodies. The Main Ring was found to consist of three ringlets, one just outside the orbit of Adrastea, one just inside the orbit of Adrastea, and one just outside the orbit of Metis. Boulder-sized clumps, consisting of a close-paired clump and a cluster of three to five clumps, were discovered in the Main Ring just inside the orbit of Adrastea. Although the origin and nonrandom distribution of the clumps remain unexplained, they are confined to a narrow belt of motion by the gravitational influence of the two innermost satellites of Jupiter. New Horizons images established a lower limit to the diameter of Jupiter's moons of 0.5 kilometer.

From observations, measurements, and numerical modeling methods applied to data collected from Voyager 2, Galileo, Cassini, New Horizons, the Hubble Space Telescope (HST), and the Keck telescope, it has been concluded that Jupiter's rings are extremely tenuous and contain significant amounts of short-lived dust. In addition to the gravitational perturbations produced by the small satellites embedded within and bounding Jupiter's ring system, the dynamics of its faint, ethereal dusty rings are dominated by effects that involve electromagnetic forces, solar radiation pressure, and various drag forces. In the process of conserving angular momentum, the rapid spin rates tend to flatten the rings. The relatively bright, narrow Main Ring has a rather sharp outer boundary that coincides with the orbit of Adrastea. Just inside this boundary is the orbit of Metis. Since the Main Ring extends only inward from these small source moonlets, it has been concluded that particles in the Main Ring must drift inward. The width of the Main Ring is approximately 6,440 kilometers, with a thickness that

varies between 30 and 300 kilometers. Dust size ranges from 0.5 to 2.5 microns in diameter.

Interior to the Main Ring lies a thick torus of particles known as the Halo Ring. Its thickness is primarily determined by Jupiter's very strong magnetic field operating on the ring's submicron dusty grains. The thickness of the Halo Ring is approximately 12,500 kilometers, while its width is about 30,500 kilometers. In visible light, the Halo Ring has a bluish color. The very faint Almathea Gossamer Ring has an estimated width of 53,000 kilometers and a thickness of 2,500 kilometers. It has been imaged from the Earth using the Keck telescope. It appears brighter near its top and bottom edges and also brightens toward Jupiter. The dust grain size in this ring is similar to that in the Main Ring. The faintest Jovian ring, the Thebe Gossamer Ring, has a width of approximately 97,000 kilometers and a thickness of 8,500 kilometers. Dust grain size varies from 0.2 to 3.0 microns. The Thebe Gossamer Ring is observed to extend beyond Thebe, which is apparently due to coupled oscillations produced by time-varying electromagnetic forces that cause the ring to extend outward. The thickness of each Jovian ring is primarily controlled by the inclination of the orbit of its embedded moonlet.

KNOWLEDGE GAINED

The existence of Jupiter's ring system was unambiguously determined in March, 1979, by Voyager 1. Until that time, most astronomers and astrophysicists were confounded as to why Saturn had a ring system but Jupiter did not. In July, 1979, more detailed images from Voyager 2 showed that the dull, diffuse ring system of Jupiter consisted of three separate rings. The ring system exists within an intense radiation belt of electrons and ions that are trapped in Jupiter's magnetic field. Resulting drag forces play an important role in determin-

The diagram displays the main components of Jupiter's ring system as well as the small inner moons, from which the dust of the rings originates. (NASA/JPL/Cornell University)

ing the motion of the ring particles.

Images obtained from the Galileo spacecraft between 1995 and 2003 provided increasing detail about Jupiter's rings. The shape, width, thickness, optical depth, and brightness of each ring were determined, as well as dust spatial densities, grain sizes, and grain collision speeds. Jupiter's faint, dark, narrow rings (albedo about 0.05) consist of submicron- to micron-sized rock fragments and dust but do not contain ice, as do Saturn's rings. The number of separate rings in Jupiter's ring system was found to be four when it was determined that the Gossamer Ring consisted of two dis-

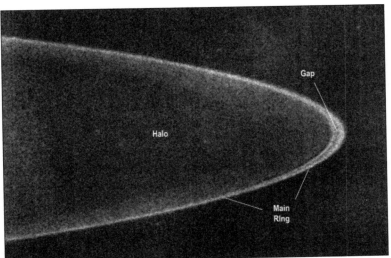

Jupiter's ring system was imaged by the New Horizons Long Range Reconnaissance Imager on February 24, 2007. (NASA/Johns Hopkins University Applied Physics Laboratory/Southwest Research Institute)

tinct rings. Further constraints on the composition, distribution, size, and shape of particles within Jupiter's rings were established in 2000 by the Cassini probe. In 2007, images from the New Horizons spacecraft revealed the fine structure of Jupiter's Main Ring. It consists of three ringlets and contains two families of boulder-sized clumps.

From the variety of measurements, observations, and analysis of collected data, Jupiter's ring system is now the best understood prototype of planetary ring systems that consist of thin, diffuse sheets of dusty debris that are primarily generated by small source moonlets. The relative motion of the dust grains within Jupiter's rings and the orientation of the orbits of the rings are primarily controlled by three processes: the spinning, asymmetric, very strong magnetic field of Jupiter; the absorption, reemission, and scattering of solar radiation energy by the dust particles, which produces momentum changes that induce orbital changes; and drag forces on the grains produced by solar radiated photons, as well as by ions and atoms that are orbiting around Jupiter. Since dust particles are continually being removed from Jupiter's rings by these processes and then replenished by dust from the four inner satellites, the dust grains existing in the rings are esti-

mated to be relatively young, probably much less than one million years old. As dust particles are ejected from the moonlets, the particles enter orbits like those of the moonlets, which causes the rings to wobble up and down as they orbit around Jupiter's equator. Micrometeoritic impacts that generate dust from the moonlets also color, chip, erode, and fragment the dust particles within the rings.

CONTEXT

Spacecraft flybys and orbiters of Jupiter and Saturn have greatly increased the scientific understanding of planetary rings. Numerical methods have been employed to simulate the physical processes occurring within Jupiter's rings by including collisional, gravitational, and electromagnetic interactions among the orbiting ring particles. Resulting models are providing keys to help guide observational strategies for future space missions.

The ring system of Jupiter provides insights into the characteristics of flattened systems of gas and colliding dust particles that are analogous to those that have eventually resulted in the formation of solar systems. In particular, Jupiter's rings offer an accessible laboratory for observing, measuring, and modeling the ongoing processes similar to those associated with the circumstellar disks that were most likely active in the solar nebula disk when the solar system containing the Earth was formed.

Further analysis, detailed examination, and numerical modeling of the data acquired by the Cassini probe and New Horizons spacecraft should provide more high-resolution maps, identify the detailed radial structure of Jupiter's ring system, and reveal invaluable time-variable features associated with the evolution of the rings. Future observations and measurements will offer new insights into the dynamic forces that shape and maintain these fascinating structures. The National Aeronautics and Space Administration (NASA) is considering a mission to Jupiter to explore the planet and its satellite-ring system in detail from a polar orbit. NASA also plans to develop small spacecraft with the capability of hovering over the rings of Jupiter and Saturn, which should provide the necessary additional data and insights for producing refined models and an advanced understanding of planetary ring structures and why they vary vastly among the gas giants.

Alvin K. Benson

FURTHER READING

Bagenal, Fran, Timothy E. Dowling, and William B. McKinnon, eds. *Jupiter: The Planet, Satellites, and Magnetosphere.* Cambridge, England: Cambridge University Press, 2004. The physical characteristics, temperature, atmospheric makeup, and satellite and ring systems of Jupiter are clearly analyzed. A description of spacecraft missions to Jupiter, numerous high-quality, full-color photographs, and many figures, tables, and diagrams elucidate a tour of the Jovian system.

Elkins-Tanton, Linda T. *Jupiter and Saturn.* New York: Chelsea House, 2006. Discusses the role that Jupiter has played in advancing our understanding of the planets that orbit the Sun. Clearly describes the discovery of Jupiter's ring system, details the Jupiter satellite-ring system, explains why the rings exist, and contrasts the ring systems of Jupiter and Saturn. An appendix lists all the known satellites of Jupiter and Saturn.

Esposito, Larry. *Planetary Rings.* Cambridge, England: Cambridge University Press, 2006. This treatise covers all aspects of planetary ring systems, including ring history, physical processes involved in ring evolution, and mathematical models used to describe them. In particular, Esposito discusses Jupiter's ring-satellite system, the age of Jupiter's rings, and the size distribution of Jupiter's rings. The text is clearly written, illustrated with many diagrams and images, and geared for a wide reading audience.

Harland, David M. *Jupiter Odyssey: The Story of NASA's Galileo Mission.* Chichester, England: Springer Praxis, 2000. A detailed account of the long trek to Jupiter by Galileo and its five years of exploration within the Jovian system. Spectacular results are presented from the observations and measurements of the satellite-ring system of Jupiter. Written for general readers, undergraduates, and faculty alike, the book is very well illustrated and includes references to many relevant Web sites.

Krüger, Harald. *Jupiter's Dust Disc: An Astrophysical Laboratory.* Aachen, Germany: Shaker-Verlag, 2003. On the basis of spacecraft observations and measurements made by Galileo, Krüger delineates the physical processes and mechanisms involved in producing Jupiter's ring system. The data clearly indicate that the Gossamer Ring material comes from Jupiter's small moonlets Thebe and Amalthea.

Miner, Ellis D., Randii R. Wessen, and Jeffrey N. Cuzzi. *Planetary Ring Systems.* New York: Springer Praxis, 2006. Provides comprehensive coverage of the scientific significance of ring systems, ring characterization and comparison, and the history of the discovery of planetary ring systems, including observations of Jupiter's ring system from Voyager 1, Voyager 2, and Galileo. Various theories for the formation of planetary ring systems are explored.

See also: Europa; Io; Jovian Planets; Jupiter's Atmosphere; Jupiter's Great Red Spot; Jupiter's Interior; Jupiter's Magnetic Field and Radiation Belts; Jupiter's Satellites; Planetary Ring Systems; Planetary Satellites; Saturn's Atmosphere; Saturn's Ring System.